Universitext

Universitext is a series of textbooks that presents material from a wide variety of mathematical disciplines at master's level and beyond. The books, often well class-tested by their author, may have an informal, personal even experimental approach to their subject matter. Some of the most successful and established books in the series have evolved through several editions, always following the evolution of teaching curricula, to very polished texts.

Thus as research topics trickle down into graduate-level teaching, first textbooks written for new, cutting-edge courses may make their way into Universitext.

More information about this series at http://www.springer.com/series/223

Felipe Linares • Gustavo Ponce

Introduction to Nonlinear Dispersive Equations

Felipe Linares
Instituto Nacional de Matemática
Pura e Aplicada (IMPA)
Rio de Janeiro
Rio de Janeiro
Brazil

Gustavo Ponce
Dept. Mathematics
University of California, Santa Barbara
College of Letters & Science
Santa Barbara
California
USA

ISSN 0172-5939 ISSN 2191-6675 (electronic)
Universitext
ISBN 978-1-4939-2180-5 ISBN 978-1-4939-2181-2 (eBook)
DOI 10.1007/978-1-4939-2181-2

Library of Congress Control Number: 2014958590

Springer New York Heidelberg Dordrecht London

Printed on acid-free paper

Springer is part of Springer Science+Business Media (www.springer.com)

Preface

The goal of this text is to present an introduction to a sampling of ideas and methods from the subject of nonlinear dispersive equations. This subject has been of great interest and has rapidly developed in the last few years. Here we will try to expose some aspects of the recent developments.

The presentation is intended to be self-contained, but we will assume that the reader has knowledge of the material usually taught in courses of theory of one complex variable and integration theory.

This text is the product of lecture notes used for mini-courses and graduate courses taught by the authors. The first version of the lecture notes was written by Gustavo Ponce with Wilfredo Urbina from the Universidad Central de Venezuela and designed to teach a mini-course at the Venezuelan School of Mathematics in Mérida, Venezuela, in 1990. A second version of those notes was presented by Gustavo Ponce at the Colombian School of Mathematics in Cali, Colombia in 1991. These notes comprise a part of the materials covered in the first six chapters of the present text. Most of the original notes were used to teach various graduate courses at IMPA and UNICAMP by Felipe Linares. During these lectures the previous versions were complemented with some new materials presented here. These notes were also used by Hebe Biagioni and Marcia Scialom from UNICAMP in their seminars and graduate courses. The idea to write the present text arose from the need for a more complete treatment of these topics for graduate students.

Before going any further we would first like to give a notion of what a dispersive type of partial differential equation is. We will do this in the one-dimensional frame. We consider a linear partial differential equation

$$F\,(\partial_x, \partial_t)\, u(x,t) = 0, \tag{1}$$

where F is a polynomial in the partial derivatives. We look for plane wave solutions of the form $u(x,t) = Ae^{i(kx-\omega t)}$ where A, k, and ω are constants representing the amplitude, the wavenumber, and the frequency, respectively. Hence u will be a solution if and only if

$$F(ik, -i\omega) = 0. \tag{2}$$

This equation is called the dispersion relation. This relation characterizes the plane wave motion. In several models we can write ω as a real function of k, namely,

$$\omega = \omega(k).$$

The phase and group velocities of the waves are defined by

$$c_p(k) = \frac{\omega}{k} \quad \text{and} \quad c_g = \frac{d\omega}{dk}.$$

The waves are called dispersive if the group velocity $c_g = \omega'(k)$ is not constant, i.e., $\omega''(k) \neq 0$. In the physical context this means that when time evolves, the different waves disperse in the medium, with the result that a single hump breaks into wave-trains.

To present the material we have chosen to study two very well-known models in the class of nonlinear dispersive equations: the Korteweg–de Vries equation

$$\partial_t v + \partial_x^3 v + v\partial_x v = 0, \tag{3}$$

where v is a real-valued function and the nonlinear Schrödinger equation

$$i\partial_t u + \Delta u = f(u, \bar{u}), \tag{4}$$

where u is a complex-valued function.

Before commenting on the theory presented in this text regarding these equations we would like to say a few words concerning the physical models described by these equations in the context of water waves.

The first model (3) goes back to the discovery of Scott Russell in 1835 of what he called a traveling wave. This equation describes the propagation of waves in shallow water and was proposed by Diederik Johannes Korteweg and Gustav de Vries in 1895 [KdV]. In the one-dimensional context, the (cubic) nonlinear Schrödinger equation (4) with $f(u, u) = |u|^2 u$ models the propagation of wave packets in the theory of water waves.

We also have to mention that there is a very well-known strong relationship between these two equations and the theory of completely integrable systems, or Soliton theory.

In many cases, we present the details of simple proof, which may not be that of the strongest result. We give several examples to illustrate the theory. At the end of every chapter we complement the theory described either with a set of exercises or with a section with comments on open problems, extensions, and recent developments.

The first three chapters attempt to review several topics in Fourier analysis and partial differential equations. These are the elementary tools needed to develop the theory in the rest of the notes.

The properties of solutions to the linear problem associated to the Schrödinger equation are discussed in Chapter 4. Then the initial value problem associated to (4) and properties of its solutions are studied in Chaps. 5 and 6. Chapters 7 and 8 are devoted to the study of the initial value problem for the generalized Korteweg–de

Vries equation. A survey of results concerning several nonlinear dispersive equations that generalize (3) and (4) as Davey–Stewartson systems, Ishimori equations, Kadomtsev–Petviashvili equations, Benjamin–Ono equations, and Zakharov systems is presented in Chapter 9. In the last chapter we present the most recent result regarding local well-posedness for the nonlinear Schrödinger equation.

We shall point out that by no means our presentation is completely exhaustive. We refer the reader to the lecture notes by Cazenave [Cz1], [Cz2] and the books by Sulem and Sulem [SS2], Bourgain [Bo2], and Tao [To7]. In these works many topics not covered in these notes are studied in detail.

Acknowledgments

The authors are indebted to several friends who made this project possible. We would like to thank Carlos Kenig, who allowed us to use part of his lecture notes regarding the material in Chapter 10; Luis Vega for useful comments and suggestions; Rafael Iório and Carlos Isnard who are great supporters of the idea of having graduate courses at IMPA in the topics discussed here and the writing of notes concerning; Hebe Biagioni, Marcia Scialom, and Jaime Angulo who gave us feedback about the former lecture notes. We also thank Daniela Bekiranov, Mahendra Panthee, Aniura Milanes, Wee Keong Lim, German Fonseca, Didier Pilod, Aida Gonzalez, José Jiménez, and Luiz Farah for reading the most part of the manuscript and for giving us many corrections and useful comments. The first author is grateful to the Mathematics Department of University of California at Santa Barbara for the support to accomplish this project. The second author was supported by an NSF grant.

Rio de Janeiro and Santa Barbara
June 2008

Felipe Linares
Gustavo Ponce

Preface to the Second Edition

In this version several errors and typos of the first edition have been corrected thanks to the comments of many friends.

Very few changes were made in the (basic) first four chapters. Some of the material presented in the other chapters has been expanded and updated. In addition, several new exercises have been added.

Despite considerable increase, the bibliography is not intended to be complete. We also thank Cynthia Flores and Derek Smith for reading some parts of the new version of this manuscript.

Rio de Janeiro and Santa Barbara
August 2014

Felipe Linares
Gustavo Ponce

Contents

Chapter 1
The Fourier Transform

In this chapter, we shall study some basic properties of the Fourier transform. Section 1.1 is concerned with its definition and properties in $L^1(\mathbb{R}^n)$. The case $L^2(\mathbb{R}^n)$ is considered in Section 1.2. The space of tempered distributions is briefly considered in Section 1.3. Finally, Sections 1.4 and 1.5 give an introduction to the study of oscillatory integrals in one dimension and some applications, respectively.

1.1 The Fourier Transform in $L^1(\mathbb{R}^n)$

Definition 1.1. The *Fourier transform* of a function $f \in L^1(\mathbb{R}^n)$, denoted by $\widehat{f}$, is defined as:

$$\widehat{f}(\xi) = \int_{\mathbb{R}^n} f(x) e^{-2\pi i (x\cdot\xi)} dx, \quad \text{for } \xi \in \mathbb{R}^n, \tag{1.1}$$

where $(x \cdot \xi) = x_1\xi_1 + \cdots + x_n\xi_n$.

We list some basic properties of the Fourier transform in $L^1(\mathbb{R}^n)$.

Theorem 1.1. *Let $f \in L^1(\mathbb{R}^n)$. Then:*

1. *$f \mapsto \widehat{f}$ defines a linear transformation from $L^1(\mathbb{R}^n)$ to $L^\infty(\mathbb{R}^n)$ with*

$$\|\widehat{f}\|_\infty \le \|f\|_1. \tag{1.2}$$

2. *$\widehat{f}$ is continuous.*
3. *$\widehat{f}(\xi) \to 0$ as $|\xi| \to \infty$ (Riemann–Lebesgue).*
4. *If $\tau_h f(x) = f(x - h)$ denotes the translation by $h \in \mathbb{R}^n$, then*

$$\widehat{(\tau_h f)}(\xi) = e^{-2\pi i (h\cdot\xi)} \widehat{f}(\xi), \tag{1.3}$$

and

$$(e^{-2\pi i (x\cdot h)} f)^{\widehat{}}(\xi) = (\tau_{-h}\widehat{f})(\xi). \tag{1.4}$$

F. Linares, G. Ponce, *Introduction to Nonlinear Dispersive Equations*, Universitext, DOI 10.1007/978-1-4939-2181-2_1

5. *If* $\delta_a f(x) = f(ax)$ *denotes a dilation by* $a > 0$, *then*

$$\widehat{(\delta_a f)}(\xi) = a^{-n}\widehat{f}(a^{-1}\xi). \tag{1.5}$$

6. *Let* $g \in L^1(\mathbb{R}^n)$ *and* $f * g$ *be the convolution of* f *and* g. *Then,*

$$\widehat{(f * g)}(\xi) = \widehat{f}(\xi)\widehat{g}(\xi). \tag{1.6}$$

7. *Let* $g \in L^1(\mathbb{R}^n)$. *Then,*

$$\int_{\mathbb{R}^n} \widehat{f}(y)g(y)dy = \int_{\mathbb{R}^n} f(y)\widehat{g}(y)dy. \tag{1.7}$$

Notice that the equality in (1.2) holds for $f \geq 0$, i.e., $\widehat{f}(0) = \|\widehat{f}\|_\infty = \|f\|_1$.

Proof. It is left as an exercise. □

Next, we give some examples to illustrate the properties stated in Theorem 1.1.

Example 1.1 Let $n = 1$ and $f(x) = \chi_{(a,b)}(x)$ (the characteristic function of the interval (a,b)). Then,

$$\begin{aligned}\widehat{f}(\xi) &= \int_a^b e^{-2\pi i x\xi}\, dx \\ &= -\frac{e^{-2\pi i b\xi} - e^{-2\pi i a\xi}}{2\pi i\xi} \\ &= -e^{-\pi i(a+b)\xi}\,\frac{\sin(\pi(a-b)\xi)}{\pi\xi}.\end{aligned}$$

Notice that $\widehat{f} \notin L^1(\mathbb{R})$ and that $\widehat{f}(\xi)$ has an analytic extension $\widehat{f}(\xi + i\eta)$ to the whole plane $\xi + i\eta \in \mathbb{C}$. In particular, if $(a,b) = (-k,k)$, $k \in \mathbb{Z}^+$, then we have

$$\widehat{\chi}_{(-k,k)}(\xi) = \frac{\sin(2\pi k\xi)}{\pi\xi}.$$

Example 1.2 Let $n = 1$ and for $k \in \mathbb{Z}^+$ define

$$g_k(x) = \begin{cases} k+1+x, & \text{if } x \in (-k-1, -k+1] \\ 2, & \text{if } x \in (-k+1, k-1) \\ k+1-x, & \text{if } x \in [k-1, k+1) \\ 0, & \text{if } x \notin (-k-1, k+1), \end{cases}$$

i.e., $g_k(x) = \chi_{(-1,1)} * \chi_{(-k,k)}(x)$. The identity (1.6) and the previous example show that

$$\widehat{g_k}(\xi) = \frac{\sin(2\pi\xi)\sin(2\pi k\xi)}{(\pi\xi)^2}.$$

Notice that $\widehat{g_k} \in L^1(\mathbb{R})$ and has an analytic extension to the whole plane $\mathbb{C}$.

Example 1.3 Let $n \geq 1$ and $f(x) = e^{-4\pi^2 t|x|^2}$ with $t > 0$. Then, changing variables $x \to x/\sqrt{t}$ and using (1.5), we can restrict ourselves to the case $t = 1$. From Fubini's theorem we write:

$$\begin{aligned}
\int_{\mathbb{R}^n} e^{-4\pi^2|x|^2} e^{-2\pi i(x\cdot\xi)}\, dx &= \prod_{j=1}^{n} \int_{-\infty}^{\infty} e^{(-4\pi^2 x_j^2 - 2\pi i \xi_j x_j)}\, dx_j \\
&= \prod_{j=1}^{n} \int_{-\infty}^{\infty} e^{(-4\pi^2 x_j^2 - 2\pi i \xi_j x_j + \xi_j^2/4)} e^{-\xi_j^2/4}\, dx_j \\
&= \prod_{j=1}^{n} e^{-\xi_j^2/4} \int_{-\infty}^{\infty} e^{-(2\pi x_j + i\xi_j/2)^2}\, dx_j \\
&= 2^{-n}\pi^{-n/2} e^{-|\xi|^2/4},
\end{aligned}$$

where in the last equality, we have employed the following identities from complex integration and calculus:

$$\int_{-\infty}^{\infty} e^{-(2\pi x + i\xi/2)^2}\, dx = \int_{-\infty}^{\infty} e^{-(2\pi x)^2}\, dx = \int_{-\infty}^{\infty} e^{-x^2}\, \frac{dx}{2\pi} = \frac{1}{2\sqrt{\pi}}.$$

Hence,

$$\widehat{e^{-4\pi^2 t|x|^2}}(\xi) = \frac{e^{-|\xi|^2/4t}}{(4\pi t)^{n/2}}. \tag{1.8}$$

Observe that taking $t = 1/4\pi$ and changing variables $t \to 1/16\pi^2 t$ we get:

$$\widehat{e^{-\pi|x|^2}}(\xi) = e^{-\pi|\xi|^2} \quad \text{and} \quad \widehat{\frac{e^{-|x|^2/4t}}{(4\pi t)^{n/2}}}(\xi) = e^{-4\pi^2 t|\xi|^2},$$

respectively.

Example 1.4 Let $n \geq 1$ and $f(x) = e^{-2\pi|x|}$. Then,

$$\hat{f}(\xi) = \frac{\Gamma[\frac{(n+1)}{2}]}{\pi^{(n+1)/2}} \frac{1}{(1+|\xi|^2)^{(n+1)/2}},$$

where $\Gamma(\cdot)$ denotes the Gamma function. See Exercise 1.1 (i).

Example 1.5 Let $n = 1$ and $f(x) = \dfrac{1}{\pi}\dfrac{1}{1+x^2}$. Using complex integration one obtains the identity:

$$\int_{-\infty}^{\infty} \frac{\cos(ax)}{x^2+b^2}\, dx = \frac{\pi}{b}\, e^{-ab}, \quad a,\, b > 0.$$

Hence,

$$\frac{1}{\pi}\widehat{\frac{1}{1+x^2}}(\xi) = \frac{1}{\pi}\int_{-\infty}^{\infty} \frac{e^{-2\pi i x\xi}}{1+x^2}\,dx$$
$$= \frac{1}{\pi}\int_{-\infty}^{\infty} \frac{\cos(2\pi|\xi|x)}{1+x^2}\,dx = e^{-2\pi|\xi|}.$$

One of the most important features of the Fourier transform is its relationship with differentiation. This is described in the following results.

Proposition 1.1. *Suppose* $x_k f \in L^1(\mathbb{R}^n)$, *where* x_k *denotes the kth coordinate of* x. *Then,* $\widehat{f}$ *is differentiable with respect to* ξ_k *and*

$$\frac{\partial \widehat{f}}{\partial \xi_k}(\xi) = (\widehat{-2\pi i x_k f(x)})(\xi). \tag{1.9}$$

In other words, the Fourier transform of the product $x_k f(x)$ is equal to a multiple of the partial derivative of $\widehat{f}(\xi)$ with respect to the kth variable.

To consider the converse result, we need to introduce a definition.

Definition 1.2. Let $1 \le p < \infty$. A function $f \in L^p(\mathbb{R}^n)$ is *differentiable* in $L^p(\mathbb{R}^n)$ with respect to the kth variable, if there exists $g \in L^p(\mathbb{R}^n)$ such that

$$\int_{\mathbb{R}^n} \left| \frac{f(x+he_k) - f(x)}{h} - g(x) \right|^p dx \to 0 \text{ as } h \to 0,$$

where e_k has kth coordinate equals 1 and 0 in the others. If such a function g exists (in this case it is unique), it is called the partial derivative of f with respect to the kth variable in the L^p-norm.

Theorem 1.2. *Let* $f \in L^1(\mathbb{R}^n)$ *and* g *be its partial derivative with respect to the kth variable in the* L^1*-norm. Then,* $\widehat{g}(\xi) = 2\pi i \xi_k \widehat{f}(\xi)$.

Proof. Properties (1.2) and (1.4) in Theorem 1.1 allow us to write

$$\left|\widehat{g}(\xi) - \widehat{f}(\xi)\frac{(1 - e^{-2\pi i h(\xi\cdot e_k)})}{h}\right|,$$

then take $h \to 0$ to obtain the result. □

From the previous theorems it is easy to obtain the formulae:

$$\begin{aligned} P(D)\widehat{f}(\xi) &= (P(-2\pi i x)f(x))^\wedge(\xi), \\ (\widehat{P(D)f})(\xi) &= P(2\pi i \xi)\widehat{f}(\xi), \end{aligned} \tag{1.10}$$

where P is a polynomial in n variables and $P(D)$ denotes the differential operator associated to P.

Now we turn our attention to the following question: Given the Fourier transform $\widehat{f}$ of a function in $L^1(\mathbb{R}^n)$, how can one recover f?

Examples 1.3–1.5 suggest the use of the formula

$$f(x) = \int_{\mathbb{R}^n} \widehat{f}(\xi) e^{2\pi i (x\cdot\xi)} d\xi.$$

Unfortunately, $\widehat{f}(\xi)$ may be nonintegrable (see Example 1.1). To avoid this problem, one needs to use the so called method of summability (Abel and Gauss) similar to those used in the study of Fourier series. Combining the ideas behind the Gauss summation method and the identities (1.4), (1.7), (1.8), we obtain the following equalities:

$$\begin{aligned} f(x) &= \lim_{t\to 0} \frac{e^{-|\cdot|^2/4t}}{(4\pi t)^{n/2}} * f(x) = \lim_{t\to 0} \int_{\mathbb{R}^n} \frac{e^{-|x-y|^2/4t}}{(4\pi t)^{n/2}} f(y) dy \\ &= \lim_{t\to 0} \int_{\mathbb{R}^n} \tau_x \frac{e^{-|y|^2/4t}}{(4\pi t)^{n/2}} f(y) dy \\ &= \lim_{t\to 0} \int_{\mathbb{R}^n} (\widehat{e^{2\pi i (x\cdot\xi)} e^{-4\pi^2 t|\xi|^2}})(y) f(y) dy \\ &= \lim_{t\to 0} \int_{\mathbb{R}^n} e^{2\pi i (x\cdot\xi)} e^{-4\pi^2 t|\xi|^2} \widehat{f}(\xi) d\xi, \end{aligned}$$

where the limit is taken in the L^1-norm.

Thus, if f and $\widehat{f}$ are both integrable, the Lebesgue dominated convergence theorem guarantees the point-wise equality. Also, if $f \in L^1(\mathbb{R}^n)$ is continuous at the point x_0, we get:

$$f(x_0) = \lim_{t\to 0} \frac{e^{-|\cdot|^2/4t}}{(4\pi t)^{n/2}} * f(x_0) = \lim_{t\to 0} \int_{\mathbb{R}^n} e^{2\pi i (x_0\cdot\xi)} e^{-4\pi^2 t|\xi|^2} \widehat{f}(\xi) d\xi.$$

Collecting this information, we get the following result.

Proposition 1.2. *Let $f \in L^1(\mathbb{R}^n)$. Then,*

$$f(x) = \lim_{t\to 0} \int_{\mathbb{R}^n} e^{2\pi i (x\cdot\xi)} e^{-4\pi^2 t|\xi|^2} \widehat{f}(\xi) d\xi,$$

where the limit is taken in the L^1-norm. Moreover, if f is continuous at the point x_0, then the following point-wise equality holds:

$$f(x_0) = \lim_{t\to 0} \int_{\mathbb{R}^n} e^{2\pi i (x_0\cdot\xi)} e^{-4\pi^2 t|\xi|^2} \widehat{f}(\xi) d\xi.$$

Let $f, \widehat{f} \in L^1(\mathbb{R}^n)$. *Then,*

$$f(x) = \int_{\mathbb{R}^n} e^{2\pi i (x\cdot\xi)} \widehat{f}(\xi) d\xi, \quad \textit{almost everywhere } x \in \mathbb{R}^n.$$

From this result and Theorem 1.1 we can conclude that

$$\wedge : L^1(\mathbb{R}^n) \longrightarrow C_\infty(\mathbb{R}^n)$$

is a linear, one-to-one (Exercise 1.6 (i)), bounded map. However, it is not surjective (Exercise 1.6 (iii)).

1.2 The Fourier Transform in $L^2(\mathbb{R}^n)$

To define the Fourier transform in $L^2(\mathbb{R}^n)$, we shall first consider that $L^1(\mathbb{R}^n) \cap L^2(\mathbb{R}^n)$ is a dense subset of $L^1(\mathbb{R}^n)$ and $L^2(\mathbb{R}^n)$.

Theorem 1.3 (Plancherel). *Let* $f \in L^1(\mathbb{R}^n) \cap L^2(\mathbb{R}^n)$. *Then,* $\widehat{f} \in L^2(\mathbb{R}^n)$ *and*

$$\|\widehat{f}\|_2 = \|f\|_2. \tag{1.11}$$

Proof. Let $g(x) = -x$. Using Young's inequality (1.39), (1.6), and Exercise 1.7 (ii), it follows that

$$f * g \in L^1(\mathbb{R}^n) \cap C_\infty(\mathbb{R}^n) \quad \text{and} \quad \widehat{(f * g)}(\xi) = \widehat{f}(\xi)\widehat{g}(\xi).$$

Since $\widehat{g} = \overline{(\widehat{f})}$, we find that $\widehat{(f * g)} = |\widehat{f}|^2 \geq 0$. Hence, $\widehat{(f * g)} \in L^1(\mathbb{R}^n)$ (see Exercise 1.7 (iii)). Proposition 1.2 shows that

$$(f * g)(0) = \int_{\mathbb{R}^n} \widehat{(f * g)}(\xi)\, d\xi,$$

and

$$\begin{aligned}
\|\widehat{f}\|_2^2 &= \int_{\mathbb{R}^n} \widehat{(f * g)}(\xi)\, d\xi = (f * g)(0) \\
&= \int_{\mathbb{R}^n} f(x) g(0 - x)\, dx = \int_{\mathbb{R}^n} f(x)\bar{f}(x)\, dx = \|f\|_2^2.
\end{aligned}$$

□

This result shows that the Fourier transform defines a linear bounded operator from $L^1(\mathbb{R}^n) \cap L^2(\mathbb{R}^n)$ to $L^2(\mathbb{R}^n)$. Indeed, this operator is an isometry. Thus, there is a unique bounded extension $\mathcal{F}$ defined in all $L^2(\mathbb{R}^n)$. $\mathcal{F}$ is called the Fourier transform in $L^2(\mathbb{R}^n)$. We shall use the notation $\widehat{f} = \mathcal{F}(f)$ for $f \in L^2(\mathbb{R}^n)$. In general, the definition $\widehat{f}$ is realized as a limit in L^2 of the sequence $\{\widehat{h}_j\}$, where $\{h_j\}$ denotes any sequence in $L^1(\mathbb{R}^n) \cap L^2(\mathbb{R}^n)$ that converges to f in the L^2-norm. It is convenient to take h_j equals f for $|x| \leq j$ and to have h_j vanishing for $|x| > j$. Then,

$$\widehat{h}_j(\xi) = \int\limits_{|x|<j} f(x)e^{-2\pi i(x\cdot\xi)}\,dx = \int\limits_{\mathbb{R}^n} h_j(x)e^{-2\pi i(x\cdot\xi)}\,dx$$

and so,

$$\widehat{h}_j(\xi) \to \widehat{f}(\xi) \quad \text{in } L^2, \text{ as } j \to \infty.$$

Example 1.6 Let $n = 1$ and $f(x) = \dfrac{1}{\pi}\dfrac{x}{1+x^2}$. Observe that $f \in L^2(\mathbb{R}) \setminus L^1(\mathbb{R})$. Differentiating the identity in the Example 1.5 with respect to a and taking $b = 1$ we get:

$$\int_{-\infty}^{\infty} \frac{x \sin(ax)}{1+x^2}\,dx = \pi e^{-a}, \quad a > 0,$$

which combined with the previous remark gives:

$$\widehat{f}(\xi) = -i\,\mathrm{sgn}(\xi)e^{-2\pi|\xi|}.$$

A surjective isometry defines a "unitary operator." Theorem 1.3 affirms that $\mathcal{F}$ is an isometry. Let us see that $\mathcal{F}$ is also surjective.

Theorem 1.4. *The Fourier transform defines a unitary operator in* $L^2(\mathbb{R}^n)$.

Proof. From the identity (1.11) it follows that $\mathcal{F}$ is an isometry. In particular, its image is a closed subspace of $L^2(\mathbb{R}^n)$. Assume that this is a proper subspace of L^2. Then, there exists $g \neq 0$ such that

$$\int\limits_{\mathbb{R}^n} \widehat{f}(y)g(y)dy = 0, \quad \text{for any } f \in L^2(\mathbb{R}^n).$$

Using formula (1.7; Theorem 1.7), which obviously extends to $f, g \in L^2(\mathbb{R}^n)$, we have that

$$\int\limits_{\mathbb{R}^n} f(y)\widehat{g}(y)dy = \int\limits_{\mathbb{R}^n} \widehat{f}(y)g(y)dy = 0, \text{ for any } f \in L^2.$$

Therefore, $\widehat{g}(\xi) = 0$ almost everywhere, which contradicts

$$\|g\|_2 = \|\widehat{g}\|_2 \neq 0.$$

□

Theorem 1.5. *The inverse of the Fourier transform* $\mathcal{F}^{-1}$ *can be defined by the formula*

$$\mathcal{F}^{-1} f(x) = \mathcal{F} f(-x), \quad \textit{for any} \quad f \in L^2(\mathbb{R}^n). \tag{1.12}$$

Proof. $\mathcal{F}^{-1}\widehat{f} = \tilde{f}$ is the limit in the L^2-norm of the sequence

$$f_j(x) = \int_{|\xi|<j} \widehat{f}(\xi) e^{2\pi i(\xi \cdot x)} d\xi.$$

First, we consider the case where $f \in L^1(\mathbb{R}^n) \cap L^2(\mathbb{R}^n)$. It suffices to verify that this agrees with $\mathcal{F}^*\widehat{f}$, where $\mathcal{F}^*$ is the adjoint operator of $\mathcal{F}$ (we recall the fact that for a unitary operator the adjoint and the inverse are equal). This can be checked as follows:

$$\tilde{f}(x) = \int_{\mathbb{R}^n} \widehat{f}(\xi) e^{2\pi i(\xi \cdot x)} d\xi = \lim_{j \to \infty} f_j(x) \text{ in } L^2(\mathbb{R}^n),$$

and

$$(g, \tilde{f}) = \int_{\mathbb{R}^n} g(x) \overline{\Big(\int_{\mathbb{R}^n} \widehat{f}(\xi)\, e^{2\pi i(\xi \cdot x)}\, d\xi \Big) dx}$$

$$= \int_{\mathbb{R}^n} \Big(\int_{\mathbb{R}^n} g(x) e^{-2\pi i(x \cdot \xi)} dx \Big) \overline{\widehat{f}(\xi)} d\xi = (\mathcal{F} g, \widehat{f})$$

for any $g \in L^1(\mathbb{R}^n) \cap L^2(\mathbb{R}^n)$. Hence $\tilde{f} = f$.

The general case follows by combining the above result and an argument involving a justification of passing to the limit. □

1.3 Tempered Distributions

From the definitions of the Fourier transform on $L^1(\mathbb{R}^n)$ and on $L^2(\mathbb{R}^n)$, there is a natural extension to $L^1(\mathbb{R}^n) + L^2(\mathbb{R}^n)$. It is not hard to see that $L^1(\mathbb{R}^n) + L^2(\mathbb{R}^n)$ contains the spaces $L^p(\mathbb{R}^n)$ for $1 \le p \le 2$. On the other hand, as we shall prove, any function in $L^p(\mathbb{R}^n)$ for $p > 2$ has a Fourier transform in the distribution sense. However, they may not be function, they are *tempered distributions*. Before studying them, it is convenient to see how far Definition 1.1 can be carried out.

Example 1.7 Let $n \ge 1$ and $f(x) = \delta_0$, the delta function, i.e., the measure of mass one concentrated at the origin. Using (1.1) one finds that

$$\widehat{\delta_0}(\xi) = \int_{\mathbb{R}^n} \delta_0(x)\, e^{-2\pi i(x \cdot \xi)} dx \equiv 1.$$

In fact, Definition 1.1 tells us that if μ is a bounded measure, then $\widehat{\mu}(\xi)$ represents a function in $L^\infty(\mathbb{R}^n)$.

Suppose that given $f(x) \equiv 1$ we want to find $\widehat{f}(\xi)$. In this case, Definition 1.1 cannot be used directly. It is necessary to introduce the notion of tempered distribution. For this purpose, we first need the following family of seminorms.

For each $(\nu, \beta) \in (\mathbb{Z}^+)^{2n}$ we denote the seminorm $\||\cdot\||_{(\nu,\beta)}$ defined as:

$$\||f\||_{(\nu,\beta)} = \|x^\nu \partial_x^\beta f\|_\infty.$$

Now we can define the Schwartz space $\mathcal{S}(\mathbb{R}^n)$, the space of the C^∞-functions decaying at infinity, i.e.,

$$\mathcal{S}(\mathbb{R}^n) = \{\varphi \in C^\infty(\mathbb{R}^n) : \||\varphi\||_{(\nu,\beta)} < \infty \text{ for any } \nu, \beta \in (\mathbb{Z}^+)^n\}.$$

Thus, $C_0^\infty(\mathbb{R}^n) \subsetneq \mathcal{S}(\mathbb{R}^n)$ (consider $f(x)$ as in Example 1.3).

The topology in $\mathcal{S}(\mathbb{R}^n)$ is given by the family of seminorms $\||\cdot\||_{(\nu,\beta)}$, $(\nu, \beta) \in (\mathbb{Z}^+)^{2n}$.

Definition 1.3. Let $\{\varphi_j\} \subset \mathcal{S}(\mathbb{R}^n)$. Then, $\varphi_j \to 0$ as $j \to \infty$, if for any $(\nu, \beta) \in (\mathbb{Z}^+)^{2n}$ one has that

$$\||\varphi_j\||_{(\nu,\beta)} \longrightarrow 0 \text{ as } j \to \infty.$$

The relationship between the Fourier transform and the function space $\mathcal{S}(\mathbb{R}^n)$ is described in the formulae (1.10). More precisely, we have the following result (see Exercise 1.13).

Theorem 1.6. *The map $\varphi \mapsto \widehat{\varphi}$ is an isomorphism from $\mathcal{S}(\mathbb{R}^n)$ into itself.*

Thus, $\mathcal{S}(\mathbb{R}^n)$ appears naturally associated to the Fourier transform. By duality, we can define the tempered distributions $\mathcal{S}'(\mathbb{R}^n)$.

Definition 1.4. We say that $\psi : \mathcal{S}(\mathbb{R}^n) \mapsto \mathbb{C}$ defines a tempered distribution, i.e., $\Psi \in \mathcal{S}'(\mathbb{R}^n)$ if:

1. Ψ is linear.
2. Ψ is continuous, i.e., if for any $\{\varphi_j\} \subseteq \mathcal{S}(\mathbb{R}^n)$ such that $\varphi_j \to 0$ as $j \to \infty$, then the numerical sequence $\Psi(\varphi_j) \to 0$ as $j \to \infty$.

It is easy to check that any bounded function f defines a tempered distribution Ψ_f, where

$$\Psi_f(\varphi) = \int_{\mathbb{R}^n} f(x)\varphi(x)dx, \quad \text{for any } \varphi \in \mathcal{S}(\mathbb{R}^n). \tag{1.13}$$

In fact, this identity allows us to see that any locally integrable function with polynomial growth at infinity defines a tempered distribution. In particular, we have the $L^p(\mathbb{R}^n)$ spaces with $1 \le p \le \infty$. The following example gives us a tempered distribution outside these function spaces.

Example 1.8 In $\mathcal{S}'(\mathbb{R})$, define the *principal value function* of $1/x$, denoted by p.v. $\frac{1}{x}$, by the expression

$$\text{p.v. } \frac{1}{x}(\varphi) = \lim_{\epsilon \downarrow 0} \int\limits_{\epsilon < |x| < 1/\epsilon} \frac{\varphi(x)}{x} dx,$$

for any $\varphi \in \mathcal{S}(\mathbb{R})$. Since $1/x$ is an odd function,

$$\text{p.v.} \frac{1}{x}(\varphi) = \int\limits_{|x|<1} \frac{\varphi(x) - \varphi(0)}{x} dx + \int\limits_{|x|>1} \frac{\varphi(x)}{x} dx. \tag{1.14}$$

Therefore,

$$\left|\text{p.v. } \frac{1}{x}(\varphi)\right| \leq 2\|\varphi'\|_\infty + 2\|x\varphi\|_\infty, \tag{1.15}$$

and consequently, $\text{p.v. } \frac{1}{x} \in \mathcal{S}'(\mathbb{R})$.

Now, given a $\Psi \in \mathcal{S}'(\mathbb{R}^n)$, its Fourier transform can be defined in the following natural form.

Definition 1.5. Given $\Psi \in \mathcal{S}'(\mathbb{R}^n)$, its Fourier transform $\widehat{\Psi} \in \mathcal{S}'(\mathbb{R}^n)$ is defined as:

$$\widehat{\Psi}(\varphi) = \Psi(\widehat{\varphi}), \quad \text{for any } \varphi \in \mathcal{S}(\mathbb{R}^n). \tag{1.16}$$

Observe that for $f \in L^1(\mathbb{R}^n)$ and $\varphi \in \mathcal{S}(\mathbb{R}^n)$, (1.7), (1.13), and (1.16) tell us that

$$\widehat{\Psi}_f(\varphi) = \Psi_f(\widehat{\varphi}) = \int\limits_{\mathbb{R}^n} f(x)\widehat{\varphi}(x)dx = \int\limits_{\mathbb{R}^n} \widehat{f}(x)\varphi(x)dx = \Psi_{\widehat{f}}(\varphi).$$

Therefore, for $f \in L^1(\mathbb{R}^n) + L^2(\mathbb{R}^n)$ one has that $\widehat{\Psi}_f = \Psi_{\widehat{f}}$. Thus, Definition 1.5 is consistent with the theory of the Fourier transform developed in Sects. 1.1 and 1.2.

Example 1.9 Let $f(x) \equiv 1 \in L^\infty(\mathbb{R}^n) \subset \mathcal{S}'(\mathbb{R}^n)$. Using the previous notation, for any $\varphi \in \mathcal{S}(\mathbb{R}^n)$ it follows that

$$\widehat{\Psi}_1(\varphi) = \Psi_1(\widehat{\varphi}) = \int\limits_{\mathbb{R}^n} 1\, \widehat{\varphi}(x)dx = \varphi(0) = \int\limits_{\mathbb{R}^n} \delta_0(x)\, \varphi(x)dx = \delta_0(\varphi).$$

Hence $\widehat{1} = \delta_0$. We recall that in Example 1.7 we already saw that $\widehat{\delta_0} = 1$.

Next we compute the Fourier transform of the tempered distribution in Example 1.8.

Example 1.10 Combining Definition 1.5, Fubini's theorem, and the Lebesgue dominated convergence theorem we have that for any $\varphi \in \mathcal{S}(\mathbb{R})$,

$$\begin{aligned}\widehat{\text{p.v. } \frac{1}{x}}(\varphi) &= \text{p.v. } \frac{1}{x}(\widehat{\varphi}) = \lim_{\epsilon \downarrow 0} \int\limits_{\epsilon < |x| < 1/\epsilon} \frac{\widehat{\varphi}(x)}{x} dx \\ &= \lim_{\epsilon \downarrow 0} \int\limits_{\epsilon < |x| < 1/\epsilon} \frac{1}{x}\Big(\int\limits_{-\infty}^{\infty} \varphi(y) e^{-2\pi i x y}\, dy\Big) dx\end{aligned}$$

$$= \lim_{\epsilon \downarrow 0} \int_{-\infty}^{\infty} \varphi(y) \Big(\int_{\epsilon < |x| < 1/\epsilon} \frac{e^{-2\pi i x y}}{x} dx \Big) \, dy$$

$$= \int_{-\infty}^{\infty} \varphi(y) \Big(\lim_{\epsilon \downarrow 0} \int_{\epsilon < |x| < 1/\epsilon} \frac{e^{-2\pi i x y}}{x} dx \Big) dy$$

$$= -i\pi \int_{-\infty}^{\infty} \operatorname{sgn}(y) \, \varphi(y) dy,$$

where a change of variables and complex integration have been used to conclude that

$$\lim_{\epsilon \downarrow 0} \int_{\epsilon < |x| < 1/\epsilon} \frac{e^{-2\pi i x y}}{x} dx = -2i \int_0^{\infty} \frac{\sin(2\pi x y)}{x} dx = -2i \operatorname{sgn}(y) \int_0^{\infty} \frac{\sin(x)}{x} dx$$

$$= -i\pi \operatorname{sgn}(y).$$

This yields the identity:

$$\widehat{\text{p.v.} \frac{1}{x}}(\xi) \; = \; - \, i\pi \operatorname{sgn}(\xi).$$

The topology in $\mathcal{S}'(\mathbb{R}^n)$ can be described in the following form.

Definition 1.6. Let $\{\Psi_j\} \subset \mathcal{S}'(\mathbb{R}^n)$. Then, $\Psi_j \to 0$ as $j \to \infty$ in $\mathcal{S}'(\mathbb{R}^n)$, if for any $\varphi \in \mathcal{S}(\mathbb{R}^n)$ it follows that $\Psi_j(\varphi) \longrightarrow 0$ as $j \to \infty$.

As a consequence of the Definitions 1.4, 1.6, we get the next extension of Theorem 1.6, whose proof we leave as an exercise.

Theorem 1.7. *The map* $\mathcal{F} : \Psi \mapsto \widehat{\Psi}$ *is an isomorphism from* $\mathcal{S}'(\mathbb{R}^n)$ *into itself.*

Combining the above results with an extension of Example 1.3 (see Exercise 1.2), we can justify the following computation related with the fundamental solution of the time-dependent Schrödinger equation.

Example 1.11 $\widehat{e^{-4\pi^2 it|x|^2}} = \lim_{\epsilon \to 0^+} \widehat{e^{-4\pi^2(\epsilon+it)|x|^2}}$ in $\mathcal{S}'(\mathbb{R}^n)$.

From Exercise 1.2, it follows that

$$(\widehat{e^{-4\pi^2(\epsilon+it)|x|^2}})(\xi) = \frac{e^{-|\xi|^2/4(\epsilon+it)}}{[4\pi(\epsilon+it)]^{n/2}}.$$

Taking the limit $\epsilon \to 0^+$, we obtain:

$$(\widehat{e^{-4\pi^2 it|x|^2}})(\xi) = \frac{e^{i|\xi|^2/4t}}{(4\pi i t)^{n/2}}. \tag{1.17}$$

As an application of these ideas, we introduce the Hilbert transform.

Definition 1.7. For $\varphi \in \mathcal{S}(\mathbb{R})$, we define its *Hilbert transform* $\mathsf{H}(\varphi)$ by

$$\mathsf{H}(\varphi)(y) \;=\; \frac{1}{\pi}\,\text{p.v.}\frac{1}{x}\;(\varphi(y-\cdot)) \;=\; \frac{1}{\pi}\,\text{p.v.}\;\frac{1}{x} * \varphi(y).$$

From (1.14) and (1.15) it is clear that $\mathsf{H}(\varphi)(y)$ is defined for any $y \in \mathbb{R}$ and it is bounded by $g(y) = a|y| + b$, with $a, b > 0$ depending on φ. In particular, we have that $\mathsf{H}(\varphi) \in \mathcal{S}'(\mathbb{R})$. Let us compute its Fourier transform.

Example 1.12 From Example 1.10 and the identity

$$\mathsf{H}(\varphi)(y) = \lim_{\epsilon \to 0} \left(\frac{1}{\pi}\,\frac{1}{x}\chi_{\{\epsilon<|x|<1/\epsilon\}} * \varphi\right)(y) \quad \text{in} \quad \mathcal{S}'(\mathbb{R})$$

it follows that

$$\lim_{\epsilon \to 0} \widehat{\left(\frac{1}{\pi}\,\frac{1}{x}\;\chi_{\{\epsilon<|x|<1/\epsilon\}} * \varphi\right)}(\xi) = -i\;\text{sgn}(\xi)\,\widehat{\varphi}(\xi).$$

This implies that

$$\widehat{\mathsf{H}(\varphi)}(\xi) = -i\;\text{sgn}(\xi)\,\widehat{\varphi}(\xi), \quad \text{for any} \quad \varphi \in \mathcal{S}(\mathbb{R}). \tag{1.18}$$

The identity (1.18) allows us to extend the Hilbert transform as an isometry in $L^2(\mathbb{R})$. It is not hard to see that

$$\|\mathsf{H}(\varphi)\|_2 = \|\varphi\|_2 \quad \text{and} \quad \mathsf{H}(\mathsf{H}(\varphi)) = -\varphi.$$

Other properties of the Hilbert transform are deduced in the exercises in Chaps. 1 and 2.

In Definition 1.7, we have implicitly utilized the following result, which is employed again in the applications at the end of this chapter.

Proposition 1.3. *Let $\varphi \in \mathcal{S}(\mathbb{R}^n)$ and $\Psi \in \mathcal{S}'(\mathbb{R}^n)$. Define*

$$\Psi * \varphi(x) = \Psi(\varphi(x-\cdot)). \tag{1.19}$$

Then,

$$\Psi * \varphi \in C^\infty(\mathbb{R}^n) \cap \mathcal{S}'(\mathbb{R}^n)$$

and

$$\widehat{\Psi * \varphi} = \widehat{\Psi}\widehat{\varphi}, \tag{1.20}$$

where $\widehat{\Psi}\widehat{\varphi} \in \mathcal{S}'(\mathbb{R}^n)$ is defined as $\widehat{\Psi}\widehat{\varphi}(\phi) = \widehat{\Psi}(\widehat{\varphi}\phi)$ for any $\phi \in \mathcal{S}(\mathbb{R}^n)$.

Proof. It is left as an exercise. □

1.4 Oscillatory Integrals in One Dimension

In many problems and applications the following question arises:

What is the asymptotic behavior of $I(\lambda)$ when $\lambda \to \infty$, where

$$I(\lambda) = \int_a^b e^{i\lambda\phi(x)} f(x)dx, \tag{1.21}$$

and ϕ is a smooth real-valued function, called the "phase function," and f is a smooth complex-valued function?

We shall see that this asymptotic behavior is determined by the points $\bar{x}$, where the derivative of ϕ vanishes, i.e., $\phi'(\bar{x}) = 0$.

Proposition 1.4. *Let $f \in C_0^\infty([a,b])$ and $\phi'(x) \neq 0$ for any $x \in [a,b]$. Then*

$$I(\lambda) = \int_a^b e^{i\lambda\phi(x)} f(x)dx = O(\lambda^{-k}), \quad \textit{as } \lambda \to \infty \tag{1.22}$$

for any $k \in \mathbb{Z}^+$.

Proof. Define the differential operator

$$\mathcal{L}(f) = \frac{1}{i\lambda\phi'}\frac{df}{dx},$$

which satisfies

$$\mathcal{L}^t(f) = -\frac{d}{dx}\left(\frac{f}{i\lambda\phi'}\right) \quad \text{and} \quad \mathcal{L}^k(e^{i\lambda\phi}) = e^{i\lambda\phi},$$

where $\mathcal{L}^t$ denotes the adjoint of $\mathcal{L}$. Using integration by parts it follows that

$$\begin{aligned}\int_a^b e^{i\lambda\phi} f dx &= \int_a^b \mathcal{L}^k(e^{i\lambda\phi}) f\, dx \\ &= (-1)^k \int_a^b e^{i\lambda\phi} (\mathcal{L}^t)^k f\, dx = O(\lambda^{-k}), \text{ as } \lambda \to \infty.\end{aligned}$$

□

Proposition 1.5. *Let $k \in \mathbb{Z}^+$ and $|\phi^{(k)}(x)| \geq 1$ for any $x \in [a,b]$ with $\phi'(x)$ monotonic in the case $k = 1$. Then,*

$$\left|\int_a^b e^{i\lambda\phi(x)} dx\right| \leq c_k \lambda^{-1/k}, \tag{1.23}$$

where the constant c_k *is independent of* a, b.

Proof. For $k = 1$, we have that

$$\int_a^b e^{i\lambda\phi}dx = \int_a^b \mathcal{L}(e^{i\lambda\phi})dx = \frac{1}{i\lambda\phi'}e^{i\lambda\phi}\Big|_a^b - \int_a^b e^{i\lambda\phi}\frac{1}{i\lambda}\frac{d}{dx}\Big(\frac{1}{\phi'}\Big)dx.$$

Clearly, the first term on the right-hand side is bounded by $2\lambda^{-1}$. On the other hand, the hypothesis of monotonicity on ϕ' guarantees that

$$\Big|\int_a^b e^{i\lambda\phi}\frac{1}{i\lambda}\frac{d}{dx}\Big(\frac{1}{\phi'}\Big)\,dx\Big| \le \frac{1}{\lambda}\int_a^b \Big|\frac{d}{dx}\Big(\frac{1}{\phi'}\Big)\Big|\,dx$$
$$= \frac{1}{\lambda}\Big|\frac{1}{\phi'(b)} - \frac{1}{\phi'(a)}\Big| \le \frac{2}{\lambda}.$$

This yields the proof of the case $k = 1$.

For the proof of the case $k \ge 2$, induction in k is used. Assuming the result for k, we shall prove it for $k+1$. By hypothesis, $|\phi^{(k+1)}(x)| \ge 1$. Let $x_0 \in [a, b]$ be such that

$$|\phi^{(k)}(x_0)| = \min_{a\le x\le b} |\phi^{(k)}(x)|.$$

If $\phi^{(k)}(x_0) = 0$, outside the interval $(x_0 - \delta,\ x_0 + \delta)$, one has that $|\phi^{(k)}(x)| \ge \delta$, with ϕ' monotonic if $k = 1$. Splitting the domain of integration and applying the hypothesis we obtain that

$$\Big|\int_a^{x_0-\delta} e^{i\lambda\phi(x)}\,dx\Big| + \Big|\int_{x_0+\delta}^b e^{i\lambda\phi(x)}\,dx\Big| \le c_k(\lambda\delta)^{-1/k}.$$

A simple computation shows that

$$\Big|\int_{x_0-\delta}^{x_0+\delta} e^{i\lambda\phi(x)}\,dx\Big| \le 2\delta.$$

Thus,

$$\Big|\int_a^b e^{i\lambda\phi(x)}\,dx\Big| \le c_k(\lambda\delta)^{-1/k} + 2\delta.$$

If $\phi^{(k)}(x_0) \ne 0$, then $x_0 = a$ or b and a similar argument provides the same bound. Finally, taking $\delta = \lambda^{-1/(k+1)}$ we complete the proof. □

Corollary 1.1 (van der Corput). *Under the hypotheses of Proposition 1.5,*

$$\Big|\int_a^b e^{i\lambda\phi(x)} f(x)\,dx\Big| \le c_k\lambda^{-1/k}\left(\|f\|_\infty + \|f'\|_1\right) \tag{1.24}$$

with c_k *independent of* a, b.

Proof. Define

$$G(x) = \int_a^x e^{i\lambda\phi(y)}\,dy.$$

By (1.23) one has that

$$|G(x)| \le c_k \lambda^{-1/k}.$$

Now using integration by parts we obtain:

$$\begin{aligned}\left|\int_a^b e^{i\lambda\phi} f\,dx\right| = \left|\int_a^b G' f\,dx\right| &\le \left|\,(Gf)\Big|_a^b\right| + \left|\int_a^b Gf'\,dx\right| \\ &\le c_k \lambda^{-1/k}\left(\|f\|_\infty + \|f'\|_1\right).\end{aligned}$$

□

Next, we shall study an application of these results.

Proposition 1.6. *Let* $\beta \in [0, 1/2]$ *and* $I_\beta(x)$ *be the oscillatory integral*

$$I_\beta(x) = \int_{-\infty}^{\infty} e^{i(x\eta+\eta^3)}\,|\eta|^\beta d\eta. \tag{1.25}$$

Then, $I_\beta \in L^\infty(\mathbb{R})$.

Proof. First, we fix $\varphi_0 \in C^\infty(\mathbb{R})$ such that

$$\varphi_0(\eta) = \begin{cases} 1, & \text{if } |\eta| > 2 \\ 0, & \text{if } |\eta| < 1. \end{cases}$$

Observe that $(1-\varphi_0)(\eta)e^{i\eta^3}|\eta|^\beta \in L^1(\mathbb{R})$, therefore its Fourier transform belongs to $L^\infty(\mathbb{R})$. Thus, it suffices to consider

$$\tilde{I}_\beta(x) = \int_{-\infty}^{\infty} e^{i(x\eta+\eta^3)}\,|\eta|^\beta \varphi_0(\eta)d\eta.$$

For $x \ge -3$, the phase function $\phi_x(\eta) = x\eta + \eta^3$, in the support of φ_0, satisfies

$$|\phi_x'(\eta)| = |x + 3\eta^2| \ge (|x| + |\eta|^2).$$

In this case, integration by parts leads to the desired result.

For $x < -3$, we consider the functions $(\varphi_1, \varphi_2) \in C_0^\infty \times C^\infty$ such that $\varphi_1(\eta) + \varphi_2(\eta) = 1$ with

$$\operatorname{supp}\varphi_1 \subset A = \left\{\eta : |x + 3\eta^2| \leq \frac{|x|}{2}\right\},$$

and

$$\varphi_2 = 0 \quad \text{in} \quad B = \left\{\eta : |x + 3\eta^2| < \frac{|x|}{3}\right\},$$

and we split the integral $\tilde{I}_\beta(x)$ in two pieces,

$$|\tilde{I}_\beta(x)| \leq |\tilde{I}_\beta^1(x)| + |\tilde{I}_\beta^2(x)|,$$

where

$$\tilde{I}_\beta^j(x) = \int_{-\infty}^{\infty} e^{i(x\eta+\eta^3)}\, |\eta|^\beta \varphi_0(\eta)\varphi_j(\eta) d\eta, \quad \text{for} \quad j = 1, 2.$$

When $\varphi_2(\eta) \neq 0$, the triangle inequality shows that

$$|\phi_x'(\eta)| = |x + 3\eta^2| \geq \frac{3}{13}(|x| + |\eta|^2).$$

Integration by parts leads to

$$|\tilde{I}_\beta^2(x)| = \left| \int_{-\infty}^{\infty} \frac{|\eta|^\beta}{\phi_x'(\eta)} \varphi_0(\eta)\varphi_2(\eta) \frac{d}{d\eta} e^{i(x\eta+\eta^3)}\, d\eta \right| \leq 100.$$

Now, if $\eta \in A$, we have that

$$\frac{|x|}{2} \leq 3\eta^2 \leq 3\frac{|x|}{2} \quad \text{and} \quad \left|\frac{d^2\phi_x}{d\eta^2}(\eta)\right| = 6|\eta| \geq |x|^{1/2}.$$

Thus (1.24) (van der Corput) and the form of φ_0, φ_1 guarantee the existence of a constant c independent of $x < -3$ such that

$$|\tilde{I}_\beta^1(x)| = \left| \int_{-\infty}^{\infty} e^{i(x\eta+\eta^3)}\, |\eta|^\beta \varphi_0(\eta)\varphi_1(\eta)\, d\eta \right| \leq c\,|x|^{-1/4} |x|^{\beta/2}.$$

□

1.5 Applications

Consider the initial value problem (IVP) for the linear Schrödinger equation:

$$\begin{cases} \partial_t u = i\,\Delta u, \\ u(x,0) = u_0(x), \end{cases} \tag{1.26}$$

$x \in \mathbb{R}^n$, $t \in \mathbb{R}$. Taking the Fourier transform with respect to the space variable x in (1.26) we obtain:

$$\begin{cases} \widehat{\partial_t u}(\xi,t) = \partial_t \widehat{u}(\xi,t) = \widehat{i\Delta u}(\xi,t) = -4\pi^2 i|\xi|^2 \widehat{u}(\xi,t) \\ \widehat{u}(\xi,0) = \widehat{u_0}(\xi). \end{cases}$$

The solution of this family of ordinary differential equations (ODE), with parameter ξ, can be written as:

$$\widehat{u}(\xi,t) = e^{-4\pi^2 it|\xi|^2}\widehat{u_0}(\xi).$$

By Proposition 1.3 it follows that

$$\begin{aligned} u(x,t) &= (e^{-4\pi^2 it|\xi|^2}\widehat{u_0}(\xi))^\vee = (e^{-4\pi^2 it|\xi|^2})^\vee * u_0(x) \\ &= \frac{e^{i|\cdot|^2/4t}}{(4\pi it)^{n/2}} * u_0(x) = e^{it\Delta}u_0(x), \end{aligned} \tag{1.27}$$

where we have introduced the notation $e^{it\Delta}$ which is justified in Chapter 4.

Next, we consider the IVP associated to the linearized Korteweg–de Vries (KdV) equation:

$$\begin{cases} \partial_t v + \partial_x^3 v = 0, \\ v(x,0) = v_0(x) \end{cases} \tag{1.28}$$

for $t, x \in \mathbb{R}$. The previous argument shows that

$$v(x,t) = S_t * v_0(x) = (e^{8\pi^3 it\xi^3}\widehat{v_0})^\vee = V(t)v_0(x), \tag{1.29}$$

where the kernel $S_t(x)$ is defined by the oscillatory integral:

$$S_t(x) = \int_{-\infty}^{\infty} e^{2\pi ix\xi} e^{8\pi^3 it\xi^3}\, d\xi. \tag{1.30}$$

After changing variables,

$$S_t(x) = \frac{1}{\sqrt[3]{3t}}\, Ai\left(\frac{x}{\sqrt[3]{3t}}\right), \tag{1.31}$$

where $Ai(\cdot)$ denotes the Airy function:

$$Ai(x) = \frac{1}{2\pi}\int_{-\infty}^{\infty} e^{i(\xi x + \xi^3/3)} d\xi. \tag{1.32}$$

By combining Proposition 1.6 (with $\beta = 0$) and a new change of variable we find that

$$\|S_t\|_\infty \leq c|t|^{-1/3}. \tag{1.33}$$

Moreover, if $\beta \in [0, 1/2]$, then

$$\|D_x^\beta S_t\|_\infty \le c|t|^{-(\beta+1)/3}. \tag{1.34}$$

Hence, using Exercise 1.6 it follows that

$$\|D_x^\beta V(t)v_0\|_\infty = \|D_x^\beta S_t * v_0\|_\infty \le c|t|^{-(\beta+1)/3}\|v_0\|_1, \tag{1.35}$$

where $D_x^\beta = D^\beta = (-\Delta)^{\beta/2}$ denotes the homogeneous fractional derivative of order β, i.e.,

$$D^\beta f(x) = [(2\pi|\xi|)^\beta \widehat{f}(\xi)]^\vee(x). \tag{1.36}$$

Notice that the derivative of the phase function in (1.32) $\phi(\xi) = \xi x + \xi^3/3$ does not vanish for $x > 0$, i.e., $|\phi'(\xi)| = |x + \xi^2| \ge |x|$, so using Proposition 1.4 one sees that $Ai(x)$ has fast decay for $x > 0$. In fact, one has (see [Ho2] or [SSS]) that

$$|Ai(x)| \le \frac{1}{(1+x_-)^{1/4}}\, e^{-cx_+^{3/2}}, \tag{1.37}$$

and

$$|Ai'(x)| \le (1+x_-)^{1/4} e^{-cx_+^{3/2}}, \tag{1.38}$$

where $x_+ = \max\{x; 0\}$ and $x_- = \max\{-x; 0\}$.

Hence, (1.34) with $\beta = 1/2$ can be seen as an interpolation between (1.37) and (1.38) and the scaling.

Remark 1.1. The relevant references used in this chapter are the books [SW], [S2], [S3], [Sa], [Du], and [Rd].

1.6 Exercises

1.1 (i) Let $n \ge 1$ and $f(x) = e^{-2\pi|x|}$. Show that

$$\hat{f}(\xi) = \frac{\Gamma[(n+1)/2]}{\pi^{(n+1)/2}} \frac{1}{(1+|\xi|^2)^{(n+1)/2}}.$$

Hint: From the formula of Example 1.5 with $a = \beta$ and $b = 1$ one sees that

$$e^{-\beta} = \frac{2}{\pi}\int_0^\infty \frac{\cos(\beta x)}{1+x^2}\,dx,$$

which, combined with the equality:

$$\frac{1}{1+x^2} = \int_0^\infty e^{-(1+x^2)\rho}\,d\rho, \quad \text{yields} \quad e^{-\beta} = \int_0^\infty \frac{e^{-\rho}}{\sqrt{\rho}}\, e^{-\beta^2/4\rho}\,d\rho.$$

Use this identity to obtain the desired result.

(ii) Let $n = 1$ and $f(x) = \dfrac{1}{\pi}\dfrac{1}{(1+x^2)^2}$. Show that

$$\hat{f}(\xi) = \frac{1}{2}e^{-2\pi|\xi|}(2\pi|\xi| + 1).$$

Hint: Differentiate the identity in Example 1.5.

1.2 (i) Prove the following extension in $\mathcal{S}'(\mathbb{R}^n)$ of formula (1.8):

$$\widehat{(e^{-a|x|^2})}(\xi) = \Big(\frac{\pi}{a}\Big)^{n/2} e^{-\pi^2|\xi|^2/a}, \quad \mathcal{R}e\, a \geq 0, \ a \neq 0,$$

where $\sqrt{a}$ is defined as the branch with $\mathcal{R}e\, a > 0$.
Hint: Use an analytic continuation argument.

(ii) Show that if $a = 1 + it$, then

$$\left\| \Big(\frac{\pi}{a}\Big)^{n/2} e^{-\pi^2|x|^2/a} \right\|_p \sim c_p (1+t)^{n(\frac{1}{p}-\frac{1}{2})}, \quad 1 \leq p \leq \infty, \ t > 0,$$

and

$$\|e^{-\pi a|\xi|^2}\|_q \sim c_q, \quad 1 \leq q \leq \infty,$$

where $f(t) \sim g(t)$, for $f, g \geq 0$, means that there exists $c > 1$ such that

$$c^{-1} f(t) \leq g(t) \leq c\, f(t), \quad \forall t > 0.$$

1.3 Prove Young's inequality: Let $f \in L^p(\mathbb{R}^n)$, $1 \leq p \leq \infty$, and $g \in L^1(\mathbb{R}^n)$. Then, $f * g \in L^p(\mathbb{R}^n)$ with

$$\|f * g\|_p \leq \|f\|_p \|g\|_1. \tag{1.39}$$

1.4 Prove the Minkowski integral inequality. If $1 \leq p \leq \infty$, then

$$\left(\int_{\mathbb{R}^n} \left| \int_{\mathbb{R}^n} f(x,y)\, dx \right|^p dy \right)^{1/p} \leq \int_{\mathbb{R}^n} \left(\int_{\mathbb{R}^n} \left| f(x,y) \right|^p dy \right)^{1/p} dx. \tag{1.40}$$

Observe that the proof of the cases $p = 1, \infty$ is immediate.

1.5 Let $f \in L^p((0,\infty))$, $1 < p < \infty$, $f \geq 0$:

(i) Prove Hardy's inequality:

$$\int_0^\infty \left(\frac{1}{x} \int_0^x f(s)\, ds \right)^p dx \leq \left(\frac{p}{p-1} \right)^p \int_0^\infty (f(x))^p\, dx. \tag{1.41}$$

(ii) Prove that equality in (1.41) holds if and only if $f = 0$, a.e., and that the constant $c_p = p/(p-1)$ is optimal in (1.41).

(iii) Prove that (1.41) fails for $p = 1$ and $p = \infty$.
Hint: Assuming $f \in C_0((0,\infty))$ define

$$F(x) = \frac{1}{x} \int_0^\infty f(s)\, ds, \text{so } x\, F' = f - F.$$

Use integration by parts and the Hölder inequality to obtain (1.41).

1.6 Consider the Fourier transform $\widehat{\ }$ as a map from $L^1(\mathbb{R}^n)$ into $L^\infty(\mathbb{R}^n)$.

(i) Prove that $\widehat{\ }$ is injective.

(ii) Prove that the image of $\widehat{\ }$, i.e., $\widehat{L^1(\mathbb{R}^n)}$, is an algebra with respect to the point-wise multiplication of functions.

(iii) Prove that $\widehat{L^1(\mathbb{R}^n)} \subsetneq C_\infty(\mathbb{R}^n)$, where $C_\infty(\mathbb{R}^n)$ denotes the space of continuous functions vanishing at infinity.
Hint: From Example 1.2 we have that $\|g_k\|_\infty = 2$ and

$$\lim_{k\uparrow\infty} \|\widehat{g}_k\|_1 = \infty.$$

Apply the open mapping theorem to get the desired result.

1.7 (i) Prove the following generalization of (1.6) in Theorem 1.1:
If $f \in L^1(\mathbb{R}^n)$ and $g \in L^p(\mathbb{R}^n)$, $1 \leq p \leq 2$, then $\widehat{(f * g)}(\xi) = \widehat{f}(\xi)\,\widehat{g}(\xi)$.

(ii) If $f \in L^p(\mathbb{R}^n)$, $g \in L^{p'}(\mathbb{R}^n)$, with $1/p + 1/p' = 1$, $1 < p < \infty$, then $f * g \in C_\infty(\mathbb{R}^n)$. What can you affirm if $p = 1, \infty$?

(iii) If $f \in L^1(\mathbb{R}^n)$, with f continuous at the point 0 and $\widehat{f} \geq 0$, then $\widehat{f} \in L^1(\mathbb{R}^n)$.
Hint: Use Proposition 1.2 and Fatou's lemma.

1.8 Show that

$$\int_0^\infty \frac{\sin^2 x}{x^2}\, dx = \frac{\pi}{2} \quad \text{and} \quad \int_0^\infty \frac{\sin^3 x}{x^3} dx = \frac{3\pi}{8}.$$

Hint: Combine the identities (1.7), (1.11), and Example 1.1.

1.9 For a given $f \in L^2(\mathbb{R}^n)$ prove that the following statements are equivalent:

(i) $g \in L^2(\mathbb{R}^n)$ is the partial derivative of $f \in L^2(\mathbb{R}^n)$ with respect to the kth variable according to Definition 1.2.

(ii) There exists $g \in L^2(\mathbb{R}^n)$ such that

$$\int_{\mathbb{R}^n} f(x)\partial_{x_k}\phi(x)\, dx = -\int_{\mathbb{R}^n} g(x)\phi(x)\, dx \tag{1.42}$$

for any $\phi \in C_0^\infty(\mathbb{R}^n)$. In general, if (1.42) holds for two distributions f, g, then one says that g is the kth partial derivative of f in the distribution sense.

(iii) There exists $\{f_j\} \subset C_0^\infty(\mathbb{R}^n)$ such that

$$\|f_j - f\|_2 \to 0 \quad \text{as} \quad j \to \infty,$$

and $\{\partial_{x_k} f_j\}$ is a Cauchy sequence in $L^2(\mathbb{R}^n)$.

(iv) $\xi_k \widehat{f}(\xi) \in L^2(\mathbb{R}^n)$.

(v)

$$\sup_{h>0} \int_{\mathbb{R}^n} \left| \frac{f(x+he_k) - f(x)}{h} \right|^2 dx < \infty.$$

For $p \neq 2$, which of the above statements are still equivalent?

1.10 (Paley–Wiener theorem) Prove that if $f \in C_0^\infty(\mathbb{R}^n)$ with support in $\{x \in \mathbb{R}^n : |x| \leq M\}$, then $\widehat{f}(\xi)$ can be extended analytically to $\mathbb{C}^n$. Moreover, if $k \in \mathbb{Z}^+$ one has that

$$|\widehat{f}(\xi + i\eta)| \leq c_k \frac{e^{2\pi M|\eta|}}{(1 + |(\xi + i\eta)|)^k} \quad \text{for any } \xi + i\eta \in \mathbb{C}^n. \tag{1.43}$$

Prove the converse, i.e., if $F(\xi + i\eta)$ is an analytic function in $\mathbb{C}^n$ satisfying (1.43), then F is the Fourier transform of some $f \in C_0^\infty(\mathbb{R}^n)$ with support in $\{x \in \mathbb{R}^n : |x| \leq M\}$.

1.11 Show that if $f \in L^1(\mathbb{R}^n)$, $f \not\equiv 0$, with compact support, then for any $\epsilon > 0$, $\hat{f} \notin L^1(e^{\epsilon|x|}dx)$.

1.12 Prove that given $k \in \mathbb{Z}^+$ and $a_\alpha \in \mathbb{R}^k$, with $\alpha = (\alpha_1, \ldots, \alpha_n) \in \mathbb{N}^n$, $|\alpha| = \alpha_1 + \cdots + \alpha_n \leq k$, there exists $f \in C_0^\infty(\mathbb{R}^n)$ such that

$$\int_{\mathbb{R}^n} x^\alpha f(x)\, dx = a_\alpha.$$

Hint: Use Exercise 1.10.

1.13 (i) Prove that if $f, g \in \mathcal{S}$, then $f * g \in \mathcal{S}$.

(ii) Prove that the Fourier transform is an isomorphism from $\mathcal{S}$ into itself.

(iii) Using the results in Section 1.3, find explicitly $\Psi = \widehat{|x|^2} \in \mathcal{S}'(\mathbb{R}^n)$.

(iv) Prove Proposition 1.3.

1.14 In this problem we shall prove that

$$\widehat{\frac{1}{|x|^\alpha}}(\xi) = c_{n,\alpha} \frac{1}{|\xi|^{n-\alpha}} \quad \text{for } \alpha \in (0, n)$$

as a tempered distribution, i.e., $\forall \varphi \in \mathcal{S}(\mathbb{R}^n)$

$$\int \frac{1}{|x|^\alpha} \widehat{\varphi}(x)\, dx = c_{n,\alpha} \int \frac{1}{|\xi|^{n-\alpha}} \varphi(\xi)\, d\xi, \tag{1.44}$$

where $c_{n,\alpha} = \pi^{\alpha - n/2}\, \Gamma(n/2 - \alpha/2)/\Gamma(\alpha/2)$.

(i) Combining the Parseval identity and Example 1.3 show that for $\delta > 0$

$$\int e^{-\pi\delta|x|^2} \widehat{\varphi}(x)\, dx = \delta^{-n/2} \int e^{-\pi|x|^2/\delta} \varphi(x)\, dx. \tag{1.45}$$

(ii) Prove the formula

$$\int_0^\infty e^{-\pi\delta|x|^2} \delta^{\beta-1}\, d\delta = \frac{c_\beta}{|x|^{2\beta}} \quad \text{for any } \beta > 0. \tag{1.46}$$

(iii) Multiply both sides of (1.45) by $\delta^{\frac{n-\alpha}{2}-1}$, integrate on δ, use Fubini's theorem and (1.46) to get (1.44).

1.15 Prove the following identities, where H denotes the Hilbert transform:

(i) $\mathsf{H}(fg) = \mathsf{H}(f)g + f\mathsf{H}(g) + \mathsf{H}(\mathsf{H}(f)\mathsf{H}(g))$.

(ii) $\mathsf{H}(\chi_{(-1,1)})(x) = \dfrac{1}{\pi}\log\left|\dfrac{x+1}{x-1}\right|$.

(iii) $\mathsf{H}\left(\dfrac{a}{x^2+a^2}\right) = \dfrac{x}{x^2+a^2}, \quad a > 0.$

1.16 Prove that if $\varphi \in \mathcal{S}(\mathbb{R})$, then $\mathsf{H}(\varphi) \in L^1(\mathbb{R})$ if and only if $\widehat{\varphi}(0) = 0$.

1.17 Consider the function $f_a(x) = \dfrac{x}{a-x^2}$.

(i) If $a \geq 0$ prove that the principal value function of $f_a(x)$,

$$\text{p.v.}\,\frac{x}{a-x^2}(\varphi) = \lim_{\epsilon\downarrow 0}\int_{\epsilon<|a-x^2|<1/\epsilon}\frac{x}{a-x^2}\,\varphi(x)\,dx,$$

with $\varphi \in \mathcal{S}(\mathbb{R})$ defines a tempered distribution. Moreover, prove that if

$$\widehat{f_a}(\xi) = \lim_{\epsilon\downarrow 0}\int_{\epsilon<|a-x^2|<1/\epsilon} e^{-2\pi i(x\cdot\xi)}\,\frac{x}{a-x^2}\,dx,$$

then

$$\|\widehat{f_a}\|_\infty \leq M, \tag{1.47}$$

where the constant M is independent of a.
Hint: Observe that if $a = 0$, $f_a(x)$ is just a multiple of the kernel $1/x$ of the Hilbert transform H. If $a > 0$, then $f_a(x)$ can be written as sum of translations of the kernel of the Hilbert transform H. Since the Hilbert transform satisfies a similar result, (1.47) follows in both cases. (See Example 1.10).

(ii) Show that (1.47) is also satisfied if $a < 0$.
Hint: Use Example 1.6.

1.18 Consider the IVP associated to the wave equation

$$\begin{cases} \partial_t^2 w - \Delta w = 0, \\ w(x,0) = f(x), \\ \partial_t w(x,0) = g(x), \end{cases} \tag{1.48}$$

$x \in \mathbb{R}^n$, $t \in \mathbb{R}$. Prove that

(i) If $f, g \in C_0^\infty(\mathbb{R}^n)$ are real-valued functions, then using the notation in (1.29), the solution can be described by the following expression:

$$w(x,t) = U'(t)f + U(t)g = \cos(Dt)f + \frac{\sin(Dt)}{D}g, \tag{1.49}$$

with $\widehat{Dh}(\xi) = 2\pi|\xi|\widehat{h}(\xi)$ (see (1.36)).

(ii) If $f,\ g$ are supported in $\{x \in \mathbb{R}^3 : |x| \le M\}$, show that $w(\cdot,t)$ is supported in $\{x \in \mathbb{R}^3 : |x| \le M + t\}$.

(iii) Assuming $n = 3$ and $f \equiv 0$, prove that

$$w(x,t) = \frac{1}{4\pi t}\int_{\{|y|=t\}} g(x+y)\,dS_y.$$

Hint: Derive and apply the following identity:

$$\int_{\{|x|=t\}} e^{2\pi i\xi\cdot x}\,dS_x = 4\pi t\,\frac{\sin(2\pi|\xi|t)}{2\pi|\xi|}.$$

If $g \in C_0^\infty(\mathbb{R}^3)$ is supported in $\{x \in \mathbb{R}^3 : |x| \le M\}$, where is the support of $w(\cdot,t)$?

(iv) Assuming $n = 3$ and $g \equiv 0$, prove that

$$w(x,t) = \frac{1}{4\pi t^2}\int_{\{|y|=t\}} [f(x+y) + \nabla f(x+y)\cdot y]\,dS_y. \tag{1.50}$$

(v) If $E(t) = \int_{\mathbb{R}^n}((\partial_t w)^2 + |\nabla_x w|^2)(x,t)dx$, then prove that for any $t \in \mathbb{R}$,

$$E(t) = E_0 = \int_{\mathbb{R}^n}(g^2 + |\nabla_x f|^2)(x)\,dx.$$

Hint: Use integration by parts and the equation.

(vi) (Brodsky [Br]) Show that

$$\lim_{t\to\infty}\int_{\mathbb{R}^n}(\partial_t w)^2(x,t)\,dx = \frac{E_0}{2}.$$

Hint: Use the Riemann–Lebesgue lemma (Theorem 1.1(3)).

1.19 Consider the IVP (1.28) with initial data $v_0 \in C_0^\infty(\mathbb{R})$. Prove that for any $t \ne 0$ $v(\cdot,t)$ does not have compact support.

Chapter 2
Interpolation of Operators: A Multiplier Theorem

In this chapter, we shall first study two basic results in interpolation of operators in L^p spaces, the Riesz–Thorin theorem and the Marcinkiewicz interpolation theorem (diagonal case). As a consequence of the former we shall prove the Hardy–Littlewood–Sobolev theorem for Riesz potentials. In this regard, we need to introduce one of the fundamental tools in harmonic analysis, the Hardy–Littlewood maximal function. In Section 2.4, we shall prove the Mihlin multiplier theorem.

The results deduced in this chapter are used frequently in these notes. In particular, in Chapter 4 the proof of Theorem 4.2 is based on the Riesz–Thorin theorem and the Hardy–Littlewood–Sobolev theorem.

2.1 The Riesz–Thorin Convexity Theorem

Let $(X, \mathcal{A}, \mu)$ be a measurable space (i.e., X is a set, $\mathcal{A}$ denotes a σ-algebra of subsets of X, and μ is a measure defined on $\mathcal{A}$). $L^p = L^p(X, \mathcal{A}, \mu)$, $1 \leq p < \infty$ denotes the space of complex-valued functions f that are μ-measurable such that

$$\|f\|_p = \left(\int_X |f(x)|^p \, d\mu \right)^{1/p} < \infty.$$

Functions in $L^p(X, \mathcal{A}, \mu)$ are defined almost everywhere with respect to μ. Similarly, we have $L^\infty(X, \mathcal{A}, \mu)$ the space of functions f that are μ-measurable, complex valued and essentially μ-bounded, with $\|f\|_\infty$ the essential supremum of f. The Riesz–Thorin convexity theorem can be obtained as a consequence of a version of the Hadamard three circles theorem, a result of the Phragmen–Lindelöf theorem, known as the *three lines theorem*.

Lemma 2.1. *Let F be a continuous and bounded function defined on*

$$S = \{z = x + iy : 0 \leq x \leq 1\}$$

which is also analytic in the interior of S. If for each $y \in \mathbb{R}$,

$$|F(iy)| \leq M_0 \quad \textit{and} \quad |F(1 + iy)| \leq M_1,$$

F. Linares, G. Ponce, *Introduction to Nonlinear Dispersive Equations*,
Universitext, DOI 10.1007/978-1-4939-2181-2_2

then for any $z = x + iy \in S$

$$|F(x+iy)| \le M_0^{1-x} M_1^x.$$

In other words, the function $\phi(x) = \log k_x$ *is convex, where* $k_x = \sup\{|F(x+iy)| : y \in \mathbb{R}\}$ *for* $x \in [0,1]$.

Proof. Without loss of generality one can assume that M_0, $M_1 > 0$. Moreover, considering the function $F(z)/M_0^{1-z}M_1^z$, the proof reduces to the case $M_0 = M_1 = 1$. Thus, we have that

$$|F(iy)| \le 1 \quad \text{and} \quad |F(1+iy)| \le 1 \quad \text{for any } y \in \mathbb{R},$$

and we want to show that $|F(z)| \le 1$ for any $z \in S$. If

$$\lim_{|y|\to\infty} F(x+iy) = 0 \quad \text{uniformly on } 0 \le x \le 1,$$

the result follows from the maximum principle. In this case, there exists $y_0 > 0$ such that $|F(x+iy)| \le 1$ for $|y| \ge y_0$ and $|F(z)| \le 1$ in the boundary of the rectangle with corners

$$iy_0, 1+iy_0, -iy_0, 1-iy_0.$$

The maximum principle guarantees the same estimate in the interior of the rectangle.

In the general case, we consider the function:

$$F_n(z) = F(z)e^{(z^2-1)/n}, \quad n \in \mathbb{Z}^+.$$

Since

$$\begin{aligned} |F_n(z)| &= |F(x+iy)|e^{-y^2/n}\, e^{(x^2-1)/n} \\ &\le |F(x+iy)|e^{-y^2/n} \to 0 \quad \text{as} \quad |y| \to \infty, \end{aligned}$$

uniformly on $0 \le x \le 1$, with $|F_n(iy)| \le 1$ and $|F_n(1+iy)| \le 1$, the previous argument proves that $|F_n(z)| \le 1$ for any $n \in \mathbb{Z}^+$. Letting $n \to \infty$, we obtain the desired estimate. □

Let T be a linear operator from $L^p(X)$ to $L^q(Y)$. If T is continuous or bounded, i.e.,

$$|||T||| = \sup_{f \ne 0} \frac{\|Tf\|_q}{\|f\|_p} < \infty, \tag{2.1}$$

we call the number $|||T|||$ the *norm of the operator* T.

Theorem 2.1 (Riesz–Thorin). *Let* $p_0 \ne p_1$, $q_0 \ne q_1$. *Let* T *be a bounded linear operator from* $L^{p_0}(X, \mathcal{A}, \mu)$ *to* $L^{q_0}(Y, \mathcal{B}, \nu)$ *with norm* M_0 *and from* $L^{p_1}(X, \mathcal{A}, \mu)$ *to* $L^{q_1}(Y, \mathcal{B}, \nu)$ *with norm* M_1. *Then,* T *is bounded from* $L^{p_\theta}(X, \mathcal{A}, \mu)$ *in* $L^{q_\theta}(Y, \mathcal{B}, \nu)$ *with norm* M_θ *such that*

$$M_\theta \le M_0^{1-\theta} M_1^\theta,$$

with

$$\frac{1}{p_\theta} = \frac{1-\theta}{p_0} + \frac{\theta}{p_1}, \quad \frac{1}{q_\theta} = \frac{1-\theta}{q_0} + \frac{\theta}{q_1}, \qquad \theta \in (0,1). \tag{2.2}$$

Proof. *(Thorin).* Combining the notation

$$\langle h, g\rangle = \int_Y h(y)g(y)\,d\nu(y)$$

and a duality argument it follows that

$$\|h\|_q = \sup\{|\langle h, g\rangle| : \|g\|_{q'} = 1\}$$

and

$$M_{pq} \equiv \sup\{|\langle Tf, g\rangle| : \|f\|_p = \|g\|_{q'} = 1\},$$

where $1/p + 1/p' = 1/q + 1/q' = 1$. Since $p < \infty$ and $q' < \infty$, we can assume that $f,\ g$ are simple functions with compact support. Thus,

$$f(x) = \sum_j a_j \chi_{A_j}(x) \ \text{ and } \ g(y) = \sum_k b_k \chi_{B_k}(y).$$

For $0 \le \mathcal{R}e\ z \le 1$, we define

$$\frac{1}{p(z)} = \frac{1-z}{p_0} + \frac{z}{p_1}, \quad \frac{1}{q'(z)} = \frac{1-z}{q_0'} + \frac{z}{q_1'},$$

$$\varphi(z) = \varphi(x,z) = \sum_j |a_j|^{p_\theta/p(z)} e^{i\arg(a_j)} \chi_{A_j}(x),$$

and

$$\psi(z) = \psi(y,z) = \sum_k |b_k|^{q_\theta'/q'(z)} e^{i\arg(b_k)} \chi_{B_k}(y).$$

Thus, $\varphi(z) \in L^{p_j}$, $\psi(z) \in L^{q_j'}$, and $T\varphi(z) \in L^{q_j}, j = 0,1$. Also, $\varphi'(z) \in L^{p_j}$, $\psi'(z) \in L^{q_j'}$, and $(T\varphi)'(z) \in L^{q_j}$, $j = 0,1$ for $0 < \mathcal{R}e\ z < 1$. Therefore, the function

$$F(z) = \langle T\varphi(z), \psi(z)\rangle$$

is bounded and continuous on $0 \le \mathcal{R}e\ z \le 1$ and analytic in the interior. Moreover,

$$\|\varphi(it)\|_{p_0} = \||f|^{p_\theta/p_0}\|_{p_0} = \|f\|_{p_\theta}^{p_\theta/p_0} = 1$$

and

$$\|\varphi(1+it)\|_{p_1} = \||f|^{p_\theta/p_1}\|_{p_1} = \|f\|_{p_\theta}^{p_\theta/p_1} = 1.$$

Similarly, $\|\psi(it)\|_{q_0'} = \|\psi(1+it)\|_{q_1'} = 1$.

From the hypotheses it follows that

$$|F(it)| \leq \|T\varphi(it)\|_{q_0} \, \|\psi(it)\|_{q_0'} \leq M_0$$

and

$$|F(1+it)| \leq \|T\varphi(1+it)\|_{q_1} \, \|\psi(1+it)\|_{q_1'} \leq M_1.$$

Since $\varphi(\theta) = f$, $\psi(\theta) = g$, and $F(\theta) = \langle Tf, g\rangle$, by the three lines theorem we obtain $|\langle Tf, g\rangle| \leq M_0^{1-\theta} M_1^{\theta}$. This completes the proof. □

Definition 2.1. An operator T is said to be *sublinear* if $T(f+g)$ is determined by the values of Tf, Tg, and

$$|T(f+g)| \leq |Tf| + |Tg|.$$

We shall say that a linear or sublinear operator T is of (strong) *type* (p,q) with constant M_{pq} if $\|Tf\|_q \leq M_{pq}\|f\|_p$ for any $f \in L^p$.

With this definition we can rephrase the statement of the Riesz–Thorin theorem. Let $p_0 \neq p_1$, $q_0 \neq q_1$, and T be a linear operator of type (p_0, q_0) with norm M_0 and of type (p_1, q_1) with norm M_1. Then T is of type (p,q) with

$$\frac{1}{p} = \frac{1-\theta}{p_0} + \frac{\theta}{p_1}, \quad \frac{1}{q} = \frac{1-\theta}{q_0} + \frac{\theta}{q_1}, \quad \theta \in (0,1),$$

with norm

$$M \leq M_0^{1-\theta} M_1^{\theta}.$$

2.1.1 *Applications*

Next we use the Riesz–Thorin theorem to establish some properties of the Fourier transform and the convolution operator. We fix $X = Y = \mathbb{R}^n$ and $\mu = \nu = dx$ the Lebesgue measure.

Theorem 2.2 (Young's inequality). *Let $f \in L^p(\mathbb{R}^n)$ and $g \in L^q(\mathbb{R}^n)$, $1 \leq p, q \leq \infty$ with $\frac{1}{p} + \frac{1}{q} \geq 1$. Then $f * g \in L^r(\mathbb{R}^n)$, where $\frac{1}{r} = \frac{1}{p} + \frac{1}{q} - 1$. Moreover,*

$$\|f * g\|_r \leq \|f\|_p \|g\|_q. \tag{2.3}$$

Proof. For $g \in L^q(\mathbb{R}^n)$, we define the operator

$$Tf(x) = \int_{\mathbb{R}^n} f(x-y)g(y)dy = (f * g)(x).$$

The Minkowski integral inequality shows

$$\|Tf\|_q \leq \|g\|_q \|f\|_1.$$

On the other hand, using Hölder's inequality one sees that

$$\|Tf\|_\infty \leq \|g\|_q \|f\|_{q'}.$$

Thus, T is of type $(1, q)$ and (q', ∞) with norm bounded by $\|g\|_q$. Hence, Theorem 2.1 (Riesz–Thorin) guarantees that T is of type (p, r), where

$$\frac{1}{p} = \frac{(1-\theta)}{1} + \frac{\theta}{q'} = 1 - \frac{\theta}{q}$$

and

$$\frac{1}{r} = \frac{(1-\theta)}{q} + 0 = \frac{1}{q} + \left(1 - \frac{\theta}{q}\right) - 1 = \frac{1}{q} + \frac{1}{p} - 1,$$

with norm less than $\|g\|_q$. □

Theorem 2.3 (Hausdorff–Young's inequality). *Let $f \in L^p(\mathbb{R}^n)$, $1 \leq p \leq 2$. Then $\widehat{f} \in L^{p'}(\mathbb{R}^n)$ with $\frac{1}{p} + \frac{1}{p'} = 1$ and*

$$\|\widehat{f}\|_{p'} \leq \|f\|_p. \tag{2.4}$$

Proof. From (1.2) and (1.11) it follows that the Fourier transform is of type $(1, \infty)$ and $(2, 2)$ with norm 1. Hence, Theorem 2.1 tells us that it is also of type (p, q) with

$$\frac{1}{p} = \frac{(1-\theta)}{1} + \frac{\theta}{2} = 1 - \frac{\theta}{2} \quad \text{and} \quad \frac{1}{q} = 0 + \frac{\theta}{2} = 1 - \frac{1}{p} = \frac{1}{p'}$$

with norm $M \leq 1^{(1-\theta)}\, 1^\theta = 1$. □

This estimate is the best possible when $p = 1$ or 2. This is not the case for $1 < p < 2$. Beckner [B] found the best constant for the Hausdorff–Young inequality. He showed that if $f \in L^p(\mathbb{R}^n)$, $1 \leq p \leq 2$, then

$$\|\widehat{f}\|_{p'} \leq (A_p)^n \|f\|_p, \quad \text{where } A_p = \left(\frac{p^{1/p}}{p'^{1/p'}}\right)^{1/2}.$$

2.2 Marcinkiewicz Interpolation Theorem (Diagonal Case)

Let $(X, \mathcal{A}, \mu)$ be a measurable space.

Definition 2.2. For a measurable function $f : X \to \mathbb{C}$, we define its distribution function as:

$$m(\lambda, f) = \mu(\{x \in X : |f(x)| > \lambda\}) = \mu(E_f^\lambda).$$

Thus, $m(\lambda, f)$ as a function of $\lambda \in [0, \infty]$ is well defined and takes values in $[0, \infty)$. Moreover, it is nonincreasing and continuous from the right.

Proposition 2.1. *For any measurable function* $f : X \to \mathbb{C}$ *and for any* $\lambda \geq 0$ *it follows that*

1. *(Tchebychev)*

$$m(\lambda, f) \leq \lambda^{-p} \int\limits_{E_f^\lambda} |f(x)|^p \, d\mu(x) \leq \lambda^{-p} \|f\|_p^p.$$

2. *If* $1 \leq p < \infty$,

$$\|f\|_p^p = -\int\limits_0^\infty \lambda^p \, dm(\lambda, f) = p \int\limits_0^\infty \lambda^{p-1} m(\lambda, f) \, d\lambda.$$

If $p = \infty$,

$$\|f\|_\infty = \inf \{\lambda : m(\lambda, f) = 0\}.$$

3. $m(\lambda, f + g) \leq m(\lambda/2, f) + m(\lambda/2, g)$.

Proof. It is left as an exercise. □

Definition 2.3. For $1 \leq p < \infty$, we denote by $L^{p*}(X, \mathcal{A}, \mu)$ (*weak* L^p*-spaces*) the space of all measurable functions $f : X \to \mathbb{C}$ such that

$$\|f\|_p^* = \sup_{\lambda > 0} \lambda (m(\lambda, f))^{1/p} < \infty.$$

Observe that $L^{\infty *} = L^\infty$.

Proposition 2.2. *If* $1 \leq p < \infty$, *then*

1. $L^p(\mathbb{R}^n) \subsetneq L^{p*}(\mathbb{R}^n)$.
2. $\|f + g\|_p^* \leq 2(\|f\|_p^* + \|g\|_p^*)$.

Proof. It is left as an exercise. □

Therefore, $L^{p*}(X, \mathcal{A}, \mu)$ is a *quasinormed vector space*

$$\|f + g\| \leq k(\|f\| + \|g\|)$$

with $k = 2$, i.e., it only satisfies a quasitriangular inequality. The spaces L^p and L^{p*} are particular cases of the *Lorentz spaces* $L^{p,q}$ (see [BeL]).

Definition 2.4. Let $(X_j, \mathcal{A}_j, \mu_j)$, $j = 1, 2$, be two measurable spaces. Let $M(X_2)$ be the space of complex-valued, measurable functions defined on X_2. A linear or sublinear operator $T : L^p(X_1) \to M(X_2)$ with $1 \leq p < \infty$ is said to be of *weak type* (p, q) if there exists a constant $c > 0$ such that for any $f \in L^p(X_1)$

$$\|Tf\|_q^* \leq c \|f\|_p.$$

If $q = \infty$, type (p, ∞) and weak type (p, ∞) agree. Tchebychev's inequality shows that if T is of type (p, q), then it is of weak type (p, q).

In the rest of this chapter, we shall consider $X_j = \mathbb{R}^n,\ j = 1, 2$.

Theorem 2.4 (Marcinkiewicz). *Let* $1 < r \leq \infty$ *and*

$$T : L^1(\mathbb{R}^n) + L^r(\mathbb{R}^n) \to M(\mathbb{R}^n)$$

be a sublinear operator (see Definition 2.1*). If* T *is of weak type* $(1,1)$ *and of weak type* (r,r)*, then* T *is of (strong) type* (p,p) *for any* $p \in (1,r)$.

Proof. First we consider the case $r = \infty$. Changing the operator T by $\|T\|^{-1}T$ one can assume that

$$\|Tf\|_\infty \leq \|f\|_\infty.$$

Given $f \in L^1(\mathbb{R}^n) + L^r(\mathbb{R}^n)$, for each $\lambda \in \mathbb{R}^+$ we define

$$f_1^\lambda(x) = \begin{cases} f(x), & \text{if } |f(x)| \geq \lambda/2 \\ 0, & \text{if } |f(x)| < \lambda/2 \end{cases}$$

and $f_2^\lambda(x) = f(x) - f_1^\lambda(x)$. Therefore,

$$|Tf(x)| \leq |Tf_1^\lambda(x)| + \lambda/2,$$

and

$$\{x \in \mathbb{R}^n : |Tf(x)| > \lambda\} \subseteq \{x \in \mathbb{R}^n : |Tf_1^\lambda(x)| > \lambda/2\}.$$

Since T is of weak type $(1,1)$, it follows that

$$\begin{aligned} |\{x \in \mathbb{R}^n : |Tf_1^\lambda(x)| > \lambda/2\}| &\leq c\left(\frac{\lambda}{2}\right)^{-1} \int_{\mathbb{R}^n} |f_1^\lambda(x)|\,dx \\ &= 2c\lambda^{-1} \int_{|f|>\lambda/2} |f(x)|\,dx, \end{aligned}$$

where $|\cdot|$ denotes the Lebesgue measure. Combining this estimate, part (2) of Proposition 2.1, and a change in the order of integration, one has:

$$\begin{aligned} \int_{\mathbb{R}^n} |Tf(x)|^p\,dx &= p\int_0^\infty \lambda^{p-1}|\{x \in \mathbb{R}^n : |Tf(x)| > \lambda\}|\,d\lambda \\ &\leq p\int_0^\infty \lambda^{p-1}\left(2c\lambda^{-1}\int_{|f|>\lambda/2} |f(x)|\,dx\right) d\lambda \\ &= 2cp\int_0^\infty \lambda^{p-2}\left(\int_{|f|>\lambda/2} |f(x)|\,dx\right) d\lambda \\ &= 2cp\int_{\mathbb{R}^n}\left(\int_0^{2|f(x)|} \lambda^{p-2}\,d\lambda\right) |f(x)|\,dx = \frac{2^p cp}{p-1}\|f\|_p^p, \end{aligned}$$

which yields the result for the case $r = \infty$.

In the case $r < \infty$, we have

$$\begin{aligned} m(\lambda, Tf) &= |\{x \in \mathbb{R}^n : |Tf(x)| > \lambda\}| \\ &\leq m(\lambda/2, Tf_1^\lambda) + m(\lambda/2, Tf_2^\lambda) \\ &\leq c_1 \left(\frac{\lambda}{2}\right)^{-1} \int_{\mathbb{R}^n} |f_1^\lambda(x)|\, dx + c_r^r \left(\frac{\lambda}{2}\right)^{-r} \int_{\mathbb{R}^n} |f_2^\lambda(x)|^r\, dx \\ &= 2c_1\lambda^{-1} \int_{|f|\geq\lambda/2} |f(x)|\, dx + (2c_r)^r \lambda^{-r} \int_{|f|<\lambda/2} |f(x)|^r\, dx. \end{aligned}$$

As in the proof of the case $r = \infty$, we have that

$$\int_0^\infty \lambda^{p-2} \left(\int_{|f|\geq\lambda/2} |f(x)|\, dx \right) d\lambda = \frac{2^{p-1}}{p-1} \|f\|_p^p.$$

A similar argument shows that

$$\int_0^\infty \lambda^{p-1-r} \left(\int_{|f|<\lambda/2} |f(x)|^r\, dx \right) d\lambda = \frac{2^{p-r}}{r-p} \|f\|_p^p.$$

Combining these inequalities and part (2) of Proposition 2.1, we find that

$$\|Tf\|_p \leq c_p \|f\|_p, \quad \text{with } c_p = 2\sqrt[p]{p} \left(\frac{c_1}{p-1} + \frac{c_r^r}{r-p} \right)^{1/p}.$$

□

2.2.1 Applications

We shall use the Marcinkiewicz interpolation theorem to study some basic properties of the Hardy–Littlewood maximal function. First, we introduce some notation.

We denote by $L^1_{\text{loc}}(\mathbb{R}^n)$ the spaces of functions $f : \mathbb{R}^n \to \mathbb{C}$ such that $\int_K |f| dx < \infty$ for any compact $K \subseteq \mathbb{R}^n$. The volume of the unit ball in $\mathbb{R}^n$ will be denoted by ω_n and $B_r(x) = \{y \in \mathbb{R}^n : \|x - y\| < r\}$ is the ball of center x and radius r.

Definition 2.5. For a given $f \in L^1_{\text{loc}}(\mathbb{R}^n)$, we define $\mathcal{M}f(x)$, the *Hardy–Littlewood maximal function* associated to f, as:

$$\begin{aligned} \mathcal{M}f(x) &= \sup_{r>0} \frac{1}{|B_r(x)|} \int_{B_r(x)} |f(y)|\, dy = \sup_{r>0} \frac{1}{\omega_n} \int_{B_1(0)} |f(x - ry)|\, dy \\ &= \sup_{r>0} \left(|f| * \frac{1}{|B_r(0)|} \chi_{B_r(0)} \right)(x). \end{aligned}$$

Proposition 2.3.

1. *$\mathcal{M}$ defines a sublinear operator, i.e.,*

$$|\mathcal{M}(f+g)(x)| \leq |\mathcal{M}f(x)| + |\mathcal{M}g(x)|, \qquad x \in \mathbb{R}^n.$$

2. *If $f \in L^\infty(\mathbb{R}^n)$, then*

$$\|\mathcal{M}f\|_\infty \leq \|f\|_\infty. \tag{2.5}$$

Proof. It is left as an exercise. □

Part (2) of Proposition 2.3 tells us that $\mathcal{M}$ is of type (∞, ∞). Next, we show that $\mathcal{M}$ is of weak type $(1, 1)$. For this purpose, we need the following result.

Lemma 2.2 (Vitali's covering lemma). *Let $E \subseteq \mathbb{R}^n$ be a measurable set such that $E \subseteq \cup_\alpha B_{r_\alpha}(x_\alpha)$ with the family of open balls $\{B_{r_\alpha}(x_\alpha)\}_\alpha$ satisfying $\sup\limits_\alpha r_\alpha = c_0 < \infty$. Then there exists a subfamily $\{B_{r_j}(x_j)\}_j$ disjoint and numerable such that*

$$|E| \leq 5^n \sum_{j=1}^{\infty} |B_{r_j}(x_j)|.$$

Proof. Choose $B_{r_1}(x_1)$ such that $r_1 \geq c_0/2$. For $j \geq 2$, take $B_{r_j}(x_j)$ such that $B_{r_j}(x_j) \cap \bigcup\limits_{k=1}^{j-1} B_{r_k}(x_k) = \emptyset$ and

$$r_j > \frac{1}{2} \sup \{r_\alpha \ : \ B_{r_\alpha}(x_\alpha) \cap B_{r_k}(x_k) = \emptyset \ \text{ for } \ k = 1, \ldots, j-1\}.$$

It is clear that the $B_{r_j}(x_j)$ are disjoint. If $\sum |B_{r_j}(x_j)| = \infty$, we have completed the proof. In the case $\sum |B_{r_j}(x_j)| < \infty$ (hence, $\lim\limits_{j\to\infty} r_j = 0$), it will suffice to show that

$$B_{r_\alpha}(x_\alpha) \subseteq \bigcup_j B_{5r_j}(x_j), \quad \text{for any } \alpha.$$

If $B_{r_\alpha}(x_\alpha) = B_{r_j}(x_j)$ for some j, there is nothing to prove. Thus, we assume that $B_{r_\alpha}(x_\alpha) \neq B_{r_j}(x_j)$ for any j. Define j_α as the smallest j such that $r_j < r_\alpha/2$. By the construction of $B_{r_j}(x_j)$, there exists $j \in \{1, \ldots, j_\alpha - 1\}$ such that $B_{r_\alpha}(x_\alpha) \cap B_{r_j}(x_j) \neq \emptyset$. Denoting by j^* this index it follows that $B_{r_\alpha}(x_\alpha) \subseteq B_{5r_{j^*}}(x_{j^*})$ since $r_{j^*} \geq r_\alpha/2$. □

Theorem 2.5 (Hardy–Littlewood). *Let $1 < p \leq \infty$. Then $\mathcal{M}$ is a sublinear operator of type (p, p), i.e., there exists c_p such that*

$$\|\mathcal{M}f\|_p \leq c_p \|f\|_p, \quad \text{for any } f \in L^p(\mathbb{R}^n). \tag{2.6}$$

Proof. We first show that $\mathcal{M}$ is of weak type $(1, 1)$, that is, there exists a constant c_1 such that for any $f \in L^1(\mathbb{R}^n)$

$$\sup_{\lambda > 0} \lambda \, m(\lambda, \mathcal{M}f) \leq c_1 \|f\|_1. \tag{2.7}$$

Once (2.7) has been established, a combination of (2.5), (2.7), and the Marcinkiewicz theorem yields (2.6).

To obtain (2.7), we define $E_f^\lambda = \{x \in \mathbb{R}^n : \mathcal{M}f(x) > \lambda\}$ for any $\lambda > 0$. Thus, if $x \in E_f^\lambda$, then there exists $B_{r_x}(x)$ such that

$$\int_{B_{r_x}(x)} |f(y)|dy > \lambda |B_{r_x}(x)|.$$

Clearly, we have that

$$E_f^\lambda \subseteq \bigcup_{x \in E_f^\lambda} B_{r_x}(x),$$

then the Vitali covering lemma guarantees the existence of a countable, disjoint subfamily $\{B_{r_{x_j}}(x_j)\}_{j\in\mathbb{Z}+}$ such that

$$|E_f^\lambda| \leq 5^n \sum_{j=1}^{\infty} |B_{r_{x_j}}(x_j)| \leq 5^n \lambda^{-1} \sum_{j=1}^{\infty} \int_{B_{r_{x_j}}(x_j)} |f(y)|\,dy \leq 5^n \lambda^{-1} \|f\|_1,$$

which implies (2.7). □

Next, we extend the estimates (2.6) and (2.7) to a large class of kernels.

Proposition 2.4. *Let $\varphi \in L^1(\mathbb{R}^n)$ be a radial, positive, and nonincreasing function of $r = \|x\| \in [0, \infty)$. Then*

$$\sup_{t>0} |\varphi_t * f(x)| = \sup_{t>0} \left| \int_{\mathbb{R}^n} \frac{\varphi(t^{-1}(x-y))}{t^n} f(y)\,dy \right| \leq \|\varphi\|_1 \mathcal{M}f(x). \tag{2.8}$$

Proof. First, we assume that, in addition to the hypotheses, φ is a simple function

$$\varphi(x) = \sum_k a_k \chi_{B_{r_k}(0)}(x), \quad \text{with} \quad a_k > 0.$$

Hence,

$$\varphi * f(x) = \sum_k a_k |B_{r_k}(0)| \frac{1}{|B_{r_k}(0)|} \chi_{B_{r_k}(0)} * f(x) \leq \|\varphi\|_1 \mathcal{M}f(x).$$

(observe that $\|\varphi\|_1 = \sum_k a_k |B_{r_k}(0)|$).

In the general case, we approximate φ by an increasing sequence of simple functions satisfying the hypotheses. Since dilations of φ satisfy the same hypotheses and preserve the L^1-norm, they verify (2.8). Finally, passing to the limit we obtain the desired result. □

Next, we shall apply these results to deduce some continuity properties of the Riesz potentials. We recall that a fundamental solution of the Laplacian Δ is given by the following formula describing the Newtonian potential

$$Uf(x) = c_n \int_{\mathbb{R}^n} \frac{f(y)}{|x-y|^{n-2}}\, dy \quad \text{for } n \geq 3.$$

The Riesz potentials generalize this expression.

Definition 2.6. Let $0 < \alpha < n$. The Riesz potential of order α, denoted by I_α, is defined as:

$$I_\alpha f(x) = c_{\alpha,n} \int_{\mathbb{R}^n} \frac{f(y)}{|x-y|^{n-\alpha}}\, dy = k_\alpha * f(x), \tag{2.9}$$

where $c_{\alpha,n} = \pi^{-n/2} 2^{-\alpha}\, \Gamma(n/2 - \alpha/2)/\Gamma(\alpha/2)$.

Since the Riesz potentials are defined as integral operators, it is natural to study their continuity properties in $L^p(\mathbb{R}^n)$.

Theorem 2.6 (Hardy–Littlewood–Sobolev). *Let* $0 < \alpha < n$, $1 \leq p < q < \infty$, *with* $\dfrac{1}{q} = \dfrac{1}{p} - \dfrac{\alpha}{n}$.

1. *If* $f \in L^p(\mathbb{R}^n)$, *then the integral (2.9) is absolutely convergent almost every* $x \in \mathbb{R}^n$.
2. *If* $p > 1$, *then* I_α *is of type (p,q), i.e.,*

$$\|I_\alpha(f)\|_q \leq c_{p,\alpha,n} \|f\|_p. \tag{2.10}$$

Proof. We split the kernel

$$k_\alpha(x) = \frac{c_{\alpha,n}}{|x|^{n-\alpha}} = k_\alpha^0(x) + k_\alpha^\infty(x)$$

as

$$k_\alpha^0(x) = \begin{cases} k_\alpha(x) & \text{if } |x| \leq \varepsilon, \\ 0 & \text{if } |x| > \varepsilon \end{cases}$$

and $k_\alpha^\infty(x) = k_\alpha(x) - k_\alpha^0(x)$, where ε is a positive constant to be determined. Thus,

$$|I_\alpha f(x)| \leq |k_\alpha^0 * f(x)| + |k_\alpha^\infty * f(x)| = I + II. \tag{2.11}$$

The integral I represents the convolution of a function $k_\alpha^0 \in L^1(\mathbb{R}^n)$ with $f \in L^p(\mathbb{R}^n)$. The integral II is the convolution of a function $f \in L^p(\mathbb{R}^n)$ with $k_\alpha^\infty \in L^{p'}(\mathbb{R}^n)$. Therefore, both integrals converge absolutely.

Also, using that

$$\int\limits_{|y|<\varepsilon} \frac{dy}{|y|^{n-\alpha}} = c_n \int\limits_0^{\varepsilon} \frac{r^{n-1}}{r^{n-\alpha}}\, dr = c_{\alpha,n}\, \varepsilon^{\alpha},$$

together with (2.8) in Proposition 2.4 we infer that

$$I \le \varepsilon^{\alpha} \left(\frac{1}{\varepsilon^{\alpha}} \chi_{\{|y/\varepsilon|<1\}}(y) \frac{1}{|y|^{n-\alpha}} * |f| \right)(x) \le c_{\alpha,n}\, \varepsilon^{\alpha} \mathcal{M} f(x). \tag{2.12}$$

On the other hand, Hölder's inequality implies that

$$\begin{aligned} II &\le c_{\alpha,n}\, \|f\|_p \left(\int\limits_{|y|\ge\varepsilon} \frac{1}{|y|^{(n-\alpha)p'}}\, dy \right)^{1/p'} \\ &= c_{\alpha,n}\, \|f\|_p \left(\int_{\varepsilon}^{\infty} \frac{r^{n-1}}{r^{(n-\alpha)p'}}\, dr \right)^{1/p'} \\ &= c_{\alpha,n}\, \varepsilon^{n/p'-n+\alpha} \|f\|_p. \end{aligned} \tag{2.13}$$

Next, we minimize the sum of the bounds in (2.12) and (2.13). Hence, we fix $\varepsilon = \varepsilon(x)$ such that

$$c\varepsilon^{\alpha} \mathcal{M} f(x) = c\varepsilon^{n/p'-n+\alpha} \|f\|_p,$$

using $n/p' - n = -n/p$. This is equivalent to

$$c\mathcal{M} f(x) = c\varepsilon^{-n/p} \|f\|_p. \tag{2.14}$$

Combining (2.11)–(2.14) we can write

$$\begin{aligned} |I_{\alpha} f(x)| &\le c\, (\|f\|_p\, (\mathcal{M} f(x))^{-1})^{\alpha p/n}\, \mathcal{M} f(x) \\ &= c\, \|f\|_p^{\alpha p/n}\, (\mathcal{M} f(x))^{1-\alpha p/n} \\ &= c\, \|f\|_p^{\theta}\, (\mathcal{M} f(x))^{1-\theta}, \qquad \theta = \alpha p/n \in (0,1). \end{aligned} \tag{2.15}$$

Finally, taking the L^q-norm in (2.15) and using (2.6) we conclude:

$$\|I_{\alpha} f\|_q \le c\|f\|_p^{\theta} \|(\mathcal{M} f)^{1-\theta}\|_q = c\|f\|_p^{\theta} \|\mathcal{M} f\|_{(1-\theta)q}^{1-\theta} \le c\|f\|_p,$$

since $(1-\theta)q = (1-\alpha p/n)q = p$, i.e., $1/q = 1/p - \alpha/n$. This completes the proof. □

2.3 The Stein Interpolation Theorem

So far we have discussed interpolation theorems for fixed linear or sublinear operators. We now have to cover the following situation: Suppose we have linear operators varying together with the indices p and q smoothly. Is it possible to extend the Riesz–Thorin theorem to this case? The answer is affirmative and we shall describe this extension next.

Let S be the strip defined in Lemma 2.1 and $z = x + iy \in S$. Suppose that for each $z \in S$ there corresponds a linear operator T_z defined on the space of simple functions in $L^1(X, \mathcal{A}, \mu)$ into measurable functions on Y in such a way that $(T_z f)g$ is integrable on Y provided f is a simple function in $L^1(X, \mathcal{A}, \mu)$ and g is a simple function in $L^1(Y, \mathcal{B}, \nu)$.

Definition 2.7. The family of operators $\{T_z\}_{z\in S}$ is called *admissible* if the mapping

$$z \mapsto \int_Y (T_z f) g d\nu$$

is analytic in the interior of S, continuous on S and there exists a constant $a < \pi$ such that

$$e^{-a|y|} \log \left| \int_Y (T_z f) g d\nu \right|$$

is uniformly bounded above in the strip S.

Theorem 2.7 (Stein). *Suppose $\{T_z\}, z \in S$, is an admissible family of linear operators satisfying*

$$\|T_{iy} f\|_{q_0} \le M_0(y) \|f\|_{p_0} \quad \text{and} \quad \|T_{1+iy} f\|_{q_1} \le M_1(y) \|f\|_{p_1}, \quad y \in \mathbb{R}^n,$$

for all simple functions f in $L^1(X, \mathcal{A}, \mu)$, where $1 \le p_j, q_j \le \infty$, $M_j(y)$, $j = 0, 1$, are independent of f and satisfy

$$\sup_{-\infty<y<\infty} e^{-b|y|} \log M_j(y) < \infty$$

for some $b < \pi$. Then, if $0 \le t \le 1$, there exists a constant M_t such that

$$\|T_t f\|_{q_t} \le M_t \|f\|_{p_t}$$

for all simple functions f, provided

$$\frac{1}{p_t} = \frac{(1-t)}{p_0} + \frac{t}{p_1} \quad \text{and} \quad \frac{1}{q_t} = \frac{(1-t)}{q_0} + \frac{t}{q_1}.$$

Proof. For the proof of this theorem, we refer the reader to [SW]. □

2.4 A Multiplier Theorem

Let $m(\cdot)$ be a bounded measurable function in $\mathbb{R}^n$. Define the operator

$$T_m f(x) = (m(\cdot)\widehat{f}(\cdot))^{\vee}(x), \qquad f \in L^1(\mathbb{R}^n) \cap L^2(\mathbb{R}^n). \tag{2.16}$$

Notice that if $\widehat{m}(x) = K(x)$, then, formally, $T_m f(x) = K * f(x)$. However, $K \in \mathcal{S}'(\mathbb{R}^n)$, i.e., a temperate distribution so $K * f$ is not necessarily defined.

As we have seen in (1.27), (1.29), and (1.49), solutions of the linear evolution equation can be written in this form.

Definition 2.8. An $m(\cdot)$ is said to be an *L^p-multiplier* if

$$\|T_m f\|_p \le c_p \|f\|_p, \quad \text{for all} \quad f \in L^2(\mathbb{R}^n) \cap L^p(\mathbb{R}^n). \tag{2.17}$$

In this case, $T_m(\cdot)$ can be extended to $L^p(\mathbb{R}^n)$. The smallest constant c_p^* in (2.17) is the operator norm of T_m in $L^p(\mathbb{R}^n)$, i.e., $|||T_m|||$ (see 2.1). Notice that if $p = 2$, one has $c_2^* = \|m\|_\infty$. Also, by duality, if $m(\cdot)$ is an L^p-multiplier, $1 < p < \infty$, then $m(\cdot)$ is an $L^{p'}$-multiplier with $\frac{1}{p} + \frac{1}{p'} = 1$, and $c_{p'}^* = c_p^*$.

Theorem 2.8 (Mihlin–Hörmander). *Let $m \in C^k(\mathbb{R}^n \setminus \{0\})$, $k \in Z^+$, $k > n/2$. If for $|\alpha| \le k$*

$$\sup_{R>0} R^{-n+|\alpha|} \int_{R<|\xi|<2R} |\partial_\xi^\alpha m(\xi)|^2 \, d\xi = A_\alpha < \infty, \tag{2.18}$$

then $m(\cdot)$ is an L^p-multiplier for any $p \in (1,\infty)$. Moreover, T_m is of weak type (1,1), i.e., for $\lambda > 0$

$$\lambda\, |\{x \in \mathbb{R}^n : |T_m f(x)| > \lambda\}| \le c\|f\|_1 \quad \textit{for all} f \in L^1(\mathbb{R}^n), \tag{2.19}$$

where $|A|$ denotes the Lebesgue measure of the set A.

Notice that if $m \in C^k(\mathbb{R}^n \setminus \{0\})$, $k \in \mathbb{Z}^+$, $k > n/2$ with

$$\sup_{x\neq 0} \sup_{|\alpha|\le k} |x|^{|\alpha|} |\partial_x^\alpha m(x)| = B_\alpha < \infty \quad \text{for} \quad |\alpha| \le k, \tag{2.20}$$

then (2.18) holds. Condition (2.20) is due to Mihlin, the weaker assumptions in (2.18) is due to Hörmander.

Combining a duality argument and the Marcinkiewicz interpolation theorem, it suffices to establish (2.19) to obtain Theorem 2.8. This is done in Appendix A.

2.5 Exercises

2.1 Prove the continuity part of Theorem 2.1 (Riesz–Thorin) in the cases $p_0 = p_1$ and $q_0 = q_1$.

2.2 Prove Proposition 2.1.

2.3 Prove Proposition 2.2.

2.4 Prove Proposition 2.3.

2.5 (i) Prove that the Fourier transform defines a continuous operator from $L^p(\mathbb{R}^n)$ to $L^q(\mathbb{R}^n)$ only if $1/p + 1/q = 1$ with $q \geq p$.

(ii) Prove that for $1 \leq p < 2$

$$\widehat{L^p(\mathbb{R}^n)} \subsetneq L^q(\mathbb{R}^n).$$

Hint: Use Exercise 1.2(ii) and the open mapping theorem.

2.6 (i) Prove the Lebesgue differentiation theorem: If $f \in L^1_{\text{loc}}(\mathbb{R}^n)$, then for almost every $x \in \mathbb{R}^n$

$$\lim_{r\to 0} \frac{1}{|B_r(x)|} \int_{B_r(x)} f(y)\,dy = f(x). \tag{2.21}$$

Hint: Without loss of generality take $f \in L^1(\mathbb{R}^n)$. Define $O(f,x)$ the oscillation of f at x as

$$O(f,x) = \left| \limsup_{r\to 0} \frac{1}{|B_r(x)|} \int_{B_r(x)} f(y)\,dy - \liminf_{r\to 0} \frac{1}{|B_r(x)|} \int_{B_r(x)} f(y)\,dy \right|.$$

Prove that (2.21) is equivalent to $O(f,x) = 0$. Use that

$$\lim_{r\to 0} \frac{1}{|B_r(0)|} \chi_{B_r(0)} * f = f \;\text{ in }\; L^1(\mathbb{R}^n);$$

therefore, there exists a sequence $\{r_j\}$ such that

$$\lim_{j\to 0} \frac{1}{|B_{r_j}(0)|} \chi_{B_{r_j}(0)} * f(x) = f(x) \;\text{ almost everywhere } x \in \mathbb{R}^n.$$

Combine (2.7), the inequality $O(f,x) \leq 2\mathcal{M}f(x)$, and a density argument to obtain the result.

(ii) Let $f \in L^1_{\text{loc}}(\mathbb{R}^n)$ and Q_j be a sequence of closed cubes in $\mathbb{R}^n$ such that $Q_1 \supseteq Q_2 \supseteq \ldots$, $|Q_1| < \infty$ and $|Q_j| = 2^n\,|Q_{j+1}|$. If $x \in \bigcap_{j=1}^{\infty} Q_j$ prove that

$$\lim_{j\to\infty} \frac{1}{|Q_j|} \int_{Q_j} f(y)\,dy = f(x). \tag{2.22}$$

Hint: Define

$$\mathcal{M}^* f(x) = \sup_{\substack{Q \text{ cube} \\ x\in Q}} \frac{1}{|Q|} \int_Q |f(y)|\,dy. \tag{2.23}$$

Show that there exist $c_n, d_n > 0$ such that

$$d_n \mathcal{M} f(x) \leq \mathcal{M}^* f(x) \leq c_n \mathcal{M} f(x),$$

and reapply the argument in (i).

2.7 Assuming to be true the case $n = 1$ of the Hardy–Littlewood–Sobolev inequality (2.10) prove the general case $n \geq 2$.
Hint: Combine the Hölder, Young, and Minkowski inequalities with the identity

$$\int_{\mathbb{R}^{n-1}} \frac{dy_1 \cdots dy_{n-1}}{|x-y|^n} = \frac{c_n}{|x_n - y_n|}.$$

2.8 Prove that the Hilbert transform (see Definition 1.7) is of type (p,q) if and only if $1 < p = q < \infty$.
Hint: (a) The identity (1.18) provides the result for the case $p = 2$. Use the formula deduced in Exercise 1.15 part (i) with $f = g$ to prove the result in the case $p = 4$. Apply the Riesz–Thorin interpolation theorem to extend the result to $2 < p < 4$. Reapply this argument to obtain the proof for $p > 2$. Finally, use duality to complete the proof.
(b) Otherwise use Theorem 2.8.
(c) Use (1.5) and part (ii) of Exercise 1.15.

2.9 Prove that the Riesz potential of order α, I_α, $\alpha \in (0,n)$ defines a bounded operator from $L^p(\mathbb{R}^n)$ to $L^q(\mathbb{R}^n)$ only if $1 < p < q < \infty$, with $1/q = 1/p - \alpha/n$.
Hint: Prove the formula $\delta_{a^{-1}} I_\alpha \delta_a = a^{-\alpha} I_\alpha$, where $\delta_a f(x) = f(ax)$. Show that the value of the norms of $\delta_a f(x)$ and $\delta_{a^{-1}} I_\alpha \delta_a f$ give the relation $1/q = 1/p - \alpha/n$. To see that the inequality does not hold for the extremal cases $p = 1$ and $q = n/(n-\alpha)$, use an approximation of the identity instead of f (case $p = 1$). For the case $q = n/\alpha$, use duality.

2.10 Prove that the multipliers

$$m_j(\xi) = \frac{i\xi_j}{|\xi|}, \; j = 1, \ldots, n, \quad \text{(the } j\text{-Riesz transform)}$$

and

$$m_y(\xi) = |\xi|^{iy}, \quad y \in \mathbb{R},$$

are L^p-multipliers with $1 < p < \infty$.
Hint: Use condition (2.20).

2.11 Let $s > 0$ and $\rho \in (0,s)$:

(i) Prove that for any $p \in (1,\infty)$

$$\|D^\rho f\|_p \leq c\, \|f\|_p^{1-\rho/s} \|D^s f\|_p^{\rho/s} \; f \in \mathcal{S}(\mathbb{R}^n). \tag{2.24}$$

(ii) More general, prove that for any $p,q,r \in (1,\infty)$

$$\|D^\rho f\|_p \leq c\, \|f\|_r^{1-\rho/s} \|D^s f\|_q^{\rho/s}, \quad f \in \mathcal{S}(\mathbb{R}^n), \tag{2.25}$$

with

$$\frac{1}{p} = \left(1 - \frac{\rho}{s}\right)\frac{1}{r} + \left(\frac{\rho}{s}\right)\frac{1}{q}. \tag{2.26}$$

(iii) Prove that the estimates (2.24) and (2.25) still hold with $\Lambda = (1-\Delta)^{1/2}$ instead of D, and that in both cases the proof for $p = q = r = 2$ is immediate. Prove that $r = \infty$ is allowed in (2.25).

Hint: For (ii) fix $f \in \mathcal{S}(\mathbb{R}^n)$, use that

$$\|D^\rho f\|_p = \sup_{\|g\|_{p'}=1} \Big| \int_{\mathbb{R}^n} D^\rho f(y) g(y) dy \Big|$$

and define

$$F_k(z) = e^{(z^2-1)/k} \int_{\mathbb{R}^n} D^{sz} f(y) \Psi(y,z)\, dy, \quad \text{for } z = x + iy \text{ with } 0 \le x \le 1,$$

where

$$\psi(y,z) = |g(y)|^{p'/q(z)} \frac{g(y)}{|g(y)|} \quad \text{and} \quad \frac{1}{q(z)} = \frac{1-z}{r'} + \frac{z}{q'}$$

with $\frac{1}{p} + \frac{1}{p'} = \frac{1}{r} + \frac{1}{r'} = \frac{1}{q} + \frac{1}{q'} = 1$. Verify that $F_k(\cdot)$ satisfies the hypotheses of Lemma 2.1 using Theorem 2.8 (see Exercise 2.10). Let k tend to infinity to get the result.

2.12 [Pi] Pitt's Theorem affirms: if $1 < p \le q < \infty$,

$$0 \le \alpha < n\left(1 - \frac{1}{p}\right), \quad 0 \le \gamma < \frac{n}{q}, \quad \alpha - \gamma = n\left(1 - \frac{1}{q} - \frac{1}{p}\right),$$

then there exists $c > 0$ such that

$$\|\widehat{f}\, |x|^{-\gamma}\|_q \le c\, \|f\, |x|^{\alpha}\|_p \tag{2.27}$$

with:

(i) Prove (2.27) in the case $\alpha = 0$ and $q \ge 2$.
(ii) Prove (2.27) in the case $\gamma = 0$ and $p \le 2$.

2.13 For the initial value problem associated to the heat equation:

$$\begin{cases} \partial_t u = \Delta u, \\ u(x,0) = f(x), \end{cases}$$

$x \in \mathbb{R}^n$, $t > 0$, prove that the solution $u(x,t) = e^{t\Delta} f(x)$ satisfies the following inequalities:

(i)

$$\|D_x^s\, u(\cdot,t)\|_p \leq c_s\, t^{-(\frac{n}{2r}+\frac{s}{2})}\, \|f\|_q, \tag{2.28}$$

for $s \geq 0$ and

$$\frac{1}{p} = \frac{1}{q} - \frac{1}{r}.$$

(ii)

$$\left(\int_0^\infty \|D_x^\rho u(\cdot,t)\|_p^\sigma\, dt \right)^{1/\sigma} \leq c\, \|f\|_q \tag{2.29}$$

with $\rho \in [0,2)$ and

$$0 < \frac{1}{\sigma} = \frac{n}{2}\left(\frac{1}{q} - \frac{1}{p}\right) + \frac{\rho}{2} \leq \frac{1}{q}, \quad \text{(see [G1])}.$$

Hint: For (i) use Example 1.3 to deduce that

$$u(x,t) = K_t * f(x) = \frac{e^{-|\cdot|^2/4t}}{(4\pi t)^{n/2}} * f(x).$$

Obtain the identity $\|D_x^s K_t\|_\infty = c_s t^{-(n/2+s/2)}$ for $s > 0$ and combine it with Young's inequality to obtain (2.28).

For (ii) define $(\Omega f)(t) = \|D_x^\rho e^{t\Delta} f\|_p$. Then by (2.28) ,$(\Omega f)(t) \leq c\, t^{-1/\sigma}\, \|f\|_q$, $t \in (0,\infty)$. Hence, the sublinear operator Ω is bounded from $L^q(\mathbb{R}^n)$ into $L^{\sigma *}((0,\infty))$, (i.e., L^σ-weak). Use Marcinkiewicz interpolation theorem to get (2.29).

2.14 Consider the initial value problem (IVP) associated to the wave equation:

$$\begin{cases} \partial_t^2 w - \Delta w = 0, \\ w(x,0) = f(x), \\ \partial_t w(x,0) = g(x), \end{cases}$$

$x \in \mathbb{R}^n$, $t \in \mathbb{R}$, prove that

(i) If $n = 1$, then

$$w(x,t) = \frac{f(x+t) + f(x-t)}{2} + \frac{1}{2}\int_{x-t}^{x+t} g(s)\, ds.$$

Hint: Use the formula deduced in Exercise 1.18(i) or the change of variables $\zeta = x + t$, $\eta = x - t$.

(ii) If $n = 3$, $f = 0$ and g is a radial function ($g(\|x\|)$), then

$$w(x,t) = w(\|x\|,t) = \frac{1}{2\|x\|} \int_{|\,\|x\|-t|}^{\|x\|+t} \rho g(\rho)\,d\rho.$$

Hint: Deduce the formula for the Laplacian of radial functions, use the change of variables

$$v(\rho,t) = \rho\, w(\rho,t) = \|x\|\, w(\|x\|,t)$$

and part (i) of this exercise.

(iii) Under the same hypotheses of part (ii) use the Hardy–Littlewood maximal function to show that

$$\left(\int_{-\infty}^{\infty} \|w(\cdot,t)\|_\infty^2\, dt \right)^{1/2} \leq c\,\|g\|_2. \tag{2.30}$$

In [KlM], it was established that (2.30) does not hold for nonradial functions g.

2.15 Let m_1, m_2 be two L^p-multipliers. Prove

(i) $T_{m_1} \circ T_{m_2} = T_{m_1 \cdot m_2}$.
(ii) $(T_{m_1})^* = T_{\overline{m}_1}$.

2.16 (i) Prove that if $n = 3$, then for any $t \neq 0$

$$m_t(\xi) = \cos(2\pi\,|\xi|t), \quad T_{m_t} f(x) = (m_t(\cdot)\widehat{f}(\cdot))^\vee(x) \tag{2.31}$$

is not an L^p-multiplier for $p \neq 2$.

(ii) Prove that if $n = 3$, then (see 3.38)

$$\|T_{m_t} f\|_\infty \leq c t^{-1}\,\|\nabla f\|_{1,2}, \quad \text{for any } t \neq 0.$$

(iii) Prove that if $n = 1$, then $m_t(\xi) = \cos(2\pi\,|\xi|t)$ for each $t \in \mathbb{R}$ is an L^p-multiplier for $1 \leq p \leq \infty$.

(Part (i) holds in any dimension $n \geq 2$. See [Lp]).

Hint: Notice that $T_{m_t} f(x) = (m_t(\cdot)\,\widehat{f}\,)^\vee(x)$ is the solution $u(x,t)$ of the IVP

$$\begin{cases} \partial_t^2 u - \Delta u = 0, \\ u(x,0) = f(x), \\ \partial_t u(x,0) = 0, \end{cases} \tag{2.32}$$

$x \in \mathbb{R}^3$, $t > 0$. So the formula (1.50) in Exercise 1.18(iv) applies. Take $f(x) = h(|x|)/|x| = h(r)/r$, with $h(\cdot)$ supported in the annulus $\{x \in \mathbb{R}^3 : \varepsilon \leq |x| \leq 2\varepsilon\}$. Check that $u(x,t) = (h(r+t) + h(r-t))/2r$, and derive the desired result.

Chapter 3
An Introduction to Sobolev Spaces and Pseudo-Differential Operators

In this chapter, we give a brief introduction to the classical Sobolev spaces $H^s(\mathbb{R}^n)$. Sobolev spaces measure the differentiability (or regularity) of functions in $L^2(\mathbb{R}^n)$ and they are a fundamental tool in the study of partial differential equations. We also list some basic facts of the theory of pseudo-differential operators without proof. This is useful to study smoothness properties of solutions of dispersive equations.

3.1 Basics

We begin by defining Sobolev spaces.

Definition 3.1. Let $s \in \mathbb{R}$. We define the *Sobolev space* of order s, denoted by $H^s(\mathbb{R}^n)$, as:

$$H^s(\mathbb{R}^n)=\left\{f \in \mathcal{S}'(\mathbb{R}^n) : \Lambda^s f(x)=((1+|\xi|^2)^{s/2}\widehat{f}(\xi))^{\vee}(x) \in L^2(\mathbb{R}^n)\right\}, \tag{3.1}$$

with norm $\|\cdot\|_{s,2}$ defined as:

$$\|f\|_{s,2} = \|\Lambda^s f\|_2. \tag{3.2}$$

Example 3.1 Let $n = 1$ and $f(x) = \chi_{[-1,1]}(x)$. From Example 1.1, we have that $\widehat{f}(\xi) = \sin(2\pi\xi)/(\pi\xi)$. Thus, $f \in H^s(\mathbb{R})$ if $s < 1/2$.

Example 3.2 Let $n = 1$ and $g(x) = \chi_{[-1,1]} * \chi_{[-1,1]}(x)$. In Example 1.2, we saw that

$$\widehat{g}(\xi) = \frac{\sin^2(2\pi\,\xi)}{(\pi\,\xi)^2}.$$

Thus, $g \in H^s(\mathbb{R})$ whenever $s < 3/2$.

Example 3.3 Let $n \geq 1$ and $h(x) = e^{-2\pi|x|}$. From Example 1.4, it follows that

$$\widehat{h}(\xi) = \frac{\Gamma[(n+1)/2]}{\pi^{(n+1)/2}}\,\frac{1}{(1+|\xi|^2)^{(n+1)/2}}. \tag{3.3}$$

F. Linares, G. Ponce, *Introduction to Nonlinear Dispersive Equations*, Universitext, DOI 10.1007/978-1-4939-2181-2_3

Using polar coordinates, it is easy to see that $h \in H^s(\mathbb{R}^n)$ if $s < n/2 + 1$. Notice that in this case s depends on the dimension.

Example 3.4 Let $n \geq 1$ and $f(x) = \delta_0(x)$. From Example 1.9, we have $\widehat{\delta_0}(\xi) = 1$. Thus, $\delta_0 \in H^s(\mathbb{R}^n)$ if $s < -n/2$.

From the definition of Sobolev spaces, we deduce the following properties.

Proposition 3.1.

1. *If* $s < s'$, *then* $H^{s'}(\mathbb{R}^n) \subsetneq H^s(\mathbb{R}^n)$.
2. $H^s(\mathbb{R}^n)$ *is a Hilbert space with respect to the inner product* $\langle \cdot, \cdot \rangle_s$ *defined as follows:*

$$\text{If } f, g \in H^s(\mathbb{R}^n), \text{ then } \langle f, g \rangle_s = \int_{\mathbb{R}^n} \Lambda^s f(\xi) \overline{\Lambda^s g(\xi)} \, d\xi.$$

We can see, via the Fourier transform, that $H^s(\mathbb{R}^n)$ *is equal to:*

$$L^2(\mathbb{R}^n; (1 + |\xi|^2)^s \, d\xi).$$

3. *For any* $s \in \mathbb{R}$, *the Schwartz space* $\mathcal{S}(\mathbb{R}^n)$ *is dense in* $H^s(\mathbb{R}^n)$.
4. *If* $s_1 \leq s \leq s_2$, *with* $s = \theta s_1 + (1 - \theta) s_2$, $0 \leq \theta \leq 1$, *then*

$$\|f\|_{s,2} \leq \|f\|_{s_1,2}^{\theta} \|f\|_{s_2,2}^{1-\theta}.$$

Proof. It is left as an exercise. □

To understand the relationship between the spaces $H^s(\mathbb{R}^n)$ and the differentiability of functions in $L^2(\mathbb{R}^n)$, we recall Definition 1.2 in the case $p = 2$.

Definition 3.2. A function f is *differentiable* in $L^2(\mathbb{R}^n)$ with respect to the kth variable, if there exists $g \in L^2(\mathbb{R}^n)$ such that

$$\int_{\mathbb{R}^n} \Big| \frac{f(x + h\, e_k) - f(x)}{h} - g(x) \Big|^2 dx \to 0 \text{ when } h \to 0,$$

where e_k has kth coordinate equal to 1 and 0 in the others.

Equivalently (see Exercise 1.9) $\xi_k \widehat{f}(\xi) \in L^2(\mathbb{R}^n)$, or

$$\int_{\mathbb{R}^n} f(x) \partial_{x_k} \phi(x) \, dx = - \int_{\mathbb{R}^n} g(x) \phi(x) \, dx$$

for every $\phi \in C_0^\infty(\mathbb{R}^n)$ ($C_0^\infty(\mathbb{R}^n)$ being the space of functions infinitely differentiable with compact support).

Example 3.5 Let $n = 1$ and $f(x) = \chi_{(-1,1)}(x)$, then $f' = \delta_{-1} - \delta_1$, where δ_x represents the measure of mass 1 concentrated in x, therefore $f' \notin L^2(\mathbb{R})$.

Example 3.6 Let $n = 1$ and g be as in Example 3.2. Then,

$$\frac{dg}{dx}(x) = \chi_{(-2,0)} - \chi_{(0,2)}, \quad \text{and so} \quad \frac{dg}{dx} \in L^2(\mathbb{R}).$$

With this definition, for $k \in \mathbb{Z}^+$ we can give a description of the space $H^k(\mathbb{R}^n)$ without using the Fourier transform.

Theorem 3.1. *If k is a positive integer, then $H^k(\mathbb{R}^n)$ coincides with the space of functions $f \in L^2(\mathbb{R}^n)$ whose derivatives (in the distribution sense, see (1.42)) $\partial_x^\alpha f$ belong to $L^2(\mathbb{R}^n)$ for every $\alpha \in (\mathbb{Z}^+)^n$ with $|\alpha| = \alpha_1 + \cdots + \alpha_n \le k$.*

In this case, the norms $\|f\|_{k,2}$ and $\sum_{|\alpha|\le k} \|\partial_x^\alpha f\|_2$ are equivalent.

Proof. The proof follows by combining the formula $\widehat{\partial_x^\alpha f}(\xi) = (2\pi i \xi)^\alpha \widehat{f}(\xi)$ (see (1.10)) and the inequalities:

$$|\xi^\beta| \le (1 + |\xi|^2)^{k/2} \le \sum_{|\alpha|\le k} |\xi^\alpha|, \qquad \beta \in (\mathbb{Z}^+)^n, \ |\beta| \le k. \qquad \square$$

Theorem 3.1 allows us to define in a natural manner $H^k(\Omega)$, the Sobolev space of order $k \in \mathbb{Z}^+$ in any subset Ω (open) of $\mathbb{R}^n$. Given $f \in L^2(\Omega)$, we say that $\partial_x^\alpha f$, $\alpha \in (\mathbb{Z}^+)^n$ is the αth partial derivative (in the distribution sense) of f, if for every $\phi \in C_0^\infty(\Omega)$

$$\int_\Omega f \, \partial_x^\alpha \phi \, dx = (-1)^{|\alpha|} \int_\Omega \partial_x^\alpha f \, \phi \, dx.$$

Then,

$$H^k(\Omega) = \{ f \in L^2(\Omega) : \partial_x^\alpha f (\text{in the distribution sense}) \in L^2(\Omega), \ |\alpha| \le k \}$$

with the norm

$$\|f\|_{H^k(\Omega)} \equiv \Big(\sum_{|\alpha|\le k} \int_\Omega |\partial_x^\alpha f(x)|^2 \, dx \Big)^{1/2}.$$

Example 3.7 For $n = 1$, $b > 0$, and $f(x) = |x|$, one has that $f \in H^1((-b, b))$ and $f \notin H^2((-b, b))$.

The next result allows us to relate "weak derivatives" with derivatives in the classical sense.

Theorem 3.2 (Embedding). *If $s > n/2 + k$, then $H^s(\mathbb{R}^n)$ is continuously embedded in $C_\infty^k(\mathbb{R}^n)$, the space of functions with k continuous derivatives vanishing at infinity. In other words, if $f \in H^s(\mathbb{R}^n)$, $s > n/2 + k$, then (after a possible modification of f in a set of measure zero) $f \in C_\infty^k(\mathbb{R}^n)$ and*

$$\|f\|_{C^k} \le c_s \, \|f\|_{s,2}. \tag{3.4}$$

Proof. Case $k = 0$: We first show that if $f \in H^s(\mathbb{R}^n)$, then $\widehat{f} \in L^1(\mathbb{R}^n)$ with

$$\|\widehat{f}\|_1 \leq c_s \|f\|_{s,2}, \text{ if } s > n/2. \tag{3.5}$$

Using the Cauchy–Schwarz inequality, we deduce:

$$\begin{aligned}\int_{\mathbb{R}^n} |\widehat{f}(\xi)| d\xi &= \int_{\mathbb{R}^n} |\widehat{f}(\xi)|(1+|\xi|^2)^{s/2} \frac{d\xi}{(1+|\xi|^2)^{s/2}} \\ &\leq \|\Lambda^s f\|_2 \left(\int_{\mathbb{R}^n} \frac{d\xi}{(1+|\xi|^2)^s} \right)^{1/2} \leq c_s \|f\|_{s,2}\end{aligned}$$

if $s > n/2$. Combining (3.5), Proposition 1.2, and Theorem 1.1, we conclude that

$$\|f\|_\infty = \|(\widehat{f})^\vee\|_\infty \leq \|\widehat{f}\|_1 \leq c_s \|f\|_{s,2}.$$

Case $k \geq 1$: Using the same argument, we have that if $f \in H^s(\mathbb{R}^n)$ with $s > n/2 + k$, then for $\alpha \in (\mathbb{Z}^+)^n$, $|\alpha| \leq k$, it follows that $\widehat{\partial_x^\alpha f} \in L^1(\mathbb{R}^n)$ and

$$\|\partial_x^\alpha f\|_\infty \leq \|\widehat{\partial_x^\alpha f}\|_1 = \|(2\pi i \xi)^\alpha \widehat{f}\|_1 \leq c_s \|f\|_{s,2}.$$

□

Corollary 3.1. *If $s = n/2 + k + \theta$, with $\theta \in (0,1)$, then $H^s(\mathbb{R}^n)$ is continuously embedded in $C^{k+\theta}(\mathbb{R}^n)$, the space of C^k functions with partial derivatives of order k Hölder continuous with index θ.*

Proof. We only prove the case $k = 0$, since the proof of the general case follows the same argument. From the formula of inversion of the Fourier transform and the Cauchy–Schwarz inequality we have:

$$\begin{aligned}|f(x+y) - f(x)| &= \Big| \int_{\mathbb{R}^n} e^{2\pi i (x\cdot\xi)} \widehat{f}(\xi)(e^{2\pi i (y\cdot\xi)} - 1)\, d\xi \Big| \\ &\leq \Big(\int_{\mathbb{R}^n} (1+|\xi|^2)^{n/2+\theta} |\widehat{f}(\xi)|^2\, d\xi \Big)^{1/2} \Big(\int_{\mathbb{R}^n} \frac{|e^{2\pi i (y\cdot\xi)} - 1|^2}{(1+|\xi|^2)^{n/2+\theta}}\, d\xi \Big)^{1/2}.\end{aligned}$$

But

$$\begin{aligned}&\int_{\mathbb{R}^n} \frac{|e^{2\pi i (y\cdot\xi)} - 1|^2}{(1+|\xi|^2)^{n/2+\theta}}\, d\xi \\ &\qquad \leq c \int_{|\xi| \leq |y|^{-1}} |y|^2 |\xi|^2 \frac{d\xi}{(1+|\xi|^2)^{n/2+\theta}} + 4 \int_{|\xi| \geq |y|^{-1}} \frac{d\xi}{(1+|\xi|^2)^{n/2+\theta}}\end{aligned}$$

$$\leq c|y|^2 \int_0^{|y|^{-1}} \frac{r^{n+1}}{(1+r)^{n+2\theta}}\, dr + 4 \int_{|y|^{-1}}^{\infty} \frac{r^{n-1}}{(1+r)^{n+2\theta}}\, dr \;\leq\; c\,|y|^{2\theta}.$$

If $|y| < 1$, we conclude that $|f(x+y) - f(x)| \leq c\,|y|^{\theta}$. This finishes the proof.□

Theorem 3.3. *If $s \in (0, n/2)$, then $H^s(\mathbb{R}^n)$ is continuously embedded in $L^p(\mathbb{R}^n)$ with $p = 2n/(n-2s)$, i.e., $s = n(1/2 - 1/p)$. Moreover, for $f \in H^s(\mathbb{R}^n)$, $s \in (0, n/2)$,*

$$\|f\|_p \leq c_{n,s}\, \|D^s f\|_2 \leq c\|f\|_{s,2}, \tag{3.6}$$

where

$$D^l f = (-\Delta)^{l/2} f = ((2\pi|\xi|)^l\, \widehat{f}\,)^{\vee}.$$

Proof. The last inequality in (3.6) is immediate, so we just need to show the first one. We define

$$D^s f = g \quad \text{or} \quad f = D^{-s} g = c_{n,s} \left(\frac{1}{|\xi|^s}\, \widehat{g} \right)^{\vee} = \frac{c_{n,s}}{|x|^{n-s}} * g, \tag{3.7}$$

where we have used the result of Exercise 1.14. Thus, by the Hardy–Littlewood–Sobolev estimate (2.10) it follows that

$$\|f\|_p = \|D^{-s} g\|_p = \|\frac{c_{n,s}}{|x|^{n-s}} * g\|_p \leq c_{n,s}\, \|g\|_2 = c\|D^s f\|_2. \tag{3.8}$$

□

We notice from Theorems 3.2 and 3.3, and Corollary 3.1 that the local regularity in H^s, $s > 0$, increases with the parameter s.

Examples 3.1 and 3.3 show that the functions in $H^s(\mathbb{R}^n)$ with $s < n/2$ or $s < n/2 + 1$, respectively, are not necessarily continuous nor C^1. Moreover, let $f \in L^2(\mathbb{R}^n)$ with

$$\widehat{f}(\xi) = \frac{1}{(1+|\xi|)^n\, \log(2+|\xi|)}$$

(which is radial, decreasing, and positive). A simple computation shows that $f \in H^{\frac{n}{2}}(\mathbb{R}^n)$, but $\widehat{f} \notin L^1(\mathbb{R}^n)$ and so $f \notin L^\infty(\mathbb{R}^n)$, since $f(0) = \int \widehat{f}(\xi) d\xi = \infty$ (see also Exercise 3.11(iii)).

To complete the embedding results of the spaces $H^s(\mathbb{R}^n)$, $s > 0$, it remains to consider the case $s = n/2$ (since for $s = k + n/2$, $k \in \mathbb{Z}^+$, the result follows from this one). So, we define the space of functions of the bounded mean oscillation or BMO, introduced by John and Nirenberg [JN].

Definition 3.3. For $f : \mathbb{R}^n \to \mathbb{C}$ with $f \in L^1_{\text{loc}}(\mathbb{R}^n)$, we say that $f \in \text{BMO}(\mathbb{R}^n)$ (f has bounded mean oscillation (BMO)) if

$$\|f\|_{\text{BMO}} = \sup_{\substack{x \in \mathbb{R}^n \\ r > 0}} \frac{1}{|B_r(x)|} \int_{B_r(x)} |f(y) - f_{B_r(x)}|\, dy < \infty, \tag{3.9}$$

where

$$f_{B_r(x)} = \frac{1}{|B_r(x)|}\int_{B_r(x)} f(y)dy.$$

Notice that $\|\cdot\|_{\mathrm{BMO}}$ is a semi-norm since it vanishes for constant functions.

BMO($\mathbb{R}^n$) is a vector space with $L^\infty(\mathbb{R}^n) \subsetneq \mathrm{BMO}(\mathbb{R}^n)$ since $\|f\|_{\mathrm{BMO}} \le 2\|f\|_\infty$ and $\log|x| \in \mathrm{BMO}(\mathbb{R}^n)$.

Theorem 3.4. *$H^{n/2}(\mathbb{R}^n)$ is continuously embedded in* BMO($\mathbb{R}^n$). *More precisely, there exists $c = c(n) > 0$ such that*

$$\|f\|_{\mathrm{BMO}} \le c\,\|D^{n/2}f\|_2.$$

Proof. Without loss of generality, we assume f real valued. Consider $x \in \mathbb{R}^n$ and $r > 0$.

Let $\phi_r \in C_0^\infty(\mathbb{R}^n)$ such that $\operatorname{supp}\phi_r \subseteq \{x \mid |x| \le \frac{2}{r}\}$ with $0 \le \phi_r(x) \le 1$ and $\phi_r(x) \equiv 1$ if $|x| < 1/r$, and define

$$f(x) = f_l + f_h = (\widehat{f}\phi_r)^\vee + (\widehat{f}(1-\phi_r))^\vee.$$

We observe that

$$\|f\|_{\mathrm{BMO}} \le \|f_l\|_{\mathrm{BMO}} + \|f_h\|_{\mathrm{BMO}}$$

and $f_l \in H^s(\mathbb{R}^n)$ for any $s > 0$; therefore,

$$f_{l,B_r(x)} = \frac{1}{|B_r(x)|}\int_{B_r(x)} f_l(y)\,dy = f_l(x_0)$$

for some $x_0 \in B_r(x)$, and so for any $y \in B_r(x)$

$$|f_l(y) - f_{l,B_r(x)}| \le 2r\,\|\nabla f_l\|_\infty.$$

Using this estimate we get:

$$\begin{aligned}
\frac{1}{|B_r(x)|}\int_{B_r(x)} \left|f_l(y) - f_{l,B_r(x)}\right| dy &\le \frac{1}{|B_r(x)|^{1/2}}\Big(\int_{B_r(x)} |f_l(y) - f_{l,B_r(x)}|^2\,dy\Big)^{1/2}\\
&\le 2r\,\|\nabla f_l\|_\infty \le 2r\,\|\widehat{\nabla f_l}\|_1\\
&\le 2r\int_{|\xi|\le 1/2r} |\xi|^{1-n/2}|\xi|^{n/2}|\widehat{f}(\xi)|\,d\xi\\
&\le 2r\Big(\int_{|\xi|\le 1/2r} |\xi|^{2-n}\,d\xi\Big)^{1/2}\|D^{n/2}f\|_2 \le c\|D^{n/2}f\|_2.
\end{aligned}$$

Also,

$$\frac{1}{|B_r(x)|}\int_{B_r(x)} |f_h(y) - f_{h,B_r(x)}| dy \le \frac{2}{|B_r(x)|^{1/2}}\|f_h\|_2$$

$$\leq \frac{2}{|B_r(x)|^{1/2}} \left(\int_{|\xi|\geq 1/2r} |\widehat{f}(\xi)|^2 \, d\xi \right)^{1/2}$$

$$= \frac{c_n}{r^{n/2}} \left(\int_{|\xi|\geq 1/2r} r^n |\xi|^n |\widehat{f}(\xi)|^2 \, d\xi \right)^{1/2} \leq \|D^{n/2} f\|_2,$$

which yields the desired result. □

We have shown that $H^s(\mathbb{R}^n)$ with $s > n/2$ is a Hilbert space whose elements are continuous functions. From the point of view of nonlinear analysis, the next property is essential.

Theorem 3.5. *If $s > n/2$, then $H^s(\mathbb{R}^n)$ is an algebra with respect to the product of functions. That is, if $f, g \in H^s(\mathbb{R}^n)$, then $fg \in H^s(\mathbb{R}^n)$ with*

$$\|fg\|_{s,2} \leq c_s \|f\|_{s,2} \|g\|_{s,2}. \tag{3.10}$$

Proof. From the triangle inequality, we have that for every $\xi, \eta \in \mathbb{R}^n$:

$$(1 + |\xi|^2)^{s/2} \leq 2^s[(1 + |\xi - \eta|^2)^{s/2} + (1 + |\eta|^2)^{s/2}].$$

Using this we deduce that

$$\begin{aligned}
|\Lambda^s(fg)| &= |(1 + |\xi|^2)^{s/2} \widehat{(fg)}(\xi)| \\
&= (1 + |\xi|^2)^{s/2} \left| \int_{\mathbb{R}^n} \widehat{f}(\xi - \eta) \widehat{g}(\eta) \, d\eta \right| \\
&\leq 2^s \int_{\mathbb{R}^n} \Big[(1 + |\xi - \eta|^2)^{s/2} \, |\widehat{f}(\xi - \eta) \widehat{g}(\eta)| \\
&\quad + (1 + |\eta|^2)^{s/2} \, |\widehat{f}(\xi - \eta) \widehat{g}(\eta)| \Big] \, d\eta \\
&\leq 2^s (|\widehat{\Lambda^s f}| * |\widehat{g}| + |\widehat{f}| * |\widehat{\Lambda^s g}|).
\end{aligned}$$

Thus, taking the L^2-norm and using (1.39) it follows that

$$\|fg\|_{s,2} = \|\Lambda^s(fg)\|_2 \leq c(\|\Lambda^s f\|_2 \|\widehat{g}\|_1 + \|\widehat{f}\|_1 \|\Lambda^s g\|_2). \tag{3.11}$$

Finally, (3.5) assures one that if $r > n/2$, then

$$\begin{aligned}
\|fg\|_{s,2} &\leq c_s(\|f\|_{s,2} \|\widehat{g}\|_1 + \|\widehat{f}\|_1 \|g\|_{s,2}) \\
&\leq c_s(\|f\|_{s,2} \|g\|_{r,2} + \|f\|_{r,2} \|g\|_{s,2}).
\end{aligned} \tag{3.12}$$

Choosing $r = s$, we obtain (3.10). □

The inequality (3.12) is not sharp as the following scaling argument shows. Let $\lambda > 0$ and

$$f(x) = f_1(\lambda x), \quad g(x) = g_1(\lambda x), \quad f_1, \, g_1 \in \mathcal{S}(\mathbb{R}^n).$$

Then, as $\lambda \uparrow \infty$ the right-hand side of (3.12) grows as λ^{s+r}, meanwhile the left-hand side grows as λ^s. This will not be the case if we replace $\|\cdot\|_{r,2}$ in (3.12) with the $\|\cdot\|_\infty$-norm to get that

$$\|fg\|_{s,2} \le c_s(\|f\|_{s,2}\,\|g\|_\infty + \|f\|_\infty\,\|g\|_{s,2}) \tag{3.13}$$

which in particular shows that for any $s > 0$, $H^s(\mathbb{R}^n) \cap L^\infty(\mathbb{R}^n)$ is an algebra under the point-wise product.

For $s \in \mathbb{Z}^+$, the inequality (3.13) follows by combining the Leibniz rule for the product of functions and the Gagliardo–Nirenberg inequality:

$$\|\partial_x^\alpha f\|_p \le c \sum_{|\beta|=m} \|\partial_x^\beta f\|_q^\theta\,\|f\|_r^{1-\theta} \tag{3.14}$$

with $|\alpha| = j$, $c = c(j,m,p,q,r)$, $1/p - j/n = \theta(1/q - m/n) + (1-\theta)1/r$, $\theta \in [j/m, 1]$. For the proof of this inequality, we refer the reader to the reference [Fm].

For the general case $s > 0$, where the usual point-wise Leibniz rule is not available, the inequality (3.13) still holds (see [KPo]). The inequality (3.13) has several extensions, for instance: Let $s \in (0,1)$, $r \in [1,\infty)$, $1 < p_j, q_j \le \infty$, $1/r = 1/p_j + 1/q_j$, $j = 1,2$. Then,

$$\|\Phi^s(fg)\|_r \le c(\|\Phi^s(f)\|_{p_1}\|g\|_{q_1} + \|f\|_{p_2}\|\Phi^s(g)\|_{q_2}),$$

with $\Phi^s = \Lambda^s$ or D^s, (for the proof of this estimate and further generalizations [KPV4], [MPTT], and [GaO]). The extension to the case $r = p_j = q_j = \infty$, $j = 1,2$ was given in [BoLi].

In many applications, the following commutator estimate is often used:

$$\begin{aligned}\sum_{|\alpha|=s} \|[\partial_x^\alpha; g]\, f\|_2 &\equiv \sum_{|\alpha|=s} \|\partial_x^\alpha(gf) - g\partial_x^\alpha f\|_2 \\ &\le c_{n,s}\Big(\|\nabla g\|_\infty \sum_{|\beta|=s-1} \|\partial_x^\beta f\|_2 + \|f\|_\infty \sum_{|\beta|=s} \|\partial_x^\beta g\|_2\Big),\end{aligned} \tag{3.15}$$

(see [Kl2]). Similarly, for $s \ge 1$ one has

$$\|[\Lambda^s; g]\, f\|_2 \le c\,(\|\nabla g\|_\infty \|\Lambda^{s-1} f\|_2 + \|f\|_\infty\,\|\Lambda^s g\|_2), \tag{3.16}$$

(see [KPo]).

There are "equivalent" manners to define fractional derivatives without relying on the Fourier transform. For instance:

Definition 3.4 (Stein [S1]). For $b \in (0,1)$ and an appropriate f define

$$\mathcal{D}^b f(x) = \Big(\int \frac{|f(x) - f(y)|^2}{|x-y|^{n+2b}}\,dy\Big)^{1/2}. \tag{3.17}$$

Theorem 3.6 (Stein [S1]). *Let $b \in (0,1)$ and $\frac{2n}{(n+2b)} \leq p < \infty$. Then $f,\ D^b f \in L^p(\mathbb{R}^n)$ if and only if $f,\ \mathcal{D}^b f \in L^p(\mathbb{R}^n)$.*

Moreover,

$$\|f\|_p + \|D^b f\|_p \sim \|f\|_p + \|\mathcal{D}^b f\|_p.$$

The case $p = 2$ was previously considered in [AS].

For other "equivalent" definitions of fractional derivatives see [Str1].

Finally, to complete our study of Sobolev spaces we introduce the localized Sobolev spaces.

Definition 3.5. Given $f : \mathbb{R}^n \to \mathbb{R}$, we say that $f \in H^s_{\text{loc}}(\mathbb{R}^n)$ if for every $\varphi \in C_0^\infty(\mathbb{R}^n)$ we have $\varphi f \in H^s(\mathbb{R}^n)$. In other words, for any $\Omega \subseteq \mathbb{R}^n$ open bounded $f|_\Omega$ coincides with an element of $H^s(\mathbb{R}^n)$.

This means that f has the sufficient regularity, but may not have enough decay to be in $H^s(\mathbb{R}^n)$.

Example 3.8 Let $n = 1$, $f(x) = x$, and $g(x) = |x|$, then $f \in H^s_{\text{loc}}(\mathbb{R})$ for every $s \geq 0$ and $g \in H^s_{\text{loc}}(\mathbb{R})$ for every $s < 3/2$.

3.2 Pseudo-Differential Operators

We recall some results from the theory of pseudo-differential operators that we need to describe the local smoothing effect for linear elliptic systems.

The class $S^m = S^m_{1,0}$ of classical symbols of order $m \in \mathbb{R}$ is defined by

$$S^m = \{p(x,\xi) \in C^\infty(\mathbb{R}^n \times \mathbb{R}^n) : |p|^{(j)}_{S^m} < \infty,\ j \in \mathbb{N}\}, \tag{3.18}$$

where

$$|p|^{(j)}_{S^m} = \sup\{\|\langle\xi\rangle^{-m+|\alpha|}\partial_\xi^\alpha \partial_x^\beta p(\cdot,\cdot)\|_{L^\infty(\mathbb{R}^n\times\mathbb{R}^n)} : |\alpha+\beta| \leq j\} \tag{3.19}$$

and $\langle\xi\rangle = (1+|\xi|^2)^{1/2}$.

The pseudo-differential operator Ψ_p associated to the symbol $p \in S^m$ is defined by

$$\Psi_p f(x) = \int_{\mathbb{R}^n} e^{2\pi i x\cdot\xi}\, p(x,\xi)\hat{f}(\xi)\, d\xi, \qquad f \in \mathcal{S}(\mathbb{R}^n). \tag{3.20}$$

Example 3.9 A partial differential operator

$$P = \sum_{|\alpha|\leq N} a_\alpha(x)\partial_x^\alpha,$$

with $a_\alpha \in \mathcal{S}(\mathbb{R}^n)$ is a pseudo-differential operator $P = \Psi_p$ with symbol

$$p(x,\xi) = \sum_{|\alpha|\leq N} a_\alpha(x)(2\pi i\xi)^\alpha \in S^N.$$

Example 3.10 The fractional differentiation operator defined in (3.1) as $\Lambda^{\rho} = \Psi_{\langle\xi\rangle^{\rho}}$ is also a pseudo-differential operator with symbol in S^{ρ}, $\rho \in \mathbb{R}$.

The collection of symbol classes S^m, $m \in \mathbb{R}$, is in some cases closed under composition, adjointness, division, and square root operations. This is not the case for polynomials in ξ, and sometimes this closure allows one to construct approximate inverses and square roots of pseudo-differential operators.

Next, we list some properties of pseudo-differential operators whose proofs can be found for instance in [Kg].

Theorem 3.7 (Sobolev boundedness). *Let $m \in \mathbb{R}$, $p \in S^m$, and $s \in \mathbb{R}$. Then, Ψ_p extends to a bounded linear operator from $H^{m+s}(\mathbb{R}^n)$ to $H^s(\mathbb{R}^n)$. Moreover, there exist $j = j(n;m;s) \in \mathbb{N}$ and $c = c(n;m;s)$ such that*

$$\|\Psi_p f\|_{H^s} \leq c\,|p|_{S^m}^{(j)}\,\|f\|_{H^{m+s}}. \tag{3.21}$$

Theorem 3.8 (Symbolic calculus). *Let $m_1, m_2 \in \mathbb{R}$, $p_1 \in S^{m_1}$, $p_2 \in S^{m_2}$. Then, there exist $p_3 \in S^{m_1+m_2-1}$, $p_4 \in S^{m_1+m_2-2}$, and $p_5 \in S^{m_1-1}$ such that*

$$\begin{aligned} \Psi_{p_1}\Psi_{p_2} &= \Psi_{p_1 p_2} + \Psi_{p_3}, \\ \Psi_{p_1}\Psi_{p_2} - \Psi_{p_2}\Psi_{p_1} &= \Psi_{-i\{p_1,p_2\}} + \Psi_{p_4}, \\ (\Psi_{p_1})^* &= \Psi_{\bar{p}_1} + \Psi_{p_5}, \end{aligned} \tag{3.22}$$

where $\{p_1, p_2\}$ denotes the Poisson bracket, i.e.,

$$\{p_1, p_2\} = \sum_{j=1}^{n} (\partial_{\xi_j} p_1\, \partial_{x_j} p_2 - \partial_{x_j} p_1\, \partial_{\xi_j} p_2), \tag{3.23}$$

and such that for any $j \in \mathbb{N}$ there exist $j' \in \mathbb{N}$ and $c_1 = c_1(n;m_1;\,m_2;j)$, $c_2 = c_2(n;m_1;j)$ such that

$$\begin{aligned} |p_3|_{S^{m_1+m_2-1}}^{(j)} + |p_4|_{S^{m_1+m_2-2}}^{(j)} &\leq c_1\,|p_1|_{S^{m_1}}^{(j')}\,|p_2|_{S^{m_2}}^{(j')} \\ |p_5|_{S^{m_1-1}}^{(j')} &\leq c_2\,|p_1|_{S^{m_1}}^{(j')}. \end{aligned}$$

Remark 3.1.

(i) (3.22) tell us that the "principal symbol" of the commutator $[\psi_{p_1}; \psi_{p_2}]$ is given by the formula in (3.23).

(ii) It is useful for our purpose to consider the class of symbols $S^{m,N} = S^{m,N}_{1,0}$ defined as $p(x,\xi) \in C^N(\mathbb{R}^n \times \mathbb{R}^n)$ such that

$$|p|_{S^m}^{(N)} < \infty, \quad \text{with } |p|_{S^m}^{(N)} \text{ defined in (3.19).} \tag{3.24}$$

For N sufficiently large the results in Theorem 3.7 extend to the class $S^{m,N}$.

3.3 The Bicharacteristic Flow

In this section, we introduce the notion of bicharacteristic flow. This plays a key role in the study of linear variable coefficients Schrödinger equations and in the well-posedness of the initial value problem (IVP) associated to the quasilinear case as we can see in the next and the last chapters.

Let $\mathcal{L} = \partial_{x_j} a_{jk}(x) \partial_{x_k}$ be an elliptic self-adjoint operator, that is, $(a_{jk}(x))_{jk}$ is a $n \times n$ matrix of functions $a_{jk} \in C_b^\infty$, real, symmetric, and positive definite, i.e., $\exists\, \nu > 0$ such that $\forall\, x, \xi \in \mathbb{R}^n$,

$$\nu^{-1} \|\xi\|^2 \leq \sum_{j,k=1}^{n} a_{jk}(x) \xi_j \xi_k \leq \nu \|\xi\|^2. \tag{3.25}$$

Let h_2 be the principal symbol of $\mathcal{L}$, i.e.,

$$h_2(x, \xi) = - \sum_{j,k=1}^{n} a_{jk}(x) \xi_j \xi_k. \tag{3.26}$$

The bicharacteristic flow is the flow of the Hamiltonian vector field:

$$H_{h_2} = \sum_{j=1}^{n} [\partial_{\xi_j} h_2 \cdot \partial_{x_j} - \partial_{x_j} h_2 \cdot \partial_{\xi_j}] \tag{3.27}$$

and is denoted by $(X(s; x_0, \xi_0), \Xi(s; x_0, \xi_0))$, i.e.,

$$\begin{cases} \dfrac{d}{ds} X_j(s; x_0, \xi_0) = -2 \sum\limits_{k=1}^{n} a_{jk}(X(s; x_0, \xi_0))\, \Xi_k(s; x_0, \xi_0), \\ \dfrac{d}{ds} \Xi_j(s; x_0, \xi_0) = \sum\limits_{k,l=1}^{n} \partial_{x_j} a_{lk}(X(s; x_0, \xi_0))\, \Xi_k(s; x_0, \xi_0)\, \Xi_l(s; x_0, \xi_0) \end{cases} \tag{3.28}$$

for $j = 1, \ldots, n$, with

$$(X(0; x_0, \xi_0), \Xi(0; x_0, \xi_0)) = (x_0, \xi_0). \tag{3.29}$$

The bicharacteristic flow exists in the time interval $s \in (-\delta, \delta)$ with $\delta = \delta(x_0, \xi_0)$, and $\delta(\cdot)$ depending continuously on (x_0, ξ_0).

The bicharacteristic flow preserves h_2, i.e.,

$$\frac{d}{ds} h_2(X(s; x_0, \xi_0), \Xi(s; x_0, \xi_0)) = 0,$$

so the ellipticity hypothesis (3.25) gives

$$\nu^{-2} \|\xi_0\|^2 \leq \|\Xi(s; x_0, \xi_0)\|^2 \leq \nu^2 \|\xi_0\|^2, \tag{3.30}$$

and hence $\delta = \infty$.

In the case of constant coefficients, $h_2(x,\xi) = -|\xi|^2$, the bicharacteristic flow is given by $(X,\ \Xi)(\xi, x_0, \xi_0) = (x_0 - 2s\xi_0, \xi_0)$.

For general symbol $h(x,\xi)$, the bicharacteristic flow is defined as:

$$\begin{cases} \dfrac{dX}{ds} = \partial_\xi\, h(X,\Xi) \\[2ex] \dfrac{d\,\Xi}{ds} = -\partial_x\, h(X,\Xi). \end{cases} \tag{3.31}$$

In applications, the notion of the bicharacteristic flow

$$t \mapsto (X(t; x_0, \xi_0),\ \Sigma(t; x_0, \xi_0)) \tag{3.32}$$

being nontrapping arises naturally.

Definition 3.6. A point $(x_0,\ \xi_0) \in \mathbb{R}^n \times \mathbb{R}^n \setminus \{0\}$ is nontrapped forward (respectively, backward) by the bicharacteristic flow if

$$\|X(t;\ x_0,\ \xi_0)\| \to \infty \text{ as } t \to \infty \quad (\text{resp, } t \to -\infty). \tag{3.33}$$

If each point $(x_0,\ \xi_0) \in \mathbb{R}^n \times \mathbb{R}^n - \{0\}$ is nontrapped forward, then the bicharacteristic flow is said to be nontrapping.

In particular, if one assumes that the "metric" $(a_{jk}(x))$ in (3.26) possesses an "asymptotic flat property," for example,

$$|\partial_x^\alpha(a_{jk}(x) - \delta_{jk})| \le \frac{c_\alpha}{|x|^{1+\epsilon(\alpha)}},\quad \epsilon(\alpha) > 0,\quad 0 \le |\alpha| \le m = m(n), \tag{3.34}$$

then it suffices to have that for each $(x_0, \xi_0) \in \mathbb{R}^n \times \mathbb{R}^n \setminus \{0\}$ and for each $\mu > 0$ there exists $\hat{t} = \hat{t}(\mu;\ x_0, \xi_0) > 0$ such that

$$\|X(\hat{t};\ x_0,\ \xi_0)\| \ge \mu$$

to guarantee that the bicharacteristic flow is nontrapping.

The next result shows that the Hamiltonian vector field is differentiation along the bicharacteristics.

Lemma 3.1. *Let $\phi \in C^\infty(\mathbb{R}^n \times \mathbb{R}^n)$. Then,*

$$(H_{h_2}\phi)(x,\xi) = \partial_s[\phi(X(s;x,\xi), \Xi(s;x,\xi))]|_{s=0} = \{h_2, \phi\}. \tag{3.35}$$

Notice that $-i\{h_2,\phi\}$ is the principal symbol of the commutator $[\psi_{h_2}, \psi_\phi]$ (see 3.22).

Proof. By the chain rule,

$$\begin{aligned} \partial_s[\phi(X(s;x,\xi), \Xi(s;x,\xi))] &= (\nabla_x\phi)(X(s;x,\xi), \Xi(s;x,\xi)) \cdot \partial_s X(s;x,\xi) \\ &\quad + (\nabla_\xi\phi)(X(s;x,\xi), \Xi(s;x,\xi)) \cdot \partial_s \Xi(s;x,\xi) \\ &= (\nabla_x\phi \cdot \nabla_\xi h_2)(X(s;x,\xi), \Xi(s;x,\xi)) \\ &\quad - (\nabla_\xi\phi \cdot \nabla_x h_2)(X(s;x,\xi), \Xi(s;x,\xi)). \end{aligned}$$

Setting $s = 0$, the lemma follows. □

3.4 Exercises

3.1 Prove that for any $k \in \mathbb{Z}^+$ and any $\theta \in (0,1)$

$$\chi_{(-1,1)} \overset{k \text{ times}}{* \cdots *} \chi_{(-1,1)}(x) \in C_0^{k-1,\theta}(\mathbb{R}) \backslash C^k(\mathbb{R}).$$

3.2 Prove Proposition 3.1.

3.3 Let $f_n : \mathbb{R}^n \to \mathbb{R}$ with $f_n(x) = e^{-2\pi|x|}$.

(i) Prove that $f_1 * f_1(x) = \dfrac{e^{-2\pi|x|}}{2\pi}(1 + 2\pi|x|)$.
Hint: Use an explicit computation or Exercise 1.1(ii).

(ii) Show that $f_1 * f_1(x) \in C^2(\mathbb{R})$, but is not in $C^3(\mathbb{R})$.

(iii) Prove that $f_n * f_n \in C_\infty^{n+1}(\mathbb{R}^n)$.

(iv) More general, prove that if $g \in H^{s_1}(\mathbb{R}^n)$ and $h \in H^{s_2}(\mathbb{R}^n)$, then $g * h \in C_\infty^{[s_1+s_2]}(\mathbb{R}^n)$ (where $[\,\cdot\,]$ denotes the greatest integer function.)

3.4 Let $\phi(x) = e^{-|x|}$, $x \in \mathbb{R}$:

(i) Prove that

$$\phi(x) - \phi''(x) = 2\delta, \tag{3.36}$$

(a) in the distribution sense, i.e., $\forall \varphi \in C_0^\infty(\mathbb{R})$,

$$\int \phi(x)(\varphi(x) - \varphi''(x))\,dx = 2\varphi(0),$$

(b) by taking the Fourier transform in (3.36).

(ii) Prove that given $g \in L^2(\mathbb{R})$ (or $H^s(\mathbb{R})$) the equation:

$$\left(1 - \frac{d^2}{dx^2}\right) f = g$$

has solution $f = \frac{1}{2} e^{-|\cdot|} * g \in H^2(\mathbb{R})$ (or $H^{s+2}(\mathbb{R})$).

3.5 Show that if $k \in \mathbb{Z}^+$ and $p \in [1, \infty)$, then

$$F_{k,p}(\mathbb{R}^n) = L_k^p(\mathbb{R}^n) \cap L^\infty(\mathbb{R}^n)$$

is a Banach algebra with respect to point-wise product of functions. Moreover, if $f, g \in F_{k,p}$, then

$$\|fg\|_{k,p} \le c_k(\|f\|_{k,p}\|g\|_\infty + \|f\|_\infty\|g\|_{k,p}). \tag{3.37}$$

Notation:

$$L_k^p(\mathbb{R}^n) = \{f : \mathbb{R}^n \to \mathbb{C} : \partial^\alpha f \text{ (distribution sense) } \in L^p,\ |\alpha| \le k\},$$

whose norm is defined as:

$$\|f\|_{k,p} = \sum_{|\alpha| \le k} \|\partial^\alpha f\|_p.$$

Observe that when $p = 2$ one has $L^p_k(\mathbb{R}^n) = H^k(\mathbb{R}^n)$.
More generally, we define

$$L^p_s(\mathbb{R}^n) = (1-\Delta)^{-s/2} L^p(\mathbb{R}^n) \text{ for } s \in \mathbb{R}, \text{ with } \|f\|_{s,p} = \|(1-\Delta)^{s/2} f\|_p. \tag{3.38}$$

Hint: From Leibniz formula and Hölder's inequality it follows that (assume $n = 1$ to simplify)

$$\|(fg)^{(k)}\|_p \leq \sum_{j=0}^{k} c_j \|f^{(k-j)}\|_{p_{j_1}} \|g^{(j)}\|_{p_{j_2}}, \text{ with } \frac{1}{p} = \frac{1}{p_{j_1}} + \frac{1}{p_{j_2}}.$$

Combine the Gagliardo–Nirenberg inequality (3.14):

$$\|h^{(k-j)}\|_{p_j} \leq c\|h^{(k)}\|_p^{\theta} \|h\|_{\infty}^{1-\theta}, \qquad \theta = \theta(n,k,j,p_j),$$

with Young's inequality (if $1/p + 1/p' = 1$ with $p > 1$, then $ab \leq a^p/p + b^{p'}/p'$) to get the desired result (3.37).

3.6 Extend the result of Theorem 3.3 to the spaces $L^p_s(\mathbb{R}^n)$, i.e., if $f \in L^p_s(\mathbb{R}^n)$, $0 < s < n/p$, then $f \in L^r(\mathbb{R}^n)$ with $s = n(\frac{1}{p} - \frac{1}{q})$, and

$$\|f\|_r \leq c_{n,s} \|D^s f\|_p \leq c_{n,s} \|f\|_{s,p}. \tag{3.39}$$

3.7 (i) Prove that if $1 < p < \infty$ and $b \in (0,1)$, then

$$\|\Lambda^b f\|_p \sim \|f\|_p + \|D^b f\|_p.$$

Hint: Use Theorem 2.8.

(ii) Given any $s \in \mathbb{R}$ find $f_s \in H^s(\mathbb{R})$ such that $f_s \notin H^{s'}(\mathbb{R})$ for any $s' > s$.
Hint:

(a) Notice that it suffices to find f_0.
(b) Show that if $g \in L^2(\mathbb{R})$ and $g \notin L^p(\mathbb{R})$ for any $p > 2$, then one can take $f_0 = g$.
(c) Use (b) to find f_0.

3.8 Show that if $f \in H^s(\mathbb{R}^n)$, $s > n/2$, with $\|f\|_{n/2,2} \leq 1$, then

$$\|f\|_\infty \leq c\,[1 + \log(1 + \|f\|_{s,2})]^{1/2}$$

with $c = c(s,n)$, see [BGa].

3.9 Prove the following inequalities:

(i) If $s > n/2$, then

$$\|f\|_\infty \leq c_{n,s} \|f\|_2^{1-n/2s} \|D^s f\|_2^{n/2s}.$$

(ii) If $s > n/p$, $1 < p < \infty$, then

$$\|f\|_\infty \le c_{n,s,p} \|f\|_p^{1-n/ps} \|D^s f\|_p^{n/ps}.$$

(iii) Prove Gagliardo–Nirenberg inequality (3.14) for p even integer, $m = 2$, $j = 2$, and $q, r \in (1, \infty)$ such that $1/q + 1/r = 2/p$.

(iv) Combine Exercises 2.10 and 2.11, and Theorem 2.6 to prove the Gagliardo–Nirenberg inequality in the general case.

3.10 ([AS]). Using Definition 3.4:

(i) Prove that for $b \in (0, 1)$

$$\|D^b f\|_2 = c_n \, \|\mathcal{D}^b f\|_2. \tag{3.40}$$

(ii) Prove that

$$\mathcal{D}^b(fg)(x) \le \|f\|_\infty \mathcal{D}^b g(x) + |g(x)| \mathcal{D}^b f(x) \tag{3.41}$$

and

$$\|\mathcal{D}^b(fg)\|_2 \le \|f \mathcal{D}^b g\|_2 + \|g \mathcal{D}^b f\|_2. \tag{3.42}$$

(iii) Let $F \in C_b^1(\mathbb{R} : \mathbb{R})$, $F(0) = 0$. Show that

$$\|D^b(F(f))\|_2 \le \|F'\|_\infty \|D^b f\|_2.$$

Hint: Apply part (i).

3.11 (i) Let $f \in L^p(\mathbb{R})$, $1 < p < \infty$, be such that $f(x_0^+)$, $f(x_0^-)$ exist and $f(x_0^+) \neq f(x_0^-)$ for some x_0. Prove that $f \notin L^p_{1/p}(\mathbb{R})$.

(ii) Let $\varphi \in C_0^\infty(\mathbb{R})$ with $\varphi(x) = 1$ if $|x| \le 1$ and $\varphi(x) = 0$ if $|x| > 2$. Let $a, b \in (0, 1)$. Prove that $|x|^a \varphi(x) \in H^b(\mathbb{R})$ if and only if $b < a + 1/2$.

(iii) Let $\alpha \in (0, 1/2)$. Prove that

$$|\log |x||^\alpha \, \chi_{\{|x| \le 1/10\}} + \frac{10}{9}(1 - |x|) \, \chi_{\{1/10 \le |x| \le 1\}} \in H^1(\mathbb{R}^2) - L^\infty(\mathbb{R}^2).$$

3.12 (Sobolev's inequality for radial functions) Let $f : \mathbb{R}^n \to \mathbb{R}$, $n \ge 3$, be a radial function, i.e., $f(x) = f(y)$ if $|x| = |y|$. Show that f satisfies

$$|f(x)| \le c_n \, |x|^{(2-n)/2} \, \|\nabla f\|_2.$$

3.13 (Hardy's inequalities (see Exercise 1.5))

(i) Let $1 \le p < \infty$. If $f \in L_1^p(\mathbb{R}^n)$, then

$$\left\| \frac{|f(\cdot)|}{|x|} \right\|_p \le \frac{p}{n - p} \, \|\nabla f\|_p. \tag{3.43}$$

(ii) Let $1 \leq p < \infty$, $q < n$, and $q \in [0, p]$. If $f \in L_1^p(\mathbb{R}^n)$, then

$$\int_{\mathbb{R}^n} \frac{|f(x)|^p}{|x|^q}\, dx \leq \left(\frac{p}{n-q}\right)^q \|f\|_p^{p-q} \|\nabla f\|_p^q. \tag{3.44}$$

Hint: Assume that $f \in C_0^\infty(\mathbb{R}^n)$. For (i), write $\||\cdot|^{-1} f\|_p^p$ in spherical coordinates, use integration by parts in the radial variable and Hölder inequality to get the result. For (ii), assume $p > q$, and apply (3.43) to $|x|^{-1} g(x)$ with $g(x) = |f(x)|^{p/q}$.

3.14 Prove Heisenberg's inequality. If $f \in H^1(\mathbb{R}^n) \cap L^2(|x|^2\, dx)$, then

$$\|f\|_2^2 \leq \frac{2}{n} \|x_j f\|_2 \|\partial_{x_j} f\|_2 = \frac{4\pi}{n} \|x_j f\|_2 \|\xi_j \widehat{f}\|_2 \leq \frac{2}{n} \|xf\|_2 \|\nabla f\|_2. \tag{3.45}$$

Hint: Use the density of $\mathcal{S}(\mathbb{R}^n)$ and integration by parts to obtain the identity

$$\|f\|_2^2 = -\frac{1}{n} \int x_j \partial_{x_j}(|f(x)|^2)\, dx.$$

3.15 Denote $u = u(x,t)$, the solution of the IVP associated to the inviscid Burgers' equation:

$$\begin{cases} \partial_t u + u \partial_x u = 0, \\ u(x,0) = u_0(x) \in C_0^\infty(\mathbb{R}), \end{cases} \tag{3.46}$$

$t, x \in \mathbb{R}$. Prove that for every $T > 0$,

$$u \in C^\infty(\mathbb{R} \times [-T,T]) \quad \text{or} \quad u \notin C^1(\mathbb{R} \times [-T,T]).$$

Hint: Combine the commutator estimate (3.16) and integration by parts to obtain the energy estimate

$$\frac{d}{dt} \|u(t)\|_{k,2} \leq c_k \|\partial_x u(t)\|_\infty \|u(t)\|_{k,2} \quad \text{for all } k \in \mathbb{Z}^+. \tag{3.47}$$

3.16 Let $P(x, \partial_x) = \sum_{|\alpha| \leq m_1} a_\alpha(x) \partial_x^\alpha$ and $Q(x, \partial_x) = \sum_{|\alpha| \leq m_2} b_\alpha(x) \partial_x^\alpha$ be two differential operators. Check the properties stated in Theorem 3.8 for P and Q.

3.17 (i) If $\Lambda = (1 - \Delta)^{1/2}$ and $y \in \mathbb{R}$, show that the symbol $p = p(\xi)$ of Λ^{iy}, $p(\xi) = (1 + |\xi|^2)^{iy/2} \in S_0$, and

$$|p|_{S^0}^j \leq c_n (1 + |y|)^j.$$

(ii) Show that if $p = p(x, \xi) \in S^0 = S_{1,0}^0$, then $e^{p(x,\xi)} \in S^0 = S_{1,0}^0$.

3.18 Prove that the bicharacteristic flow in (3.28) $(X(s; x_0, \xi_0), \Xi_k(s; x_0, \xi_0))$ satisfies

(i) $X(s; x_0, \rho\,\xi_0) = X(\rho\, s; x_0, \xi_0)$,
(ii) $\Xi_k(s; x_0, \rho\,\xi_0) = \rho\, \Xi_k(\rho\, s; x_0, \xi_0)$.

Hint: Use the homogeneity of $h_2(x, \xi) = -a_{jk}(x)\,\xi_j\xi_k$.

3.19 Prove that if Ψ_p is a pseudo-differential operator with symbol $p \in S^0$, then for any $b \in \mathbb{R}$,

$$\|\Psi_p f\|_{L^2(\langle x\rangle^b\,dx)} \leq c_{k,n}\, \|f\|_{L^2(\langle x\rangle^b\,dx)}, \tag{3.48}$$

where

$$||g||_{L^2(\langle x\rangle^b\,dx)} = \left(\int |g(x)|^2 \langle x\rangle^b\,dx\right)^{1/2}$$

and

$$\langle x\rangle = (1 + |x|^2)^{1/2}. \tag{3.49}$$

Hint:

(i) Follow an argument similar to that given in the proof of Theorem 2.1 to show that it suffices to establish (3.48) for $b = 4k$, $k \in \mathbb{Z}$.
(ii) Consider the case $b = -4k$, $k \in \mathbb{Z}^+$, and show that (3.48) is equivalent to

$$\left\| \frac{1}{\langle x\rangle^{2k}}\, \Psi_p\, (\langle x\rangle^{2k} g) \right\|_2 \leq c\, \|g\|_2. \tag{3.50}$$

(iii) Obtain (3.50) by combining integration by parts, Theorems 3.7 and 3.8.
(iv) Finally, prove the case $b = 4k$, $k \in \mathbb{Z}^+$, by duality.

3.20 Let $a, b > 0$. Assume that $\Lambda^a f = (1 - \Delta/4\pi^2)^{a/2} f \in L^2(\mathbb{R}^2)$ (i.e., $f \in H^a(\mathbb{R}^n)$) and $\langle x\rangle^b f \in L^2(\mathbb{R}^n)$ (see 3.49). Prove that for any $\theta \in (0, 1)$,

$$||\Lambda^{(1-\theta)a}(\langle x\rangle^{\theta b} f)||_2 \leq c_{a,b,n}\, ||\langle x\rangle^b f||_2^{\theta} ||\Lambda^a f||_2^{1-\theta}.$$

Hint: Combine the three lines theorem, Exercises 3.17 part (i) and (3.19).

Chapter 4
The Linear Schrödinger Equation

In this chapter, we study the smoothing properties of solutions of the initial value problem:

$$\begin{cases} \partial_t u = i\Delta u + F(x,t), \\ u(x,0) = u_0(x), \end{cases} \tag{4.1}$$

$x \in \mathbb{R}^n$, $t \in \mathbb{R}$. These properties are fundamental tools in the next chapters. In Section 4.1, we present some general basic results concerning the initial value problem (4.3). The global smoothing properties of solutions of (4.3) described by estimates of the type $L^q(\mathbb{R} : L^p(\mathbb{R}^n))$ are discussed in Section 4.2. In Section 4.3, we derive the local smoothing arising from estimates of type $L^2_{\text{loc}}(\mathbb{R} : H^{1/2}_{\text{loc}}(\mathbb{R}^n))$. We end the chapter with some remarks and comments regarding the issues discussed in the previous sections.

4.1 Basic Results

We begin by recalling the notation (see (1.27))

$$e^{it\Delta} u_0 = \frac{e^{-|x|^2/4it}}{(4\pi i t)^{n/2}} * u_0 = \left(e^{-4\pi^2 it|\xi|^2} \widehat{u_0}\right)^{\vee}. \tag{4.2}$$

The identity (4.2) describes the solution $u(x,t)$ of the linear homogeneous initial value problem (IVP)

$$\begin{cases} \partial_t u = i\Delta u, \\ u(x,0) = u_0(x). \end{cases} \tag{4.3}$$

$x \in \mathbb{R}^n$, $t \in \mathbb{R}$. In the following examples, we illustrate some of the properties exhibited by solutions of IVP (4.3).

F. Linares, G. Ponce, *Introduction to Nonlinear Dispersive Equations*,
Universitext, DOI 10.1007/978-1-4939-2181-2_4

Example 4.1 Consider the Gaussian function $u_0(x) = e^{-\pi|x|^2}$. Using Examples 1.3, 1.11, and Exercise 1.2 we find that the solution of the IVP (4.3) is given by

$$\begin{aligned} u(x,t) &= \left(e^{-4\pi^2 it|\xi|^2}\widehat{u_0}(\xi)\right)^\vee \\ &= \left(e^{-(1+4\pi it)\pi\,|\xi|^2}\right)^\vee \\ &= \frac{1}{(1+4\pi it)^{n/2}}\exp\left(\frac{-\pi|x|^2}{1+4\pi it}\right) \\ &= (1+4\pi it)^{-n/2}\exp\left(-\frac{\pi|x|^2}{1+16\pi^2t^2}\right)\exp\left(\frac{4\pi^2 it|x|^2}{1+16\pi^2t^2}\right). \end{aligned} \tag{4.4}$$

Notice that when $t \gg 1$ and $|x| < t$, the absolute value of the solution is bounded below by $c_n\,t^{-n/2}$ and the solution oscillates for $|x| > t^{1/2}$. Furthermore, if $|x| > t$ the absolute value of the solution decays exponentially. Moreover,

$$C\,t^{-n/2}\chi_{\{|x|<t\}}(x) \le |u(x,t)| \le c\,t^{-n/2}, \tag{4.5}$$

which is the expected behavior of the solution in order to have its $L^2(\mathbb{R}^n)$-norm independent of t.

Example 4.2 We can write the solution of the IVP (4.3) as

$$\begin{aligned} u(x,t) &= \left(e^{-4\pi^2 it|\xi|^2}\widehat{u_0}\right)^\vee(x) = \int_{\mathbb{R}^n}\frac{e^{i|x-y|^2/4t}}{(4\pi it)^{n/2}}\,u_0(y)\,dy \\ &= \frac{e^{i|x|^2/4t}}{(4\pi it)^{n/2}}\int_{\mathbb{R}^n} e^{-2ix\cdot y/4t}\,e^{i|y|^2/4t}\,u_0(y)\,dy \\ &= \frac{e^{i|x|^2/4t}}{(4\pi it)^{n/2}}\left(\widehat{e^{i|\cdot|^2/4t}\,u_0}\right)\left(\frac{x}{4\pi t}\right). \end{aligned} \tag{4.6}$$

Thus, if $c_t = (4\pi it)^{n/2}$,

$$c_t\,e^{-i|x|^2/4t}\,u(x,t) = \left(\widehat{e^{i|\cdot|^2/4t}\,u_0}\right)\left(\frac{x}{4\pi t}\right). \tag{4.7}$$

Notice that if $u_0 \in C_0(\mathbb{R}^n)$ from (4.7) we deduce that for any $t \in \mathbb{R}\setminus\{0\}$ and any $\epsilon > 0$, $u(\cdot,t) \notin L^1(e^{\epsilon|x|}dx)$. In particular, if $t \neq 0$, $u(x,t)$ has an analytic extension to $\mathbb{C}^n$ (see Exercise 4.5).

Example 4.3 This example describes the propagation of oscillatory pulses. Now we take $u_0(x) = e^{ix\cdot x_0}e^{-\pi|x|^2}$, $x_0 \in \mathbb{R}^n$. From Examples 1.3 and 1.4 we have $\widehat{u_0}(\xi) = e^{-\pi|\xi - x_0/2\pi|^2}$. Thus, using Example 4.1 we obtain

$$\begin{aligned} u(x,t) &= \left(e^{-4\pi^2 it(|\xi-x_0/2\pi|^2+2(\xi-x_0/2\pi)\cdot x_0/2\pi+|x_0|^2/4\pi^2)}\,e^{-\pi|\xi-x_0/2\pi|^2}\right)^\vee \\ &= \left(\tau_{x_0/2\pi}\left(e^{-4\pi^2 it(|\xi|^2+2\xi\cdot x_0/2\pi+|x_0|^2/4\pi^2)}\,e^{-\pi|\xi|^2}\right)\right)^\vee \end{aligned}$$

$$
\begin{aligned}
&= \left(\tau_{x_0/2\pi}(e^{-i2t\xi\cdot x_0}\, e^{-it|x_0|^2}\, e^{-(1+4\pi it)\pi|\xi|^2})\right)^{\vee} \\
&= e^{ix_0\cdot x}\tau_{2x_0t}(e^{-it|x_0|^2}\, e^{-(1+4\pi it)\pi|\xi|^2})^{\vee} \\
&= e^{ix_0\cdot x}\, e^{-it|x_0|^2}(1+4\pi it)^{-n/2}\, e^{\frac{-\pi|x-2tx_0|^2}{(1+4\pi it)}},
\end{aligned}
\tag{4.8}
$$

where τ is the translation operator (see (1.4)). In other words, the solution of the IVP (4.3) with data $u_0(x) = e^{ix_0\cdot x}\, e^{-\pi|x|^2}$ is given by

$$
u(x,t) = e^{ix\cdot x_0}\, e^{-i|x_0|^2 t}\, \mathbf{u}(x - 2t\, x_0, t), \tag{4.9}
$$

where $\mathbf{u}$ denotes the solution of the IVP (4.3) given in Example 4.1.

In the next proposition, we list several invariance properties of solutions of the equation in (4.3).

Proposition 4.1. *If $u = u(x,t)$ is a solution of the equation in (4.3), then*

$u_1(x,t) = e^{i\theta} u(x,t), \quad \theta \in \mathbb{R}$ *fixed,*

$u_2(x,t) = u(x - x_0, t - t_0)$, *with* $x_0 \in \mathbb{R}^n,\ t_0 \in \mathbb{R}$ *fixed,*

$u_3(x,t) = u(Ax,t)$, *withA any orthogonal matrix* $n \times n$,

$u_4(x,t) = u(x - 2x_0\, t, t)\, e^{i(x\cdot x_0 - |x_0|^2 t)}$, *with* $x_0 \in \mathbb{R}^n$ *fixed,*

$u_5(x,t) = \lambda^{n/2} u(\lambda x, \lambda^2 t),\ \lambda > 0,$

$$
u_6(x,t) = \frac{1}{(\alpha+\omega t)^{n/2}} \exp\left[\frac{i\omega|x|^2}{4(\alpha+\omega t)}\right] u\left(\frac{x}{\alpha+\omega t}, \frac{\gamma+\theta t}{\alpha+\omega t}\right), \quad \alpha\theta - \omega\gamma = 1,
$$

$u_7(x,t) = \overline{u(x,-t)},$

also satisfy the equation in (4.3).

In (4.2), we have used an exponential formula to describe the solution of the IVP (4.3). To justify this formula, we state next some properties of the family of operators $\{e^{it\Delta}\}_{t=-\infty}^{\infty}$.

Proposition 4.2.

1. *For all* $t \in \mathbb{R}$, $e^{it\Delta} \colon L^2(\mathbb{R}^n) \mapsto L^2(\mathbb{R}^n)$ *is an isometry, which implies*

$$
\|e^{it\Delta} f\|_2 = \|f\|_2.
$$

2. $e^{it\Delta} e^{it'\Delta} = e^{i(t+t')\Delta}$ *with* $(e^{it\Delta})^{-1} = e^{-it\Delta} = (e^{it\Delta})^*$.

3. $e^{i0\Delta} = 1$.

4. *Fixing* $f \in L^2(\mathbb{R}^n)$, *the function* $\Phi_f \colon \mathbb{R} \mapsto L^2(\mathbb{R}^n)$, *where* $\Phi_f(t) = e^{it\Delta} f$ *is a continuous function, i.e., it describes a curve in* $L^2(\mathbb{R}^n)$.

Proof. The proof is left as an exercise. □

In general, a family of operators $\{T_t\}_{t=-\infty}^{\infty}$ defined on a Hilbert space H which satisfies properties (1)–(4) in Proposition 4.2 is called a *unitary group of operators.*

Example 4.4 Let $L_t : L^2(\mathbb{R}) \mapsto L^2(\mathbb{R})$ be the one parameter family of translation operators $L_t(u_0)(x) = u_0(x+t)$. It is easy to see that $\{L_t\}_{t=-\infty}^{\infty}$ is a unitary group of operators, which describes the solution $u(x,t) = L_t(u_0)(x)$ of the problem

$$\begin{cases} \partial_t u = \partial_x u, \\ u(x,0) = u_0(x), \end{cases}$$

$t,\ x \in \mathbb{R}$.

The next result of M. H. Stone, characterizes the unitary group of operators.

Theorem 4.1 (M. H. Stone). *The family of operators $\{T_t\}_{t=-\infty}^{\infty}$ defined on the Hilbert space H is a unitary group of operators if and only if there exists a self-adjoint operator A (not necessarily bounded) on H such that*

$$T_t = e^{itA} \tag{4.10}$$

in the following sense: Consider $D(A)$ the domain of the operator A, which is a dense subspace of H; if $f \in D(A)$, then we have

$$\lim_{t \to 0} \frac{T_t f - f}{t} = iAf. \tag{4.11}$$

In other words, if $f \in D(A)$, then the curve Φ_f defined in Proposition 4.2 (4) is differentiable at $t = 0$ with derivative iAf.

For a proof of this theorem, we refer the reader to [Yo].

The operator A in Theorem 4.1 is called the *infinitesimal generator* of the unitary group. In (4.2), the operator A is the Laplacian Δ with $D(A) = H^2(\mathbb{R}^n)$. In Example 4.4, we have $A = -i\frac{d}{dx}$ and in this case, formula (4.10) can be interpreted as a generalized Taylor series.

Now we establish the properties of the group $\{e^{it\Delta}\}_{t=-\infty}^{\infty}$ in the $L^p(\mathbb{R}^n)$ spaces.

Lemma 4.1. *If $t \neq 0$, $1/p + 1/p' = 1$ and $p' \in [1,2]$, then $e^{it\Delta}\colon L^{p'}(\mathbb{R}^n) \mapsto L^p(\mathbb{R}^n)$ is continuous and*

$$\left\| e^{it\Delta} f \right\|_p \leq c|t|^{-n/2(1/p'-1/p)} \|f\|_{p'}. \tag{4.12}$$

Proof. From Proposition 4.2 it follows that

$$e^{it\Delta}\colon L^2(\mathbb{R}^n) \mapsto L^2(\mathbb{R}^n)$$

is an isometry, that is,

$$\|e^{it\Delta} f\|_2 = \|f\|_2.$$

Using Young's inequality (1.39), we have

$$\|e^{it\Delta} f\|_\infty = \left\| \frac{e^{i|\cdot|^2/4t}}{\sqrt{(4\pi i t)^n}} * f \right\|_\infty$$

$$\leq \left\| \frac{e^{i|\cdot|^2/4t}}{\sqrt{(4\pi i t)^n}} \right\|_{\infty} \|f\|_1 \leq c|t|^{-n/2}\|f\|_1. \tag{4.13}$$

A combination of these inequalities with the Riesz–Thorin theorem (Theorem 2.1) lead to

$$e^{it\Delta} : L^{p'}(\mathbb{R}^n) \mapsto L^p(\mathbb{R}^n) \ \text{ with } \ \frac{1}{p} + \frac{1}{p'} = 1,$$

and

$$\|e^{it\Delta} f\|_p \leq (c|t|^{-n/2})^{1-\theta} \|f\|_{p'} = c|t|^{-n/2(1/p'-1/p)}\|f\|_{p'},$$

where

$$\frac{1}{p} = \frac{\theta}{2} \quad \text{and} \quad 1-\theta = 1 - \frac{2}{p} = \frac{1}{p'} - \frac{1}{p}, \quad \theta \in [0,1].$$

Thus, the lemma follows. □

This result indicate that if $f \in L^2(\mathbb{R}^n)$ decreases fast enough when $|x| \to \infty$ such that $f \in L^1(\mathbb{R}^n)$, $e^{it\Delta} f$, $t \neq 0$, is bounded (and so more regular than f). In general, decay on the initial data f is translated into smoothing property of the solution $e^{it\Delta} f$ (see Exercise 4.4).

Note that $e^{it\Delta}$ with $t \neq 0$ is not a bounded operator from $L^p(\mathbb{R}^n)$ in $L^p(\mathbb{R}^n)$ if $p \neq 2$, i.e., $m(\xi) = e^{-4\pi^2 it|\xi|^2}$ is not an L^p multiplier for $p \neq 2$ (see Definition 2.8). In fact, if it were bounded for $p \neq 2$ it would be bounded also for p' by duality. Then, without loss of generality, we can assume $p > 2$. Using (4.12), we would have that for all $f \in L^{p'}(\mathbb{R}^n) \cap L^p(\mathbb{R}^n) \subseteq L^2(\mathbb{R}^n)$,

$$\|f\|_p = \|e^{it\Delta} e^{-it\Delta} f\|_p \leq c_0 \|e^{-it\Delta} f\|_p \leq c_0\, c(t) \|f\|_{p'},$$

which is a contradiction.

Next proposition help us to understand the regularizing effects present in the group $\{e^{it\Delta}\}_{t=-\infty}^{\infty}$.

Proposition 4.3.

1. *Given $t_0 \neq 0$ and $p > 2$, there exists $f \in L^2(\mathbb{R}^n)$ such that $e^{it_0\Delta} f \notin L^p(\mathbb{R}^n)$.*

2. *Let $s' > s > 0$ and $f \in H^s(\mathbb{R}^n)$ such that $f \notin H^{s'}(\mathbb{R}^n)$. Then, for all $t \in \mathbb{R}$, $e^{it\Delta} f \in H^s(\mathbb{R}^n)$ and $e^{it\Delta} f \notin H^{s'}(\mathbb{R}^n)$.*

Proof. To show (1), it is enough to choose $g \in L^2(\mathbb{R}^n)$ such that $g \notin L^p(\mathbb{R}^n)$ and take $f = e^{-it_0\Delta} g$.
The statement (2) follows from the fact that $\{e^{it\Delta}\}_{t=-\infty}^{\infty}$ is a unitary group in $H^s(\mathbb{R}^n)$ for all $s \in \mathbb{R}$ since

$$\|e^{it\Delta} f\|_{s,2} = \|\Lambda^s(e^{it\Delta} f)\|_2 = \|e^{it\Delta}(\Lambda^s f)\|_2 = \|\Lambda^s f\|_2 = \|f\|_{s,2}.$$

Therefore, if $e^{it\Delta} f \in H^{s_0}(\mathbb{R}^n)$, then $f = e^{-it\Delta}(e^{it\Delta}) f \in H^{s_0}(\mathbb{R}^n)$. □

4.2 Global Smoothing Effects

The next theorem describes the *global smoothing* property of the group $\{e^{it\Delta}\}_{t=-\infty}^{\infty}$.

Theorem 4.2. *The group* $\{e^{it\Delta}\}_{t=-\infty}^{\infty}$ *satisfies:*

$$\left(\int_{-\infty}^{\infty} \|e^{it\Delta} f\|_p^q \, dt\right)^{1/q} \le c\, \|f\|_2, \tag{4.14}$$

$$\left(\int_{-\infty}^{\infty} \Big\| \int_{-\infty}^{\infty} e^{i(t-t')\Delta} g(\cdot,t')\, dt' \Big\|_p^q \, dt\right)^{1/q} \le c \left(\int_{-\infty}^{\infty} \|g(\cdot,t)\|_{p'}^{q'} \, dt\right)^{1/q'}, \tag{4.15}$$

$$\Big\| \int_{-\infty}^{\infty} e^{it\Delta} g(\cdot,t)\, dt \Big\|_2 \le c \left(\int_{-\infty}^{\infty} \|g(\cdot,t)\|_{p'}^{q'} \, dt\right)^{1/q'}, \tag{4.16}$$

and

$$\left(\int_{-\infty}^{\infty} \Big\| \int_{0}^{t} e^{i(t-t')\Delta} g(\cdot,t')\, dt' \Big\|_p^q \, dt\right)^{1/q} \le c \left(\int_{-\infty}^{\infty} \|g(\cdot,t)\|_{p'}^{q'} \, dt\right)^{1/q'}, \tag{4.17}$$

with

$$\left.\begin{array}{ll} 2 \le p < \frac{2n}{n-2} & \text{if } n \ge 3 \\ 2 \le p < \infty & \text{if } n = 2 \\ 2 \le p \le \infty & \text{if } n = 1 \end{array}\right\} \quad \text{and} \quad \frac{2}{q} = \frac{n}{2} - \frac{n}{p}, \tag{4.18}$$

where $c = c(p,n)$ *is a constant depending only on* p *and* n.

From here on, we always use the notation

$$\frac{1}{p} + \frac{1}{p'} = \frac{1}{q} + \frac{1}{q'} = 1.$$

Proof. First, we shall prove that (4.14), (4.15), and (4.16) are equivalent.

Fubini's theorem gives us that

$$\int_{-\infty}^{\infty} \int_{\mathbb{R}^n} (e^{it\Delta} f)(x) g(x,t)\, dx dt = \int_{\mathbb{R}^n} f(x) \left(\int_{-\infty}^{\infty} e^{it\Delta} g(x,t)\, dt\right) dx.$$

Therefore, using duality,

$$\left(\int_{-\infty}^{\infty} \|h(\cdot,t)\|_p^q\,dt\right)^{1/q}$$

$$= \sup\left\{\left|\int_{-\infty}^{\infty}\int_{\mathbb{R}^n} h(x,t)w(x,t)\,dx\,dt\right| : \left(\int_{-\infty}^{\infty} \|w(\cdot,t)\|_{p'}^{q'}\,dt\right)^{1/q'} = 1\right\}$$

it follows that (4.14) and (4.16) are equivalent. An argument due to P. Tomas implies that

$$\left\|\int_{-\infty}^{\infty} e^{it\Delta}g(\cdot,t)\,dt\right\|_2^2 = \int_{\mathbb{R}^n}\left(\int_{-\infty}^{\infty} e^{it\Delta}g(\cdot,t)\,dt\right)\overline{\left(\int_{-\infty}^{\infty} e^{it'\Delta}g(\cdot,t')dt'\right)}\,dx$$

$$= \int_{\mathbb{R}^n}\int_{-\infty}^{\infty} g(x,t)\left(\int_{-\infty}^{\infty} e^{i(t-t')\Delta}\overline{g(\cdot,t')}\,dt'\right)dt\,dx. \tag{4.19}$$

From these identities we obtain (applying again an argument of duality and Hölder's inequality) the equivalence between (4.15) and (4.16).

Next we shall establish (4.15). Minkowski's inequality (1.40) and Lemma 4.1 give

$$\begin{aligned}\left\|\int_{-\infty}^{\infty} e^{i(t-t')\Delta}g(\cdot,t')\,dt'\right\|_p &\le \int_{-\infty}^{\infty}\|e^{i(t-t')\Delta}g(\cdot,t')\|_p\,dt' \\ &\le c\int_{-\infty}^{\infty}\frac{1}{|t-t'|^{\alpha}}\|g(\cdot,t')\|_{p'}\,dt'\end{aligned} \tag{4.20}$$

with $\alpha = (n/2)\left(1/p' - 1/p\right)$. Inequality (4.20) and Theorem 2.6 (Hardy–Littlewood–Sobolev) imply

$$\left(\int_{-\infty}^{\infty}\left\|\int_{-\infty}^{\infty} e^{i(t-t')\Delta}g(\cdot,t')\,dt'\right\|_p^q dt\right)^{1/q}$$

$$\le c\left\|\int_{-\infty}^{\infty}\frac{1}{|t-t'|^{\alpha}}\|g(\cdot,t')\|_{p'}\,dt'\right\|_q \le c\left(\int_{-\infty}^{\infty}\|g(\cdot,t)\|_{p'}^{q'}\,dt\right)^{1/q'}$$

with $1/q' = 1/q + (1-\alpha)$ and $0 < 1-\alpha < 1$, that is, $n/2 = 2/q + n/p$, where

$$\begin{cases} 2 \le p < \dfrac{2n}{n-2} & \text{if } n \ge 3, \\ 2 \le p < \infty & \text{if } n = 2, \\ 2 \le p \le \infty & \text{if } n = 1. \end{cases}$$

Finally, we turn to the proof of (4.17). This is a consequence of (4.15) and the following result due to Christ-Kiselev [CrK].

Lemma 4.2. *Let*

$$Tf(t) = \int_{-\infty}^{\infty} K(t,s)\, f(s)\, ds \tag{4.21}$$

be a bounded map from $L^r(\mathbb{R})$ to $L^l(\mathbb{R})$ with $1 < r < l < \infty$. Then the map

$$\tilde{T} f(t) = \int_{s<t} K(t,s)\, f(s)\, ds \tag{4.22}$$

also maps $L^r(\mathbb{R})$ into $L^l(\mathbb{R})$.

For the proof of this lemma, see Appendix B. □

In particular, this theorem tells us that if $f \in L^2(\mathbb{R}^n)$, then $e^{it\Delta} f \in L^p(\mathbb{R}^n)$, for any fixed $p \in (2,\ p(n))$ for almost all time $t \in \mathbb{R}$, with $p(n)$ depending on the dimension. In particular, if $n = 1$, $p(1) = \infty$, and $q = 4$, then for $f \in L^2(\mathbb{R})$ we have

$$\left(\int_{-\infty}^{\infty} \| e^{it\partial_x^2} f \|_\infty^4 \, dt \right)^{1/4} \le c\, \| f \|_2,$$

which implies that $e^{it\partial_x^2} f \in L^\infty(\mathbb{R})$ for almost every t. Indeed, in this case, one has that for almost every $t \in \mathbb{R}$, $e^{it\partial_x^2} f$ is continuous in $\mathbb{R}$ (see Exercise 4.9). Note that this fact does not contradict Proposition 4.3.

Corollary 4.1. *Let (p_0, q_0), $(p_1, q_1) \in \mathbb{R}^2$ satisfying the condition (4.18) in Theorem 4.2. Then, for all $T > 0$ we have*

$$\left(\int_0^T \left\| \int_0^t e^{i(t-t')\Delta} g(\cdot, t')\, dt' \right\|_{p_1}^{q_1} dt \right)^{1/q_1} \le c \left(\int_0^T \| g(\cdot, t) \|_{p_0'}^{q_0'} \, dt \right)^{1/q_0'},$$

with $c = c(n, p_0, p_1)$.

Proof. By hypothesis, the points $(1/p_0, 1/q_0)$ and $(1/p_1, 1/q_1)$ are in the segment of the line connecting $P = (1/2, 0)$ with $Q = (1/p(n), n/4 - n/2\, p(n))$. So $p(n) = \infty$ if $n = 1, 2$, and $p(n) = 2n/(n-2)$ if $n \ge 3$. Therefore, without loss of generality we can assume $p_0 \in [2, p_1)$. An application of the inequalities (4.16) and (4.17) in Theorem 4.2 provides the following estimates:

$$\left(\int_0^T \left\| \int_0^t e^{i(t-t')\Delta} g(\cdot, t')\, dt' \right\|_{p_1}^{q_1} dt \right)^{1/q_1} \le c \left(\int_0^T \| g(\cdot, t) \|_{p_1'}^{q_1'} \, dt \right)^{1/q_1'},$$

and

$$\sup_{[0,T]} \Big\| \int_0^t e^{i(t-t')\Delta} g(\cdot,t')\,dt' \Big\|_2 = \sup_{[0,T]} \Big\| e^{it\Delta} \int_0^t e^{-it'\Delta} g(\cdot,t')\,dt' \Big\|_2$$

$$= \sup_{[0,T]} \Big\| \int_0^t e^{-it'\Delta} g(\cdot,t')\,dt' \Big\|_2 \le c \left(\int_0^T \|g(\cdot,t)\|_{p_1'}^{q_1'}\,dt \right)^{1/q_1'}.$$

These estimates and Hölder's inequality lead to

$$\left(\int_0^T \Big\| \int_0^t e^{i(t-t')\Delta} g(\cdot,t')\,dt' \Big\|_{p_0}^{q_0} dt \right)^{1/q_0} \le c \left(\int_0^T \|g(\cdot,t)\|_{p_1'}^{q_1'}\,dt \right)^{1/q_1'}.$$

To finish the proof, an argument of duality allows us to write the inequality

$$\left(\int_0^T \Big\| \int_0^t e^{i(t-t')\Delta} g(\cdot,t')\,dt' \Big\|_{p_1}^{q_1} dt \right)^{1/q_1} \le c \left(\int_0^T \|g(\cdot,t)\|_{p_0'}^{q_0'}\,dt \right)^{1/q_0'}.$$

This yields the result. □

4.3 Local Smoothing Effects

In this section, we study the local smoothing effects of the group $\{e^{it\Delta}\}_{t=-\infty}^{\infty}$.

Theorem 4.3. *If $n = 1$, then*

$$\sup_x \int_{-\infty}^{\infty} |D_x^{1/2} e^{it\Delta} f(x)|^2\,dt \le c\|f\|_2^2. \tag{4.23}$$

If $n \ge 2$, then for all $j \in \{1,\dots,n\}$

$$\sup_{x_j} \int_{\mathbb{R}^n} |D_{x_j}^{1/2} e^{it\Delta} f(x)|^2 dx_1 \cdots dx_{j-1} dx_{j+1} \cdots dx_n\,dt \le c\|f\|_2^2, \tag{4.24}$$

where $D_{x_j}^{1/2} g(x,t) = ((2\pi|\xi_j|)^{1/2} \widehat{g}(\xi,t))^{\vee}(x,t)$ denotes the homogeneous fractional derivative of order $1/2$ in the variable x_j.

Proof. We begin considering the case $n = 1$. So,

$$\begin{aligned} D_x^{1/2} e^{it\Delta} f &= c\big(|\xi|^{1/2} e^{-4\pi^2 it|\xi|^2} \widehat{f}(\xi)\big)^{\vee} \\ &= c\big(|\xi|^{1/2} e^{-4\pi^2 it|\xi|^2} \widehat{f_+}(\xi)\big)^{\vee} + c\big(|\xi|^{1/2} e^{-4\pi^2 it|\xi|^2} \widehat{f_-}(\xi)\big)^{\vee}, \end{aligned}$$

where $\widehat{f_\pm}(\xi) = \chi_{\mathbb{R}^\pm}\widehat{f}(\xi)$. Thus, it is enough to show (4.23) with f_+ replacing f. A combination of the change of variables $2\pi\xi^2 = r$, Plancherel's theorem (1.11) and the inverse change of variables $\xi = +\sqrt{r/2\pi}$ produce the following identities:

$$\int_{-\infty}^{\infty} |D_x^{1/2} e^{it\Delta} f_+|^2(x)\,dt = c\int_{-\infty}^{\infty}\Big|\int_{-\infty}^{\infty} |\xi|^{1/2} e^{2\pi i x\xi} e^{-4\pi^2 it\xi^2}\widehat{f_+}(\xi)\,d\xi\Big|^2 dt$$

$$= c\int_{-\infty}^{\infty}\Big|\int_{0}^{\infty} r^{1/4} e^{-2\pi itr} e^{ix\sqrt{2\pi r}}\widehat{f_+}\left(\sqrt{\frac{r}{2\pi}}\right)\frac{dr}{r^{1/2}}\Big|^2 dt$$

$$= c\int_{0}^{\infty}\Big|e^{ix\sqrt{2\pi r}}\widehat{f_+}\left(\sqrt{\frac{r}{2\pi}}\right)\frac{1}{r^{1/4}}\Big|^2 dr = c\int_{-\infty}^{\infty}|\widehat{f_+}(\xi)|^2\,d\xi = c\,\|f_+\|_2^2,$$

which gives (4.23). Moreover, when $\widehat{f}$ has support in $[0,\infty)$ or $(-\infty,0]$, inequality (4.23) becomes an equality.

To obtain (4.24), we fix $j = 1$ to simplify the notation. We then define $\widehat{f_\pm}(\xi) = \chi_{\mathbb{R}^\pm}(\xi_1)\widehat{f}(\xi)$. Without the loss of generality , we prove (4.24) with f_+ replacing f.

Denote $\bar{x} = (x_2,\dots,x_n)$ and $\bar{\xi} = (\xi_2,\dots,\xi_n)$. The change of variables

$$(\xi_1,\xi_2,\dots,\xi_n) = (\xi_1,\bar{\xi}) \overset{\Phi}{\to} (2\pi(\xi_1^2+\cdots+\xi_n^2),\bar{\xi}) = (r,\bar{\xi}),$$

$$d\xi_1 d\bar{\xi} = \left|\begin{pmatrix} \frac{\partial r}{\partial\xi_1} & \frac{\partial r}{\partial\xi_2} & \cdots & \frac{\partial r}{\partial\xi_n} \\ 0 & 1 & \cdots & 0 \\ \vdots & \vdots & \ddots & \vdots \\ 0 & 0 & \cdots & 1 \end{pmatrix}\right|^{-1} dr d\bar{\xi} = \frac{1}{4\pi|\xi_1|}\,dr\,d\bar{\xi},$$

Plancherel's identity (1.11) and the change of variables Φ^{-1} yield

$$\|D_{x_1}^{1/2} e^{it\Delta} f_+\|_{L^2_{\bar{x}t}}^2 = c\,\Big\|\int_{\mathbb{R}^n} e^{2\pi x\cdot\xi}\,|\xi_1|^{1/2} e^{-4\pi^2 it|\xi|^2}\widehat{f_+}(\xi)d\xi\,\Big\|_{L^2_{\bar{x}t}}^2$$

$$= c\,\Big\|\int_{\mathbb{R}^n} e^{2\pi i(\bar{x}\cdot\bar{\xi}+rt)}\frac{1}{|\xi_1|^{1/2}}\,e^{2\pi x_1\sqrt{\frac{|r-2\pi|\bar{\xi}|^2|}{2\pi}}}\,\widehat{f_+}(r,\bar{\xi})\,dr d\bar{\xi}\,\Big\|_{L^2_{\bar{x}t}}$$

$$= c\int_{\mathbb{R}^n}\frac{1}{|\xi_1|}|\widehat{f_+}(r,\bar{\xi})|^2\,dr d\bar{\xi} = c\,\|\widehat{f_+}\|_{L^2_\xi}^2 = c\,\|f_+\|_{L^2_x}^2,$$

which leads to (4.24). □

Corollary 4.2.

$$\left(\int_{-\infty}^{\infty}\int_{\{|x|\le R\}} |D_x^{1/2} e^{it\Delta} f|^2(x)\,dx\,dt\right)^{1/2} \le c\,R^{1/2}\,\|f\|_2, \tag{4.25}$$

where $D_x^{1/2} v(x,t) = ((2\pi|\xi|)^{1/2}\widehat{v}(\xi,t))^{\vee}$.

Notice that from this result and the translation invariance property of the solution one gets

$$\sup_{x_0\in\mathbb{R}^n,\ R>0}\left(\frac{1}{R}\int_{-\infty}^{\infty}\int_{B_R(x_0)} |D_x^{1/2} e^{it\Delta} f(x)|^2\,dxdt\right)^{1/2} \le c\|f\|_2.$$

Proof. If $n=1$, inequality (4.25) follows from (4.23).

Consider the case $n\ge 2$. Defining $D_j = \{\xi\in\mathbb{R}^n : |\xi_j| > \frac{1}{\sqrt{2n}}|\xi|\}$, with $j=1,\dots,n$. It is easy to see that $\bigcup_{j=1}^{n} D_j = \mathbb{R}^n - \{0\}$. Let $\{\phi_j\}_{j=1}^n$ be a partition of unity subordinate to the covering $\{D_j\}_{j=1}^n$ (the ϕ_j can be defined in the sphere $\mathbb{S}^{n-1}$ and extended such that they are homogeneous of order zero). Using linearity it suffices to show that

$$\int_{-\infty}^{\infty}\int_{\{|x|\le R\}} |e^{it\Delta} f(x)|^2\,dx\,dt \le c\,R\|D_x^{-1/2} f\|_2^2 = cR\||\xi|^{-1/2}\widehat{f}\|_2^2.$$

From (4.24), we obtain for all $j=1,\dots,n$,

$$\int_{-\infty}^{\infty}\int_{\{|x|\le R\}} |e^{it\Delta} g(x)|^2\,dx\,dt \le c\,R\|D_{x_j}^{-1/2} g\|_2^2.$$

Therefore, using the notation $\widehat{f_j} = \widehat{f}\phi_j$, $j=1,\dots,n$, it follows that

$$\begin{aligned}
\int_{-\infty}^{\infty}\int_{\{|x|\le R\}} |e^{it\Delta} f|^2(x)\,dxdt &\le c\sum_{j=1}^{n}\int_{-\infty}^{\infty}\int_{\{|x|\le R\}} |e^{it\Delta} f_j|^2(x)\,dxdt\\
&\le c\,R\sum_{j=1}^{n}\|D_{x_j}^{-1/2} f_j\|_2^2 = c\,R\sum_{j=1}^{n}\big\||\xi_j|^{-1/2}\widehat{f_j}\big\|_2^2\\
&= c\,R\sum_{j=1}^{n}\big\||\xi_j|^{-1/2}\widehat{f}\phi_j\big\|_2^2 \le c\,R\big\||\xi|^{-1/2}\widehat{f}\big\|_2^2\\
&= c\,R\,\|D_x^{-1/2} f\|_2^2.
\end{aligned}$$

□

From Corollary 4.2 and the group properties, we deduce that if $f \in L^2(\mathbb{R}^n)$, then $e^{it\Delta} f \in L^2_{\text{loc}}(\mathbb{R} : H^{1/2}_{\text{loc}}(\mathbb{R}^n))$ and thus $e^{it\Delta} f \in H^{1/2}_{\text{loc}}(\mathbb{R}^n)$ for almost every $t \in \mathbb{R}$.

On the other hand, from (4.23) (case $n = 1$) using duality we have

$$\left\| D_x^{1/2} \int_{-\infty}^{\infty} e^{it\Delta} F(\cdot, t)\, dt \right\|_2 \le c \int_{-\infty}^{\infty} \|F(x,\cdot)\|_2\, dx. \tag{4.26}$$

Similarly, from (4.24) we obtain the corresponding inequality for the case $n \geq 2$.

For solutions of the inhomogeneous problem:

$$\begin{cases} \partial_t u = i\Delta u + F(x,t), \\ u(x,0) = 0, \end{cases} \tag{4.27}$$

$x \in \mathbb{R}^n,\ t \in \mathbb{R}$, we observe that the gain of derivatives doubles that obtained in the homogeneous case.

Theorem 4.4. *If $u(x,t)$ is the solution of problem (4.27), then, when $n = 1$ it satisfies*

$$\sup_x \left(\int_{-\infty}^{\infty} |\partial_x u(x,t)|^2\, dt \right)^{1/2} \le c \int_{-\infty}^{\infty} \left(\int_{-\infty}^{\infty} |F(x,t)|^2 dt \right)^{1/2} dx, \tag{4.28}$$

and in the case $n \geq 2$

$$\sup_{x_j} \left(\int_{\mathbb{R}^n} |\partial_{x_j} u(x,t)|^2 d\mu_j dt \right)^{1/2} \le c \int_{-\infty}^{\infty} \left(\int_{\mathbb{R}^n} |F(x,t)|^2 d\mu_j dt \right)^{1/2} dx_j, \tag{4.29}$$

where $d\mu_j = dx_1 \cdots dx_{j-1}\, dx_{j+1} \cdots dx_n$. Therefore in the case $n \geq 2$ we have that

$$\sup_\alpha \left(\int_{Q_\alpha} \int_{-\infty}^{\infty} |\partial_x u(x,t)|^2\, dt\, dx \right)^{1/2} \le c \sum_\alpha \left(\int_{Q_\alpha} \int_{-\infty}^{\infty} |F(x,t)|^2 dt\, dx \right)^{1/2}, \tag{4.30}$$

where $\{Q_\alpha\}_{\alpha \in \mathbb{Z}^n}$ denotes a family of disjoint unit cubes with sides parallel to the axes and covering $\mathbb{R}^n$.

Proof. We only sketch the proof in the case $n = 1$. Using Exercise 4.16 in this chapter, we deduce that

$$\begin{aligned} \partial_x u(x,t) &= \int_{-\infty}^{\infty} \int_{\mathbb{R}^n} \frac{2\pi i \xi}{4\pi^2 i |\xi|^2 + 2\pi i \tau} (e^{2\pi i \tau t} - e^{-4\pi^2 i |\xi|^2 t}) e^{2\pi i x\cdot\xi} \widehat{F}(\xi,\tau) d\xi d\tau \\ &= \int_{-\infty}^{\infty} \int_{\mathbb{R}^n} \frac{2\pi i \xi\ e^{2\pi i \tau t}}{4\pi^2 i |\xi|^2 + 2\pi i \tau} e^{2\pi i x\cdot\xi} \widehat{F}(\xi,\tau)\, d\xi d\tau \end{aligned} \tag{4.31}$$

$$-\int_{-\infty}^{\infty}\int_{\mathbb{R}^n} \frac{2\pi i\xi\, e^{-4\pi^2 i|\xi|^2 t}}{4\pi^2 i|\xi|^2 + 2\pi i\tau} e^{2\pi i x\cdot\xi}\widehat{F}(\xi,\tau)\,d\xi d\tau$$

$$= \partial_x u_1(x,t) + \partial_x u_2(x,t),$$

where $\widehat{F}(\xi,\eta)$ represents the Fourier transform with respect to the variables x,t. Since the numerator in the first integrand vanishes on the zeros of its denominator, the integrals in the second equality are understood in the principal value sense. From Exercise 1.17, we have that

$$\left(\text{p.v.}\,\frac{2\pi i\xi}{4\pi^2 i|\xi|^2 + 2\pi i\tau}\right)^{\vee(\xi)} = K(x,\tau) \in L^\infty(\mathbb{R}^2).$$

Plancherel's identity (1.11), Young's and Minkowski's inequalities, (1.39) and (1.40), respectively, imply that for all $x \in \mathbb{R}$,

$$\left(\int_{-\infty}^{\infty} |\partial_x u_1(x,t)|^2\,dt\right)^{1/2} = c\left\| \int_{-\infty}^{\infty} e^{2\pi i\tau t}\int_{-\infty}^{\infty} K(x-y,\tau)\widehat{F}^{(t)}(y,\tau)\,dy d\tau\right\|_{2(t)}$$

$$= c\left\|\int_{-\infty}^{\infty} K(x-y,\tau)\widehat{F}^{(t)}(y,\tau)\,dy\right\|_{2(\tau)}$$

$$\leq c\int_{-\infty}^{\infty} \|\widehat{F}^{(t)}(y,\cdot)\|_{2(\tau)}\,dy \leq c\int_{-\infty}^{\infty} \|F(y,\cdot)\|_{2(t)}\,dy,$$

which proves

$$\sup_x \left(\int_{-\infty}^{\infty} |\partial_x u_1(x,t)|^2\,dt\right)^{1/2} \leq c\int_{-\infty}^{\infty}\left(\int_{-\infty}^{\infty} |F(x,t)|^2\,dt\right)^{1/2} dx.$$

On the other hand, we have that

$$\partial_x u_2(x,t) = D_x^{1/2} e^{it\Delta} G(x),$$

where

$$\widehat{G}(\xi) = c\int_{-\infty}^{\infty} \frac{\text{sgn}(\xi)\,|\xi|^{1/2}\,\widehat{F}(\xi,\tau)}{4\pi^2 i|\xi|^2 + 2\pi i\tau}\,d\tau.$$

A simple computation and (1.18) shows that

$$\left(\text{p.v.}\,\frac{1}{4\pi^2 i|\xi|^2 + 2\pi i\tau}\right)^{\vee(\tau)} = \int_{-\infty}^{\infty} \frac{e^{-2\pi i\tau t}}{4\pi^2 i|\xi|^2 + 2\pi i\tau}\,d\tau = c\,\text{sgn}(t)\,e^{-4\pi^2 i|\xi|^2 t}.$$

Therefore, using (4.23), (4.26), and Plancherel's identity (1.11), we infer that

$$\sup_x \left(\int_{-\infty}^{\infty} |\partial_x u_2(x,t)|^2 \, dt \right)^{1/2} \leq c \left\| \int_{-\infty}^{\infty} \frac{\operatorname{sgn}(\xi)\, |\xi|^{1/2}\, \widehat{F}(\xi,\tau)}{4\pi^2 i |\xi|^2 + 2\pi i \tau} \, d\tau \right\|_{2(\xi)}$$

$$= c \left\| \int_{-\infty}^{\infty} e^{-4\pi^2 i|\xi|^2 t} \operatorname{sgn}(\xi)\, |\xi|^{1/2} \widehat{F}^{(x)}(\xi,t) \operatorname{sgn}(t)\, dt \right\|_{2(\xi)}$$

$$= c \left\| \left(\int_{-\infty}^{\infty} e^{it\Delta} D_x^{1/2} \mathsf{H} F(\cdot,t) \operatorname{sgn}(t)\, dt \right)^{\vee} \right\|_{2(\xi)}$$

$$= c \left\| D_x^{1/2} \int_{-\infty}^{\infty} e^{it\Delta} \mathsf{H} F(\cdot,t) \operatorname{sgn}(t)\, dt \right\|_2$$

$$\leq c \int_{-\infty}^{\infty} \left(\int_{-\infty}^{\infty} |F(x,t)|^2 \, dt \right)^{1/2} dx,$$

where H denotes the Hilbert transform (see Definition 1.7). This leads to the result. □

4.4 Comments

The first result concerning smoothing effects for the particular group $\{e^{it\Delta}\}_{t=-\infty}^{\infty}$ or for general group of unitary operators was obtained by Kato in [K1]. In this work on theory of operators, Kato introduced the notion of A-regular and A-super regular operators.

Let A be a self-adjoint operator (not necessarily bounded) defined on a Hilbert space H such that the resolvent of A, $R(\lambda) = (\lambda I - A)^{-1}$, exists for all $\lambda \in \mathbb{C}$ with $\mathcal{I}m\, \lambda \neq 0$ and let L be an operator of closed graph with domain $D(L)$ dense in H.

Definition 4.1. We say that the operator L is *A-regular* (respectively, A-super regular) if for all $x \in D(L^*)$ and for all $\lambda \in \mathbb{C}$ with $\mathcal{I}m\ \lambda \neq 0$,

$$|\, \mathcal{I}m\ < R(\lambda)L^*x, L^*x > | \leq c\,\pi\, \|x\|^2$$

(respectively, $|\langle R(\lambda)L^*x, L^*x\rangle| \leq c\pi\, \|x\|^2$), where the constant c is independent of x and λ.

The following theorems establish the relationship between the notion of A-regular operator and the type of results described in this chapter .

Theorem 4.5 ([K1]). *The operator L is A-regular if and only if for all $x \in H$*

$$\int_{-\infty}^{\infty} \|Le^{itA}x\|\,dt \le c\|x\|.$$

In particular, $e^{itA}x \in D(L)$ for almost every $t \in \mathbb{R}$.

Theorem 4.6 ([KY]). *Let $L = L_h$ be an operator of multiplication by h with $h \in L^n(\mathbb{R}^n)$ and $n \ge 3$. Then, L_h is Δ-super regular.*

Theorem 4.7 ([KY] see also [BKl]). *Let $\tilde{L}$ be the operator*

$$(1+|x|^2)^{-1/2}\Lambda^{1/2} = (1+|x|^2)^{-1/2}(1-\Delta)^{1/4}$$

with domain $C_0^\infty(\mathbb{R}^n)$ and $n \ge 3$. Then, the closure of $\tilde{L}$ is Δ-super regular.

Combining Theorems 4.5 and 4.6, we have that if $f \in L^2(\mathbb{R}^n)$ with $n \ge 3$, then $e^{it\Delta}f \in D(L_h)$ for almost every $t \in \mathbb{R}$. When $h \notin L^\infty(\mathbb{R}^n)$, then $D(L_h)$ is a set of first category in $L^2(\mathbb{R}^n)$. These results neither imply nor are consequence of the estimate (4.14) in Theorem 4.2.

Later on, Strichartz [Str3], motivated by the work of Segal [Se], studied special properties of the Fourier transform. He proved that

$$\left(\int_{-\infty}^{\infty}\int_{\mathbb{R}^n} |e^{it\Delta}f|^{2(n+2)/n}\,dx\,dt\right)^{n/2(n+2)} \le c\|f\|_2. \tag{4.32}$$

In his proof, he employed previous results of Tomas [Tm] and Stein [S2] regarding restriction theorems (and extension) of the Fourier transform. More precisely, Strichartz used the fact that

$$\begin{aligned} e^{it\Delta}f(x) &= \int_{\mathbb{R}^n} e^{2\pi i x\cdot\xi} e^{-4\pi^2 it|\xi|^2}\widehat{f}(\xi)\,d\xi \\ &= \int_{\mathbb{R}^{n+1}} e^{2\pi i <(x,t);(\xi,\tau)>} g(\xi,\tau)\,d\sigma(\xi,\tau) = \widehat{g\,d\sigma}, \end{aligned}$$

where g is a measure supported on the hypersurface $M_\sigma \subset \mathbb{R}^{n+1}$, where the symbol $\sigma(\xi,\eta) = \eta + 2\pi|\xi|^2$ vanishes, i.e.,

$$M_\sigma = \{(\xi,\tau) \in \mathbb{R}^n \times \mathbb{R} \;:\; \sigma(\xi,\tau) = 0\} \tag{4.33}$$

(in this case, $\sigma(\xi,\eta) = \eta + 2\pi|\xi|^2$), with density $\widehat{f}(\xi)$ and $d\sigma(\tilde{\xi}) = d\xi$.

Similarly,

$$\widehat{e^{it\Delta}f}(\xi,\tau) = \int_{-\infty}^{\infty} e^{-2\pi it\tau} e^{-4\pi^2 it|\xi|^2}\widehat{f}(\xi)\,dt = \widehat{f}(\xi)\,\delta(\tau + 2\pi|\xi|^2),$$

where $\widehat{\ }$ on the left-hand side denotes the Fourier transform with respect to both variables: space x and time t. In other words, the Fourier transform in the variables (x,t) of the solution $e^{it\Delta}f(x)$ is a distribution with support on the parabola $\tau = -2\pi|\xi|^2$. Thus, inequality (4.14) can be seen as a result on the extension of the Fourier transform of measures with support on this parabola. Similarly, we can see (4.16) as a result of restriction because using the Fubini theorem and the Plancherel identity (1.11) we have,

$$\begin{aligned}
\left\| \int_{-\infty}^{\infty} e^{it\Delta} g(\cdot,t)\,dt \right\|_2 &= \left\| \int_{-\infty}^{\infty} \left(\int_{\mathbb{R}^n} e^{2\pi i x\cdot\xi} e^{-4\pi^2 it|\xi|^2} \widehat{g}(\xi,t)\,d\xi \right) dt \right\|_2 \\
= \left\| \int_{\mathbb{R}^n} e^{2\pi i x\cdot\xi} \left(\int_{-\infty}^{\infty} e^{-4\pi^2 it|\xi|^2} \widehat{g}(\xi,t)\,dt \right) d\xi \right\|_2 &= \| \widehat{g}(\xi,-2\pi|\xi|^2)\|_2.
\end{aligned} \tag{4.34}$$

The proof presented in Section 4.2 is due to J. Ginibre and G. Velo [GV1] (see also [M], [P1]).

The main point in the argument is the *curvature* of the hypersurface M_σ defined by the symbol σ as in (4.33) and not the ellipticity of Δ. In particular, the same inequalities (4.14), (4.17) hold when we replace Δ by

$$\mathfrak{L}_j = \partial_{x_1}^2 + \cdots + \partial_{x_j}^2 - \partial_{x_{j+1}}^2 - \cdots \partial_{x_n}^2, \quad \text{for some} \quad j \in \{1,\ldots,n\}. \tag{4.35}$$

The curvature of hypersurface M_σ for the symbol $\sigma = \tau + 2\pi|\xi|^2$ is reflected on the decay estimates (4.12) in Lemma 4.1. In fact, the results in Theorem 4.2 are true for any unitary group satisfying decay estimates of the type described in Lemma 4.1. Thus, in particular for the linear problem associated to the KdV equation (1.28), we have that the unitary group $V(t)v_0 = (e^{i8\pi^3\xi^3 t}\widehat{v_0})^\vee$ describing the solutions satisfies for any $(\theta,\alpha) \in [0,1] \times [0,1/2]$

$$\left\| D^{\alpha\theta/2} V(t) v_0 \right\|_{L^{2/1-\theta}} \le c\,|t|^{-\theta(\alpha+1)/3} \|v_0\|_{L^{2/1+\theta}}. \tag{4.36}$$

Therefore, the argument used in Theorem 4.2 shows that for any $(\theta,\alpha) \in [0,1] \times [0,1/2]$,

$$\left\| D^{\alpha\theta/2} V(t) v_0 \right\|_{L^q(\mathbb{R}:\,L^p(\mathbb{R}))} \le c\,\|v_0\|_2, \tag{4.37}$$

where $(q,p) = (6/\theta(\alpha+1), 2/(1-\theta))$. Notice that in (4.37) there is a possible gain of $1/4$ derivatives. Roughly speaking, in general this gain is equal to $(m-2)/4$, where m is the order of the dispersive operator (see [KPV2]).

In the case of the IVP associated wave equation:

$$\begin{cases} \partial_t^2 w = \Delta w, \\ w(x,0) = 0, \\ \partial_t w(x,0) = g(x), \end{cases} \tag{4.38}$$

$x \in \mathbb{R}^n$, $t \in \mathbb{R}^+$, whose solution

$$w(x,t) = U(t)g = \left(\frac{\sin(2\pi|\xi|t)}{2\pi|\xi|}\,\widehat{g}(\xi)\right)^{\vee}$$

(see (1.49)) is associated to the unitary group $M(t) = (e^{i2\pi|\xi|t}\,\widehat{g})^{\vee}$, we have the decay estimate:

$$\|U(t)g\|_{L^p(\mathbb{R}^n)} \leq c\,t^{(n-1)(\frac{1}{2}-\frac{1}{p'})}\|D^{\alpha}g\|_{L^{p'}(\mathbb{R}^n)}, \tag{4.39}$$

with

$$\alpha = \frac{n-1}{2} - \frac{n+1}{p'}, \quad 2 \leq p < \infty, \quad n \geq 2.$$

From this, we can deduce the equivalent to Theorem 4.2:

$$\left\|(-\Delta)^{(1-b)/4}\,U(t)g\right\|_{L^q(\mathbb{R}:\,L^p(\mathbb{R}^n))} \leq c\,\|g\|_2, \tag{4.40}$$

where

$$2 < q < \infty, \quad \frac{1}{2} - \frac{2}{(n-1)q} = \frac{1}{p}, \quad \text{and} \quad b = \frac{n-1}{2} - \frac{n+1}{p}$$

(see [M], [P1]).

As we mentioned above, the decay estimates (4.12), (4.36), and (4.39) are related to the "curvature" of the hypersurfaces M_{σ_j}, $j = 1, 2, 3$, which described the zero set of the symbols $\sigma_1 = \tau + 2\pi|\xi|^2$, $\sigma_2 = \tau - 4\pi^2\xi^3$, and $\sigma_3 = \tau \pm |\xi|$, respectively. In the case σ_1 and σ_3, we observe that the hypersurfaces M_{σ_1} and M_{σ_3} have nonvanishing curvature in n and $n-1$ directions (rank of the Hessian), respectively.

In the limiting case, the inequality (4.14) in dimension $n = 2$ (i.e., $(q, p) = (2, \infty)$) fails (see [MSm]). Similarly, the limiting case of the estimate (4.40) for the wave equation in dimension $n = 3$ (i.e., $(q, p) = (2, \infty)$) fails (see [KlM]) although both hold in the radial case; see [To1] for the Schrödinger equation and [KlM] for the wave equation. Moreover, in the case of the Schrödinger equation ($(n, p, q) = (2, \infty, 2)$) one has the following generalization of the radial result, see [To1]

$$\|e^{it\Delta}u_0\|_{L^2_t(\mathbb{R}:\,L^\infty_r L^2_\theta(\mathbb{R}^2))} \leq c\,\|u_0\|_2,$$

where

$$\|f\|_{L^\infty_r L^2_\theta} = \sup_{r>0}\left(\frac{1}{2\pi}\int_{-\pi}^{\pi}|f(re^{i\theta})|^2\,d\theta\right)^{1/2}.$$

In [KT1], the limiting cases in higher dimension were shown to hold in both cases, i.e., the Schrödinger equation (4.14) holds for $n \geq 3$, $(q, p) = (2, 2n/(n-2))$, as well as the wave equation in (4.40) hold for $n \geq 4$, $(q, p) = (2, 2(n-1)/(n-3))$.

The problem of finding the best constant for the Strichartz estimate (4.14):

$$c(n;p;q) = \sup_{\|u_0\|_2=1} \left(\int_{-\infty}^{\infty} \|e^{it\Delta}u_0\|_p^q dt \right)^{1/q} \tag{4.41}$$

as well as its maximizers, i.e., the $u_0 \in L^2(\mathbb{R}^n)$ for which the equality (4.41) holds with (p,q) as in (4.18) has been studied in several works. In [Kz], it was proved the existence of a maximizer for $n = 1$ and $p = q = 6$. In [Fs] and [HuZ], it was established that for the case $n = 1,2$ and $p = q = 2 + 4/n$, one has $c(1;6;6) = 12^{-1/12}$ and $c(2;4;4) = 2^{-1/2}$ with the maximizer, up to the invariant of the Schrödinger equation (see Proposition 4.1), equal to $c_n e^{-|x|^2}$, $n = 1,2$. Also in [Fs], the same problem was settled for the case of the wave equation (4.38) in dimension $n = 2,3$ with $p = q = 2 + 4/(n-1)$. The value $c(1;8;4) = 2^{-1/4}$ in (4.41) was computed in [BBCH] and [Car].

Corollary 4.1 was proved in [CzW1]. For further results in this directio, we refer to [Vi1].

Concerning the decay of the free Schrödinger equation, on one hand, one has that if $u_0 \in C_0^\infty(\mathbb{R}^n)$ with $u_0 \not\equiv 0$, then for any $t \neq 0$ and any $\epsilon > 0$, $e^{it\Delta}u_0 \in \mathcal{S}(\mathbb{R}^n)\backslash L^1(e^{\epsilon|x|}dx)$ (see Exercises 4.4 and 4.5). On the other hand, Example 4.2 tells us that solutions corresponding to Gaussian data exhibits a global Gaussian decay. In [EKPV1], it was shown that given $u_0 \in \mathcal{S}'(\mathbb{R}^n)$ the following conditions are equivalent:

(i) There are two different real numbers t_1 and t_2, such that $e^{it_j\Delta}u_0 \in L^2(e^{a_j|x|^2}dx)$ for some $a_j > 0$, $j = 1,2$.

(ii) $u_0 \in L^2(e^{b_1|x|^2}dx)$ and $\widehat{u}_0 \in L^2(e^{b_2|x|^2}dx)$, for some $b_j > 0$, $j = 1,2$.

(iii) There is $\nu : [0,+\infty) \longrightarrow (0,+\infty)$, such that $e^{it\Delta}u_0 \in L^2(e^{\nu(t)|x|^2}dx)$, for all $t \geq 0$.

(iv) $u_0(x+iy)$ is an entire function such that $|u_0(x+iy)| \leq N\, e^{-a|x|^2+b|y|^2}$ for some constants N, a, $b > 0$.

(v) There exist δ, $\epsilon > 0$, and $h \in L^2(e^{\epsilon|x|^2}dx)$ such that $u_0(x) = e^{\delta\Delta}h(x)$.

It was also established in [EKPV1] that if one of the above conditions holds then for appropriate values $\alpha, \beta > 0$ the function

$$f(t) = \left\| e^{\frac{|x|^2}{(\alpha t+\beta)^2}} e^{it\Delta}u_0 \right\|_2$$

is logarithmically convex. In particular, one has that

$$f(t) \leq f(0)^{\theta(t)}\, f(T)^{1-\theta(t)},$$

with $\theta(t) = \beta(T-t)/(T(\alpha t+\beta))$ for all $t \in [0,T]$.

In [EKPV1], the constants used above were described in a precise manner as a consequence of (4.7) and the following result due to Hardy for $n = 1$ [H] and its extension to higher dimension given in [SS]: if $f(x) = O(e^{-\pi A|x|^2})$ and $\hat{f}(\xi) = O(e^{-\pi B|\xi|^2})$, with $A > 0$, $B > 0$, and $AB > 1$, then $f \equiv 0$.

Extensions of these results to the case of Schrödinger equation, with potential (in an appropriate class) as (4.42) below depending on x or on (x,t), i.e., $V = V(x,t)$ (as well as application to unique continuation properties of semilinear Schrödinger equations) were given in [EKPV2].

Consider the IVP associated to the Schrödinger equation with a potential V:

$$\begin{cases} i\partial_t u = \Delta u - V(x)u, \\ u(x,0) = u_0(x). \end{cases} \tag{4.42}$$

Assume first that the potential $V = V(x)$ is real and regular enough such that $L = -\Delta + V(x)$ is self-adjoint.

A natural question is whether or not the unitary group $e^{itL} = e^{it(-\Delta+V)}$ satisfies the $L^\infty - L^1$ estimate in (4.13) as in the free case $V \equiv 0$, i.e., there exists $c > 0$ for every $f \in L^2(\mathbb{R}^n)$ such that

$$\|e^{itL} f\|_\infty \leq c\, t^{-n/2} \|f\|_1. \tag{4.43}$$

If L has an eigenvalue (with eigenfunction $f \in L^1(\mathbb{R}^n) \cap L^2(\mathbb{R}^n)$), (4.43) fails. Similarly, if zero is a resonance of L. So, one reformulates the inequality (4.43) as

$$\|e^{itL} P_{\mathrm{ac}}(L) f\|_\infty \leq c|t|^{-n/2} \|f\|_1, \tag{4.44}$$

where $P_{\mathrm{ac}}(L)$ defines the projection onto the absolutely continuous spectrum of L.

The following conditions on the decay of V have been shown to be sufficient for (4.43) to hold: $n = 1$ and $(1 + |x|)\, V \in L^1(\mathbb{R})$ [GSch], $n = 2$ and $|V(x)| \leq c\,(1 + |x|)^{-3-\epsilon}$ [Scl1], $n = 3$ and $V \in L^{3/2-\epsilon}(\mathbb{R}^3)$ [Gb].

In [JSS], for $n \geq 3$ sufficient conditions on the decay and regularity on the potential $V(x)$ which guarantees (4.43) were deduced. In [GVi], it was shown that for $n > 3$ decay assumptions alone do not imply the estimate (4.43). More precisely, it was proved that (4.43) fails for any potential V with compact support such that $\sum_{|\alpha| \leq \frac{n-3}{2}} \|\partial^\alpha V\|_\infty \leq 1$.

The cases of time-dependent potentials have been also studied (see for instance [RS]). Also, decay estimates of the type in (4.43) with electromagnetic potentials were obtained in [FFFP].

For conditions on the potential V that guarantee the extension of the local smoothing effect described in Corollary 4.2 to solutions of the IVP (4.42) see [RV], [BRV].

Local-in-time extensions of Strichartz estimates to the variable coefficients' case, where the Laplacian Δ is replaced by an elliptic operator of the form:

$$L = \partial_{x_k} a_{jk}(x,t)\partial_{x_j} + \partial_{x_l} b_l(x,t) + b_l(x,t)\partial_{x_l} + V(x,t) \tag{4.45}$$

have been considered in several works. In [StTa], Staffilani and Tataru established these estimates under the assumptions: $b_l = V = 0$, $(a_{jk}(x,t))$ a compactly supported perturbation of the Laplacian and a nontrapping condition on the bicharacteric flow. Extensions of this result under appropriate hypotheses on the "asymptotic flatness" and the nontrapping condition of the coefficients a_{jk} were given in [MMTa1], [RZ], [Td]. The one-dimensional case was considered in [Sl].

Next, we briefly treat the periodic case:

$$\begin{cases} i\,\partial_t u = \partial_x^2 u, \\ u(x,0) = u_0(x), \end{cases} \tag{4.46}$$

$x \in \mathbb{S}^1 \times \cdots \times \mathbb{S}^1$, $t \in \mathbb{S}^1$.

Theorem 4.8 ([Z]).

$$\Big\| \sum_{k=-\infty}^{\infty} a_k e^{i(tk^2+kx)} \Big\|_{L^4(\mathbb{T}^2)} \le c \left(\sum_{k=-\infty}^{\infty} |a_k|^2 \right)^{1/2}, \tag{4.47}$$

where $(x,t) \in \mathbb{S}^1 \times \mathbb{S}^1 = \mathbb{T}^2$.

Note that $u(x,t) = \sum_k a_k e^{i(tk^2+kx)}$ is the solution of the periodic problem (4.46) for $n = 1$ with $u_0(x) = \sum_k a_k e^{ikx}$.

Proof. If $u(x,t) = \sum_k a_k e^{i(tk^2+kx)}$, then $\|u\|^2_{L^4(\mathbb{T}^2)} = \|u \cdot \bar{u}\|_{L^2(\mathbb{T}^2)}$. It is easy to see that

$$u\bar{u} = \sum_k |a_k|^2 + \sum_{k_1 \neq k_2} a_{k_1} \bar{a}_{k_2} e^{i((k_1-k_2)x+(k_1^2-k_2^2)t)}.$$

If we fix $l_1 = k_1 - k_2$ and $l_2 = k_1^2 - k_2^2$ we have at most one pair (k_1, k_2) of solutions of these equations. So, we can conclude that

$$\begin{aligned} \|u \cdot \bar{u}\|_2 &= \sum_k |a_k|^2 + \left(\sum_{k_1 \neq k_2} |a_{k_1} \bar{a}_{k_2}|^2 \right)^{1/2} \\ &\le \sum_k |a_k|^2 + \left(\sum_{k_1} |a_{k_1}|^2 \sum_{k_2} |a_{k_2}|^2 \right)^{1/2} = 2 \sum_k |a_k|^2. \end{aligned}$$

□

We observe that for the case $n = 1$, the corresponding inequality to (4.32) in $\mathbb{R}$ is true with $p = 6$. So, the next question is natural: Is the inequality (4.47) still true if we substitute 4 by 6? The answer is negative. In fact, one has that

$$\Big\| \sum_{k=1}^{N} e^{i(kx+k^2t)} \Big\|_{L^6(\mathbb{T}^2)} \gtrsim (\log N)^{1/6} N^{1/2}. \tag{4.48}$$

So, if $\phi = \sum_{k=1}^{N} e^{ikx}$, then $\|\phi\|_2 = N^{1/2}$, which combined with (4.48) implies that

$$\big\| e^{it\Delta} \phi \big\|_{L^{\frac{2(n+2)}{n}}(\mathbb{T}^{n+1})} \le c\, \|\phi\|_2 \tag{4.49}$$

fails for $n = 1$.

Nevertheless, Bourgain [Bo1] proved that there exists a constant $c_0 > 0$ such that for all $\epsilon > 0$ and $N \in \mathbb{Z}^+$ we have

$$\Big\| \sum_{|k|\leq N} a_k e^{i(tk^2+kx)} \Big\|_{L^6(\mathbb{T}^2)} \leq c_0\, N^{\epsilon} \left(\sum_{|k|\leq N} |a_k|^2 \right)^{1/2}. \tag{4.50}$$

It is an *open problem* to determine if the inequality can be obtained in the interval $(4,6)$. More precisely, it was conjectured in [Bo2] that

$$\|e^{it\Delta}\phi\|_{L^q(\mathbb{T}^{n+1})} \leq c\,\|\phi\|_2 \ \text{ if } q < \frac{2(n+2)}{n}, \tag{4.51}$$

and assuming $\operatorname{supp}\widehat{\phi} \subset B(0,N)$

$$\|e^{it\Delta}\phi\|_{L^q(\mathbb{T}^{n+1})} \ll N^{\frac{n}{2}-\frac{n+2}{q}+\epsilon}\|\phi\|_2 \ \text{ if } q \geq \frac{2(n+2)}{n} \tag{4.52}$$

hold. In this direction, some partial results are gathering in the next proposition.

Proposition 4.4 **([Bo2]).**

1. *For* $n = 1, 2$, *inequality (4.52) holds.*
2. *For* $n \geq 3$, *inequality (4.52) holds for* $q \geq 4$.

For details, see [Bo1] and [Bo2].

The extension of Theorem 4.8 to other compact manifolds (i.e., L^p–L^q estimates for the Schrödinger flow on manifolds) has been studied by Burq, Gerard and Tzvetkov [BGT3].

In the particular case of the two-dimensional sphere $\mathbb{S}^2$, they proved that

$$\left(\int_I \left(\int_{\mathbb{S}^2} |e^{it\Delta}u_0(x)|^q\, dx \right)^{p/q} dt \right)^{1/q} \leq c_I\, \|u_0\|_{1/p,2}, \tag{4.53}$$

where I is a finite time interval and $\|\cdot\|_{1/p,2}$ is defined as in (3.38), for every admissible pair in (4.18) Theorem 4.2 with $n = 2$, i.e.,

$$\frac{1}{p} = \frac{1}{2} - \frac{1}{q}.$$

Roughly, (4.53) gives a gain of $1/2$ derivatives with respect to the Sobolev embedding (Theorem 3.3),

$$\|u_0\|_q \leq c\|u_0\|_{1/r,2} \quad \text{with} \quad \frac{1}{r} = n\left(\frac{1}{2} - \frac{1}{q}\right).$$

The local smoothing effect studied in Section 4.3 was first established by T. Kato [K2] for solutions of the Korteweg–de Vries equation:

$$\begin{cases} \partial_t u + \partial_x^3 u + u\partial_x u = 0, \\ u(x,0) = u_0(x). \end{cases} \tag{4.54}$$

$t,\ x \in \mathbb{R}$. More precisely, Kato proved the following inequality:

$$\left(\int_{-T}^{T}\int_{-R}^{R} |\partial_x u(x,t)|^2\,dx\,dt\right)^{1/2} \leq c(T,R)\|u_0\|_2, \tag{4.55}$$

which was the main ingredient in his proof of existence of the global weak solutions of (4.54) with initial data $u_0 \in L^2(\mathbb{R})$ (see [K2]). In [KF], Kruzhkov and Faminskii independently obtained a similar result to that described in (4.55). Later on and simultaneously, Constantin and Saut [CS], Sjölin [Sj], and Vega [V] showed that the estimates of the type in (4.55) are intrinsic properties of linear dispersive equations. Let $P(\xi)$ be the real symbol associated to the operator $P(D)$. Suppose that at infinity $P(\xi) \sim |\xi|^{\alpha}$, for α a real positive number, and $u(x,t) = e^{itP(D)}u_0(x)$, then

$$\left(\int_{-T}^{T}\int_{|x|\leq R} \left|(-\Delta)^{(\alpha-1)/4}u(x,t)\right|^2 dx\,dt\right)^{1/2} \leq c(T,R)\|u_0\|_2. \tag{4.56}$$

In particular, inequality (4.56) implies that if $u_0 \in L^2(\mathbb{R}^n)$, then the solutions $e^{itP(D)}u_0 \in H^{(\alpha-1)/2}_{\text{loc}}(\mathbb{R}^n)$ for almost all t. Notice that this gain of derivatives is a pure dispersive phenomenon, which cannot hold in hyperbolic problems.

The version of the homogeneous smoothing effect given here (Theorem 4.3) is taken from [KPV3] (see also [LP]). The inhomogeneous smoothing effect version described in Theorem 4.4 was first established in [KPV3]. Observe that the gain of derivatives here doubles from that in the homogeneous case. Also, one has that the result in Theorem 4.4 still holds with $\mathfrak{L}_j$ as in (4.35) instead of the Laplacian.

It is interesting to note that in [CS] the authors extended Kato's result (4.55) to linear dispersive equations. In contrast, in [Sj] and [V] inequality (4.56) with $\alpha = 2$ appears implicitly in the study of the following problem introduced by L. Carleson: Determine the minimum value of s which guarantees that if $u_0 \in H^s(\mathbb{R}^n)$, then

$$\lim_{t\downarrow 0} e^{it\Delta}u_0(x) = u_0(x) \text{ for almost every } x \in \mathbb{R}^n. \tag{4.57}$$

In the one-dimensional case $n = 1$, we have that $s \geq 1/4$ implies (4.57) (see [C]) and this is the best possible result (see [DK], [KR]). For the case $n = 2$, the best result asserting (4.57) is $s > 3/8$ obtained in [Le] (improving previous results of [Sj], [V], $s > 1/2$, [Bo3], $s > 1/2 - \epsilon$, [MVV2], $s > (164 + \sqrt{2})/339$, [TV] $s > 15/32$). In [Bo11], it was shown that in any dimension n the statement (4.57) holds if $s > 1/2 - 1/4n$ (improving previous results of [Sj], [V], $s > 1/2$). Moreover, it was also established in [Bo11] that for $n > 4$ the condition $s \geq \frac{n-2}{2n}$ is necessary for (4.57) to hold.

The original Kato's proof of the smoothing effect (4.55) was based on an energy estimate argument. Let us consider the linear problem (4.54) with data $u_0 \in L^2(\mathbb{R})$. Then multiplying the equation by $u(x,t)\varphi(Rx) = u(x,t)\varphi_R(x)$, $\varphi \in C^{\infty}(\mathbb{R})$, $(\varphi(x) =$

1 for $x > 2$, $\varphi(x) = 0$ for $x < -2$, with $\varphi'(x) > 0$ for $-1 < x < 1$ and $R > 0$), we obtain after integration by parts that

$$\frac{1}{2}\frac{d}{dt}\int u^2\varphi_R\,dx + \frac{3}{2}\int (\partial_x u)^2\,\varphi_R'\,dx - \frac{1}{2}\int u^2\varphi_R^{(3)}\,dx = 0.$$

Thus, integrating in the time interval $[0, T]$ and using that the L^2-norm of the solution is preserved we get (4.55).

The extension of the estimate (4.56) to general dispersive linear models (with constant coefficients) given in [CS] was based on a Fourier transform argument. In nonlinear problems and in linear ones with variable coefficients (where the Fourier transform does not provide the result) it may be useful to obtain the result via "energy estimates."

For example, consider the IVP:

$$\begin{cases} \partial_t u = i\,Au, \\ u(x,0) = u_0(x), \end{cases} \tag{4.58}$$

$x \in \mathbb{R}^n$, $t \in \mathbb{R}$, where A has a real symbol $a = a(x,\xi)$ of order m (for instance, $A = \partial_{x_j}(a_{jk}(x)\partial_{x_k})$, $i\partial_x^3$, Δ, and $i\mathsf{H}\partial_x^2$). By integration by parts, we have that the solutions $u(\cdot,t)$ preserve the L^2-norm, i.e., $\|u(\cdot,t)\|_2 = \|u_0\|_2$. Now to establish the corresponding local smoothing effect (4.55), we follow the argument in [CKS]. First, one applies an operator B of order zero with real symbol $b(x,\xi)$ to our equation to get:

$$\partial_t B\,u = i\,AB\,u + i[B;A]\,u. \tag{4.59}$$

By multiplying the equation (4.59) by $\bar{u}$ and the conjugate of equation (4.58) by Bu, adding the results and integrating in the x-variable, and then in the time interval $[0, T]$, it follows that

$$\int_0^T\int_{\mathbb{R}} i\,[B;A]\,u\,\bar{u}\,dx\,dt \le c_0\,(T;B)\|u_0\|_2. \tag{4.60}$$

Let $C = i\,[B;A] = -i\,[A;B]$. The operator C has order $m-1$ and its symbol $c(x,\xi)$ is given by

$$c(x,\xi) = -\{a,b\} = -\frac{d}{ds}b\,(\varphi(s;x,\xi))\big|_{s=0} = H_a(b)(x,\xi), \tag{4.61}$$

(where $\varphi(s;x,\xi)$ denotes the bicharacteristic flow associated to the symbol of A, that is, $a(x,\xi)$, and $H_a(b)$ is defined as in (3.27)). The aim is to find an operator B such that $C > 0$. By quadrature,

$$b(x,\xi) = \int_0^\infty c(\varphi(s;x,\xi))\,ds. \tag{4.62}$$

Thus, if $A = \Delta/8\pi^2$, $a(x,\xi) = |\xi|^2/2$, and $\varphi(s;x,\xi) = (x + s\xi, \xi)$. Taking

$$c(x,\xi) = -\frac{f'(|x_j|)\,\xi_j^2}{\langle\xi\rangle^2}, \tag{4.63}$$

with $f \in L^1([0,\infty) : \mathbb{R}^+)$, f decreasing, and $\langle\xi\rangle = (1 + |\xi|)^{1/2}$, we have $C > 0$ of order 1.

By (4.61) one gets:

$$b(x,\xi) = \frac{f(|x_j|)\,\xi_j}{\langle\xi\rangle} \quad \text{(nonlocal operator of order zero).}$$

Now, from (4.60), (4.61), (4.63) it follows that

$$\begin{aligned}
\int_0^T\int_{\mathbb{R}} C u\,\bar{u}\,dx\,dt &= \int_0^T\int_{\mathbb{R}} -f'(|x_j|)\,\Lambda^{-1}\partial_{x_j}^2 u\,\bar{u}\,dxdt \\
&= \int_0^T\int_{\mathbb{R}} \partial_{x_j}\Lambda^{-1/2}(-\,f'(|x_j|)\,\Lambda^{-1/2}\partial_{x_j}u)\,\bar{u}\,dxdt \\
&\quad + \int_0^T\int_{\mathbb{R}} \underbrace{[-\,f'(|x_j|);\partial_{x_j}\Lambda^{-1/2}]\,\Lambda^{-1/2}\partial_{x_j}}_{\text{zero order operator}} u\,\bar{u}\,dxdt.
\end{aligned} \tag{4.64}$$

From (4.60) combined with (4.64) and the choice of f, one basically has that

$$\begin{aligned}
\int_0^T\int_{|x|\le R} |D^{1/2}u(x,t)|^2\,dx\,dt &\lesssim \int_0^T\int_{\mathbb{R}} \partial_{x_j}\Lambda^{-1/2}(-f'(|x_j|)\,\Lambda^{-1/2}\partial_{x_j}u)\,\bar{u}\,dx\,dt \\
&\le c_0(R;f;T)\|u_0\|_2.
\end{aligned} \tag{4.65}$$

Repeating the argument for $A = i\partial_x^3$ and taking $c(x,\xi) = \varphi'(x)\xi^2$ with $\varphi'(x) = 1$ if $|x| \le R$ and $\varphi'(x) = 0$ and if $|x| \ge 2R$, even, C^∞, nonincreasing for $x > 0$, we obtain $b(x,\xi) = \varphi(x)$ (local operator as in Kato's approach). Similarly, for $A = i\mathsf{H}\partial_x^2$ (the dispersive operator associated to the Benjamin–Ono equation) with the same choice of $c(x,\xi) = \varphi'(x)\,\xi^2$, we get the same $b(x,\xi) = \varphi(x)$, again a local operator so the result can be obtained by standard integration by parts.

For the variable coefficients case $A = \partial_{x_j}(a_{jk}(x)\partial_{x_k})$, we need several hypotheses that guarantee the appropriate behavior of the bicharacteristic flow at infinity as well as the integrability of $l(s) = c(\varphi(s;x,\xi))$ in (4.62). In this regard, we find the following result due to Doi [Do1].

Let $A(x) = (a_{jk}(x))$ be a real and symmetric $n \times n$ matrix of functions $a_{jk} \in C_b^\infty$. Assume that

$$|\nabla a_{jk}(x)| = o(|x|^{-1}) \quad \text{as } |x| \to \infty, \quad j,k = 1,\ldots,n, \tag{4.66}$$

and that $A(x)$ is positive definite, so the operator $\partial_{x_j}(a_{jk}(x)\partial_{x_k})$ is elliptic as in (3.25). Assume that the bicharacteristic flow is nontrapped in one direction, which means that the set

$$\{X(s; x_0, \xi_0) \; : \; s \in \mathbb{R}\}$$

is unbounded in $\mathbb{R}^n$ for each $(x_0, \xi_0) \in \mathbb{R}^n \times \mathbb{R}^n - \{0\}$.

Lemma 4.3. *Let $A(x)$ and its bicharacteristic flow satisfy the assumptions above. Suppose $\lambda \in L^1([0,\infty)) \cap C([0,\infty))$ is strictly positive and nonincreasing. Then, there exist $c > 0$ and a real symbol $p \in S^0$, both depending on h_2 and λ, such that*

$$H_{h_2} p = \{h_2, p\}(x,\xi) \geq \lambda(|x|)\,|\xi| - c, \quad \forall\, (x,\xi) \in \mathbb{R}^n \times \mathbb{R}^n. \tag{4.67}$$

Extensions and refinements as well as different proofs of the estimates in Theorems 4.2 and 4.3 have been deduced in connection with specific problems. To simplify the exposition we shall only mention some of them.

In [Bo5], Bourgain showed that there exists $c_0 > 0$ such that if $u_1, u_2 \in L^2(\mathbb{R}^2)$, $0 < M_1 \leq M_2$ satisfying that

$$u_j(x) = P_{M_j} u_j = \int\limits_{M_j/2 \leq |\xi| \leq 2M_j} e^{2\pi x\cdot\xi}\, \widehat{u}_j(\xi)\, d\xi, \quad j = 1, 2, \tag{4.68}$$

then

$$\left\| (e^{it\Delta} u_1)(e^{-it\Delta} u_2) \right\|_{L^2(\mathbb{R}^2_x \times \mathbb{R}_t)} \leq c_0 \left(\frac{M_1}{M_2} \right)^{1/2} \|u_1\|_2 \|u_2\|_2. \tag{4.69}$$

Inequality (4.69) measures the interaction of a pair of solutions corresponding to data with localized support in the frequency space.

Notice that for $M_1 \sim M_2$ (4.69) yields the case $p = q = 4 = 2 + 2/n$ of Theorem 4.2.

In [OT1], Ozawa and Tsutsumi studying the bilinear form:

$$(u_0, v_0) \to \partial_x (e^{it\partial_x^2} u_0)(e^{-it\partial_x^2} \bar{v}_0)$$

established the following identity: there exists $c_0 > 0$ such that for any $u_0, v_0 \in L^2(\mathbb{R})$

$$\left\| D_x^{1/2} [(e^{it\partial_x^2} u_0)(e^{-it\partial_x^2} \bar{v}_0)] \right\|_{L^2(\mathbb{R}_x \times \mathbb{R}_t)} = c_0 \|u_0\|_2 \|v_0\|_2. \tag{4.70}$$

The estimate (4.70) resembles the gain of $1/2$ derivative in Theorem 4.3 as well as (after Sobolev embedding) the limit case ($p = \infty$, $q = 4$, $n = 1$, $u_0 = v_0$) of Theorem 4.2.

In higher dimensions, Lions and Perthame [LP] applied the Winger transformation to obtain a different proof of (4.23) in Theorem 4.3. They also showed that for $\alpha \in (0,\infty)$,

$$\left(\int_{-\infty}^{\infty} \int_{\mathbb{R}^n} \frac{|\nabla e^{it\Delta} u_0(x)|^2}{1 + |x|^{1+\alpha}}\, dx dt \right)^{1/2} \leq c_{n,\alpha} \left\| D_x^{1/2} u_0 \right\|_2. \tag{4.71}$$

Finally, we shall briefly discuss the L^2-well-posedness of the IVP

$$\begin{cases} \partial_t u = i\Delta u + b_j(x)\partial_{x_j} u + d(x)u + f(x,t), \\ u(x,0) = u_0(x), \end{cases} \tag{4.72}$$

where the coefficients b_j and d and their derivatives are assumed to be bounded.

The problem (4.72) is said to be L^2-well-posed if for any $u_0 \in L^2(\mathbb{R}^n)$ and $f \in C_0([0,\infty) : L^2(\mathbb{R}^n))$ (where C_0 stands for the set of continuous functions with compact support) there exist $T > 0$ and a unique solution $u \in C([0,T] : L^2(\mathbb{R}^n))$ of (4.72) such that for $t \in [0,T]$

$$\sup_{[0,t]} \|u(\cdot,s)\|_2 \le c(t) \left\{ \|u_0\|_2 + \int_0^t \|f(\cdot,s)\|_2 \, ds \right\}.$$

Notice that if the b_j take real values the result follows by integration by parts. Also, if $b_j(x) = b_{0j}$ is a constant then the assumption $\mathcal{I}m\ b_{0j} = 0$ for all j is a necessary and sufficient condition. In the one-dimensional case, Takeushi [Ta1] proved that the condition

$$\sup_{\ell \in \mathbb{R}} \left| \int_0^\ell \mathcal{I}m\, b(s)\, ds \right| < \infty \tag{4.73}$$

is sufficient for the L^2-well-posedness of (4.72). In [Mz] Mizohata showed that in any dimension n the condition

$$\sup_{\widehat{w} \in \mathbb{S}^{n-1}} \sup_{\substack{x \in \mathbb{R}^n \\ \ell \in \mathbb{R}}} \left| \int_0^\ell \mathcal{I}m\, b_j(x + s \cdot \widehat{w}) \cdot \widehat{w}_j \, ds \right| < \infty \tag{4.74}$$

is necessary. (4.74) is an integrability condition on the coefficients $b = (b_1, \ldots, b_n)$ of the first order term along the bicharacteristic. In fact, Ichinose [I] extended (4.74) to the case where the Laplacian Δ in (4.72) is replaced by the elliptic variable coefficients $A = \partial_{x_j}(a_{jk}(x)\partial_{x_k})$ by deducing that

$$\sup_{\widehat{w} \in \mathbb{S}^{n-1}} \sup_{\substack{x \in \mathbb{R}^n \\ \ell \in \mathbb{R}}} \left| \int_0^\ell \mathcal{I}m\, b_j(X(s;x,\widehat{w})) \cdot \Xi(s;x,\widehat{w})\, ds \right| < \infty \tag{4.75}$$

is a necessary condition for the L^2-well-posedness (to the IVP associated to the equation $\partial_t u = iAu + b_j(x)\partial_{x_j} u + d(x)u + f(x,t)$), where $s \to (X(s;x,\hat{w}), \Sigma(s;x,\hat{w}))$ denotes the bicharacteristic flow associated to A (see 3.28).

Notice that the notion of nontrapping for the bicharacteristic flow associated is essential in the hypothesis (4.75) for $b_j(\cdot)$, even in $C_0^\infty(\mathbb{R}^n)$. We will return to this in Chapter 10, where the above results are further studied.

4.5 Exercises

4.1 Prove Proposition 4.1

4.2 Prove Proposition 4.2.

4.3 Prove that if $1 < p \leq q < \infty$, $0 \leq \gamma < n/q$, $0 \leq \alpha < n(1 - 1/p)$, and $\alpha - \gamma = n(1 - \frac{1}{q} - \frac{1}{p})$, then there exists $c > 0$ such that for all $t \in \mathbb{R}\backslash\{0\}$

$$\left\| e^{it\Delta} u_0 \, |x|^{-\gamma} \right\|_q \leq c \, |t|^{-(\alpha+\gamma)/2 - n/2(1/p - 1/q)} \left\| u_0 \, |x|^{\alpha} \right\|_p. \tag{4.76}$$

Notice that the exponent in (4.76) satisfies

$$-\frac{\alpha+\gamma}{2} - \frac{n}{2}\Big(\frac{1}{p} - \frac{1}{q}\Big) = -\frac{n}{2} + 1 - \frac{1}{p} - \alpha = -\frac{n}{2} + \frac{1}{q} - \gamma.$$

Hint: Combine the formula (4.7) and Pitt's theorem (Exercise 2.12).

4.4 Define the operators:

$$\Gamma_j = x_j + 2it\partial_{x_j}, \qquad j = 1, \ldots, n.$$

(i) Prove that for any $\alpha \in (\mathbb{Z}^+)^n$ (with multi-index notation),

$$\Gamma^{\alpha} f(x,t) = e^{i|x|^2/4t} \, (2it\partial_x)^{\alpha} \, e^{-i|x|^2/4t} \, f = e^{it\Delta} \, x^{\alpha} \, e^{-it\Delta} \, f.$$

(ii) Prove that Γ_j commutes with $\mathcal{O}_s = \partial_t - i\Delta$.

(iii) If $u_0 \in L^2$ and $x^{\alpha} \, u_0 \in L^2(\mathbb{R}^n)$, show that $\Gamma^{\alpha} u \in C(\mathbb{R} : L^2(\mathbb{R}^n))$ and so

$$\partial_x^{\alpha} \big(e^{i|x|^2/4t} \, e^{it\Delta} \, u_0\big) \in C\big(\mathbb{R} \setminus \{0\} : L^2(\mathbb{R}^n)\big).$$

In particular, $\partial_x^{\alpha} \, e^{it\Delta} \, u_0 \in L^2_{\text{loc}}(\mathbb{R}^n)$ for $t \neq 0$.

(iv) If $u_0 \in H^s(\mathbb{R}^n)$, $s \in \mathbb{Z}^+$, and $x^{\alpha} \, u_0 \in L^2$, $|\alpha| \leq s$, prove that

$$u = e^{it\Delta} u_0 \in C\left(\mathbb{R} : H^s \cap L^2(|x|^s \, dx)\right).$$

(v) If $u_0 \in \mathcal{S}(\mathbb{R}^n)$ show that $e^{it\Delta} u_0 \in \mathcal{S}(\mathbb{R}^n)$.

4.5 (i) Prove that if u_0, $x^{\alpha} \, u_0 \in L^2(\mathbb{R}^n)$, and $\partial_x^{\alpha} \, u_0 \notin L^2(\mathbb{R}^n)$, then $x^{\alpha} \, e^{it\Delta} \, u_0 \notin L^2(\mathbb{R}^n)$ for any $t \neq 0$.

(ii) Show that if $u_0 \in C_0(\mathbb{R}^n)$, then for any $t \in \mathbb{R} \setminus \{0\}$ and any $\epsilon > 0$, $e^{it\Delta} u_0 \notin L^1(e^{\epsilon|x|} dx)$, and that $e^{it\Delta} u_0$ has an analytic extension to $\mathbb{C}^n$ for $t \neq 0$.

Hint: Use formula (4.7).

4.6 Using the notation in Definition 3.4.

(i) Prove that for $t > 0$, and $b \in (0,1)$

$$\mathcal{D}^b \, (e^{it|x|^2}) \leq c_{n,b} \, (t^{b/2} + t^b \, |x|^b).$$

(ii) Prove that for $b \in (0,1)$

$$\left\| \, |x|^b \right\|_2 \leq c\big(t^{b/2} \, \|u_0\|_2 + t^b \, \|D^b u_0\|_2 + \| \, |x|^b u_0\|_2\big).$$

(iii) Prove that if $s \geq b/2$, $b \in (0,2)$ and

$$u_0 \in H^s(\mathbb{R}^n) \cap L^2(|x|^b\,dx) \equiv \mathcal{F}_b^s,$$

then

(a) $e^{it\Delta}u_0 \in \mathcal{F}_b^s$, for all $t \neq 0$.
(b) Moreover, $e^{it\Delta}u_0 \in C(\mathbb{R} : \mathcal{F}_b^s)$ (see [NhPo1]).

Hint: For (ii) combine part (i) and Exercise 3.10 inequality 3.42.

4.7 Check that for the group of translations

$$L_t : L^2(\mathbb{R}^n) \mapsto L^2(\mathbb{R}^n)$$

defined by $L_t(u_0)(x) = u_0(x+t)$ the inequalities (4.12) and (4.14) are not true.

4.8 Prove that there do not exist p, q, t with $1 \leq q < p < \infty$, $t \in \mathbb{R}\setminus\{0\}$ such that

$$e^{it\Delta} : L^p(\mathbb{R}^n) \mapsto L^q(\mathbb{R}^n) \;\text{ is continuous.}$$

This is a particular case of Hörmander's theorem in [Ho2].
Hint:

(i) Verify that $e^{it\Delta}$ commutes with translations. That is, if $\tau_h f(x) = f(x-h)$, then $\tau_h(e^{it\Delta}f(x)) = e^{it\Delta}\tau_h f(x)$.
(ii) Show that if $f \in L^{p_0}(\mathbb{R}^n)$, $1 \leq p_0 < \infty$, then

$$\lim_{|h|\to\infty} \|f + \tau_h f\|_{p_0} = 2^{1/p_0}\|f\|_{p_0}.$$

(iii) Using (ii) deduce that if T commutes with translations and $\|Tf\|_q \leq c\|f\|_p$, then

$$\|Tf\|_q \leq c\,2^{(1/p-1/q)}\|f\|_p,$$

which leads to a contradiction because $q < p$.

4.9 (i) Prove that if $f \in L^2(\mathbb{R})$, then $e^{it\Delta}f$ is continuous in $\mathbb{R}$ for almost every $t \in \mathbb{R}$.
Hint: Combine Strichartz estimate (4.14) with $(p,q) = (\infty, 4)$ and a density argument.
(ii) Prove that inequality (4.14) is not true when the pair (p,q) does not satisfy the condition $2/q = n/2 - n/p$ in (4.18) Theorem 4.2.
Hint: Use the fact that if $u(x,t)$ is a solution of the linear Schrödinger equation, then for all $\lambda > 0$, $\lambda\, u(\lambda x, \lambda^2 t)$ is also a solution.

4.10 Given a sequence of times $A = \{t_j \in \mathbb{R} : j \in \mathbb{Z}^+\}$ converging to t_0, prove that there exists $f \in L^2(\mathbb{R})$ such that $e^{it\Delta}f \notin L^\infty(\mathbb{R})$ if $t \in A$ (compare this result with the inequality (4.14)).
Hint: For all $t \in A$ choose $a_t g_t \in L^1(\mathbb{R}) \cap L^2(\mathbb{R})$ such that $g_t \notin L^\infty(\mathbb{R})$ and where the constants a_t are fixed and such that if $f_t = e^{-it\Delta}a_t g_t$ then $f = \sum_{t\in A} f_t$ satisfies the statement (use Lemma 4.1).

4.11 Prove that

$$\|e^{it\Delta}u_0\|_{L^{3r}_{xt}(\mathbb{R}^2)} = \left(\int_{-\infty}^{\infty}\int_{-\infty}^{\infty} |e^{it\Delta}u_0(x)|^{3r}\,dxdt\right)^{1/3r} \leq c\,\|\widehat{u_0}\|_{r'} \tag{4.77}$$

with $1/r + 1/r' = 1$ and $2 \leq r < \infty$. (The inequality (4.77) holds for $4/3 < r < \infty$ see [Ff]).

4.12 Prove that if $f \in L^2(\mathbb{R}^n)$, then

$$\lim_{t\to\pm\infty} \left\| e^{it\Delta}f - \frac{e^{i|\cdot|^2/4t}}{\sqrt{(4\pi i t)^n}}\widehat{f}(\,\cdot\,/4\pi t)\right\|_2 = 0. \tag{4.78}$$

Hint:

(i) Verify that for all $t \neq 0$,

$$U(t)f(x) = (4\pi i t)^{-n/2} e^{i|x|^2/4t}\,\widehat{f}(x/4\pi t)$$

defines a unitary operator. Hence, it is enough to prove (4.78) assuming $f \in \mathcal{S}(\mathbb{R}^n)$.

(ii) Prove that

$$e^{it\Delta}f(x) - U(t)f(x) = \frac{e^{i|x|^2/4t}}{\sqrt{(4\pi i t)^{n/2}}}\widehat{F_t}(x/4\pi t),$$

with $F_t(y) = (e^{i|y|^2/4t} - 1)f(y)$.

(iii) Use the estimate $|e^{i|x|^2/4t} - 1| \leq c\frac{|x|^2}{4t}$ to complete the proof (see [Dl]).

4.13 Prove that if $u_0 \in H^1(\mathbb{R}) \cap L^2(|x|^2dx)$, then

$$\|x\, e^{it\partial_x^2}u_0\|_2 \geq 2|t|\,\|\partial_x u_0\|_2 - \|xu_0\|_2.$$

4.14 Show that the initial value problem:

$$\begin{cases} \partial_t u = i\Delta\bar{u}, \\ u(x,0) = u_0(x), \end{cases} \tag{4.79}$$

$x \in \mathbb{R}^n$, $t > 0$, is ill-posed.
Hint: Differentiate equation (4.79) with respect to the variable t, then use the conjugate of equation (4.79) to obtain an equation in terms of second-order derivatives with respect to t and the bi-Laplacian.

4.15 (Duhamel's principle) Prove that the solution $u(x,t)$ of the inhomogeneous IVP:

$$\begin{cases} \partial_t u = i\Delta u + F(x,t), \\ u(x,0) = u_0(x), \end{cases} \tag{4.80}$$

$x \in \mathbb{R}^n$, $t \in \mathbb{R}$, with $F \in C(\mathbb{R} : \mathcal{S}(\mathbb{R}^n))$ is given by the formula:

$$u(x,t) = e^{it\Delta}u_0 + \int_0^t e^{i(t-t')\Delta}F(\cdot,t')\,dt'. \tag{4.81}$$

4.16 Prove that if $F \in \mathcal{S}(\mathbb{R}^{n+1})$, then the solution $u(x,t)$ of problem (4.80) can be written as:

$$u(x,t)=e^{it\Delta}u_0+\int_{-\infty}^{\infty}\int_{\mathbb{R}^n}\frac{e^{2\pi i\tau t}-e^{-4\pi^2 i|\xi|^2 t}}{4\pi^2 i|\xi|^2+2\pi i\tau}e^{2\pi i x\cdot\xi}\widehat{F}(\xi,\tau)\,d\xi\,d\tau, \tag{4.82}$$

where $\widehat{F}$ represents the Fourier transform of F with respect to the variables x, t.

4.17 [AvHe]

(i) Show that if $u(x,t)$ is a solution of the IVP for the Schrödinger equation with Stark potential:

$$\begin{cases} \partial_t u = i(\Delta u + (\nu\cdot x)u), \\ u(x,0) = u_0(x) \end{cases} \tag{4.83}$$

with $\nu \in \mathbb{R}^n$ for $(x,t) \in \mathbb{R}^n \times \mathbb{R}$, then

$$w(x,t) = u(x+t^2\nu,t)\,e^{-it\nu\cdot x-it^3|\nu|^2/3}$$

solves the linear Schrödinger equation with the same data, i.e., $w(x,t) = e^{it\Delta}u_0$.

(ii) Do the estimates (4.14) and (4.24) hold for the solution $u(x,t)$ of (4.83)?

4.18 Prove inequality (4.37).

4.19 Prove that $m(\xi) = e^{8\pi^3 it\xi^3}$ is not an L^p-multiplier for $p \neq 2$.

4.20 Using the estimates (4.23), (4.24) from Theorem 4.3, prove that:

(i) If $n > 2$, $\alpha > 1/2$, and $f \in L^2(\mathbb{R}^n)$, then $(1+|x|)^{-\alpha}D_x^{1/2}e^{it\Delta}f \in L^2(\mathbb{R}^n)$, a.e. $t \in \mathbb{R}$.

(ii) If $n = 1$ the result in (i) is not true.

(iii) What can be said in the case $n = 2$? (See [KY]).

4.21 Use the commutator estimates in (3.16) to show that operator defined in (4.64), i.e.,

$$\left[-f'(|x_j|);\partial_{x_j}\Lambda^{-1/2}\right]\Lambda^{-1/2}\partial_{x_j},$$

is in fact of order zero.

Chapter 5
The Nonlinear Schrödinger Equation: Local Theory

In this chapter, we shall study local well-posedness of the nonlinear initial value problem (IVP):

$$\begin{cases} i\,\partial_t u = -\Delta u - \lambda |u|^{\alpha-1}u, \\ u(x,0) = u_0(x), \end{cases} \tag{5.1}$$

$t \in \mathbb{R}$, $x \in \mathbb{R}^n$, where λ and α are real constants with $\alpha > 1$.

The equation (5.1) appears as a model in several physical problems (see references [GV1], [N], [SCMc], [ZS]).

Formally solutions of problem (5.1) satisfy the following *conservation laws,* that is, if $u(x,t)$ is solution of (5.1), then for all $t \in [0,T]$, the L^2-norm

$$M(u_0) = \|u(\cdot,t)\|_2^2 = \|u_0\|_2^2, \tag{5.2}$$

the energy

$$\begin{aligned} E(u_0) &= \int\limits_{\mathbb{R}^n} \left(|\nabla_x u(x,t)|^2 - \frac{2\lambda}{\alpha+1}|u(x,t)|^{\alpha+1} \right) dx \\ &= \|\nabla u_0\|_2^2 - \frac{2\lambda}{\alpha+1}\|u_0\|_{\alpha+1}^{\alpha+1}, \end{aligned} \tag{5.3}$$

the momentum

$$\mathcal{I}m \int\limits_{\mathbb{R}^n} \nabla u(x,t)\,\bar{u}(x,t)\,dx = \mathcal{I}m \int\limits_{\mathbb{R}^n} \nabla u_0(x)\,\bar{u}_0(x)\,dx, \tag{5.4}$$

and the so-called quasiconformal law [GV1]

$$\begin{aligned} &\|(x+2it\nabla)u(t)\|_2^2 - \frac{8\lambda t^2}{\alpha+1}\|u(t)\|_{\alpha+1}^{\alpha+1} \\ &\quad = \|xu_0\|_2^2 - 4\lambda\frac{(4-n(\alpha-1))}{\alpha+1}\int_0^t \left(\int\limits_{\mathbb{R}^n} |u(x,s)|^{\alpha+1}\,dx \right) s\,ds. \end{aligned} \tag{5.5}$$

F. Linares, G. Ponce, *Introduction to Nonlinear Dispersive Equations*, Universitext, DOI 10.1007/978-1-4939-2181-2_5

We will use these identities in the next chapter.

We shall say that equation in (5.1) is focusing if $\lambda > 0$ (attractive nonlinearity) and defocusing if $\lambda < 0$ (repulsive nonlinearity).

In any dimension, the equation in (5.1) in the focusing case $\lambda > 0$ has solutions of the form

$$u(x,t) = e^{it}\,\varphi(x), \tag{5.6}$$

called *standing waves*. The *ground state* φ is closely related to the elliptic problem

$$-\Delta v = f(v), \tag{5.7}$$

which have been extensively studied. In our case, $f(v) = -v + |v|^{\alpha-1}v$, with $\lambda = 1$. Indeed, the problem is to find $\varphi \in H^1(\mathbb{R}^n)$, positive, such that

$$-\Delta\varphi + \varphi = |\varphi|^{\alpha-1}\varphi. \tag{5.8}$$

Hence, for any $\omega > 0$,

$$u_\omega(x,t) = e^{i\omega t}\omega^{1/(\alpha-1)}\varphi(\sqrt{\omega}\,x) = e^{i\omega t}\varphi_\omega(x) \tag{5.9}$$

is a solution of the equation in (5.1) with $\lambda = 1$.

The existence of solutions of the equation (5.8) in dimension $n \geq 3$ was established by Strauss [Sr2] and Berestycki and Lions [BLi] (see also [BLiP]). The bidimensional case was considered in [BGK] by Berestycki, Gallouët and Kavian Regarding the uniqueness of solutions of (5.8), Kwong [Kw1] showed that positive solutions of the problem (5.7) with $f(v) = -v + v^p$ are unique up to translations. We summarize these results in the next theorem.

Theorem 5.1. *Let $n \geq 2$ and $1 < \alpha < (n+2)/(n-2)$ $(1 < \alpha < \infty,\ n = 2)$. Then, there exists a unique positive, spherically symmetric solution of* (5.8) *$\varphi \in H^1(\mathbb{R}^n)$. Moreover, φ and its derivatives up to order 2 decay exponentially at infinity.*

Remark 5.1. The restriction on α comes from Pohozaev's identity (5.83) since we want to have H^1-solutions of (5.8) (see Exercise 5.3).

Remark 5.2. There are infinitely many radially symmetric solutions under the hypothesis of Theorem 5.1 without the positivity assumption (see [BLi], [E], [JK]).

As given below, once we have a solution of (5.1), we can use the invariance of the equation to generate other solutions. Thus, if $u = u(x,t)$ is a solution of the equation in (5.1), then the following are also solutions:

$$\begin{aligned} &\text{(i)} \quad u_\mu(x,t) = \mu^{\frac{2}{\alpha-1}}\,u(\mu x, \mu^2 t),\ \mu \in \mathbb{R},\ \text{with initial data given by} \\ &\qquad u_{0\mu}(x) = \mu^{\frac{2}{\alpha-1}}\,u_0(\mu x). \end{aligned} \tag{5.10}$$

$$\text{(ii)} \qquad u_\theta(x,t) = e^{i\theta}u(x,t), \theta \in \mathbb{R}.$$

(iii) $$u_A(x,t) = u(Ax,t), \quad A \text{ any } n\times n \text{ orthogonal matrix.}$$

(iv) $$u_{a,b}(x,t) = u(x-a, t-b), \; a\in\mathbb{R}^n, \; b\in\mathbb{R}.$$

(v) $$\begin{aligned} &u_c(x,t) = e^{ic\cdot x}e^{-i|c|^2 t}u(x-2tc,t) \text{ for } c\in\mathbb{R}^n, \text{ with initial data} \\ &u_c(x,0) = e^{ic\cdot x}\,u_0(x). \end{aligned} \tag{5.11}$$

(vi) In addition, if $\alpha = 4/n+1$, then [GV1]

$$u_\omega(x,t) = \frac{1}{(\alpha+\omega t)^{n/2}}\exp\left(\frac{i\omega|x|^2}{4(\alpha+\omega t)}\right)$$
$$\times\, u\left(\frac{x}{\alpha+\omega t}, \frac{\gamma+\theta t}{\alpha+\omega t}\right), \quad \alpha\theta - \omega\gamma = 1,$$

(vii) $$u_7(x,t) = \overline{u(x,-t)}.$$

Property (i) is called scaling, property (v) Galilean invariance, and property (vi) pseudo-conformal invariance.

Hence, gathering this information, one gets the multiparametric family of solutions $R = R(\nu,\omega,\theta,x_0)$ with ν, $x_0\in\mathbb{R}^n$, $\omega>0$, and $\theta\in\mathbb{R}$.

$$R(x,t) = e^{i(\nu\cdot x - |\nu|^2 t + \omega t + \theta)}\,\varphi_\omega(x - x_0 - 2\nu t) \tag{5.12}$$

of (5.1) with $\lambda = 1$ (focusing case), where $\varphi(\cdot)$ is the positive solution of (5.8) and $\varphi_\omega(\cdot)$ is defined in (5.9). Notice that the solitary wave in (5.12) moves on the line $x = x_0 + 2\nu t$. In the one-dimensional (1-D) case, the equation (5.8) becomes an ordinary differential equation (ODE) and one has that

$$\varphi_\omega(x) = \left\{\frac{(\alpha+1)}{2}\omega\,\mathrm{sech}^2\left(\frac{\alpha-1}{2}\sqrt{\omega}\,x\right)\right\}^{1/(\alpha-1)}. \tag{5.13}$$

Thus, for all $t\in\mathbb{R}$ and $p\in[1,\infty]$

$$\|u(\cdot,t)\|_p = \|u_0\|_p = K(\alpha,\omega). \tag{5.14}$$

From the nonlinear differential equations point of view, the existence of the solitary wave describes a perfect balance between the nonlinearity and the dispersive character of its linear part. More precisely, although the solutions of the linear problem $e^{it\Delta}u_0$ with $u_0\in L^1(\mathbb{R}^n)\cap L^2(\mathbb{R}^n)$ decay as $t\to\infty$ (see (4.12) for the case $u_0\in L^1(\mathbb{R}^n)$ and (4.14) for $u_0\in L^2(\mathbb{R}^n)$), the solutions of (5.1) neither decay nor develop singularities. The latter situation is addressed in the next chapter.

5.1 L² Theory

We consider the integral equation (see Exercise 4.15)

$$u(t) = e^{it\Delta}u_0 + i\lambda \int_0^t e^{i(t-t')\Delta}(|u|^{\alpha-1}u)(t')\,dt'. \tag{5.15}$$

The difference between this equation and the one in (5.1) is that (5.15) does not require any differentiability of the solution. Using the properties described in Proposition 4.2, it is easy to see that if u is a solution of the differential equation in (5.1), then it is also a solution of (5.15). We shall prove in Section 5.3 that under some hypotheses on α and n, if $u_0 \in H^2(\mathbb{R}^n)$, the solution of (5.15) also satisfies the differential equation in (5.1).

We say that the integral equation (5.15) is *locally well-posed* in X, where X is a function space, if for every $u_0 \in X$ there exist $T > 0$ and a unique solution $u \in C([0,T) : X) \cap \ldots$ of (5.15) for $(x,t) \in \mathbb{R}^n \times [0,T)$. Moreover, the map data solution, i.e., $u_0 \mapsto u(\cdot,t)$, locally defined from X to $C([0,T) : X)$, is continuous. Therefore, our notion of well-posedness includes existence, uniqueness, and persistence (the solution $u(t)$ belongs to the same space as the initial data and its time trajectory describes a curve on it). Thus, the solution flow of (5.15) defines a dynamical system in X. In the case that T can be taken arbitrarily large, we shall say that (5.15) is *globally well-posed in X*.

As we shall see below in the subcritical case, one has that $T = T(\|u_0\|_X) > 0$ and in the critical case that $T = T(u_0) > 0$. These definitions of local and global well-posedness also apply to the initial value problem (IVP) (5.1).

Our first result indicate that under some restriction on the power of the nonlinearity, $\alpha \in (1, 1+4/n)$, problem (5.15) is locally well-posed in L^2.

Theorem 5.2 (Local theory in L^2). *If $1 < \alpha < 1 + 4/n$, then for each $u_0 \in L^2(\mathbb{R}^n)$ there exist $T = T(\|u_0\|_2, n, \lambda, \alpha) > 0$ and a unique solution u of the integral equation* (5.15) *in the time interval $[-T,T]$ with*

$$u \in C([-T,T]\colon L^2(\mathbb{R}^n)) \cap L^r([-T,T]\colon L^{\alpha+1}(\mathbb{R}^n)), \tag{5.16}$$

where $r = 4(\alpha+1)/n(\alpha-1)$.

Moreover, for all $T' < T$ there exists a neighborhood V of u_0 in $L^2(\mathbb{R}^n)$ such that

$$\mathbb{F} : V \mapsto C([-T',T']\colon L^2(\mathbb{R}^n)) \cap L^r([-T',T']\colon L^{\alpha+1}(\mathbb{R}^n)), \quad \tilde{u}_0 \mapsto \tilde{u}(t),$$

is Lipschitz.

As we shall see in the proof of Theorem 5.2 (see (5.24)) and in Exercise 5.5, one can give a precise estimate for the life span of the solution according to the size of the data in L^2-norm. This fact holds whenever the problem is "subcritical" and the scaling of the norm of the initial data is homogeneous, i.e., in our case, if $u = u(x,t)$ is a solution of (5.1) or (5.15), then

$$u_\mu(x,t) = \mu^{2/(\alpha-1)}\, u(\mu x, \mu^2 t),$$

is also a solution with data $u_\mu(x,0) = \mu^{2/(\alpha-1)}\, u_0(\mu x)$ so that

$$\|u_\mu(0)\|_2 = \mu^{2/(\alpha-1)-n/2}\, \|u_0\|_2.$$

If, in addition to the hypothesis of Theorem 5.2, one has that $u_0 \in H^s(\mathbb{R}^n)$, $s > 0$, and $\alpha \geq [s]+1$, $[\cdot]$ denoting the greatest integer function, then

$$u \in C([0,T] : H^s(\mathbb{R}^n)) \cap L^r([-T,T] : L_s^{\alpha+1}(\mathbb{R}^n)), \tag{5.17}$$

with T as in the theorem. This fact holds in any subcritical case with a regular enough nonlinearity, since by taking s derivatives the problem becomes linear in this variable.

The proof of Theorem 5.2 is based on the contraction mapping principle. This has the advantage that it also shows that if the nonlinearity is smooth, i.e., α is an odd integer, then the map data-solution $u_0 \mapsto u(t)$ is smooth (see Corollary 5.6).

Corollary 5.1. *The solution* u *of equation (5.15) obtained in Theorem 5.2 belongs to* $L^q([-T,T] \colon L^p(\mathbb{R}^n))$ *for all* (p,q) *defined by condition (4.18) of Theorem 4.2, that is:*

$$\left.\begin{array}{ll} 2 \leq p < \dfrac{2n}{n-2} & \text{if } n \geq 3 \\ 2 \leq p < \infty & \text{if } n = 2 \\ 2 \leq p \leq \infty & \text{if } n = 1 \end{array}\right\} \quad \textit{and} \quad \frac{2}{q} = \frac{n}{2} - \frac{n}{p}. \tag{5.18}$$

In the proof of Theorem 5.2, we use the following notation: For all positive constants T and a, we define

$$E(T,a) = \Big\{ v \in C([-T,T] \colon L^2(\mathbb{R}^n)) \cap L^r([-T,T] \colon L^{\alpha+1}(\mathbb{R}^n)) :$$
$$|||v|||_T \equiv \sup_{[-T,T]} \|v(t)\|_2 + \left(\int_{-T}^{T} \|v(t)\|_{\alpha+1}^r dt \right)^{1/r} \leq a \Big\} \tag{5.19}$$

with $1 < \alpha < 1 + 4/n$ and $r = 4(\alpha+1)/n(\alpha-1)$. Note that $E(T_0, a)$ is a complete metric space.

Proof of Theorem 5.2 For appropriate values of a and $T > 0$, we shall show that

$$\Phi_{u_0}(u)(t) = \Phi(u)(t) = e^{it\Delta} u_0 + i\lambda \int_0^t e^{i\Delta(t-t')} (|u|^{\alpha-1} u)(t')\, dt' \tag{5.20}$$

defines a contraction map on $E(T,a)$.

Without loss of generality we consider only the case $t > 0$. Using (4.14), (4.17), and Hölder's inequality combined with the definition $\Phi(\cdot)$ in (5.20), we obtain:

$$\left(\int_0^T \|\Phi(u)(t)\|_{\alpha+1}^r dt\right)^{1/r} \leq c\|u_0\|_2 + c|\lambda| \left(\int_0^T \| |u(t)|^\alpha \|_{(\alpha+1)/\alpha}^{r'} dt\right)^{1/r'}$$
$$\leq c\|u_0\|_2 + c|\lambda| \left(\int_0^T \|u(t)\|_{(\alpha+1)}^{\alpha r'} dt\right)^{1/r'}. \tag{5.21}$$

By hypothesis ($1 < \alpha < 1 + 4/n$), we have that $\alpha r' < r$, that is,

$$\alpha \frac{r}{r-1} < r \quad \text{or} \quad \alpha < r - 1 = \frac{4(\alpha+1)}{n(\alpha-1)} - 1.$$

Therefore, from (5.21) we deduce that

$$\left(\int_0^T \|\Phi(u)(t)\|_{\alpha+1}^r dt\right)^{1/r} \leq c\|u_0\|_2 + c|\lambda| T^\theta \left(\int_0^T \|u\|_{\alpha+1}^r dt\right)^{\alpha/r} \tag{5.22}$$

with $\theta = 1 - n(\alpha-1)/4 > 0$. Then, if $u \in E(T,a)$ we have

$$\left(\int_0^T \|\Phi(u)(t)\|_{\alpha+1}^r dt\right)^{1/r} \leq c\|u_0\|_2 + c|\lambda|\, T^\theta\, a^\alpha.$$

Using 4.16 and the unitary group properties in expression (5.20), we obtain that if $u \in E(T,a)$, then

$$\sup_{[0,T]} \|\Phi(u)(t)\|_2 \leq c\|u_0\|_2 + c|\lambda| \left(\int_0^T \| |u|^\alpha \|_{(\alpha+1)/\alpha}^{r'} dt\right)^{1/r'}$$
$$\leq c\|u_0\|_2 + c|\lambda|\, T^\theta\, a^\alpha, \tag{5.23}$$

where the constant c depends only on α and the dimension n. Hence,

$$|||\Phi(u)|||_T \leq c\,\|u_0\|_2 + c|\lambda| T^\theta a^\alpha.$$

If we fix $a = 2c\|u_0\|_2$ and take $T > 0$ such that

$$2^\alpha c^\alpha\, |\lambda|\, T^\theta\, \|u_0\|_2^{\alpha-1} < 1, \tag{5.24}$$

it follows that the application Φ is well defined on $E(T,a)$. Now, if $u, v \in E(T,a)$,

$$(\Phi(v) - \Phi(u))(t) = i\lambda \int_0^t e^{i(t-t')\Delta}(|v|^{\alpha-1}v - |u|^{\alpha-1}u)(t')\,dt'.$$

The same argument as in (5.21) and (5.22) shows that

$$\begin{aligned}
&\left(\int_0^T \|(\Phi(v) - \Phi(u))(t)\|_{\alpha+1}^r\,dt\right)^{1/r} \\
&\leq c|\lambda| \left(\int_0^T \|\,|v|^{\alpha-1}v - |u|^{\alpha-1}u\|_{(\alpha+1)/\alpha}^{r'}\,dt\right)^{1/r'} \\
&\leq c_\alpha|\lambda| \left(\int_0^T (\|v\|_{\alpha+1}^{\alpha-1} + \|u\|_{\alpha+1}^{\alpha-1})^{r'}\,\|v-u\|_{\alpha+1}^{r'}(t)\,dt\right)^{1/r'} \\
&\leq c_\alpha|\lambda|\,T^\theta \left\{\left(\int_0^T \|v\|_{(\alpha+1)}^r\,dt\right)^{(\alpha-1)/r} + \left(\int_0^T \|u\|_{(\alpha+1)}^r\,dt\right)^{(\alpha-1)/r}\right\} \\
&\quad\times \left(\int_0^T \|v(t) - u(t)\|_{(\alpha+1)}^r\,dt\right)^{1/r} \\
&\leq 2c_\alpha\,|\lambda|\,T^\theta\,a^{\alpha-1} \left(\int_0^T \|v(t) - u(t)\|_{(\alpha+1)}^r\,dt\right)^{1/r}.
\end{aligned}$$

Combining (4.16) with the unitary group properties and the arguments used in (5.21) and (5.22), we see as in (5.23) that

$$\sup_{[0,T]} \|(\Phi(v) - \Phi(u))(t)\|_2 \leq 2c_\alpha|\lambda|\,T^\theta\,a^{\alpha-1} \left(\int_0^T \|v(t) - u(t)\|_{\alpha+1}^r\,dt\right)^{1/r}.$$

Finally, it follows from the choice of a, $a \leq 2c\|u_0\|_2$, and inequality (5.24) that

$$2c\,|\lambda|\,T^\theta\,a^{\alpha-1} \leq 2^\alpha c^\alpha|\lambda|\,T^\theta\,\|u_0\|_2^{\alpha-1} < 1.$$

Hence,

$$T \simeq \|u_0\|_2^\beta, \quad \text{with } \beta = \frac{4(1-\alpha)}{4-n(\alpha-1)}. \tag{5.25}$$

Thus, we have proved the existence and uniqueness in an appropriate class of the solution of equation (5.15). To prove the continuous dependence of $\Phi(u(t)) = \Phi_{u_0}(u(t))$ with respect to u_0, note that if u, v are the corresponding solutions of (5.15) with initial data u_0, v_0, respectively, then

$$u(t) - v(t) = e^{it\Delta}(u_0 - v_0) + i\lambda \int_0^t e^{i(t-t')\Delta}(|u|^{\alpha-1}u - |v|^{\alpha-1}v)(t')dt'.$$

Therefore, the same argument used in (5.21) and (5.22) implies

$$\left(\int_0^T \|u(t) - v(t)\|_{\alpha+1}^r \, dt\right)^{1/r} \leq c\|u_0 - v_0\|_2$$

$$+ K_\alpha |\lambda| \, T^\theta \left(\|u_0\|_2^{\alpha-1} + \|v_0\|_2^{\alpha-1}\right) \left(\int_0^T \|u(t) - v(t)\|_{\alpha+1}^r \, dt\right)^{1/r}.$$

As a consequence, if $\|u_0 - v_0\|_2$ is small enough (see (5.24)), then

$$\left(\int_0^T \|u(t) - v(t)\|_{\alpha+1}^r \, dt\right)^{1/r} \leq \widetilde{K}\|u_0 - v_0\|_2.$$

Analogously we can prove that

$$\sup_{[0,T]} \|u(t) - v(t)\|_2 \leq \widetilde{K}\|u_0 - v_0\|_2,$$

which completes the proof. □

Proof of Corollary 5.1 The proof is obtained by combining Corollary 4.1 with inequality (5.21). That is, taking (p, q) in Corollary 4.1 instead of $(\alpha + 1, r)$ on the left-hand side of (5.21) and then using the argument in the proof of Theorem 5.2. The details of this proof are left as an exercise to the reader. □

Remark 5.3. Observe that in the proof of Theorem 5.2 we only used the hypothesis on the growth of the nonlinear term but not its particular form.

Next, we show how to extend the argument used in the proof of Theorem 5.2 to the critical case $\alpha = 1 + 4/n$.

Proposition 5.1. *Let (p, q) be a pair satisfying condition* (5.18) *in Corollary* 5.1. *Given $u_0 \in L^2(\mathbb{R}^n)$ and $\epsilon > 0$, there exist $\delta > 0$ and $T > 0$ such that if $\|v_0 - u_0\|_2 < \delta$, then*

$$\left(\int_0^T \|e^{it\Delta}v_0\|_p^q \, dt\right)^{1/q} < \epsilon. \tag{5.26}$$

Proof. If we take $\delta < \epsilon/2c$, then it suffices to show that

$$\left(\int_0^T \|e^{it\Delta}u_0\|_p^q\,dt\right)^{1/q} < \epsilon/2. \tag{5.27}$$

We choose $\widetilde{u}_0 \in \mathcal{S}(\mathbb{R}^n)$ such that $\|u_0 - \widetilde{u}_0\|_2 < \epsilon/4c$ and then combining Theorem 4.2 (inequality (4.14)), the fact that $\{e^{it\Delta}\}$ defines a unitary group in $H^s(\mathbb{R}^n)$ and Sobolev's inequality (Theorem 3.3), we have

$$\begin{aligned}\left(\int_0^T \|e^{it\Delta}u_0\|_p^q\,dt\right)^{1/q} &\leq \left(\int_0^T \|e^{it\Delta}(\widetilde{u}_0 - u_0)\|_p^q\,dt\right)^{1/q} + \left(\int_0^T \|e^{it\Delta}\widetilde{u}_0\|_p^q\,dt\right)^{1/q}\\ &\leq c\|\widetilde{u}_0 - u_0\|_2 + cT^{1/q}\,\|\widetilde{u}_0\|_{s,2},\end{aligned}$$

where $s \geq n(1/2 - 1/p)$. Fixing T such that $cT^{1/q}\,\|\widetilde{u}_0\|_{s,2} < \epsilon/4$, we obtain (5.27). □

Theorem 5.3 (Critical case, $\alpha = 1 + 4/n$ in $L^2(\mathbb{R}^n)$). *If $\alpha = 1 + 4/n$, then for each $u_0 \in L^2(\mathbb{R}^n)$ there exist $T = T(u_0, \lambda, \alpha) > 0$ and a unique solution u of the integral equation* (5.15) *in the time interval $[-T, T]$ with*

$$u \in C([-T,T]: L^2(\mathbb{R}^n)) \cap L^\sigma([-T,T]: L^\sigma(\mathbb{R}^n)), \tag{5.28}$$

where $\sigma = 2 + 4/n$.

Moreover, for all $T' < T$ there exists a neighborhood V of u_0 in $L^2(\mathbb{R}^n)$ such that

$$\mathbb{F}: V \mapsto C([-T',T']: L^2(\mathbb{R}^n)) \cap L^\sigma([-T',T']: L^\sigma(\mathbb{R}^n)), \quad \tilde{u}_0 \mapsto \tilde{u}(t),$$

is Lipschitz.

Remark 5.4. Notice that the time of existence in Theorem 5.2 depends only on the size of u_0 (that is, on $\|u_0\|_2$); meanwhile, in Theorem 5.3, the time of existence depends on the position of u_0, and not only on its size.

Proof. We shall show that $\Phi_{u_0} = \Phi$ in (5.20) defines a contraction in:

$$\widetilde{E}(T,a) = \Big\{v \in C([-T,T]: L^2(\mathbb{R}^n)) \cap L^\sigma([-T,T]: L^\sigma(\mathbb{R}^n)) :$$

$$\|v\|_T \equiv \sup_{[-T,T]} \|v(t) - e^{it\Delta}u_0\|_2 + \left(\int_{-T}^T \|v(t)\|_\sigma^\sigma\,dt\right)^{1/\sigma} \leq a\Big\}.$$

First, from (5.20) it follows that

$$\sup_{[0,T]} \|\Phi(u)(t) - e^{it\Delta}u_0\|_2 \leq \sup_{[0,T]} \|\int_0^t e^{i\Delta(t-t')}\lambda|u|^\alpha(t')\,dt'\|_2$$

$$\leq c|\lambda| \left(\int_0^T \| |u(t)|^\alpha \|_{\sigma'}^{\sigma'} dt \right)^{1/\sigma'} \tag{5.29}$$
$$\leq c|\lambda| \left(\int_0^T \|u(t)\|_\sigma^\sigma dt \right)^{\alpha/\sigma}.$$

On the other hand, it is easy to see that the pair (σ, σ) satisfies the condition (5.18) of Corollary 5.1. Then, combining the integral equation (5.20), estimates (4.14), and (5.26) with $(p, q) = (\sigma, \sigma)$, with the argument used on (5.21), we obtain:

$$\left(\int_0^T \|\Phi(u)(t)\|_\sigma^\sigma dt \right)^{1/\sigma} \leq c\epsilon + c|\lambda| \left(\int_0^T \| |u(t)|^\alpha \|_{\sigma'}^{\sigma'} dt \right)^{1/\sigma'} \tag{5.30}$$
$$\leq c\epsilon + c|\lambda| \left(\int_0^T \|u(t)\|_\sigma^\sigma dt \right)^{\alpha/\sigma},$$

because $\alpha\sigma' = (1 + 4/n)\,((2n + 4)/(n + 4)) = 2 + 4/n = \sigma$.

From Proposition 5.1, inequalities (5.29) and (5.30), we obtain that given $\epsilon > 0$, there exists $T > 0$ such that if $u \in \widetilde{E}(T, a)$, then

$$|||\Phi(u)|||_T \leq c\epsilon + c|\lambda|\, a^\alpha.$$

Therefore, if

$$c\epsilon + c\,|\lambda|\, a^\alpha < a, \tag{5.31}$$

we get that $\Phi(\widetilde{E}(T, a)) \subseteq \widetilde{E}(T, a)$. The argument used in the proof of Theorem 5.2 yields:

$$\left(\int_0^T \|(\Phi(v) - \Phi(u))(t)\|_\sigma^\sigma dt \right)^{1/\sigma} \leq 2c|\lambda|\, a^{\alpha-1} \left(\int_0^T \|v(t) - u(t)\|_\sigma^\sigma dt \right)^{1/\sigma}.$$

Thus, for

$$2c\,|\lambda|\, a^{\alpha-1} < 1/2, \tag{5.32}$$

we have that $\Phi(\cdot)$ is a contraction. Now, fixing $\epsilon > 0$ such that

$$c|\lambda|\, \epsilon^{\alpha-1} < 1/2$$

we see that both (5.31) and (5.32) are verified. This basically completes the proof, the remainder of the proof follows using the same argument employed to show Theorem 5.2. □

Corollary 5.2. *There exists* $\epsilon_0 > 0$ *depending on* λ *and* n *such that for all* $u_0 \in L^2(\mathbb{R}^n)$ *with* $\|u_0\|_2 \leq \epsilon_0$, *the results of Theorem* 5.3 *extends to any time interval* $[0, T]$, *i.e,*

$$u \in C(\mathbb{R} : L^2(\mathbb{R}^n)) \cap L^\sigma(\mathbb{R} : L^\sigma(\mathbb{R}^n)), \quad \sigma = 2 + 4/n. \tag{5.33}$$

Proof. It is enough to note that if $\|u_0\|_2$ is sufficiently small, then taking $\epsilon = \|u_0\|_2$ and $a = 2\|u_0\|_2$ (both independent of T), and

$$c\,|\lambda|\,\|u_0\|_2^{\alpha-1} < 1/2,$$

we see that (5.31) and (5.32) hold. □

Combining the results in Corollary 5.2 and those in Exercise 6.2 (concerning the scattering of the solutions obtained in Corollary 5.2), one should expect that the constant ϵ in Corollary 5.2 be given by $\|\varphi\|_2$, where φ is the positive solution of equation (5.8), with $\omega = 1$ and $\alpha = 1 + 4/n$. This has been proved in the radial case and for dimension $n = 2$ in [KTV].

5.2 H^1 Theory

We consider the integral equation (5.15) with $u_0 \in H^1(\mathbb{R}^n)$ with the nonlinearity α satisfying

$$\begin{cases} 1 < \alpha < \dfrac{n+2}{n-2}, & \text{if } n > 2 \\ 1 < \alpha < \infty, & \text{if } n = 1, 2. \end{cases} \tag{5.34}$$

Theorem 5.4 (Local theory in H^1). *If* α *satisfies hypothesis* (5.34), *then for all* $u_0 \in H^1(\mathbb{R}^n)$ *there exist* $T = T(\|u_0\|_{1,2}, n, \lambda, \alpha) > 0$ *and a unique solution* u *of the integral equation* (5.15) *in the time interval* $[-T, T]$ *with*

$$u \in C([-T, T] \colon H^1(\mathbb{R}^n)) \cap L^r([-T, T] \colon L_1^\rho(\mathbb{R}^n)), \tag{5.35}$$

where $(\rho, r) = \left(\dfrac{n(\alpha+1)}{n+\alpha-1}, \dfrac{4(\alpha+1)}{(n-2)(\alpha-1)}\right)$ *for* $n \geq 3$, *and* (ρ, r) *satisfies* (5.18) *for* $n = 1, 2$, *and* L_1^ρ *is defined as in (3.38).*

Moreover, for all $T' < T$ *there exists a neighborhood* W *of* u_0 *in* $H^1(\mathbb{R}^n)$ *such that the function*

$$\mathbb{F} \colon W \mapsto C([-T', T'] \colon H^1(\mathbb{R}^n)) \cap L^r([-T', T'] \colon L_1^\rho(\mathbb{R}^n)), \quad \tilde{u}_0 \mapsto \tilde{u}(t),$$

is Lipschitz.

If in addition to the hypothesis of Theorem 5.4 one has that $u_0 \in H^s(\mathbb{R}^n)$, $s > 1$, and $\alpha \geq [s] + 1$, $[\cdot]$ denoting the greatest integer function, then

$$u \in C([0, T] : H^s(\mathbb{R}^n)) \cap L^r([0, T] : L_s^{\alpha+1}(\mathbb{R}^n)), \tag{5.36}$$

where $[0,T]$ is the same time interval given for $s=1$. As in (5.17), the problem becomes linear in $D_x^s u$ once one takes D_x^s in the equation and the result follows by reapplying the argument in the proof of Theorem 5.4 in this linear equation whose coefficients (depending on u) have sufficient regularity to get the desired result.

As we shall see in the next chapter, in the critical case, a similar result was quite difficult to establish.

Corollary 5.3. *The solution of the integral equation* (5.15) *obtained in Theorem* 5.4 *belongs to* $u \in L^q([-T,T]: L_1^p(\mathbb{R}^n))$ *for all pair* (p,q) *defined by condition* (5.18) *in Corollary* 5.1. *Moreover, in these spaces, the solution depends continuously on the initial data.*

The proof of this theorem is similar to the one given in the previous section for the L^2 case; therefore, we can give only a sketch of it.

Proof of Theorem 5.4 We will show the theorem in the case $n \geq 3$. We first define

$$E^1(T,a)=\Big\{v \in C([-T,T]: H^1)\cap L^r([-T,T]: L_1^{\rho}) : \|\!|v|\!\|_T \equiv \sup_{[-T,T]} \|v(t)\|_{1,2} + \left(\int_{-T}^{T} (\|v(t)\|_\rho^r + \|\nabla_x v(t)\|_\rho^r)\,dt\right)^{1/r} \leq a\Big\}. \tag{5.37}$$

Notice that the pair (ρ,r) is an admissible pair (see Corollary 4.1).

We prove that there exist positive constants T and a such that the operator defined in (5.20) is a contraction on $E^1(T,a)$.

Combining Hölder's inequality and the Sobolev inequality (Theorem 3.3) it follows that

$$\||u|^{\alpha-1}\nabla u\|_{\rho'} \leq c\,\||u|^{\alpha-1}\|_l \|\nabla u\|_\rho \leq c\,\|u\|_{(\alpha-1)l}^{\alpha-1}\|\nabla u\|_\rho \leq c\,\|\nabla u\|_\rho^\alpha.$$

Thus,

$$\||u|^{\alpha-1}u\|_{1,\rho'} \leq c\,\|u\|_{1,\rho}^\alpha, \tag{5.38}$$

with $1/\rho' = 1/l + 1/\rho$. Then,

$$\frac{1}{l} = 1 - \frac{2}{\rho} \text{ and } \frac{1}{(\alpha-1)l} = \frac{1}{\rho} - \frac{1}{n}, \text{ i.e., } \frac{1}{l} = \frac{\alpha-1}{\rho} - \frac{\alpha-1}{n}.$$

Therefore, $(\alpha+1)/\rho = (n+\alpha-1)/n$.

Using Corollary 4.1, (5.20), and (5.38), we have

$$\begin{aligned}\|\!|\Phi(u)|\!\|_T &\leq c\,\|u_0\|_{1,2} + c\left(\int_0^T \||u|^{\alpha-1}u(t)\|_{1,\rho'}^{r'}\,dt\right)^{1/r'}\\ &\leq c\,\|u_0\|_{1,2} + c\left(\int_0^T \|u(t)\|_{1,\rho}^{\alpha r'}\,dt\right)^{1/r'}\end{aligned} \tag{5.39}$$

$$\leq c\,\|u_0\|_{1,2} + c\,T^{\delta}\left(\int_0^T \|u(t)\|_{1,\rho}^r\,dt\right)^{\alpha/r},$$

with $\delta = 1 - (\alpha+1)/r = 1 - (n-2)(\alpha-1)/4$. Hence, taking $a = 2c\,\|u_0\|_{1,2}$ in (5.37), we get from (5.39) that

$$\begin{aligned}|||\Phi(u)|||_T &\leq c\,\|u_0\|_{1,2} + c\,T^{\delta}\,|||u|||_T^{\alpha}\\ &\leq \frac{a}{2} + c\,T^{\delta}\,\frac{a^{\alpha}}{(2c)^{\alpha}} \leq a\end{aligned}$$

if T is sufficiently small, i.e.,

$$\frac{c\,T^{\delta}}{(2c)^{\alpha}}\,a^{\alpha-1} \leq \frac{1}{2}.$$

Thus,

$$T \lesssim a^{(1-\alpha)/\delta}. \tag{5.40}$$

To complete the proof of existence and uniqueness of the solution, it is enough to show that the operator Φ is a contraction. The proof of this as well as the continuous dependence is similar to the one given in the previous section, so it will be omitted.□

Remark 5.5. As we commented in the previous section, in the proof of this (local) result, we did not use the particular structure of the nonlinear term.

Theorem 5.5 (Critical case, $\alpha = (n+2)/(n-2)$, $n > 2$, in $H^1(\mathbb{R}^n)$). *Let $n > 2$ and $\alpha = (n+2)/(n-2)$. Given $u_0 \in H^1(\mathbb{R}^n)$, there exist $T = T(u_0, n, \lambda, \alpha) > 0$ and a unique solution u of the integral equation* (5.15) *in the time interval $[-T, T]$ with*

$$u \in C([-T,T]\colon H^1(\mathbb{R}^n)) \cap L^r([-T,T]\colon L_1^{\rho}(\mathbb{R}^n)),$$

where $r = 2n/(n-2)$, $\rho = 2n^2/(n^2-2n+4)$ and L_1^p is defined as in (3.38).

Moreover, for all $T' < T$ there exists a neighborhood W of u_0 in $H^1(\mathbb{R}^n)$ such that the function

$$\mathbb{F}\colon W \to C([-T',T']\colon H^1(\mathbb{R}^n)) \cap L^r([-T',T']\colon L_1^{\rho}(\mathbb{R}^n)), \quad \tilde{u}_0 \to \tilde{u}(t),$$

is Lipschitz.

Remark 5.6. We notice that the time of existence depends on the initial data. In Theorem 5.4, it depends only on the size of u_0, that is, on $\|u_0\|_{1,2}$. In Theorem 5.5, the interval of existence depends on the position of u_0, and not only on its size.

Proof. Observe that the pair $(r,\rho) = (2n/(n-2), 2n^2/(n^2-2n+4))$ satisfies condition (5.18) of Corollary 5.1. First, we have that

$$\left(\int_0^T \|\nabla_x(|u|^{\alpha-1}u)\|_{\rho'}^{r'}\right)^{1/r'} \leq c\left(\int_0^T \|\nabla_x u\|_{\rho}^{r}\right)^{1/r}\left(\int_0^T \||u|^{\alpha-1}\|_{\nu}^{l}\right)^{1/l} \tag{5.41}$$

$$\leq c \left(\int_0^T \|\nabla_x u\|_\rho^r \right)^{1/r} \left(\int_0^T \|u\|_{\nu(\alpha-1)}^{l(\alpha-1)} \right)^{1/l},$$

where $1/r + 1/r' = 1/\rho + 1/\rho' = 1$, $1/\rho' = 1/\rho + 1/\nu$, and $1/r' = 1/r + 1/l$. Since $(\alpha, r, \rho) = ((n+2)/(n-2), 2n/(n-2), 2n^2/(n^2-2n+4))$, we have $l(\alpha-1) = r$ and $\nu(\alpha-1) = 2n^2/(n-2)^2$. Then by Gagliardo–Nirenberg's inequality (3.14) it follows:

$$\|u\|_{\nu(\alpha-1)} \leq c\, \|u\|_{1,\rho} = c\,(\|u\|_\rho + \|\nabla_x u\|_\rho). \tag{5.42}$$

Combining (5.41), (5.42), Proposition 5.1, Theorem 4.2, and the notation in the proof of Theorem 5.4, we obtain that for any $\varepsilon > 0$ fixed there exists $T > 0$ such that

$$\begin{aligned}
&\left(\int_0^T \|\Phi(u)(t)\|_{1,\rho}^r \, dt \right)^{1/r} \\
&\leq c \left(\int_0^T \|\Phi(u)(t)\|_\rho^r \, dt \right)^{1/r} + \left(\int_0^T \|\nabla_x \Phi(u)(t)\|_\rho^r \, dt \right)^{1/r} \\
&\leq c\epsilon + c|\lambda| \left(\int_0^T \|u\|_\rho^r \, dt \right)^{1/r} \\
&\quad + c|\lambda| \left(\int_0^T \|\nabla_x u\|_\rho^r \, dt \right)^{1/r} \left(\int_0^T \|u\|_{1,\rho}^r \, dt \right)^{(\alpha-1)/r} \\
&\leq c\epsilon + c|\lambda| \left(\int_0^T \|u\|_{1,\rho}^r \, dt \right)^{\alpha/r}.
\end{aligned} \tag{5.43}$$

On the other hand, we have that

$$\sup_{[0,T_0]} \|\Phi(u)(t) - e^{it\Delta} u_0\|_{1,2} \leq c|\lambda| \left(\int_0^T \|u\|_{1,\rho}^r dt \right)^{\alpha/r}. \tag{5.44}$$

Therefore, defining

$$\begin{aligned}
\widetilde{E}^1(T,a) = \Big\{ & v \in C([0,T] : H^1(\mathbb{R}^n)) \cap L^r([0,T] : L_1^\rho(\mathbb{R}^n)) : \\
& |||v|||_T \equiv \sup_{[0,T_0]} \|v(t) - e^{it\Delta} u_0\|_{1,2} + \left(\int_0^T \|v\|_{1,\rho}^r \, dt \right)^{1/r} \leq a \Big\},
\end{aligned}$$

and applying (5.44) and (5.43), we have that for all $\epsilon > 0$ there exists $T > 0$ such that if $u \in \widetilde{E}^1(T,a)$, then

$$\|\!|\Phi(u)(t)|\!\| \leq c\epsilon + c|\lambda|\, a^{\alpha}. \tag{5.45}$$

Once inequality (5.45) is established, the remainder of the proof follows an argument given previously, so it will be omitted. □

Corollary 5.4. *There exists* $\epsilon_0 > 0$ *depending on* λ *and* n *such that for all* $u_0 \in H^1(\mathbb{R}^n)$ *with* $\|u_0\|_{1,2}$ *small, the results of Theorem* 5.5 *extend to all time intervals* $[0,T]$*, so*

$$u \in C(\mathbb{R} : H^1(\mathbb{R}^n)) \cap L^r(\mathbb{R} : L_1^{\rho}(\mathbb{R}^n)) \tag{5.46}$$

with (r,ρ) *as in Theorem* 5.5.

Proof. Once Theorem 5.5 is established, we follow the argument used in the proof of Corollary 5.2. □

5.3 H² Theory

Consider again the integral equation (5.15) with $u_0 \in H^2(\mathbb{R}^n)$.

Assume that the nonlinearity α satisfies

$$\begin{cases} 2 \leq \alpha < \dfrac{n}{n-4}, & \text{if } n \geq 5 \\ 2 \leq \alpha < \infty, & \text{if } n \leq 4. \end{cases} \tag{5.47}$$

Theorem 5.6 (Local theory in $H^2(\mathbb{R}^n)$). *If* α *satisfies* (5.47)*, then for all* $u_0 \in H^2(\mathbb{R}^n)$ *there exist* $T = T(\|u_0\|_{2,2}, n, \lambda, \alpha) > 0$ *and a unique solution* u *of the integral equation* (5.15) *in the interval of time* $[-T,T]$ *with*

$$u \in C([-T,T]\colon H^2(\mathbb{R}^n)) \cap L^q([-T,T]\colon L_2^p(\mathbb{R}^n)) \tag{5.48}$$

for all pairs (p,q) *defined by condition (4.18) of Corollary* 5.1.

Moreover, for all $T' < T$ *there exists a neighborhood* W *of* u_0 *in* $H^2(\mathbb{R}^n)$ *such that for all pairs* (p,q) *in (4.18) the function*

$$\mathbb{F} : W \mapsto C([-T',T']\colon H^2(\mathbb{R}^n)) \cap L^q([-T',T']\colon L_2^p(\mathbb{R}^n)), \quad \tilde{u}_0 \mapsto \tilde{u}(t),$$

is Lipschitz.

The proof of this result is similar to the one exposed to establish Theorem 5.2 and Corollary 5.1, so it is left to the reader to complete the details.

As a consequence of Theorem 5.5 we obtain the following relation between the differential equation (5.1) and integral equation (5.15).

Corollary 5.5. *If* u *is the solution of equation* (5.15) *obtained in Theorem* 5.6, *then for all pair* (p,q) *which verifies condition* (5.18) *of Corollary* 5.1, *we have*

$$\partial_t u \in L^q([-T,T]: L^p(\mathbb{R}^n)).$$

Moreover, u *is the (unique) solution of the differential equation* (5.1) *in the time interval* $[-T,T]$.

Proof. Using Theorem 3.3 and hypothesis (5.47) on the nonlinearity, it is easy to see that $u \in C([-T,T]: H^2)$ implies that $|u|^{\alpha-1}u \in C([-T,T]: L^2)$. Combining Theorem 5.6, which guarantees $\Delta u \in C([-T,T]: L^2)$ with the previous results and the integral equation (5.15), we see that $\partial_t u \in C([-T,T]: L^2)$, and that the differential equation in (5.1) is realized in the space $C([-T,T]: L^2)$.

The end of the proof is left as an exercise to the reader. □

In the next chapter, we will use the identities (5.2) and (5.3) to establish global solutions. To justify them, we present the following result in H^2.

Theorem 5.7.

1. *Let* $u \in C([-T,T]: L^2(\mathbb{R}^n)) \cap L^q([-T,T]: L^p(\mathbb{R}^n))$ *be the solution of integral equation* (5.15) *obtained in Section 5.1. If* $u_0 \in H^1(\mathbb{R}^n)$, *then*

$$u \in C([-T,T]: H^1(\mathbb{R}^n)) \cap L^q([-T,T]: L^p_1(\mathbb{R}^n)). \tag{5.49}$$

2. *Let* $u \in C([-T,T]: H^1) \cap L^q([-T,T]: L^p_1)$ *be the solution of the integral equation* (5.15) *obtained in Section 5.2. If* $u_0 \in H^2(\mathbb{R}^n)$ *and* $\alpha \geq 2$, *then* $u \in C([-T,T]: H^2)$ *and satisfies the differential equation* (5.1) *and estimates* (5.2) *and* (5.3).

Proof. We prove only part 1 of the theorem. Given $u_0 \in H^1(\mathbb{R}^n)$, we know by Theorem 5.4 that there exists $T' > 0$ such that $u \in C([-T',T']: H^1(\mathbb{R}^n))$. If $T' > T$ it is easy to see that the solution in L^2 can be extended to the interval $[-T',T']$. Thus, we assume that $T' < T$. To get the desired result, it is enough to prove that

$$\sup_{[0,T']} \|\nabla_x u(t)\|_2 \leq K\|u(0)\|_{1,2}$$

with K depending only on T and $M = \sup\{\|u(t)\|_2 : t \in [0,T]\}$.

Differentiate the integral equation (5.15) and use the notation $v_j = \partial_{x_j} u$, $j = 1,\dots,n$, to have that

$$v_j(t) = e^{it\Delta}v_j(0) + i\lambda\alpha \int_0^t e^{i(t-t')\Delta}(|u|^{\alpha-1}v_j)(t')\,dt', \tag{5.50}$$

which is a linear integral equation, because $u(\cdot)$ is known in the time interval $[0,T]$. With the same method used in the proof of Theorem 5.2, it is easy to see that this new integral equation (5.50) has unique solution on $[0,\Delta T]$, where ΔT depends on α, λ, n, and M, which remains constant in the interval $[0, T]$. Combining this result with an iterative argument, we obtain (5.49), which leads to the result. □

Now, we explain how to use Theorems 5.6 and 5.7 to justify the use of identities (5.2) and (5.3), respectively, in the proof of theorems (global).

Assume that $u_0 \in L^2(\mathbb{R}^n)$ and $\alpha \in (2, 1 + 4/n)$, we choose $\{u_0^k\}_{k=1}^{\infty}$ in $H^2(\mathbb{R}^n)$ such that $\|u_0^k - u_0\|_2 = o(1)$ when $k \to \infty$. Combining Theorems 5.2, (5.6), and (5.7), we see that for all $T > 0$ there exist $u^k \in C([-T, T]: H^2(\mathbb{R}^n))$, $k = 1, \ldots,$ a solution of (5.1) and (5.15) with initial data u_0^k. Since it satisfies the differential equation in (5.1), we infer that for all $t \in [-T, T]$,

$$\|u^k(t)\|_2 = \|u_0^k\|_2,$$

i.e., identity (5.2). From Theorem 5.2 (continuous dependence on the initial data), we have that $\sup_{[-T',T']} \|u^k(t) - u(t)\|_2 = o(1)$ when $k \to \infty$, where $T' < T$. Thus,

$$\|u(t)\|_2 = \|u_0\|_2 \qquad \text{for all } t \in [-T', T']. \tag{5.51}$$

This identity allows us to reapply Theorem 5.2 and extend the solution to the interval $[-(T' + \Delta T'),\ T' + \Delta T']$, where (using the same argument) identity (5.51) still holds. By successive applications of this step, we obtain the desired result (identity (5.42) in any time interval).

Finally, the case $\alpha \in (1, 2)$ requires some changes: For initial data $u_0^k \in H^2(\mathbb{R}^n)$ we will have the nonlinear term $\rho_k * (|\rho_k * u|^{\alpha-1} \rho_k * u)$, where $\rho_k(\cdot) = k^n \rho(\cdot\, / k)$, with $\rho(\cdot)$ an approximation of the identity. In this case it will be necessary to prove the stability of the solution in L^2 with respect to initial data and the nonlinear term.

As we remarked at the end of Theorem 5.2 all the previous existence proofs are based on the contraction principle. This approach has the advantage that it also shows that for smooth nonlinearity the map data-solution is smooth.

This general fact follows from the implicit function theorem. However, to simplify the exposition we will sketch the details in the case of Theorem 5.2.

Corollary 5.6. *Assume the same hypotheses of Theorem 5.2. Suppose $F(u, \bar{u}) = i\lambda|u|^{\alpha-1}u$ is smooth (i.e., $\alpha - 1$ is an even integer). Then there exists a neighborhood $\widetilde{V}$ of $u_0 \in L^2(\mathbb{R}^n)$ such that the map $\mathbb{F} : u_0 \mapsto u(t)$ from $\widetilde{V}$ into $E(T, a)$ is smooth.*

Proof. Define for $F(u, \bar{u}) = i\lambda|u|^{\alpha-1}u$

$$\begin{aligned} H : V \times E(T,a) &\mapsto E(T,a) \\ (v_0, v(t)) &\mapsto v(t) - \Phi_{v_0}(v)(t) \\ &= v(t) - \Big(e^{it\Delta} v_0 + \int_0^t e^{i(t-t')\Delta}\, F(v, \bar{v})(t')\, dt'\Big). \end{aligned}$$

Thus, H is smooth, $H(u_0,\ u(t)) = 0$, and

$$D_v H(u_0, u(t))v(t) = v(t) + \int_0^t e^{i(t-t')\Delta}\, [\partial_v F(u, \bar{u})\, v + \partial_{\bar{v}} F(u, \bar{u})\bar{v}](t')\, dt'.$$

Hence,

$$D_v H(u_0, u(t)) = I + L.$$

From the proof of Theorem 5.2 it is easy to see that

$$|||Lv||| \leq c|\lambda|T^{\theta}a^{\alpha-1} < 1$$

for any choice of a in (5.24). Then,

$$D_u H(u_0, u(t)) : E(T,a) \to E(T,a)$$

is invertible, i.e., one-to-one and onto. Thus, by the implicit function theorem there exists $h : \widetilde{V} \to E(T,a)$ smooth ($\widetilde{V} \subset V$ neighborhood of $u_0 \in L^2(\mathbb{R}^n)$) such that

$$H(v_0, h(v_0)) = 0, \quad \forall v_0 \in \widetilde{V},$$

so,

$$h(v_0) = e^{it\Delta}v_0 + \int_0^t e^{i(t-t')\Delta} F(h(v_0), \overline{h(v_0)})(t')\,dt'$$

is a solution of (5.15) with data v_0 (instead of u_0). □

Remark 5.7. The same argument shows that if $F(u,\bar{u}) = i\lambda|u|^{\alpha-1}u$ is $C^{[\alpha]}$ (when $\alpha - 1$ is not an even integer), then the map $\mathbb{F} : u_0 \mapsto u(t)$ from $\widetilde{V}$ into $E(T,a)$ is $C^{[\alpha]}$.

5.4 Comments

The L^2 theory exposed on Section 5.1 was obtained by Y. Tsutsumi [T1] in the case $\alpha \in (1, 1+4/n)$. The critical case L^2 $(\alpha = 1+4/n)$ was established by Cazenave and Weissler [CzW3]. The results of Section 5.2 were taken from references [CzW2], [GV1], [K1], and [T2]. Finally, the H^2 theory can be found in [K2].

It is important to note that Theorems 5.2, 5.4, and 5.6 prove that under some conditions on the power of the nonlinearity α, the solutions of the integral equation possess, at least locally in time, the same smoothing properties as the Strichartz type (discussed in Section 4.2, Theorem 4.2) that the solution of the associated linear problem.

From the proof of Theorem 5.3, one sees that the conditions on the data u_0 in the existence results can be significantly weaker. To simplify the exposition, let us concentrate on the results in Theorem 5.3: Instead of $u_0 \in L^2(\mathbb{R}^n)$, one can take $u_0 \in \mathcal{S}'(\mathbb{R}^n)$ such that

$$\|e^{it\Delta}u_0\|_{L^{\sigma}(\mathbb{R}^n_x \times \mathbb{R}_t)} < \infty, \quad \sigma = 2 + \frac{4}{n} \tag{5.52}$$

to get the same local result, or

$$\|e^{it\Delta}u_0\|_{L^\sigma(\mathbb{R}^n_x\times\mathbb{R}_t)} \ll 1 \tag{5.53}$$

to obtain a global one in the function space $u(\cdot)$ with

$$u - e^{it\Delta}u_0 \in C([0,T] : L^2(\mathbb{R}^n)) \text{ and } u \in L^\sigma([0,T] : L^\sigma(\mathbb{R}^n)). \tag{5.54}$$

Several methods to construct $u_0 \in \mathcal{S}'(\mathbb{R}^n)$ such that (5.52) or (5.53) are satisfied (or simply, $u_0 \in L^2(\mathbb{R}^n)$ with $\|u_0\|_2 \gg 1$ such that (5.53) holds) have been developed.

Let us consider first the last problem. Without loosing generality, assume the 1-D case. We will use Examples 4.1–4.3 Chapter 4 to obtain $u_0 \in L^2(\mathbb{R})$ with $\|u_0\|_2 \gg 1$ such that (5.53) holds.

Let $\varphi \in C_0^\infty(\mathbb{R})$ with $\operatorname{supp}\varphi \subseteq B_1(0)$ and $\|\varphi\|_2 = 1$. Let $N \in \mathbb{Z}^+$ and define

$$u_0^N(x) \equiv \sum_{j=1}^N \varphi(x - \nu_j)e^{2\pi i\mu_j x} = \sum_{j=1}^N \varphi_j(x), \tag{5.55}$$

where $\nu_1, \dots, \nu_N$ and $\mu_1, \dots, \mu_N$ are numbers chosen such that for $t > 0$ the "cones" containing most of the mass of $u_j(x,t) = e^{it\Delta}\varphi_j$, i.e., for $t_0 \gg 1$ fixed

$$c_j = \left\{(x,t) : \frac{(2N-1)t_0 + 1}{t_0}t - 1 \le x \le \frac{(2N+1)t_0 - 1}{t_0}t + 1,\ 0 \le t \le t_0\right\},$$

do not overlap. Thus,

$$\|u_0^N\|_2 = \sqrt{N} \tag{5.56}$$

and using that outside c_j, $u_j(x,t)$ decays exponentially (for a t fixed and $|x| \to \infty$), one can show (for a similar computation see [Vi2]):

$$\begin{aligned}\|e^{it\Delta}u_0^N\|_{L^6(\mathbb{R}\times\mathbb{R}^+)} &= \Big\|\sum_{j=1}^N e^{it\Delta}\varphi_j\Big\|_{L^6(\mathbb{R}\times\mathbb{R}^+)} \\ &\simeq \left(\sum_{j=1}^N \int_0^\infty\int_{-\infty}^\infty |e^{it\Delta}\varphi_j(x)|^6\,dxdt\right)^{1/6} \simeq N^{1/6}\end{aligned} \tag{5.57}$$

(since $\|e^{it\Delta}\varphi_j\|_{L^6}^6 \le \|\varphi_j\|_{L^2}^6 = 1$). So by taking $v_0^N = u_0^N/\sqrt{N}$, we get a sequence of data with $\|v_0^N\|_{L^2} = 1$ and

$$\|e^{it\Delta}v_0\|_{L^6(\mathbb{R}\times\mathbb{R}^+)} \le cN^{-1/3} \ll 1 \quad \text{for } N \text{ large}.$$

For the same problem, Bourgain [Bo3] introduced the following norm (two-dimensional case, $n = 2$)

$$|||u_0|||_{X_p} = \left(\sum_{j=1}^\infty\sum_{k=1}^\infty 2^{-4j}\left(\frac{1}{2^{-2j}}\int_{Q_k^j}|u_0(x)|^p\,dx\right)^{4/p}\right)^{1/4}, \tag{5.58}$$

where $\{Q_k^j\}_{k\in\mathbb{Z}^+}$ denotes a grid of squares with disjoint interior of side 2^{-j} parallel to the axes.

First one notices that the norm $\|\!|\cdot|\!\|_{X_p}$ scales like the $L^2(\mathbb{R}^2)$-norm, i.e., $\|\!| f_\lambda |\!\|_{X_p}$ with $f_\lambda(x) = \lambda f(\lambda x)$ is independent of λ (see Exercise 5.5). In [MVV1], [MVV2] Moyua, Vargas and Vega (improving and extending results in [Bo3]) showed that

$$\|e^{it\Delta}u_0\|_{L^4(\mathbb{R}^2_x\times\mathbb{R}_t)} \leq c\|\!|u_0|\!\|_{X_p} \tag{5.59}$$

for $12/7 \leq p \leq 2$ for any $u_0 \in L^1_{loc}(\mathbb{R}^2)$, and for $4(\sqrt{2}-1) \leq p < 2$ if u_0 is the characteristic function of a measurable set. Moreover, they showed that $p > 4(\sqrt{2}-1)$ is sharp.

Using (5.58) and (5.59), one can find $u_0 \in L^1_{loc}(\mathbb{R}^2) \setminus L^2(\mathbb{R}^2)$ such that (5.53) holds.

Let

$$u_{0j}(x,y) = \chi_{\{[0,2^{-j}]\times[0,2^j]\}}(x,y), \quad j \in \mathbb{Z}^+. \tag{5.60}$$

It is not hard to see that $\|\!|u_{0j}|\!\|_{X_p} \leq 2^{-j/4}$ (Exercise 5.7) while $\|u_{0j}\|_2 \equiv 1$. Then taking

$$u_0(x,y) = \epsilon\sum_{j=1}^{\infty} u_{0j}((x,y)-(j,0)), \quad \epsilon > 0 \tag{5.61}$$

it follows that $u_0 \notin L^2(\mathbb{R}^2)$ and

$$\|e^{it\Delta}u_0\|_{L^4(\mathbb{R}^2\times\mathbb{R})} \leq \|\!|u_{0j}|\!\|_{X_p} \leq c\epsilon.$$

It is not difficult to show that solutions of (5.15) also enjoy the local regularity property described in Section 4.3. For instance, we see that the solution $u(\cdot)$ of (5.15) obtained in Theorem 5.2 satisfies

$$u \in L^2([-T,T]: H^{1/2}_{loc}(\mathbb{R}^n)). \tag{5.62}$$

In fact, writing the equation (5.15) in the form:

$$u(t) = e^{it\Delta}\Big(u_0 + \int_0^t e^{-it'\Delta}(|u|^{\alpha-1}u)(t')\,dt'\Big)$$

and using (4.25) (or (4.23) when $n = 1$) and (4.16) we have that

$$\begin{aligned}\left(\int_{\{|x|\leq R\}}\int_{-T}^{T}|D_x^{1/2}u(x,t)|^2\,dtdx\right)^{1/2} &\leq c\,R\left(\|u_0\|_2 + \sup_{[-T,T]}\Big\|\int_0^t e^{-it'\Delta}(|u|^{\alpha-1}u)(t')\,dt'\Big\|_2\right)\\ &\leq c\,R\left(\|u_0\|_2 + \left(\int_0^T \|\,|u|^{\alpha}(t)\|^{r'}_{(\alpha+1)/\alpha}\,dt\right)^{1/r'}\right),\end{aligned}$$

where $r = 4(\alpha+1)/n(\alpha-1)$. Combining (5.21) and (5.22) with Corollary 5.1, we obtain (5.62).

As we have seen along this chapter, the results concerning local existence are a consequence of the estimates obtained in Theorem 4.2. Thus, the method of proof applied can be extended to any group satisfying Theorem 4.2 (even locally). In particular, we obtain the same local theorems for the nonlinear Schrödinger (NLS) equation with real potential

$$\partial_t u = i\Delta u + V(x)u + \lambda |u|^{\alpha-1}u,$$

under appropriate conditions on V (see the references [C], [Y]).

Theorems 5.6 and 5.7 are concerned with the regularity of solution measured in Sobolev spaces. One can also ask whether the decay properties of the data are preserved by the solution. To simplify the matter, consider the case where α is an odd integer (or, where the nonlinearity has the form $f(|u|^2)u$ with $f(\cdot)$ smooth). In [HNT1], [HNT2], [HNT3], Hayashi, Nakamitsu, and Tsutsumi showed that if $u_0 \in H^m(\mathbb{R}^n) \cap L^2(|x|^k\,dx)$ with $m \geq k$, then there exists $T = T(\|u_0\|_{H^l})$, $l = \min\{m; n/2^+\}$ such that the IVP (5.1) has a unique solution

$$u \in C([0,T] : H^m(\mathbb{R}^n) \cap L^2(|x|^k\,dx)) \cap L^q([0,T] : L^p_k(\mathbb{R}^n) \cap L^p(|x|^k\,dx))$$

with p, q as in Theorem 4.2 and where $L^p_k(\mathbb{R}^n)$ is defined as in (3.38).

In the case $k \geq m$ they showed that the solution u does not belong to $L^2(|x|^k\,dx)$ but possesses a further regularity property, roughly speaking $\partial_x^\alpha u(\cdot,t) \in L^2_{\text{loc}}(\mathbb{R}^n)$, $t \neq 0$, for $|\alpha| \leq k$, (see [HNT1], [HNT2]).

In particular, one has that if $u_0 \in \mathcal{S}(\mathbb{R}^n)$, then the solution $u(\cdot)$ of the IVP (5.1) (with α an odd integer) belongs to $C([0,T] : \mathcal{S}(\mathbb{R}^n))$, and that if $u_0 \in H^1(\mathbb{R}^n)$ with compact support, α an odd integer, and $1 + 4/n < \alpha < 1 + 4/(n-2)$, then $u \in C^\infty(\mathbb{R}^n \times \mathbb{R} - \{0\})$.

The proofs given in [HNT1]–[HNT3] are based on the properties of the operators $\Gamma_j = x_j + 2it\partial_{x_j}$, $j = 1, \dots, n$, deduced there.

In particular, using that for $\Gamma = (\Gamma_1, \dots, \Gamma_n)$,

$$\Gamma^\alpha u = e^{i|x|^2/4t}(2it)^{|\alpha|}\partial_x^\alpha(e^{-i|x|^2/4t}u) \text{ for } \alpha \in \mathbb{Z}^+ \tag{5.63}$$

and

$$x^\alpha e^{it\Delta}u_0 = e^{it\Delta}\,\Gamma^\alpha u_0 \tag{5.64}$$

(see Exercise 4.4), they developed a calculus of inequalities for the operators Γ_j similar to that in (3.15) for the operators ∂_{x_j}. For instance, for $n = 1$ they showed that

$$\|\Gamma^m(|v|^{2\alpha}v)(t)\|_{L^2} \leq c_m \|v(t)\|^{2\alpha}_{L^\infty} \|\Gamma^m v(t)\|_{L^2}$$

and

$$\|v(t)\|_{L^\infty} \leq t^{-1/2}\,\|\Gamma v(t)\|^{1/2}_{L^2}\|v(t)\|^{1/2}_{L^2}$$

(compare with (3.14), (3.15), and (3.16) in Chapter 3) which have been essential tools in the study of the asymptotic behavior of solution of (5.1). The extension of these weighted results to $L^2(|x|^k\,dx)$ with $k \geq 0$ (not necessarily an integer) was obtained in [NhPo1].

To simplify the exposition, we have presented local well-posedness results in Sobolev spaces with integer indexes, i.e., $H^s(\mathbb{R}^n)$, $s = 0, 1, 2$. Concerning the local existence theory in fractional Sobolev spaces, $H^s(\mathbb{R}^n)$, $s \geq 0$, we have the following result due to Cazenave and Weissler [CzW4].

Theorem 5.8. *Let* $1 + 4/n \leq \alpha < \infty$ *and* $s > s_\alpha = n/2 - 2/(\alpha - 1)$, *with* $[s] < \alpha - 1$ *if* $\alpha - 1$ *is not an even integer. Given* $v_0 \in H^s(\mathbb{R}^n)$, *there exist* $T = T(\|v_0\|_{s,2}; s) > 0$ *and a unique strong solution* $v(\cdot)$ *of the IVP* (5.1) *satisfying*

$$v \in C([-T, T] : H^s(\mathbb{R}^n)) \cap W^T_{s,n}. \tag{5.65}$$

Moreover, given $T' \in (0, T)$ *there exist a constant* $r = r(\|v_0\|_{s,2}; s; T') > 0$ *and a continuous, nondecreasing function* $G(\cdot) = G(\|v_0\|_{s,2})$ *with* $G(0) = 0$ *such that*

$$\sup_{[0,T']} \|(v - \widetilde{v})(t)\|_{s,2} \leq G(\|v_0\|_{s,2})\|v_0 - \widetilde{v}_0\|_{s,2} \tag{5.66}$$

for any $\widetilde{v}_0 \in H^s(\mathbb{R}^n)$ *with* $\|v_0 - \widetilde{v}_0\|_{s,2} < r$, *i.e., the map data-solution is locally Lipschitz.*

The space $W^T_{s,n}$ in (5.65) is related to the Strichartz estimates, and its precise definition will not be needed in the discussion below. We recall that for $1 < \alpha < 1 + 4/n$ the problem is locally well-posed in $L^2(\mathbb{R}^n)$.

From the scaling argument, i.e., if $u(x,t)$ is a solution of the IVP (5.1), then

$$u_\mu(x,t) = \mu^{2/(\alpha-1)} u(\mu x, \mu^2 t), \qquad \mu > 0, \tag{5.67}$$

is also a solution with data $u_\mu(x,0) = \mu^{2/(\alpha-1)} u_0(\mu x)$, for which one has that

$$\|D^s_x u_\mu(\cdot, 0)\|_2 = c\mu^{2/(\alpha-1)} \mu^{s-n/2} \|u_0\|_2.$$

To have results invariant by rescaling, one needs to consider data $u_0 \in \dot{H}^s(\mathbb{R}^n)$ $(= (-\Delta)^{-s/2} L^2(R^n))$, with $s(\alpha) = s_c = n/2 - 2/(\alpha - 1)$ which is called the *critical case*. The case $s > s_c = n/2 - 2/(\alpha - 1)$ is called *subcritical case*. Notice that Theorem 5.8 above corresponds to the subcritical case and Theorem 5.3 to the critical case in $L^2(\mathbb{R}^n)$ $(s = 0)$.

So the following question arises. Are the results in Theorem 5.8 optimal? This seems to be the case. First, let us consider the "focusing case," i.e., for $\lambda > 0$ in (5.1), the following result was obtained in [BKPSV].

Theorem 5.9. *If* $4/n + 1 \leq \alpha < \infty$, *then the IVP* (5.1) *with* $\lambda > 0$ *is ill-posed in* $H^{s_c}(\mathbb{R}^n)$ *with* $s_c = n/2 - 2/(\alpha - 1)$, *in the sense that the time of existence T and the continuous dependence cannot be expressed in terms of the size of the data in the* H^{s_c}*-norm. More precisely, there exists* $c_0 > 0$ *such that for any* δ, $t > 0$ *small there exist data* u_1, $u_2 \in \mathcal{S}(\mathbb{R}^n)$ *such that*

$$\|u_1\|_{s,2} + \|u_2\|_{s,2} \leq c_0, \quad \|u_1 - u_2\|_{s,2} \leq \delta, \quad \|u_1(t) - u_2(t)\|_{s,2} > c_0/2,$$

where $u_j(\cdot)$ *denotes the solution of the IVP* (5.1) *with data* u_j, $j = 1, 2$.

Proof. For simplicity, we shall only consider the case $0 < s_c < 1$ and fix $\lambda = 1$. We consider the one-parameter family of ground states:

$$v_\mu(x,t) = e^{i\mu t}\varphi_\mu(x) = e^{i\mu t}\mu^{1/(\alpha-1)}\varphi(\sqrt{\mu}x),$$

where the function $\varphi(\cdot) = \varphi_1(\cdot)$ solves the nonlinear elliptic eigenvalue problem (5.8) with $0 < \alpha < 4/(n-2)$, if $n > 2$. The idea is to estimate

$$\|D_x^{s_c}(v_{\mu_1} - v_{\mu_2})(t)\|_2^2$$

and

$$\|D_x^{s_c}(\mu_1^{1/(\alpha-1)}\varphi_1(\sqrt{\mu}_1\cdot) - \mu_2^{1/(\alpha-1)}\varphi_1(\sqrt{\mu}_2\cdot))\|_2^2.$$

Choosing $\mu_1 = (N+1)^2$ and $\mu_2 = N^2$ so that $\mu_1 - \mu_2 > 2N$, we have that

$$\begin{aligned}
&\|D_x^{s_c}(v_{\mu_1} - v_{\mu_2})(t)\|_2^2\\
&= \|D_x^{s_c}v_{\mu_1}(t)\|_2^2 + \|D_x^{s_c}v_{\mu_2}(t)\|_2^2 - 2\mathcal{R}e\left\{e^{it(\mu_1-\mu_2)}\langle v_{\mu_1}(t), v_{\mu_2}(t)\rangle_{s_\alpha}\right\}\\
&= \Psi(\mu_1,\mu_2)(t).
\end{aligned}$$

Given any $T > 0$ there exist $N > c(T)$ and $t \in (0,T)$ such that

$$\mathcal{R}e\ \{e^{it(\mu_1-\mu_2)}\langle v_{\mu_1}(t), v_{\mu_2}(t)\rangle_{s_c}\} = 0,$$

hence,

$$\sup_{[0,T]} \Psi(\mu_1,\mu_2)(t) = 2\|D_x^{s_c}\varphi_1\|_2^2.$$

On the other hand,

$$\begin{aligned}
\lim_{N\to\infty} \|D_x^{s_c}(v_{\mu_1} - v_{\mu_2})(0)\|_2^2 &= \|D_x^{s_c}v_{\mu_1}(0)\|_2^2 + \|D_x^{s_c}v_{\mu_2}(0)\|_2^2\\
&\quad - 2\mathcal{R}e\ \{\langle v_{\mu_1}, v_{\mu_2}\rangle_{s_c}\} = 0
\end{aligned}$$

by using that $\mu_1/\mu_2 \to 1$ as $N \to \infty$ and so

$$\lim_{N\to\infty} \mathcal{R}e\ \{\langle v_{\mu_1}, v_{\mu_2}\rangle_{s_c}\} = \|D_x^{s_c}\varphi_1\|_2^2.$$

Therefore, for any $T > 0$

$$\lim_{N\to\infty}\sup_{[0,T]} \|D_x^{s_c}(v_{\mu_1} - v_{\mu_2})(t)\|_2 = \sqrt{2}\|D_x^{s_c}\varphi_1\|_2,$$

while

$$\lim_{N\to\infty} \|D_x^{s_c}(\mu_1^{1/(\alpha-1)}\varphi(\sqrt{\mu_1}\cdot) - \mu_2^{1/(\alpha-1)}\varphi(\sqrt{\mu}_2\cdot))\|_2 = 0,$$

which essentially proves the result. □

Christ, Colliander and Tao [CrCT1] have shown that the results in Theorem 5.9 extend to the defocusing case $\lambda < 0$. Moreover, the following stronger ill-posedness result in *norm inflation* concerning the IVP (5.1) in both the focusing and defocusing cases was established in [CrCT3]:

Theorem 5.10. *Given $s \in (0, s_c)$ $\exists\{u_0^m : m \in \mathbb{Z}^+\} \subset \mathcal{S}(\mathbb{R}^n)$ and $\{t_m : t_m > 0\}$ with $\|u_0^m\|_{s,2} \to 0$, $t_m \to 0$ as $m \uparrow \infty$ such that the corresponding solution u^m of the IVP* (5.1) *with $\lambda \neq 0$, and initial data $u^m(x,0) = u_0^m(x)$ satisfies that*

$$\|u_m(\cdot, t_m)\|_{s,2} \to \infty, \quad as \quad m \uparrow \infty. \tag{5.68}$$

In the case $\alpha \geq 3$, this result has been strengthened in [AlCa] by showing:

Theorem 5.11. *Given $\alpha \geq 3$ and $s \in (0, s_c)$ there exist $\{u_0^m : m \in \mathbb{Z}^+\} \subset \mathcal{S}(\mathbb{R}^n)$ and $\{t_m : t_m > 0\}$ with $\|u_0^m\|_{s,2} \to 0$, $t_m \to 0$ as $m \uparrow \infty$ such that the corresponding solution u^m of the IVP* (5.1)*, in the defocusing case $\lambda < 0$, with initial data $u^m(x,0) = u_0^m(x)$, satisfies that*

$$\|u_m(\cdot, t_m)\|_{l,2} \to \infty, \quad as \quad m \uparrow \infty, \forall l \in \left(\frac{2s}{2 + (\alpha - 1)(s_c - s)}, s\right). \tag{5.69}$$

All the existence results for the IVP (5.1) discussed so far are restricted to Sobolev spaces with nonnegative index, i.e., in $H^s(\mathbb{R}^n)$, $s \geq 0$, even in the cases when the scaling argument tells us that the critical value is negative, that is, $s(\alpha) = s_c = n/2 - 2/(\alpha - 1) < 0$. Thus, for example, we can ask whether for the IVP for the cubic 1-D Schrödinger equation:

$$\begin{cases} i\partial_t v + \partial_x^2 v + \lambda |v|^2 v = 0, \\ v(x,0) = v_0(x), \end{cases} \tag{5.70}$$

$t \in \mathbb{R}$, $x \in \mathbb{R}$, $\lambda \in \mathbb{R}$, for which $s_c = 1/2 - 2/(\alpha - 1) = -1/2$, one can obtain a local existence result in $H^s(\mathbb{R})$, with $s < 0$ (we recall that Theorem 5.2 provides the result in $H^s(\mathbb{R})$, with $s \geq 0$). In this regard, we have the following result found in [KPV4].

Theorem 5.12. *If $s \in (-1/2, 0)$, then the mapping data-solution $u_0 \mapsto u(t)$, where $u(t)$ solves the IVP* (5.70) *with $\lambda > 0$ (focusing case), is not uniformly continuous.*

In [VV] and [Gr3], Vargas and Vega, and Grünrock found spaces which scale is below the one from L^2 but above that of $\dot{H}^{-1/2}(\mathbb{R})$, i.e., spaces whose norm is invariant by $\lambda^\theta\, u_0(\lambda x)$ with $\theta \in (-1/2, 0)$, for which the IVP (5.70) is locally and globally well-posed.

Remark 5.8. The result in Theorem 5.12 can be extended to higher dimensions. More precisely, it applies to the IVP

$$\begin{cases} i\partial_t u + \Delta u + |u|^{\rho-1} u = 0, \\ u(x,0) = u_0(x), \end{cases} \tag{5.71}$$

$t \in \mathbb{R}$, $x \in \mathbb{R}^n$, with $u_0 \in H^s(\mathbb{R}^n)$, $n < 4/(\rho - 1)$, and $s \in (n/2 - 2/(\rho - 1), 0)$.

For the IVP (5.70), on one hand, one has that the map data solution fails to be continuous, i.e., there exist data $u_0 \in \mathcal{S}(\mathbb{R})$ with arbitrary small H^s-norm for $s \leq -1/2$, whose corresponding solution $u(t)$ provided by Theorem 5.2 has arbitrary large H^s-norm at an arbitrary small time (see (5.68)). On the other hand, Theorem 5.12 for the focusing case and the results in [CrCT1] for the defocusing case, shows that the map data-solution is not uniformly continuous in H^s for $s < 0$.

In [KTa3], Koch and Tataru obtained the following a priori estimates for solutions of the IVP (5.70), improving a previous result found in [CrCT2] and [KTa2].

Theorem 5.13 ([KTa3]). *Let $u \in C(\mathbb{R}^+ : L^2(\mathbb{R})) \cap L^4(\mathbb{R} \times [0, T])$ for all $T > 0$ be the global solution of the IVP* (5.70) *(see Theorem* 5.2*). Then for all $T > 0$ there exists $\alpha(T) > 0$ such that*

$$\sup_{t\in[0,T]} \|u(t)\|_{H^{-1/4}_{\alpha(T)}} \leq 1,$$

where

$$\|f\|^2_{H^{-1/4}_{\alpha(T)}} = \int_{-\infty}^{\infty} \frac{|\widehat{f}(\xi)|^2}{(\alpha + \xi^2)^{1/4}}\, d\xi.$$

This a priori estimate allows one to establish the existence of an appropriate class of global weak solution of (5.70) (see [CrCT2] and [KTa3]).

Proof of Theorem 5.12 As in the previous proof, consider the one-parameter family of standing wave solutions (with $n = 1$ in this case)

$$v_\omega(x, t) = e^{it\omega^2} \varphi_\omega(x),$$

where $\varphi_\omega(x) = \omega\varphi(\omega x)$ and $\varphi(x) = \varphi_1(x)$ solves the nonlinear equation in (5.8) with $\omega = 1$. Using the Galilean invariance (5.11), we obtain the two-parameter family of solutions:

$$u_{N,\omega}(x, t) = e^{-itN^2+iNx}\, v_\omega(x - 2tN, t) = e^{-it(N^2-\omega^2)} e^{iNx}\, \varphi_\omega(x - 2tN).$$

We fix s such that $s \in (-1/2, 0)$ and take $\omega = N^{-2s}$ and $N_1, N_2 \simeq N$.

First, we calculate

$$\|u_{N_1,\omega}(0) - u_{N_2,\omega}(0)\|^2_{s,2}\,.$$

Observing that $\widehat{\varphi}_\omega(\xi) = \widehat{\varphi}(\xi/\omega)$ so that $\widehat{\varphi}_\omega(\cdot)$ concentrates in $B_\omega(0)= \{\xi \in \mathbb{R} : |\xi| < \omega\}$. From the choice of ω and $s > -1/2$, if $\xi \in B_\omega(\pm N)$, then $|\xi| \simeq N$. Then, a straight calculation yields

$$\|u_{N_1,\omega}(0) - u_{N_2,\omega}(0)\|^2_{s,2}$$
$$\leq cN^{2s} \frac{|N_1 - N_2|}{\omega^2} \left(\int_{\eta+N_2}^{\eta+N_1} d\xi\right) \int_{-\infty}^{\infty} |\widehat{\varphi}'_\omega(\eta)|^2 d\eta$$

$$\leq cN^{2s}(N_1 - N_2)^2 \frac{1}{\omega^2}\,\omega = c(N^{2s}(N_1 - N_2))^2,$$

and that

$$\|u_{N_j,\omega}(0)\|_{s,2}^2 \simeq cN^{2s}\omega = c, \quad j = 1, 2.$$

Now, we consider the solutions $u_{N_1,\omega}(t), u_{N_2,\omega}(t)$ at time $t = T$, and compute

$$\|u_{N_1,\omega}(T) - u_{N_2,\omega}(T)\|_{s,2}.$$

Note first that

$$\|u_{N_j,\omega}(T)\|_{s,2}^2 \simeq c, \quad j = 1, 2.$$

In fact,

$$\|u_{N_j,\omega}(T)\|_{s,2}^2 = \|u_{N_j,\omega}(0)\|_{s,2}^2 \simeq c, \quad j = 1, 2.$$

Note that the frequencies of both $u_{N_j,\omega}(T)$, $j = 1, 2$, are localized around $|\xi| \simeq N$, and hence,

$$\|u_{N_1,\omega}(T) - u_{N_2,\omega}(T)\|_{s,2}^2 \simeq N^{2s}\|u_{N_1,\omega}(T) - u_{N_2,\omega}(T)\|_2^2. \tag{5.72}$$

Next, we observe that

$$u_{N_j,\omega}(x, T) = e^{-i(TN_j^2 - N_j x - T\omega^2)}\omega\varphi(\omega(x - 2TN_j)), \quad j = 1, 2.$$

Thus, the support of $u_{N_j,\omega}(T)$ is concentrated in $B_{\omega^{-1}}(2TN_j)$, $j = 1, 2$. Therefore, if for T fixed, N_1, N_2 are chosen such that

$$T(N_1 - N_2) \gg \omega^{-1} = N^{2s},$$

then there is not interaction and

$$\|u_{N_1,\omega}(T) - u_{N_2,\omega}(T)\|_2^2 \simeq \|u_{N_1,\omega}(T)\|_2^2 + \|u_{N_2,\omega}(T)\|_2^2 \simeq \omega.$$

The above estimate combined with (5.72) yields

$$\|u_{N_1,\omega}(T) - u_{N_2,\omega}(T)\|_{s,2}^2 \geq cN^{2s}\omega = c. \tag{5.73}$$

Take now

$$N_1 = N \quad \text{and} \quad N_2 = N - \frac{\delta}{N^{2s}},$$

so that

$$\begin{cases} c(N^{2s}(N_1 - N_2))^2 = c\delta^2, \\ T(N_1 - N_2) = T\frac{\delta}{N^{2s}} \gg N^{2s}, \quad \text{i.e.,} \quad T \gg \frac{N^{4s}}{\delta}. \end{cases} \tag{5.74}$$

Since $s < 0$, given $\delta,\ T > 0$, we can choose N so large that (5.74) is valid, and from this we see that (5.73) violates the uniform continuity. □

Well-posedness for some particular cases of the IVP (5.71) has been studied in other spaces. In [Pl], the problem was considered in Besov spaces, in [CVV] in L^{p*} (L^p-weak spaces) and in [Gr3] in the spaces $\widehat{H^s_r}(\mathbb{R}^n)$ defined as:

$$f \in \widehat{H^s_r}(\mathbb{R}^n) \quad \text{if} \quad \|f\|_{\widehat{H^s_r}} = \|(1+|\xi|^2)^{s/2}\widehat{f}\|_{L^{r'}_\xi} < \infty. \tag{5.75}$$

The idea was to find larger spaces than $H^s(\mathbb{R}^n)$ or other ones which scale closer to the critical homogeneity given by the equation.

In [KPV12], the study of the IVP:

$$\begin{cases} i\partial_t u \pm \Delta u + N_k(u,\overline{u}) = 0, \\ u(x,0) = u_0(x), \end{cases} \tag{5.76}$$

$x \in \mathbb{R}^n, t \in \mathbb{R}$, where

$$N_k(z_1,z_2) = \sum_{a+b=k} C_k\, z_1^a\, z_2^b, \tag{5.77}$$

was first considered. Even though it may not have a physical interpretation in general, the main purpose of this study was motivated to test new local estimates based on $X_{s,b}$ spaces (see Definition 7.1) and variant of them and their relation with the geometry of the nonlinearity N_k.

We summarize next some local results obtained for the IVP (5.76). First, we will consider the 1-D situation. For the nonlinearity $N_2(u,\overline{u}) = u^2$, Bejenaru and Tao [BTo] obtained a sharp local well-posedness result in $H^s(\mathbb{R})$ for $s \geq -1$. In [Ki3], Kishimoto established a similar result for the quadratic nonlinearity $N_2(u,\ \overline{u}) = (\overline{u})^2$. In [KiT], Kishimoto and Tsugawa showed the local well-posedness for the case $N_2(u,\ \overline{u}) = u\overline{u} = |u|^2$ in $H^s(\mathbb{R})$ for $s > -1/2$. In each case, these results improve by one fourth the previous ones obtained in [KPV12]. Grünrock [Gr1] has shown that the IVP (5.76) is locally well-posed in $H^s(\mathbb{R})$ with $s > -5/12$ for $N_3(u,\overline{u}) = (\bar{u})^3$ and $N_3(u,\overline{u}) = u^3$ and with $s > -2/5$ for $N_3(u,\overline{u}) = u(\bar{u})^2$. Notice that all these nonlinearities have the same homogeneity, but only $N_3(u,\overline{u}) = |u|^2u$ is Galilean invariant. For higher powers in (5.77), the results known are due to Grünrock [Gr1]. He proved local well-posedness for the IVP (5.76) when the nonlinearity $N_4(u,\overline{u})$ has either of the following forms: $(\overline{u})^4$, u^4, $u^3\overline{u}$, and $\overline{u}^3u$ in $H^s(\mathbb{R})$, $s > -1/6$, and for $N_4(u,\overline{u}) = |u|^4$ in $H^s(\mathbb{R})$, $s > -1/8$.

In dimension $n = 2$, Bejenaru and De Silva [BeDS] showed local well-posedness for the IVP (5.76) in $H^s(\mathbb{R}^2)$, $s > -1$ when $N_2(u,\overline{u}) = u^2$ and a similar result was obtained by Kishimoto [Ki2] for $N_2(u,\overline{u}) = (\overline{u})^2$. These results improved by one fourth the previous ones found in [CDKS]. In the later work, local well-posedness for the nonlinearity $N_2(u,\overline{u}) = u\bar{u}$ was established in $H^s(\mathbb{R}^2)$, $s > -1/4$. In the three-dimensional case, Tao [To3] proved that the IVP (5.76) is locally well-posed in $H^s(\mathbb{R}^3)$, $s > -1/2$ for either $N_2(u,\overline{u}) = u^2$ or $N_2(u,\overline{u}) = \overline{u}^2$, and in $H^s(\mathbb{R}^3)$, $s > -1/4$ for the nonlinearity $N_2(u,\overline{u}) = u\overline{u}$.

Next, we deal with the existence and uniqueness question for the IVP associated to the cubic Schrödinger equation with the delta function as initial datum:

$$\begin{cases} i\partial_t u + \partial_x^2 u \pm |u|^2 u = 0, \\ u(x,0) = \delta(x). \end{cases} \tag{5.78}$$

$t > 0, x \in \mathbb{R}$.

Theorem 5.14 ([KPV5]). *Either there is no weak solution u for the IVP* (5.78) *in the class*

$$u,\ |u|^2 u \in L^\infty([0,\infty) : \mathcal{S}'(\mathbb{R})) \quad \textit{with} \quad \lim_{t\downarrow 0} u(\cdot,t) = \delta \tag{5.79}$$

or there is more than one.

Consider now the local and global well-posedness of the periodic problem:

$$\begin{cases} i\partial_t u = -\Delta u \pm |u|^{\alpha-1} u, \\ u(x,0) = u_0(x). \end{cases} \tag{5.80}$$

$x \in \mathbb{T}^n, t \in \mathbb{R},\ \alpha > 1$.

For $n = 1$, Bourgain [Bo1] established local well-posedness for (5.80) in $H^s(\mathbb{T})$, $s \in [0, 1/2)$ for $\alpha \in (1, 1 + 4/(1-2s))$. This combined with the conservation law $\|u(t)\|_{L^2} = \|u_0\|_{L^2}$ yields the corresponding global well-posedness result.

In the defocussing cubic NLS case ((+) in (5.80)) it was shown in [BGT2], [CrCT1] that the problem (5.80) is ill-posed (the map data-solution is not uniformly continuous) in $H^s(\mathbb{T})$, $s < 0$.

For $n = 3$, local well-posedness with $\alpha = 3$ was proved in [Bo1] for $u_0 \in H^s(\mathbb{T}^3)$, $s > 1/2$. For $n \geq 2$, local well-posedness was established in [Bo1] for $\alpha \in [3, 4/(n-2s))$ and $s > 3n/n + 4$.

The problem (5.80) in an n-dimensional nonflat compact manifold M^n has been studied by Burq, Gerard and Tzvetkov [BGT2], [BGT3]. Among other results, for the case of the two-dimensional sphere $\mathbb{S}^2$ they have shown that the IVP (5.80) in the cubic defocusing case (i.e., $\alpha = 3$ and positive sign in front of the nonlinearity) is locally well-posed in $H^s(\mathbb{S}^2)$ for $s > 1/4$ and ill-posed for $s < 1/4$.

The IVP problem (5.76) can also be considered in the periodic setting. We list next some results regarding the local well-posedness for this IVP in this situation. In the 1-D case, Bourgain [Bo1] established local well-posedness in $L^2(\mathbb{T})$ for any nonlinearity in (5.76) such that $a + b = k \leq 4$. Kenig, Ponce and Vega [KPV12] established the local well-posedness theory in $H^s(\mathbb{T})$, $s > -1/2$, for $N_2(u,\overline{u}) = u^2$, and for $N_2(u,\overline{u}) = \overline{u}^2$. Also, in the 1-D case Grünrock [Gr1] proved local well-posedness for $N_3(u,\overline{u}) = \overline{u}^3$ and $N_4(u,\overline{u}) = \overline{u}^4$ in $H^s(\mathbb{T})$, with $s > -1/3$ and $s > -1/6$, respectively. In the two-dimensional case, Grünrock [Gr1] showed local well-posedness in $H^s(\mathbb{T}^2)$, $s \geq -1/2$, for $N_2(u,\overline{u}) = \overline{u}^2$, and in dimension three that the IVP (5.76) is locally well-posed in $H^s(\mathbb{T}^3)$, $s \geq -3/10$, for $N_2(u,\overline{u}) = \overline{u}^2$. For further well-posedness results in the spaces

$$\mathcal{H}_{s,p}(\mathbb{T}) = \left(\sum_{n\in\mathbb{Z}} (1+n^2)^{ps/2} \, |\widehat{f}(n)|^p \right)^{1/p},$$

we refer to [Cr] and [Th].

5.5 Exercises

5.1 (i) Prove that if $u = u(x,t)$ satisfies

$$i\partial_t u = -\Delta u + |u|^{4/n} u \tag{5.81}$$

(u is the solution of the equation in (5.1) with $\lambda = 1$ and the critical power $\alpha = 4/n + 1$ in $L^2(\mathbb{R}^n)$) then:

$$\begin{aligned}
&u_1(x,t) = e^{i\theta} u(x,t),\\
&u_2(x,t) = u(x - x_0, t - t_0), \text{ with} x_0 \in \mathbb{R}^n,\ t_0 \in \mathbb{R} \text{ fixed},\\
&u_3(x,t) = u(Ax,t), \text{ with A any orthogonal matrix} n \times n,\\
&u_4(x,t) = u(x - 2x_0\, t, t)\, e^{i(x\cdot x_0 - |x_0|^2 t)}, \text{ with} x_0 \in \mathbb{R}^n \text{ fixed},\\
&u_5(x,t) = \mu^{n/2} u(\mu x, \mu^2 t),\ \mu \in \mathbb{R} \text{ fixed},\\
&u_6(x,t) = \frac{1}{(\alpha + \omega t)^{n/2}} \exp\left[\frac{i\omega |x|^2}{4(\alpha+\omega t)}\right] u\left(\frac{x}{\alpha+\omega t}, \frac{\gamma + \theta t}{\alpha + \omega t}\right),\\
&\qquad \alpha\theta - \omega\gamma = 1,\\
&u_7(x,t) = \overline{u(x,-t)},
\end{aligned}$$

also satisfy equation (5.81).

(ii) Prove that u_1, u_2, u_3, u_4, u_5 (with different powers in μ) and u_7 still satisfy the equation (5.81) for general nonlinearity $\pm\, |u|^{\alpha-1} u$ in (5.81).

5.2 Let $u \in H^1(\mathbb{R}^n)$ solve $-\Delta u + au = b|u|^\alpha u$, where $a > 0$ and $b \in \mathbb{R}$. Show that u satisfies

(i)

$$\int_{\mathbb{R}^n} |\nabla u|^2 \, dx + a \int_{\mathbb{R}^n} |u|^2 \, dx = b \int_{\mathbb{R}^n} |u|^{\alpha+2} \, dx. \tag{5.82}$$

(ii) Pohozaev's identity:

$$(n-2) \int_{\mathbb{R}^n} |\nabla u|^2 \, dx + na \int_{\mathbb{R}^n} |u|^2 \, dx = \frac{2nb}{\alpha+2} \int_{\mathbb{R}^n} |u|^{\alpha+2} \, dx. \tag{5.83}$$

5.3 Use Pohozaev's identity to show that a necessary condition to have solution in $H^1(\mathbb{R}^n)$ of problem (5.8) is that the nonlinearity satisfies $1 < \alpha < (n+2)/(n-2)$ $(1 < \alpha < \infty,\ n = 1, 2)$.

5.4 (i) Show that a formal scaling argument yields the estimate:

$$T = T(\|u_0\|_2) = c\,\|u_0\|_2^{-\beta}, \quad \beta = \frac{4(\alpha-1)}{4-n(\alpha-1)}, \tag{5.84}$$

for the life span of the L^2-local solution as a function of the size of the data given in Theorem 5.2.

(ii) Review the proof of Theorem 5.2 to obtain the estimate (5.84).

5.5 Consider the IVP for the 1-D NLS equation

$$\begin{cases} \partial_t u = i\,\partial_x^2 u \pm i\lambda\,|u|^{\alpha-1}u, \\ u(x,0) = u_0(x), \end{cases} \tag{5.85}$$

$\lambda \in \mathbb{R}$, $\alpha > 1$.

(i) Prove that if $\alpha \in (1,5)$ and $u_0 \in L^2(\mathbb{R})$, then the solution $u(\cdot,t)$ of the IVP (5.85) provided by Theorem 5.2 satisfies that

$$u(\cdot,t) \in C(\mathbb{R}) \quad \text{a.e.} \quad t \in [-T,T].$$

(ii) Can the result in (i) be extended to the case $\alpha = 5$?
Hint: Combine the idea of the proof of Exercise 4.9(ii) with Theorem 5.2, Corollary 5.1 and Theorem 5.3.

5.6 Let $f_\mu(x) = \mu f(\mu x)$. Show that $\|\!|f_\mu\|\!|_{X_p}$ is independent of μ, where $\|\!|\cdot\|\!|_{X_p}$ was defined in (5.58).

5.7 Let

$$u_{0j}(x,y) = \chi_{\{[0,1/2^j]\times[0,2^j]\}}(x,y) \quad j \in \mathbb{Z}^+.$$

Prove that

$$\|\!|u_{0j}\|\!|_{X_p} \le 2^{-j/4}.$$

5.8 Show that

$$u(x,t) = e^{it}\left\{1 - \frac{4(1+2it)}{1+2x^2+4t^2}\right\}$$

solves the IVP associated to

$$i\partial_t u + \partial_x^2 u + |u|^2 u = 0,$$

with datum

$$u(x,0) = 1 - \frac{4}{1+2x^2}.$$

5.9 (i) Prove that for any $\omega > 0$ the function:

$$u_\omega(x,t) = e^{it\omega}\omega^{1/(\alpha-1)}\varphi(\sqrt{\omega}\,x) = e^{it\omega}\varphi_\omega(x), \quad x \in \mathbb{R}^n,\ t \in \mathbb{R}, \tag{5.86}$$

where $\varphi(\cdot)$ is the unique positive, spherical symmetric solution of (5.8), satisfies the equation in (5.1) with $\lambda = 1$ (focussing case).

(ii) Show that

$$\frac{d}{d\omega}\|u_\omega\|_2 = \frac{d}{d\omega}\|\varphi_\omega\|_2 \begin{cases} > 0, & \text{if} \quad \dfrac{1}{\alpha-1} > \dfrac{n}{4}, \\[2ex] = 0, & \text{if} \quad \dfrac{1}{\alpha-1} = \dfrac{n}{4}, \\[2ex] < 0, & \text{if} \quad \dfrac{1}{\alpha-1} < \dfrac{n}{4}. \end{cases}$$

5.10 [Generalized pseudo-conformal transformation] Let $u(x,t)$ be a solution of the equation

$$i\partial_t u + \mathfrak{L}_j u \pm |u|^{\alpha-1}\,u = 0, \quad \alpha > 1 \tag{5.87}$$

with

$$\mathfrak{L}_j = \partial_{x_1}^2 + \cdots + \partial_{x_j}^2 - \partial_{x_{j+1}}^2 - \cdots - \partial_{x_n}^2, \quad j \in \{1,\ldots,n\}.$$

Prove that for $\nu, \theta, \omega, \gamma \in \mathbb{R}$ such that $\nu\theta - \omega\gamma = 1$,

$$v(x,t) = \frac{e^{i\omega Q_j(x)/4(\nu+\omega t)}}{(\nu+\omega t)^{n/2}}\, u\left(\frac{x}{\nu+\omega t}, \frac{\gamma+\theta t}{\nu+\omega t}\right)$$

with

$$Q_j(x) = x_1^2 + \cdots + x_j^2 - x_{j+1}^2 - \cdots - x_n^2$$

verifies the equation:

$$i\partial_t v + \mathfrak{L}_j v \pm (\nu+\omega t)^{(\alpha-1)n/2-2}|v|^{\alpha-1}\,v = 0. \tag{5.88}$$

In particular, if $\alpha - 1 = 4/n$ (critical L^2-case) (5.87) and (5.88) are equal, see [GV1].

5.11 Let $u \in C([0,T] : L^2(\mathbb{R})) \cap L^4([0,T] : L^\infty(\mathbb{R}))$ be the local solution of the IVP

$$\begin{cases} \partial_t u = i(\partial_x^2 u \pm |u|^2 u), \\ u(x,0) = u_0(x), \end{cases} \tag{5.89}$$

$x, t \in \mathbb{R}$, provided by Theorem 5.2.

(i) Prove that if $x\, u_0,\, x\, u(\cdot, T) \in L^2(\mathbb{R})$, then

$$u \in C([0,T] : H^1(\mathbb{R}) \cap L^2(|x|^2\, dx)) = C([0,T] : \mathcal{F}_2^1).$$

(ii) Extend the result in (i) for any $m \in \mathbb{Z}^+$, i.e. If $|x|^m\, u_0,\, |x|^m\, u(\cdot, T) \in L^2(\mathbb{R})$, then

$$u \in C([0,T] : H^m(\mathbb{R}) \cap L^2(|x|^{2\,m}\, dx)) = C([0,T] : \mathcal{F}_{2\,m}^m).$$

(iii) Prove that if $u_0 \in H^s(\mathbb{R}) \cap L^2(|x|^{2b}\, dx) = \mathcal{F}_{2b}^s$ with $s \geq b \in \mathbb{Z}^+$, then $u \in C([0,T] : \mathcal{F}_{2b}^s)$.

Chapter 6
Asymptotic Behavior of Solutions for the NLS Equation

In this chapter, we shall study the longtime behavior of the local solutions of the initial value problem (IVP)

$$\begin{cases} i\partial_t u + \Delta u + \lambda\, |u|^{\alpha-1}u = 0, \\ u(x,0) = u_0(x), \end{cases} \tag{6.1}$$

$t \in \mathbb{R}$, $x \in \mathbb{R}^n$, obtained in the previous chapter.

In the first section, we shall present results that under appropriate conditions involving the dimension n, the nonlinearity α, the sign of λ (focusing $\lambda > 0$, defocusing $\lambda < 0$), and the size of the data u_0 guarantee that these local solutions extend globally in time, i.e., to any time interval $[-T, T]$ for any $T > 0$.

In the second section, we shall see that when these conditions are not satisfied, the local solution should blowup in finite time.

6.1 Global Results

We shall start with the L^2 case. Theorem 5.2 (subcritical case) tells us that the initial value problem (IVP) (6.1) is locally well-posed in $L^2(\mathbb{R}^n)$ for $\alpha \in (1, 1+4/n)$ in a time interval $[0, T]$ with $T = T(\|u_0\|_2) > 0$. Multiplying the equation in (6.1) by $\bar{u}$, integrating the result in the space variables, and taking the imaginary part we get that the mass is conserved:

$$M(u) = \|u(t)\|_2^2 = \|u_0\|_2^2, \tag{6.2}$$

(to justify this procedure one needs to use continuous dependence, approximate the data u_0 by a sequence in $H^2(\mathbb{R}^n)$, and take the limit). The conservation law (6.2) allows us to reapply Theorem 5.2 as many times as we wish, preserving the length of the time interval to get a global solution.

Theorem 6.1 (Global L^2-solution, subcritical case). *If the nonlinearity power $\alpha \in (1, 1+4/n)$, then for any $u_0 \in L^2(\mathbb{R}^n)$ the local solution $u = u(x,t)$ of the*

F. Linares, G. Ponce, *Introduction to Nonlinear Dispersive Equations*, Universitext, DOI 10.1007/978-1-4939-2181-2_6

initial value problem (IVP) (6.1) extends globally with

$$u \in C([0,\infty) : L^2(\mathbb{R}^n)) \cap L^q_{\text{loc}}([0,\infty) : L^p(\mathbb{R}^n)),$$

where (p,q) *satisfies the condition (4.18) in Theorem 4.2.*

The situation for the L^2-critical case $\alpha = 1 + 4/n$,

$$i\,\partial_t u + \Delta u + \lambda\,|u|^{4/n} u = 0 \tag{6.3}$$

with $u(x,0) = u_0(x) \in L^2(\mathbb{R}^n)$, whose solutions are given by Theorem 5.3, is quite different. In this case, the local result shows the existence of solution in a time interval depending on the data u_0 itself and not on its norm. So, the conservation law (6.2) does not guarantee the existence of a global solution. The problem of the longtime behavior of the L^2-solution of the equation (6.3) has received considerable attention.

The progress on this problem can be roughly described as follows:

(i) $||u_0||_2$ is small enough (Corollary 5.2); and
(ii) For the "defocusing" case, i.e., $\lambda < 0$ in (6.1), with $u_0 \in H^s(\mathbb{R}^n)$, $s > 4/7$, [CKSTT2], and for $s \geq 1/2$ for $n = 2$ [FGr] or under the decay assumption $|x|^l\, u_0 \in L^2(\mathbb{R}^n)$, $l > 3/5$ [Bo4].
In this case, it was also proved [Bo2] that if the local L^2-solution provided by Theorem 6.1 cannot be extended beyond the time interval $[0, T_*)$, then at least in the two-dimensional case ($n = 2$), the following L^2-concentration phenomenon of the L^2 mass occurs: There exists $c > 0$ such that

$$\limsup_{t \uparrow T_*} \sup_{Q \subset \mathbb{R}^2 \,:\, |Q| = (T_* - t)^{1/2}} \int_Q |u(x,t)|^2\, dx \geq c, \tag{6.4}$$

where Q denotes a square in $\mathbb{R}^2$ and $|Q|$ the size of its side. The result in (6.4) holds in both the defocusing case $\lambda < 0$ and the focusing case $\lambda > 0$ in which, as we see, blowup takes place but in the H^1-norm.

(iii) If the initial data $u_0(\cdot)$ are assumed to be radial, then
- In the defocusing case ($\lambda = -1$), the global existence and scattering results were established in [TVZ] for dimension $n \geq 3$ and in [KTV] for dimension $n = 2$.
- In the focusing case ($\lambda = 1$) for initial radially symmetric data u_0 satisfying

$$\|u_0\|_2 < \|\varphi\|_2,$$

where φ is the positive solution of the elliptic equation (5.8) with $\alpha = 1+4/n$, it was proved in [KTV] and [KVZ] that the corresponding local solution extends globally and scattering results hold (this is sharp).

(iv) Finally, in [D1]–[D3] Dobson removed the radial assumptions on the result described in (iii). More precisely, in [D1]–[D3] global well-posedness and scattering results were established in the defocussing case for any data $u_0 \in L^2(\mathbb{R}^n)$ and the focussing case for any data $u_0 \in L^2(\mathbb{R}^n)$ with $\|u_0\|_2 < \|\varphi\|_2$, with φ being the positive solution of the elliptic equation (5.8) with $\alpha = 1 + 4/n$.

Definition 6.1. A global solution u of the IVP (6.1) is said to scatter in the space X to a free solution as $t \to \pm\infty$, if there exists $u_\pm \in X$ such that

$$\lim_{t\to\pm\infty} \|e^{it\Delta}u_\pm - u(\cdot,t)\|_X = 0. \tag{6.5}$$

Notice that for the IVP associated to the L^2-critical equation (6.3) focusing case ($\lambda = 1$), scattering cannot occur for all L^2-data in the ball with center as the origin and radius R, $R > \|\varphi\|_2$ (with φ as in (5.8)).

Let us consider now the extension problem of the H^1-local solution proved in Chapter 5. We first examine the subcritical case (Theorem 5.4), i.e., $\alpha \in (1, 1 + 4/(n-2))$, $n \geq 3$, or $1 < \alpha < \infty$, if $n = 1, 2$, where the time of existence T depends on the size of the data, i.e., $T = T(\|u_0\|_{H^1})$. In this case, if u is a solution in the interval $[0, T]$, then multiplying the equation by $-\partial_t \bar{u}$, integrating the result in the space variables, taking its real part and using integration by parts, one gets that for $t \in [0, T]$

$$\frac{d}{dt}E(u(t)) = \frac{d}{dt}\int_{\mathbb{R}^n}\left(|\nabla_x u(x,t)|^2 - \frac{2\lambda}{\alpha+1}|u(x,t)|^{\alpha+1}\right)dx = 0.$$

So, $E(u(t))$ is constant and $E(u(t)) = E(u_0)$ or

$$E(u_0) = \int_{\mathbb{R}^n}\left(|\nabla_x u(x,t)|^2 - \frac{2\lambda}{\alpha+1}|u(x,t)|^{\alpha+1}\right)dx. \tag{6.6}$$

Therefore, if $\lambda < 0$ (defocusing case) it follows that

$$\sup_{[0,T]}\int_{\mathbb{R}^n}|\nabla u(x,t)|^2 dx \leq E(u_0),$$

which combined with (6.2) gives

$$\sup_{[0,T]}\|u(t)\|_{1,2}^2 \leq E(u_0) + \|u_0\|_2^2.$$

This allows us to reapply Theorem 5.4 to extend the local solution u to any time interval.

In the focusing case $\lambda > 0$, using the Gagliardo–Nirenberg inequality, see (3.14), we have that for $t \in [0, T]$

$$\|u(t)\|_{\alpha+1} \leq c\|\nabla_x u(t)\|_2^{\theta}\, \|u(t)\|_2^{1-\theta} \leq c\,\|\nabla_x u(t)\|_2^{\theta}\, \|u_0\|_2^{1-\theta}, \tag{6.7}$$

with

$$\frac{1}{\alpha+1} = \theta\Big(\frac{1}{2} - \frac{1}{n}\Big) + \frac{1-\theta}{2} \quad \text{or} \quad \theta = \frac{n(\alpha-1)}{2(\alpha+1)}.$$

Then,

$$\|u(t)\|_{\alpha+1}^{\alpha+1} \leq c\,\|u_0\|_2^{[(\alpha+1)-n(\alpha-1)/2]}\,\|\nabla_x u(t)\|_2^{n(\alpha-1)/2}.$$

This combined with (6.6) proves that if $E(u_0) < \infty$, then

$$\|\nabla_x u(t)\|_2^2 \leq |E(u_0)| + c_\alpha|\lambda|\,\|u_0\|_2^{[(\alpha+1)-n(\alpha-1)/2]}\,\|\nabla_x u(t)\|_2^{n(\alpha-1)/2}. \tag{6.8}$$

Assume first that $\alpha \in (1, 1+4/n)$, so $n(\alpha-1)/2 < 2$. Then, from (6.8) and the notation $y = y(t) = \|\nabla_x u(t)\|_2$, one gets

$$y^2 \leq E(u_0) + c\, \|u_0\|_2^{[(\alpha+1)-n(\alpha-1)/2]}\, y^{2-\gamma}, \tag{6.9}$$

with $\gamma = 2 - n(\alpha-1)/2 \in (0,2)$. Therefore, there exists $M = M(\|u_0\|_{1,2}; n; \alpha; \lambda) > 0$ independent of T such that

$$\sup_{[0,T]} \|\nabla_x u(t)\|_2 \leq M.$$

Thus, the same argument used above allow us to reapply Theorem 5.4 to extend the local solution u to any time interval.

In the case $\alpha = 1 + 4/n$, the inequality (6.9) becomes

$$y^2 \leq E(u_0) + c\, \|u_0\|_2^{4/n}\, y^2. \tag{6.10}$$

Hence, there exists $c_0 > 0$ such that if $\|u_0\|_2 < c_0$, then the local solution u provided by Theorem 5.4 extends to any time interval.

Finally, we consider the case $\alpha \in (1+4/n, (n+2)/(n-2))$. In this case, using the notation $\delta = \|u_0\|_2$, the inequality (6.9) becomes

$$y^2(t) \leq E(u_0) + c\, \delta^{[(\alpha+1)-n(\alpha-1)/2]}\, y^{2+\nu}(t), \tag{6.11}$$

with $\nu = n(\alpha-1)/2 - 2 \geq 0$. For $\|u_0\|_{1,2} = \|u_0\|_2 + \|\nabla u_0\|_2 \leq \rho$ sufficiently small, it follows from (6.11), evaluated at $t = 0$, that $E(u_0) > 0$. Also, from (6.11), one gets that there exists $M > 0$ such that $y(t) = \|\nabla_x u(t)\|_2 \leq M$, which combined with (6.2) allows us to extend the local solution to any interval of time as in the previous case.

Summarizing, we have the following result:

Theorem 6.2. *Under any of the following set of hypotheses the local solution of the IVP (6.1) with $u_0 \in H^1(\mathbb{R}^n)$ provided by Theorem 5.4 extends globally in time, if*

(i) $\lambda < 0$,
(ii) $\lambda > 0$ *and* $1 < \alpha < 1 + 4/n$,
(iii) $\lambda > 0$, $\alpha = 1 + 4/n$, *and* $\|u_0\|_2 < c_0$,
(iv) $\lambda > 0$, $\alpha > 1 + 4/n$, *and* $\|u_0\|_{1,2} \simeq \|u_0\|_2 + \|\nabla u_0\|_2 \leq \rho$, *for ρ sufficiently small.*

The size assumption on the data in (iii), i.e., $\alpha = 1 + 4/n$, can be made precise. In [W3], Weinstein showed that

$$J_n(f) = \inf_{f \in H^1} \frac{\|\nabla f\|_2^2\, \|f\|_2^{4/n}}{\|f\|_{2+4/n}^{2+4/n}} = \frac{\|\varphi\|_2^{4/n}}{1 + 2/n}, \tag{6.12}$$

and the infimum is attained at φ, where φ is the unique positive solution up to translation of the elliptic problem (5.8) (for details see Exercise 6.6). From (6.12), it follows that $E(\varphi) = 0$ and that if $u_0 \in H^1(\mathbb{R}^n)$ with $\|u_0\|_2 < \|\varphi\|_2$, then the corresponding solution of the IVP (6.1) with $\alpha = 1 + 4/n$ extends globally in time, i.e., $c_0 = \|\varphi\|_2$ in part (iii) of Theorem 6.2. We shall return to this point after Theorem 6.4.

Next, we consider the extension problem of the local solution of the IVP (6.1) with $u_0 \in H^1(\mathbb{R}^n)$ in the critical case $\alpha = (n+2)/(n-2)$. As we shall see in the next section, in the focusing case $\lambda > 0$, local solutions of this problem may blow up. So, we first consider the defocusing case $\lambda < 0$. Under these assumptions, one may ask if the local solution provided by Theorem 5.5 extends to all time and there is scattering. For this problem, the first result known is due to Bourgain [Bo7], who gave a positive answer in the case of dimensions $n = 3, 4$ for radial data, i.e., $u_0(x) = \phi(|x|)$ (see also [Gl1]). In [To5], Tao extended Bourgain's result to any dimension. For any data, not necessarily radial in dimension $n = 3$, Colliander, Keel, Staffilani, Takaoka and Tao [CKSTT7] established global well-posedness and scattering results. Ryckman and Visan [RVi] showed the corresponding result in dimension $n = 4$ and Visan [Vs] obtained it for dimension $n \geq 5$.

A similar problem for the semilinear wave equation:

$$\partial_t^2 w - \Delta w + |w|^{4/(n-2)} w = 0, x \in \mathbb{R}^n, \ t > 0, \tag{6.13}$$

with $(w(0), \partial_t w(0)) = (f, g) \in H^1(\mathbb{R}^n) \times L^2(\mathbb{R}^n)$ was previously solved by Struwe [Stw] in the radial case and $n = 3$, and by Grillakis [Gl2], [Gl3] for general data in dimensions $n = 3, 4, 5$ (see [ShS] for a simplified proof and an extension to the cases $n = 6, 7$).

In both cases, one reviews the local existence theory to deduce what happens if the local solution is assumed not to extend beyond the time interval $[0, T^*)$. In this case, a "concentration of energy" in small sets must occur as $t \uparrow T^*$. In the radial case, this should only take place at the origin. Roughly speaking, to exclude this possibility in the case of the wave equation one combines the Morawetz estimate and the finite propagation speed of the solution. The case of the Schrödinger equation is more involved. The corresponding local Morawetz estimate (appropriate truncated version) [LS] (see Exercise 6.3) is significantly more difficult to establish and even in the radial case requires an inductive argument in the accumulation of energy to disprove the possible concentration.

In [KM1], assuming that the $\dot{H}^{1/2}(\mathbb{R}^3)$-norm of the solution of the defocusing

$$i\partial_t u + \Delta u - |u|^2 u = 0$$

remains bounded, Kenig and Merle showed that the above global results apply.

Next, consider the $\dot{H}^1$ critical focusing case ($\lambda = 1$):

$$i\partial_t u + \Delta u + |u|^{4/(n-2)} u = 0, \tag{6.14}$$

assuming that the data u_0 are spherically symmetric and $3 \leq n \leq 5$, Kenig and Merle [KM2] established a sharp condition for the global existence and blow-up results. Let Φ be the solution of the elliptic problem:

$$\Delta \Phi + |\Phi|^{4/(n-2)} \Phi = 0 \tag{6.15}$$

(so-called Aubin–Talenti solution), where

$$\Phi(x) = \left(1 + \frac{|x|^2}{n(n-2)}\right)^{-(n-2)/2},$$

i.e., the solution of the associated stationary problem to (6.15).

For $u_0 \in H^1(\mathbb{R}^n)$, radial:

(i) If $E(u_0) < E(\Phi)$ and $\|\nabla u_0\|_2 < \|\nabla \Phi\|_2$, then the local solution extends to a global one and scatters as $t \to \pm\infty$.
(ii) If $E(u_0) < E(\Phi)$ and $\|\nabla u_0\|_2 > \|\nabla \Phi\|_2$, then the local solution blows up in finite time in both directions.

In [KV1], Killip and Visan extended the result in (i) to dimension $n \geq 5$ without the radial assumption on the data. They also extended the result in (ii) to all dimension $n \geq 6$ under the radial assumption on the data.

The case $E(u_0) = E(\Phi)$ for radial solutions for the equation (6.14) with $n = 3, 4, 5$ was studied in [DM]. To describe these results, we need to introduce the following notation:

Given $u = u(x,t)$, a radial solution of the IVP associated to (6.14) define Ω_u^{rad} the set of its radial symmetries:

$$\Omega_u^{\text{rad}} = \{e^{i\theta}\, \lambda^{\frac{n-2}{2}}\, u(\lambda x,\, \lambda^2 t) \;:\; \theta \in \mathbb{R}, \lambda > 0\}.$$

Then in [DM] it was shown that if $u_0 \in H^1(\mathbb{R}^n)$, radial, with $E(u_0) = E(\Phi)$ the corresponding radial solutions $u = u(x,t)$ of (6.14) verify the next threshold:

(i) If $\|\nabla u_0\|_2 < \|\nabla \Phi\|_2$, then the local solution extends to a global one in $\mathbb{R}$.
(ii) If $\|\nabla u_0\|_2 = \|\nabla \Phi\|_2$, then $u \in \Omega_\Phi^{\text{rad}}$.
(iii) If $\|\nabla u_0\|_2 > \|\nabla \Phi\|_2$, then either $u \in \Omega_{w^+}^{\text{rad}}$ (for some fixed radial solution w^+) or $u(t)$ blows up in both directions.

Above, we have considered the equation (6.3) critical in $L^2(\mathbb{R}^n)$ and (6.14) critical in $\dot{H}^1(\mathbb{R}^n)$. The critical problem in $\dot{H}^s(\mathbb{R}^n)$ with $s = s_c \in (0,1)$ was studied by Holmer and Roudenko in [HR1]. For the case $n = 3$, $s_c = 1/2$, i.e.,

$$\begin{cases} i\partial_t u + \Delta u + |u|^2 u = 0, \\ u(x,0) = u_0(x), \end{cases} \tag{6.16}$$

(denoting by $\varphi(x)$ the solution of (5.8) with $\omega = 1$) they proved:

Let $u_0 \in H^1(\mathbb{R}^3)$ radial such that

$$M(u_0)\, E(u_0) < M(\varphi)\, E(\varphi).$$

(i) If $\|u_0\|_2\, \|\nabla u_0\|_2 < \|\varphi\|_2\, \|\nabla \varphi\|_2$, then for all t, $u(t)$ satisfies

$$\|u_0\|_2\, \|\nabla u(t)\|_2 < \|\varphi\|_2\, \|\nabla \varphi\|_2$$

and is globally defined and scatters.

(ii) If $\|u_0\|_2 \, \|\nabla u_0\|_2 > \|\varphi\|_2 \, \|\nabla \varphi\|_2$, then the local solution blows up in finite time.

The radial assumption in (i) was removed by Duyckaerts, Holmer and Roudenko in [DHR]. The method in [HR1] and [DHR] follows some of the ideas introduced in [KM1].

In [NSc], Nakanishi and Schlag obtained the following:

There exists $\epsilon > 0$ such that for any $u_0 \in X_\epsilon$ where

$$X_\epsilon = \{f \in H^1(\mathbb{R}^3) : f \text{ radial}, \; M(f)\, E(f) < M(\varphi)(E(\varphi) + \epsilon^2) \text{ and } \|f\|_2 = \|\varphi\|_2\},$$

the corresponding local solution $u(t)$ of the IVP associated to the equation in (6.14) as $t \to +\infty$ (and $t \to -\infty$) satisfies one of the next three possibilities:

(i) Scatters,
(ii) Finite time blowup,
(iii) After sometime it remains close in H^1 to $B = \{e^{i\theta} \, \varphi(\cdot) \; : \; \theta \in \mathbb{R}\}$.

Considering $t \to \pm\infty$, this gives nine possibilities.

Moreover, the subset of X_ϵ having the behavior in (i) and (ii) (four possibilities) is open and B describes "their boundaries."

The case $\|u_0\|_2 \, \|\nabla u_0\|_2 = \|\varphi\|_2 \, \|\nabla \varphi\|_2$ for equation (6.16) was studied in [DRu], where they obtained results in the direction of [DM] for the $\dot{H}^1$ critical case (see above).

The extension of the results in [HR1] and [DHR] to the energy subcritical range $1 + \dfrac{4}{n} < \alpha < 1 + \dfrac{4}{n-2}$, for $n \geq 3$, and $1 + \dfrac{4}{n} < \alpha < \infty$, for $n = 1, 2$ was considered by Fang, Xie and Cazenave [FXC], and Guo [Gq].

The problem of the longtime behavior of the local solution for the supercritical H^1- case, i.e., $\alpha > 1 + 4/(n-2)$, $n \geq 3$, remains largely open. For some techniques and results in this direction, see [KM1] and [KV2].

6.2 Formation of Singularities

In this section, we prove that the global results in the previous section are optimal. We shall see that if (i)–(iii) in Theorem 6.2 do not hold, then there exists $u_0 \in H^1(\mathbb{R}^n)$ and $T^* < \infty$ such that the corresponding solution u of the IVP (6.1) satisfies

$$\lim_{t \uparrow T^*} \|\nabla u(t)\|_2 = \infty. \tag{6.17}$$

To simplify, the exposition we shall assume $\lambda = 1$. In the proof of (6.17), we need the following identities.

Proposition 6.1. *If $u(t)$ is a solution in $C([-T, T] : H^1(\mathbb{R}^n))$ of the IVP (6.1) with $\lambda = 1$ obtained in Theorems 5.4 and 5.5, then*

$$\frac{d}{dt} \int_{\mathbb{R}^n} |x|^2 |u(x,t)|^2 \, dx = 4 \, \mathcal{I}m \int_{\mathbb{R}^n} r \, \overline{u} \, \partial_r u \; dx, \tag{6.18}$$

with $r = |x|$ and $\mathcal{I}m\,(\cdot) =$ imaginary part of $(\cdot)$, and

$$\frac{d}{dt}\mathcal{I}m \int_{\mathbb{R}^n} r\overline{u}\,\partial_r u\,dx = 2\int_{\mathbb{R}^n} |\nabla u(x,t)|^2\,dx + \Big(\frac{2n}{\alpha+1} - n\Big)\int_{\mathbb{R}^n} |u(x,t)|^{\alpha+1}\,dx. \quad (6.19)$$

Proof. To obtain (6.18) we multiply the equation in (6.1) by $2\overline{u}$ and take the imaginary part to get

$$\mathcal{I}m\,(2i\,\partial_t u\,\overline{u}) = \partial_t|u|^2 = -\mathcal{I}m\,(2\Delta u\,\overline{u}) = -2\,\mathrm{div}\,(\mathcal{I}m\,(\nabla u\,\overline{u})).$$

Multiplying this identity by $|x|^2$, integrating in $\mathbb{R}^n$, using integration by parts and that $r\partial_r u = x_j\partial_{x_j}u$ (with summation convention) it follows that

$$\begin{aligned}\frac{d}{dt}\int |x|^2|u|^2\,dx &= \int |x|^2\partial_t|u|^2\,dx = -2\int \mathrm{div}\,(\mathcal{I}m\,(\,\overline{u}\,\nabla u))|x|^2dx\\ &= 2\int \mathcal{I}m\,(\,\overline{u}\,\partial_{x_j}u)\,2x_j\,dx = 4\int \mathcal{I}m\,(r\,\overline{u}\,\partial_r u)\,dx,\end{aligned}$$

which proves (6.18).

For (6.19), we multiply the equation in (6.1) by $2r\partial_r\overline{u}$, integrate in $\mathbb{R}^n$, and take the real part of this expression to get

$$\begin{aligned}\mathcal{R}e\,\Big(2i\int r\partial_r\,\overline{u}\,\partial_t u\,dx\Big) &= i\int r(\partial_r\overline{u}\,\partial_t u - \partial_r u\partial_t\,\overline{u})\,dx\\ &= -2\mathcal{R}e\int r\partial_r\,\overline{u}\,\Delta u dx - 2\mathcal{R}e\int r\partial_r\,\overline{u}\,|u|^{\alpha-1}u\,dx. \quad (6.20)\end{aligned}$$

By integration by parts and the equation in (6.1) it follows that

$$\begin{aligned}i\int r\,(\partial_r\overline{u}\,\partial_t u - \partial_r u\partial_t\overline{u})\,dx &= i\int x_j\,(\partial_{x_j}\overline{u}\,\partial_t u - \partial_{x_j}u\partial_t\overline{u})\,dx\\ &= i\int x_j\,(\partial_t(\partial_{x_j}\overline{u}\,u) - \partial_{x_j}(u\partial_t\overline{u}))\,dx\\ &= i\frac{d}{dt}\int r\,u\,\partial_r\overline{u}\,dx + n\,i\int u\partial_t\overline{u}\,dx \quad (6.21)\\ &= \frac{d}{dt}\Big(i\int r\,u\,\partial_r\overline{u}\,dx\Big) + n\Big(\int u(\Delta\overline{u} + |u|^{\alpha-1}\overline{u})\,dx\Big)\\ &= \frac{d}{dt}\Big(i\int ru\partial_r\overline{u}\,dx\Big) - n\int |\nabla u|^2\,dx + n\int |u|^{\alpha+1}\,dx.\end{aligned}$$

Similarly, we see that

$$
\begin{aligned}
2\,\mathcal{R}e\left(\int r\partial_r\overline{u}\Delta u\,dx\right) &= 2\,\mathcal{R}e\left(\int x_j\,\partial_{x_j}\overline{u}\,\partial^2_{x_k}u\,dx\right)\\
&= 2\,\mathcal{R}e\left(-\int |\nabla u|^2\,dx - \int x_j\partial_{x_k}u\,\partial_{x_k}\partial_{x_j}\overline{u}\,dx\right)\\
&= -2\int |\nabla u|^2\,dx - \int x_j\partial_{x_k}u\,\partial_{x_k}\partial_{x_j}\overline{u}\,dx\\
&\quad - \int x_j\,\partial_{x_k}\overline{u}\,\partial_{x_k}\partial_{x_j}u\,dx \qquad (6.22)\\
&= -2\int |\nabla u|^2\,dx + n\int |\nabla u|^2\,dx\\
&\quad + \int x_j\partial_{x_k}\partial_{x_j}u\,\partial_{x_k}\overline{u}\,dx - \int x_j\partial_{x_k}\partial_{x_j}u\,\partial_{x_k}\overline{u}\,dx\\
&= (n-2)\int |\nabla u|^2 dx.
\end{aligned}
$$

Also,

$$
\begin{aligned}
2\,\mathcal{R}e\left(\int |u|^{\alpha-1}ru\,\partial_r\overline{u}\,dx\right) &= 2\,\mathcal{R}e\left(\int |u|^{\alpha-1}u\,x_j\partial_{x_j}\overline{u}\,dx\right)\\
&= \int x_j\,(|u|^2)^{(\alpha-1)/2}(\partial_{x_j}u\,\overline{u} + u\partial_{x_j}\overline{u})\,dx \qquad (6.23)\\
&= \frac{2}{\alpha+1}\int x_j\partial_{x_j}[(|u|^2)^{(\alpha+1)/2}]\,dx\\
&= -\frac{2n}{\alpha+1}\int |u|^{\alpha+1}\,dx.
\end{aligned}
$$

Collecting the information in (6.21)–(6.23) we can rewrite (6.20) as:

$$
\frac{d}{dt}\mathcal{I}m\left(\int r\overline{u}\partial_r u\,dx\right) = 2\int |\nabla u|^2\,dx + \left(\frac{2n}{\alpha+1} - n\right)\int |u|^{\alpha+1}\,dx,
$$

which yields (6.19). □

In the last proof, we used implicitly the following result commented on at the end of Chapter 4.

Proposition 6.2 ([HNT2]). *If u is a solution of the IVP (6.1) in $C([-T,T]: H^1(\mathbb{R}^n))$ provided by Theorems 5.4 and 5.5 such that $x_ju_0 \in L^2(\mathbb{R}^n)$ for some $j = 1,\dots,n$, then*

$$x_j\,u(\cdot,t) \in C([-T,T]: L^2(\mathbb{R}^n)).$$

Thus, if $u_0 \in L^2(\mathbb{R}^n, |x|^2\,dx)$, then

$$u(\cdot,t) \in C([-T,T]: H^1 \cap L^2(|x|^2\,dx)).$$

Now, we shall prove one of the main results in this section.

6.2.1 Case $\alpha \in (1 + 4/n, 1 + 4/(n-2))$

Theorem 6.3. *Let u be a solution in $C([0,T]: H^1(\mathbb{R}^n) \cap L^2(|x|^2\,dx))$ of the IVP (6.1) with $\lambda = 1$ provided by Theorems 5.4 and 5.5 and Proposition 6.2. Assume that the initial data u_0 and the nonlinearity α satisfy the following assumptions:*

$$\text{(i)} \quad \int \Big(|\nabla u_0|^2 - \frac{2}{\alpha+1}|u_0|^{\alpha+1}\Big)\,dx = E(u_0) = E_0 < 0,$$

$$\text{(ii)} \quad \alpha \in (1 + 4/n, 1 + 4/(n-2));$$

then there exists $T^* > 0$ such that

$$\lim_{t \uparrow T^*} \|\nabla u(t)\|_2 = \infty. \tag{6.24}$$

We observe that condition (i) implies that $\|u_0\|_{1,2}$ is not arbitrarily small. In particular, for any $u_0 \in H^1(\mathbb{R}^n)$ one has that $E_0(\nu u_0) < 0$ for $\nu > 0$ sufficiently large.

In the proof, we just need $\alpha > 1+4/n$, therefore, the theorem extends to solutions $u \in C([0,T] : H^2(\mathbb{R}^n) \cap L^2(|x|^2\,dx))$, $\alpha < \infty$ for $n \leq 4$ and $\alpha \leq n/(n-4)$ for $n \geq 5$.

Proof. We first assume that $\mathcal{I}m\,\big(\int r\bar{u}_0\,\partial_r\,u_0\,dx\big) < 0$. We define

$$f(t) = -\mathcal{I}m \int r(\partial_r u\,\bar{u})(x,t)\,dx.$$

By hypothesis, $f(0) > 0$. Using identities (6.19) and (6.6) it follows that

$$\begin{aligned}
f'(t) &= -2\int |\nabla u(x,t)|^2\,dx - \Big(\frac{2n}{\alpha+1} - n\Big)\int |u(x,t)|^{\alpha+1}\,dx \\
&= -2\int |\nabla u(x,t)|^2\,dx + n\Big(\frac{\alpha+1}{2} - 1\Big)\frac{2}{\alpha+1}\int |u(x,t)|^{\alpha+1}\,dx \\
&= -2\int |\nabla u(x,t)|^2\,dx + n\Big(\frac{\alpha+1}{2} - 1\Big)\Big(\int |\nabla u(x,t)|^2\,dx - E_0\Big) \qquad (6.25)\\
&= -\Big[2 - n\Big(\frac{\alpha+1}{2} - 1\Big)\Big]\int |\nabla u(x,t)|^2\,dx - n\Big(\frac{\alpha+1}{2} - 1\Big)E_0 \\
&\geq M\,\|\nabla u(t)\|_2^2,
\end{aligned}$$

since by hypothesis $E_0 < 0$, $\alpha > 1$ implies that $(\alpha+1)/2 - 1 > 0$, and $\alpha > 1+4/n$ implies that $n((\alpha+1)/2 - 1) - 2 = M > 0$.

From (6.25), $f(t)$ is an increasing function, so $f(t) \geq f(0) > 0$ for all $t > 0$.

Now we use (6.18) to see that

$$\frac{d}{dt}\int |x|^2|u(x,t)|^2\,dx = 4\,\mathcal{I}m\int r(\overline{u}\,\partial_r u)(x,t)\,dx = -4\,f(t) < 0.$$

Thus, $h(t) = \int |x|^2|u(x,t)|^2\,dx$ is a decreasing function with

$$h(t) \le \int |x|^2|u_0(x)|^2\,dx = h(0).$$

The Cauchy–Schwarz inequality tells us that

$$\begin{aligned}|f(t)| = f(t) &= -\mathcal{I}m\int r(\overline{u}\,\partial_r u)(x,t)\,dx\\ &\le \Big(\int r^2|u|^2(x,t)\,dx\Big)^{1/2}\Big(\int |\partial_r u|^2(x,t)\,dx\Big)^{1/2}\\ &\le (h(0))^{1/2}\|\nabla u(t)\|_2,\end{aligned}$$

which combined with (6.25) proves that $f(t)$ satisfies the differential inequality:

$$\begin{cases} f'(t) \ge \dfrac{M}{h(0)}\,(f(t))^2,\\ f(0) > 0.\end{cases}$$

Hence,

$$(h(0))^{1/2}\|\nabla u(t)\|_2 \ge f(t) \ge \frac{h(0)f(0)}{h(0)-Mf(0)t}. \tag{6.26}$$

Defining

$$T_0 = \frac{h(0)}{Mf(0)} > 0, \tag{6.27}$$

we obtain (6.24) with $T^* = T_0$.

Next, we consider the case $\mathcal{I}m\,\big(\int r\overline{u}_0\,\partial_r\,u_0\,dx\big) \ge 0$. From (6.25), it follows that

$$\frac{d}{dt}\mathcal{I}m\int r\overline{u}\,\partial_r\,u(x,t)\,dx = 2E_0 + \Big(\frac{2(n+2)}{\alpha+1} - n\Big)\int |u(x,t)|^{\alpha+1}\,dx \le 2\,E_0$$

because $\alpha > 1 + 4/n$. Hence, since $E_0 < 0$ there exists $\hat{t} > 0$ such that

$$\mathcal{I}m\int r\overline{u}\,\partial_r\,u(x,\hat{t})\,dx < 0$$

and we are in the case previously considered. □

The antecedently result gives us an upper bound on the life span of the local solution in H^1 since we have shown that the existence of the interval of time $[0, T^*)$ implies (6.24). This only tells us that the time of life span T^* of the solution is less than or equal to T_0 as above. It is easy to see that the L^p-norm with $p \ge \alpha + 1$ of the solution u, that is $\|u(t)\|_p$, also satisfies an estimate of the type described in (6.24).

6.2.2 Case $\alpha = 1 + 4/n$

In this case, $n(1 - 2/(\alpha + 1)) = 4/(\alpha + 1)$, (6.19) can be rewritten as:

$$\frac{d}{dt}\mathcal{I}m \int r\,\bar{u}\,\partial_r u\,dx = 2\Big(\int |\nabla u(x,t)|^2\,dx - \frac{2}{\alpha+1}\int |u(x,t)|^{\alpha+1}\,dx\Big) = 2\,E_0.$$

Integrating this equality we see that

$$\mathcal{I}m \int r\,\bar{u}\,\partial_r u\,dx = \mathcal{I}m \int r\,\bar{u}_0\,\partial_r u_0\,dx + 2t\,E_0,$$

which combined with (6.18) tells us that

$$\frac{d}{dt}\int |x|^2|u(x,t)|^2\,dx = 4\,\mathcal{I}m \int r\,\bar{u}_0\,\partial_r u_0\,dx + 8t\,E_0.$$

Integrating again, we obtain the identity:

$$\int |x|^2|u(x,t)|^2\,dx = \|\,|x|\,u_0\|_2^2 + 4t\,\mathcal{I}m \int r\,\bar{u}_0\,\partial_r u_0\,dx + 4t^2\,E_0. \tag{6.28}$$

Assume first that either (i) $E_0 < 0$ or (ii) $E_0 \leq 0$ (with E_0 as in Theorem 6.3 (i)) and $\mathcal{I}m \int r\,\bar{u}_0\,\partial_r u_0\,dx < 0$ or (iii) $E_0 > 0$ and $\mathcal{I}m \int r\,\bar{u}_0\,\partial_r u_0\,dx < -\sqrt{E_0}\,\|\,|x|\,u_0\|_2$.

Suppose that the desired result (6.24) does not hold, i.e., the H^1-solution can be extended globally.

Our assumptions and (6.28) allow us to deduce that there exists T^* such that

$$\lim_{t\uparrow T^*} \|\,|x|\,u(\cdot,t)\|_2 = 0. \tag{6.29}$$

Now, we recall Weyl–Heisenberg's inequality (see Exercise 3.14): For any $f \in H^1(\mathbb{R}^n) \cap L^2(|x|^2\,dx)$,

$$\|f\|_2^2 \leq \frac{2}{n}\|\,|x|\,f\|_2\,\|\nabla f\|_2. \tag{6.30}$$

Notice that (6.22) still holds when one substitutes x by $x - a$ for any fixed $a \in \mathbb{R}^n$.

Combining (6.2) and (6.30), it follows that

$$0 < \|u_0\|_2^2 = \|u(t)\|_2^2 \leq \frac{2}{n}\|\,|x|\,u(\cdot,t)\|_2\,\|\nabla u(\cdot,t)\|_2,$$

which together with (6.29) leads to a contradiction. Therefore, it follows that

$$\lim_{t\uparrow T^*} \|\nabla u(t)\|_2 = \infty.$$

Thus, we have proved the following theorem.

Theorem 6.4. *Let $u \in C([-T,T]: H^1(\mathbb{R}^n) \cap L^2(|x|^2\,dx))$ be the solution of the IVP (6.1) with $\alpha = 1 + 4/n$ obtained in Theorem 5.4 and Proposition 6.2 such that the initial data $u_0 \in H^1(\mathbb{R}^n) \cap L^2(|x|^2\,dx)$ satisfy*

$$(i)\ \ E_0 < 0,$$

$$(ii)\ \ E_0 \leq 0 \ \text{ and } \ \mathcal{I}m \int r\,\overline{u}_0\,\partial_r u_0\,dx < 0,$$

or

$$(iii)\ \ E_0 > 0 \text{ and } \mathcal{I}m \int r\,\overline{u}_0\,\partial_r u_0\,dx \leq -\sqrt{E_0}\,\|\,|x|\,u_0\|_2,$$

where E_0 was defined in Theorem 6.3(i). Then there exists T^* for which identity (6.24) holds.

It is important to notice that (6.29) has not been proved as part of Theorem 6.3, since the singularity in (6.24) could form before time T^*, i.e., $T_0 < T^*$, T_0 being the time when the inequality (6.24) occurs since we assume the existence in the time interval $[0, T_0]$. However, when T_0 in (6.24) and T^* in (6.29) coincide then (6.24), (6.2), and (6.29) ensure that

$$|u(\cdot,t)|^2 \to c\delta(\cdot) \qquad (\text{"concentration"}), \tag{6.31}$$

when $t \uparrow T^*$ in the distribution sense.

In the critical case $\alpha = 1 + 4/n$, the pseudo-conformal invariance tells us that if $u = u(x,t)$ is a solution of the equation in (6.1) with $\alpha = 1+4/n$ and $\lambda = \pm 1$, then

$$v(x,t) = \frac{e^{i|x|^2/4t}}{|t|^{n/2}}\,u\left(\frac{x}{t}, \frac{1}{t}\right) \tag{6.32}$$

solves the same equation for $t \neq 0$, with $v(\cdot,t) \in H^1(\mathbb{R}^n) \cap L^2(|x|^2\,dx)$. In particular, in the focusing case $\lambda = 1$, if $u(x,t) = e^{i\omega t}\varphi(x)$ is the standing wave solution of the equation in (6.1) see Chapter 5 (5.7) and (5.8), to simplify the notation we fix $\omega = 1$. Then,

$$z(x,t) = \frac{e^{i(|x|^2-4)/4t}}{|t|^{n/2}}\,\varphi\left(\frac{x}{t}\right) \tag{6.33}$$

is also a solution in $C(\mathbb{R} - \{0\} : H^1(\mathbb{R}^n) \cap L^2(|x|^2\,dx))$, which blows up at time $t = 0$, i.e.,

$$\lim_{t \uparrow 0} \|\nabla z(t)\|_2 = \infty.$$

Moreover, $\|\nabla z(t)\|_2 \sim c/t$.

The next result tells us that this is the "unique" minimal mass blow up solution. Observe that $\|z(t)\|_2 = \|\varphi\|_2$ and as it was commented before, if $\|u_0\|_2 < \|\varphi\|_2$, then the corresponding H^1-solution extends globally in time.

Theorem 6.5 ([Me3]). *Let u_1 be a solution of the IVP (6.1) with $\lambda = 1, \alpha = 1+4/n$, and data $u_{1,0} \in H^1(\mathbb{R}^n)$ with*

$$\|u_{1,0}\|_2 = \|\varphi\|_2,$$

where φ is the unique positive solution of the elliptic problem (5.8). Assume that u_1 blows up at time $T > 0$, i.e.,

$$\lim_{t \uparrow T} \|\nabla u_1(t)\|_2 = \infty. \tag{6.34}$$

Then,

$$u_1(x,t) = \left(\frac{1}{T-t}\right)^{n/2} e^{i(|x|^2-4)/4(T-t)} \varphi\left(\frac{x}{T-t}\right)$$

up to the invariance of the equation (see (5.10) and (5.11)).

Next, we consider the IVP (6.1) as in Theorem 6.5, i.e., $\lambda = 1, \alpha = 1+4/n$, and $u_0 \in H^1(\mathbb{R}^n)$, $n = 1, 2$ (so that the nonlinearity is smooth). Assuming that for some $\delta > 0$ sufficiently small

$$\|u_0\|_2 = \|\varphi\|_2 + \delta$$

and that (6.34) occurs, Bourgain and Wang [BoW] have shown that the corresponding solution u can be written as:

$$u(x,t) = u_1(x,t) + u_2(x,t),$$

with u_1 as in Theorem 6.5 and where u_2 remains smooth after the blow-up time T, i.e., for some $\rho > 0$,

$$\partial_t u_2 + \Delta u_2 + |u_2|^{4/n} u_2 = 0, \quad t \in (T-\rho, T+\rho),$$

with $u_2(x,T) = \phi(x)$, where ϕ is smooth, with fast decay at infinity and vanishes at 0 to sufficiently high order.

In particular, this result tells us that at the blow-up time the solution does not need to absorb all the L^2-mass.

The following result is concerned with the concentration phenomenon in the blow up solutions.

Theorem 6.6 ([Me2]). *Given $T > 0$ and a set of points $\{x_1, \ldots, x_k\} \subset \mathbb{R}^n$, there exists an initial datum u_0 such that the corresponding solution of the IVP (6.1) with $\lambda = 1$ and $\alpha = 1 + 4/n$ blows up exactly at time T with the total L^2-mass concentrating at the points $\{x_1, \ldots, x_k\}$.*

Next, we comment on the blow-up rates. As a consequence of the H^1-local existence theorem (Theorem 5.8), we have

Corollary 6.1 ([CzW4]). *If the solution of the IVP (6.1) satisfies*

$$\lim_{t \uparrow T^*} \|\nabla u(t)\|_2 = \infty, \tag{6.35}$$

then

$$\|\nabla u(t)\|_2 \geq c_0 (T^* - t)^{-(1/(\alpha-1)-(n-2)/4)}. \tag{6.36}$$

We recall that (6.35) can only occur in the focusing case $\lambda = 1$ with $\alpha \geq 1 + 4/n$.

Proof. For $t_0 < T^*$, we consider the IVP (6.1) for time $t > t_0$ with data $u(t_0)$. By hypothesis, the solution cannot be extended in H^1 beyond the interval $[0, T^*)$. From the proof of Theorem 5.4 (estimates (5.38) and (5.39)), it follows that if for some $M > c\,\|u(t_0)\|_{1,2}$ one has that

$$c\,\|u(t_0)\|_{1,2} + c(T - t_0)^{\delta}\, M^{\alpha} \leq M, \quad \delta = 1 - \frac{(n-2)(\alpha-1)}{4}, \tag{6.37}$$

then $T < T^*$. Therefore, for all $M > c\,\|u(t_0)\|_{1,2}$,

$$c\,\|u(t_0)\|_{1,2} + c\,(T^* - t_0)^{\delta}\, M^{\alpha} \geq M. \tag{6.38}$$

Choosing $M = 2c\|u(t_0)\|_{1,2}$, it follows that

$$(T^* - t_0)^{\delta}\, \|u(t_0)\|_{1,2}^{\alpha-1} \geq c_0. \tag{6.39}$$

Since $\|u(t)\|_2 = \|u_0\|_2$, it follows that

$$\|\nabla_x u(t_0)\|_2 \geq c_0 (T^* - t_0)^{-\delta/(\alpha-1)} = c_0 (T^* - t_0)^{-(1/(\alpha-1)-(n-2)/4)}.$$

□

Thus, on the one hand we have that, in the critical case $\alpha = 1 + 4/n$, Corollary 6.1 gives the following estimate for the lower bound for the blow-up rate:

$$\|\nabla_x u(t)\|_2 \geq c_0\,(T^* - t)^{-1/2}.$$

On the other hand, numerical simulations in [LPSS] suggested the existence of solutions with blow-up rates as:

$$\|\nabla_x u(t)\|_2 \sim \left(\frac{\ln|\ln|T^* - t||}{T^* - t}\right)^{1/2}. \tag{6.40}$$

The constructions of the two previous blow-up solutions imply the following: there are at least two blow-up dynamics for (6.1) with two different rates, one which is continuation of the explicit $z(x,t)$ blow-up dynamic with the $1/(T - t)$ rate (6.36), and which is expected to be unstable; another one with the log–log rate (6.40), which has been conjectured to be stable.

In the one-dimensional case ($n = 1$), Perelman [Pe1] established the existence of a solution blowing up at the rate described in (6.40).

In [MeRa1], [MeRa2], Merle and Raphael have obtained general upper bound results for the blow up rate. More precisely, they characterize a set of data, i.e.,

$$\mathcal{B}_{\alpha^*} = \{u_0 \in H^1(\mathbb{R}^n) : \int \varphi^2 \leq \int |u_0|^2 \leq \int \varphi^2 + \alpha^*\}, \tag{6.41}$$

where α^* is a small enough parameter and φ is a ground state solution of (6.1), see (5.6)–(5.8), with $\alpha = 1 + 4/n$ and $\lambda = 1$, satisfying

$$E_G(u) = E(u) - \frac{1}{2}\left(\frac{\mathcal{I}m\,(\int \nabla_x u\,\overline{u})}{\|u\|_{L^2}}\right)^2 < 0, \tag{6.42}$$

blow up with an upper rate of the form:

$$\|\nabla_x u(t)\|_2 \lesssim \Big(\frac{\ln|\ln|T^* - t||}{T^* - t}\Big)^{1/2}. \tag{6.43}$$

Regarding the dynamics of the blow up solutions Raphael [Ra1] established the following result:

Theorem 6.7 ([Ra1]). *Let $n = 1, 2, 3, 4$. There exist universal constants C^*, C_1^* > 0 such that the following is satisfied:*

(i) Rigidity of blow-up rate: Let $u_0 \in \mathcal{B}_{\alpha^*}$ with

$$E_G(u_0) > 0,$$

and assume the corresponding solution $u(t)$ to (6.1) blows up in finite time $T < \infty$; then, there holds for t close to T either

$$\|\nabla_x u(t)\|_2 \leq C^* \Big(\frac{\log|\log(T - t)|}{T - t}\Big)^{1/2} \tag{6.44}$$

or

$$\|\nabla_x u(t)\|_2 \geq \frac{C_1^*}{(T - t)\sqrt{E_G(u_0)}}.$$

(ii) Stability of the log–log law: Moreover, the set of initial data $u_0 \in \mathcal{B}_{\alpha^*}$ such that $u(t)$ blows up in finite time with upper bound (6.44) is open in H^1.

6.3 Comments

The results shown in Section 6.1 are due to Glassey [G2], based on previous ideas of Zakharov and Shabat [ZS]. Section 6.2 was built on the works of Tsutsumi [Ts] and of Nawa and Tsutsumi [NT]. Proposition 6.1, crucial in the proof of Theorem 6.2, is known as "the pseudoconformal invariant property" and was proved by Ginibre and Velo [GV1]. Observe that all these blow up results apply to local solution $u \in C([0, T] : H^1(\mathbb{R}^n) \cap L^2(|x|^2\,dx))$. In [OgT], the one-dimensional case $n = 1$, critical case $\alpha = 5$, the weighted condition $xu_0 \in L^2(\mathbb{R})$ was removed. The formation of singularities in solutions of the problem associated to the equation in (6.1) in the case of boundary and periodic values was studied in [Ka].

We recall that the existence of solutions in L^2 for the critical power $\alpha = 1+4/n$ was established in Theorem 5.2 (see [CzW2]). Using this result, we have that an extension is only possible when the L^2-$\lim_{t\uparrow T_0} u(t)$ exists. The identity (6.2) assures the existence of the limit in the weak topology of L^2. It was proved in [MT] that the strong limit does not exist and moreover that the same extension does not exist. This shows that if a solution that corresponds to radial data and dimension $n \geq 2$ develops singularities, then this solution satisfies (6.24) and (6.29).

In [Ra2], Raphael studied the dynamical structure of the blow up for the quintic nonlinear Schrödinger (NLS) in two dimensions, i.e., the equation in (6.1) with $\alpha = 5$, $\lambda = 1$, and $n = 2$, which is supercritical in L^2. Among other results he showed the existence of H^1 radial initial data for which the associated solutions blow up in finite time on a sphere of strictly positive radius.

In this setting, one has the equation:

$$i\,\partial_t u + \partial_r^2 u + \frac{1}{r}\,\partial_r u + |u|^4 u = 0.$$

Removing the term $\partial_r u/r$ above, one gets the one-dimensional equation with the critical L^2-power ($\alpha = 1 + 4/n$). Roughly proving that the contribution of the term $\partial_r u/r$ is negligible for the analysis and using Theorem 6.6, with one point $x = 1$, one gets the idea of the form of the result in [Ra2].

This approach to get blow up results which concentrate in a surface of $\mathbb{R}^n$ has also been obtained and extended in [HR2], [HR3], [HPR], and [MRS].

Corollary 6.1 was taken from Cazenave and Weissler [CzW4]. There they also showed that the IVP (6.1) with data $u_0 \in H^s(\mathbb{R}^n)$, $s \in (0, 1)$, and $\|(-\Delta)^{s/2}u_0\|_2$ sufficiently small and nonlinearity $\alpha = 1 + 4/(n - 2\,s)$ has a unique global solution (notice that the cases $s = 0$, $s = 1$ were covered in Corollaries 5.2, 5.4, respectively). This result holds in both the focusing and defocusing case. In fact, it is just based on the homogeneity of the nonlinearity, so it applies to any nonlinear term of the form $f(u, \overline{u})$ with $f(\lambda u, \lambda \overline{u}) = \lambda^\alpha f(u, \overline{u})$.

Based on a pioneering idea of Bourgain [Bo5], one can obtain a global solution below the "energy norm," which in this case is H^1. The argument in [Bo5] has been significantly refined in a sequence of works of Colliander, Keel, Staffilani, Takaoka and Tao [CKSTT1], [CKSTT2], [CKSTT3]. For the IVP (6.1), with $\lambda < 0$, they have shown that in the cases $(n, \alpha) = (1, 5), (2, 3), (3, 3)$, $u_0 \in H^s(\mathbb{R}^n)$, $s > 1/2, 1/2, 4/5$, respectively, suffice for the global existence. In the last case $(n, \alpha) = (3, 3)$ with radial initial data the condition is lower: $s > 5/7$. Notice that in the above cases, the IVP (6.1) is locally well-posed in $H^s(\mathbb{R}^n)$, with $s > 0, 0, 1/2$. So, it is unclear whether these results are optimal.

Now, we regard the problem of the rate of growth of the higher Sobolev norm. Consider the IVP (6.1) defocusing case $\lambda < 0$ with $\alpha \in (1 + 4/n, (n+2)/(n-2))$. Theorem 5.4 provides the global solution for data $u_0 \in H^1(\mathbb{R}^n)$ with

$$\sup_{t \in \mathbb{R}} \|u(t)\|_{1,2} < \infty.$$

Assuming that $u_0 \in H^s(\mathbb{R}^n)$, $s > 1$, and the nonlinearity is sufficiently smooth, one can ask what are the best possible bounds for $\|u(t)\|_{s,2} \sim \|(-\Delta)^{s/2}u(t)\|_2$.

Standard energy estimates give an exponential upper bound. In [Bo6], Bourgain showed that if $n = 3$ and $\alpha < 5$, then

$$\|u(t)\|_{s,2} \leq c\,|t|^{c(s-1)}$$

for some constant c. In [Sta1], Staffilani refined the arguments seen in [Bo6] and among other results in the case $(n, \alpha) = (1, 3)$, she showed that

$$\|u(t)\|_{s,2} \leq c|t|^{\mu} \qquad \text{as} \quad t \to \infty,$$

with $\mu = 2/3(s-1)^+$.

Concerning the asymptotic behavior of the H^1-global solution of the IVP (6.1) obtained in Theorem 6.2, one has the following L^p-norm decay result due to Ginibre and Velo [GV2].

Theorem 6.8. *Assume*

$$\lambda < 0, \quad \alpha \in \left(1 + \frac{4}{n},\, 1 + \frac{4}{n-2}\right), \quad \textit{and } n \geq 3. \tag{6.45}$$

Then, for each $u_0 \in H^1(\mathbb{R}^n)$ the corresponding global solution $u(t)$ of the IVP (6.1) provided by Theorem 6.2(i) satisfies

$$\lim_{t \to \pm\infty} \|u(t)\|_p = 0 \quad \textit{for } p \in \left(2, \frac{2n}{n-2}\right). \tag{6.46}$$

In addition, in [GV2] and [GV3], Ginibre and Velo proved the following theorems:

Theorem 6.9. *Under assumption (6.45), for each $u_0 \in H^1(\mathbb{R}^n)$ there exists a unique $u_0^{\pm} \in H^1(\mathbb{R}^n)$ such that*

$$\lim_{t \to \pm\infty} \|e^{it\Delta}u_0^{\pm} - u(t)\|_{1,2} = 0, \tag{6.47}$$

with

$$\|u_0^{\pm}\|_2 = \|u_0\|_2, \textit{ and } \|\nabla u_0^{\pm}\|_2^2 = E(u_0).$$

Theorem 6.10. *Under assumption (6.45) for each $u_0^{\pm} \in H^1(\mathbb{R}^n)$, there exists a unique $u_0 \in H^1(\mathbb{R}^n)$ such that (6.47) holds.*

Theorems 6.9 and 6.10 were used in [GV2] and [GV3] to define (continuous) maps $W^{\pm}$ (asymptotic states) and $\Omega^{\pm}$ (wave operators) in $H^1(\mathbb{R}^n)$, respectively, as $W^{\pm}(u_0) = u_0^{\pm}$ and $\Omega^{\pm}(u_0^{\pm}) = u_0$. Hence, $W^{\pm}\Omega^{\pm} = I$ on $H^1(\mathbb{R}^n)$ and one has the scattering operator $S = W^+\Omega^-$, with $S(u_0^-) = u_0^+$. For extensions of these results, see [GV2], [GV3]; for results in the cases $n = 1, 2$, see [Na].

Regarding global well-posedness for the periodic problem:

$$\begin{cases} i\partial_t u = -\Delta u \pm |u|^{\alpha-1}u, \\ u(x,0) = u_0(x), \end{cases} \tag{6.48}$$

$x \in \mathbb{T}^n$, $t \in \mathbb{R}$, $\alpha > 1$, we have the following results:

For $n = 2$, Bourgain [Bo1] showed that (6.48) with $\alpha \geq 3$ in the defocussing case ((–) sign) is globally well-posed. A similar result holds for the focusing case

((+) sign) under the additional assumption of $\|u_0\|_{L^2}$ being small enough or for $\alpha > 3$ assuming that $\|u_0\|_{1,2}$ is sufficiently small. In [BGT1], Burq, Gerard and Tzvetkov proved the existence of finite time ($T < \infty$) H^1 blow-up solutions, with data close to the ground state, i.e., $\lim_{t\uparrow T} \|u(t)\|_{1,2} = \infty$ for (6.48) with $\alpha = 3$ in the focusing case. Moreover, they found the precise rate of the blowup showing that $(T - t)\,\|u(t)\|_{1,2} \sim c_0$.

Finally, we describe some results concerning the stability and instability of standing waves. Before doing that we introduce the following notation:

$$A = \{\phi \in H^1(\mathbb{R}^n);\ \phi \neq 0 \text{ and } -\Delta\phi + \omega\phi = |\phi|^{\alpha-1}\phi\} \tag{6.49}$$

and

$$G = \{\phi \in A;\ S(\phi) \leq S(v) \text{ for all } v \in A\}, \tag{6.50}$$

where

$$S(\phi) = \frac{1}{2}\int_{\mathbb{R}^n} |\nabla\phi|^2\,dx - \frac{1}{\alpha+2}\int_{\mathbb{R}^n} |\phi|^{\alpha+1}\,dx - \frac{\omega}{2}\int_{\mathbb{R}^n} |\phi|^2\,dx.$$

The functions in the first set are called *ground states* and $u(x,t) = e^{i\omega t}\phi(x)$ bound states or *standing waves* or solitary waves.

If we require the following conditions be satisfied:

(i) $\alpha = 1 + 4/n,\ \omega > 0$ and $\varphi \in A$ or

(ii) $1 + 4/n < \alpha < (n+2)(n-2)$, $(1 + 4/n < \alpha < \infty,\ n = 1,2)$, $\omega > 0$ and $\varphi \in G$,

then $u(x,t) = e^{i\omega t}\,\varphi(x)$ is an unstable solution of (6.1), in the sense that there is a sequence $\{\varphi_m\}_{m\in\mathbb{N}} \subset H^1(\mathbb{R}^n)$ such that

$$\varphi_m \to \varphi \quad \text{in} \quad H^1(\mathbb{R}^n)$$

and such that the corresponding maximal solution u_m of (6.1) blows up in a finite time for both $t > 0$ and $t < 0$.

The result in case (i) was established by Weinstein [W3] and case (ii) was proved by Berestycki and Cazenave [BC2]. The argument of proof involves variational methods.

On the other hand, if we let $1 < \alpha < 1 + 4/n,\ \omega > 0$, and $\varphi \in G$, then the solution $u(x,t) = e^{i\omega t}\,\varphi(x)$ is a stable solution of (6.1), in the sense that for every $\epsilon > 0$ there exists $\delta(\epsilon) > 0$ such that if $\psi \in H^1(\mathbb{R}^n)$ verifies $\|\varphi - \psi\|_{H^1} < \delta(\epsilon)$, then the corresponding maximal solution v of (6.1) with data ψ verifies

$$\sup_{t\in\mathbb{R}} \inf_{\theta\in\mathbb{R}} \inf_{y\in\mathbb{R}^n} \|v(\cdot,t) - e^{i\theta}\varphi(\cdot - y)\|_{H^1} \leq \epsilon.$$

This result shows orbital stability in the subcritical case. It was established by Cazenave and Lions [CzL]. Extensions of this result to other dispersive equations can be found in [AL].

The interaction of solitary waves:

$$R(x,t) = e^{i(\nu\cdot x - |\nu|^2 t + \omega t + \theta)}\,\varphi_\omega(x - 2\nu\,t - x_0), \tag{6.51}$$

with $\nu,\ x_0 \in \mathbb{R}^n$, $\omega,\ \theta \in \mathbb{R}$, $\omega > 0$, and φ_ω a solution of (5.8), is not yet well understood. For example, the detailed description of solutions of the IVP (6.1) with $\lambda > 0$ and data:

$$u_0(x,0) = \sum_{j=1}^{N} R_j(x,0) = \sum_{j=1}^{N} e^{i(\nu_j\cdot x + \theta_j)}\varphi_\omega(x - x_0),\quad N \geq 2 \tag{6.52}$$

such that $\exists\ j,\ k \in \{1,\ldots,N\}$, $j \neq k$, and $\widehat{t} > 0$ with

$$|2(\nu_j - \nu_k)\widehat{t} - (x_{0_j} - x_{0_k})| \ll 1$$

(i.e., at $t = \widehat{t}$ the solitary waves $R_j(x,t)$ and $R_k(x,t)$ interact) is quite open. In the integrable case $n = 1$, $\alpha = 3$, the scattering theory [ZS] describes the solution $u(x,t)$ in terms of the data as a nearly perfect *elastic* interaction between solitary waves (see Section 8.3). In the nonintegrable case, numerical simulations predict a similar behavior which has not been rigorously established. However, some results are known: In [MM7], Martel and Merle for the L^2-subcritical case ($1 < \alpha < 1 + 4/n$) proved the existence of multisolitary waves. More precisely, for

$$R(x,t) = \sum_{j=1}^{N} e^{i(\nu_j\cdot x - |\nu_j|^2 t + \omega_j t + \theta_j)}\varphi_\omega(x - 2\,\nu_j\,t - x_0), \tag{6.53}$$

the sum of the N-traveling waves in (6.51) with $\nu_j \neq \nu_k$ if $j \neq k$, they showed that there exists $u \in C([0,\infty) : H^1(\mathbb{R}^n))$ solution of the equation in (6.1) with $\lambda > 0$ such that for all $t > 0$

$$\|R(\cdot,t) - u(\cdot,t)\|_{1,2} \leq c\,e^{-\alpha_0 t} \ \text{ for some } c,\ \alpha_0 > 0. \tag{6.54}$$

Notice that in the L^2-subcritical case the solitary waves are stable (see [CzL]), (for other results in this direction see [Pe2].)

In the same vein as in [HoZ], [HMZ], and [DH], the time evolution of the solution of the IVP:

$$\begin{cases} i\,\partial_t u + \partial_x^2 u - q\,\delta_0(x)\,u + |u|^2 u = 0, \\ u(x,0) = e^{i\nu x}\text{sech}(x - x_0), \qquad x_0 \ll -1, \end{cases} \tag{6.55}$$

$q \in \mathbb{R}$, has been studied. Notice that if $q = 0$, the solution of (6.55) is the soliton:

$$u(x,t) = e^{i\nu x}\,e^{-i\nu^2 t}\,\text{sech}(x - 2\nu\,t - x_0), \tag{6.56}$$

and that for $q \neq 0$ the "soliton" should interact with the localized potential at time $t \sim |x_0|/\nu$. In [HoZ], it was shown that for $|q| \ll 1$ the "soliton solution" of

(6.55) remains "intact." In the repulsive case ($q > 0$), for high velocity ($\nu \gg 1$), it was proven in [HMZ] that the incoming solution split into transmitted and reflected components (traveling with velocity ν to the right and to the left, respectively). The attractive case ($q < 0$) was studied in [DH].

In the one-dimensional L^2-supercritical case:

$$i\partial_t u + \partial_x^2 u + |u|^{\alpha-1}u = 0, \qquad \alpha > 5,$$

it is known that the traveling wave solution in (5.14) is unstable. In [KrS], a kind of "finite dimensional version" of the stable manifold for the ordinary differential equation (ODE) system was constructed. For further developments in this direction see [Scl2] and [Bc].

The "soliton resolution conjecture" claims that any "reasonable" solution to a nonlinear dispersive equation eventually resolve into a radiation component that behaves as a linear solution plus a localized component that behaves as a finite sum of special solutions (traveling waves, standing waves, breathers, ...). This conjecture is largely open (see Section 8.3). In the case of the NLS (6.1) for the defocusing case ($\lambda < 0$ in (6.1)) (where no nontrivial special solutions exist) is known in some cases. In [To8], a weak form of this conjecture was established for energy-subcritical and mass-supercritical global $H^1(\mathbb{R}^n)$ radial solutions in higher dimension $n \geq 5$.

Finally, we return to the formula (4.7) (and the comment after it). This affirms that if $u_0 \in C_0(\mathbb{R}^n)$, then for any $t \neq 0$ and $\epsilon > 0$, $e^{it\Delta}u_0 \notin L^1(e^{\epsilon|x|}dx)$. So, one can ask if a similar result holds for solutions of the IVP (5.1). The following unique continuation principle established in [KPV15] answers this question:

Consider the equation:

$$\partial_t u = i(\Delta u + \lambda\,|u|^{\alpha-1}), \lambda \in \mathbb{R} - \{0\}, \tag{6.57}$$

with α such that $[\alpha] > n/2$ if α is not an odd integer.

Let $u_1, u_2 \in C([-T,T] : H^k(\mathbb{R}^n))$, $k > n/2$, $T > 0$, be two solutions of the equation (6.57) such that

$$\text{supp}(u_1(\cdot,0) - u_2(\cdot,0)) \subset \{x \in \mathbb{R}^n : x_1 \leq a_1\}, a_1 < \infty.$$

If for some $t \in (-T,T) - \{0\}$ and for some $\epsilon > 0$,

$$u_1(\cdot,t) - u_2(\cdot,t) \in L^2(e^{\epsilon|x_1|}\,dx),$$

then $u_1 \equiv u_2$.

Notice that fixing $u_2 \equiv 0$ one settles the above question. Moreover, by taking $u_1(x,0) = \varphi(x)$ as in (5.8) and $u_2(x,0) = \varphi(x) + \phi(x)$, with $\phi \in H^s(\mathbb{R}^n)$, $s > n/2$, with compact support, one can see that

$$u_2(\cdot,t) \notin L^2(e^{\epsilon|x|}\,dx) \text{ for any } \epsilon > 0,$$

for any $t \neq 0$ and α as in (6.57). This follows by combining Theorem 5.1 and the above unique continuation result. In other words, regardless of the stability of the standing wave a compact perturbation of it destroys its exponential decay.

6.4 Exercises

6.1 (i) Let $\alpha \in (1, 1 + 4/n]$. Prove that the local L^2-solution of the IVP (5.1) provided by Theorems 5.2 and 5.3 satisfies the identity (5.2).

(ii) Let $\alpha \in (1, 1 + 4/(n-2)]$. Show that the local H^1-solution of the IVP (5.1) provided by Theorems 5.4 and 5.5 satisfies the identities (5.2) and (5.3).

(iii) If in addition to the hypothesis in (ii), assuming that $|x|u_0 \in L^2(\mathbb{R}^n)$, prove (5.5).

(iv) Assume $\lambda < 0$ (defocusing case) and $\alpha \geq 1 + 4/n$ with the hypotheses in (iii), prove the decay estimate:

$$\|u(t)\|_{\alpha+1} \leq c\, t^{-2/(\alpha+1)}.$$

6.2 Consider the IVP (6.1) with $\alpha = 1+4/n$. Let $u(t)$ be its global L^2-solution corresponding to a datum $u_0 \in L^2(\mathbb{R}^n)$ with $\|u_0\|_2 < \epsilon$ provided by Theorem 6.2 (iii).

(i) Prove that there exist $u_0^{\pm} \in L^2(\mathbb{R}^n)$ such that

$$u(t) = e^{\pm it\Delta}\, u_0^{\pm} + R_{\pm}(t) \text{ with } \lim_{t\to\pm\infty} \|R_{\pm}(t)\|_2 = 0. \tag{6.58}$$

(ii) Prove that (6.58) fails for arbitrary $u_0 \in L^2(\mathbb{R}^n)$.
Hint: (i) Using Theorem 4.2, inequality (4.16), and Corollary 5.2 prove that

$$u_0^{\pm} = u_0 + \int_0^{\pm\infty} e^{-it'\Delta}\, |u|^{4/n}\, u(t')\, dt' \in L^2(\mathbb{R}^n).$$

(ii) Use the standing wave solutions in (5.8), (5.9), and (5.13) if $n = 1$.

6.3 (**Morawetz's estimate**) Consider the IVP (6.1) in the defocusing case $\lambda = -1$, with $\alpha < 1+4/(n-2)$, and $n \geq 3$. Let $u \in C([-T_0, T_1] : H^1(\mathbb{R}^n))$ be the local solution of this problem provided by Theorem 5.4.

(i) Prove the following estimates:

$$\text{(a) } \mathcal{R}e \int i\partial_t u\big(\partial_r \bar{u} + \frac{(n-1)}{2r}\bar{u}\big)\, dx = \mathcal{R}e\, \frac{1}{2}\frac{d}{dt}\int iu\partial_r u\, dx.$$

$$\text{(b) } \mathcal{R}e \int \Delta u\, \big(\partial_r \bar{u} + \frac{(n-1)}{r}\bar{u}\big)\, dx \leq 0.$$

$$\text{(c) } \mathcal{R}e \int -|u|^{\alpha-1} u\big(\partial_r \bar{u} + \frac{(n-1)}{r}\bar{u}\big)\, dx = -\frac{\alpha-1}{2(\alpha+1)}\int \frac{(n-1)}{2r}|u|^{\alpha+1}\, dx$$

where $r = |x|$.

(ii) Using part (i) and the equation in (6.1) show that

$$\frac{1}{2}\frac{d}{dt}\int iu\partial_r\bar{u}\,dx \geq \frac{(\alpha-1)(n-1)}{4(\alpha+1)}\int \frac{|u(x,t)|^{\alpha+1}}{|x|}\,dx. \tag{6.59}$$

(iii) Integrate (6.59) in the interval $(t_1, t_2) \subset [-T_0, T_1]$ to get

$$\int_{t_1}^{t_2}\int_{\mathbb{R}^n} \frac{|u(x,t)|^{\alpha+1}}{|x|} \leq c\,(\|u(t_1)\|_{1,2} + \|u(t_2)\|_{1,2}). \tag{6.60}$$

(iv) Use Theorem 6.2 to conclude that if in addition $\alpha < 1 + 4/n$, then the global solution $u \in C(\mathbb{R} : H^1(\mathbb{R}^n))$ satisfies

$$\int_{-\infty}^{\infty}\int_{\mathbb{R}^n} \frac{|u(x,t)|^{\alpha+1}}{|x|}\,dxdt < \infty. \tag{6.61}$$

(Notice that $|x|^{-1}$ is not integrable around the origin, so (6.61) gives information over the movement in time of the solution around the origin.)

6.4 For the IVP (6.1) with $\lambda = 1$ (focussing case), with the notation in Exercise 5.9, prove that

$$\begin{aligned} E(u_\omega(x,t)) &= \int_{\mathbb{R}^n} \left(|\nabla_x u_\omega(x,t)|^2 - \frac{2}{\alpha+1}\,|u_\omega(x,t)|^{\alpha+1}\right) dx \\ &= E(\varphi_\omega) = \int_{\mathbb{R}^n} \left(|\nabla_x \varphi_\omega(x)|^2 - \frac{2}{\alpha+1}\,|\varphi_\omega(x)|^{\alpha+1}\right) dx \\ &= \left(\frac{n(\alpha-1)-4}{2(\alpha+1)}\right)\int_{\mathbb{R}^n} |\varphi_\omega(x)|^{\alpha+1}\,dx. \end{aligned} \tag{6.62}$$

Thus, $E(\varphi_\omega) > 0$ if $\alpha - 1 > 4/n$, $E(\varphi_\omega) = 0$ if $\alpha - 1 = 4/n$ and $E(\varphi_\omega) < 0$ if $\alpha - 1 < 4/n$.
Hint: Combine the fact that $\varphi_\omega(x) = \omega^{1/(\alpha-1)}\varphi(\sqrt{\omega}\,x)$ is the solution of (5.8), with the identity (5.83).

6.5 Prove that for any $\alpha > 1$, there exist u_0^+, u_0^-, $u_0^0 \in H^1(\mathbb{R}^n)$ such that if

$$E(u_0) = \int_{\mathbb{R}^n} \left(|\nabla_x u_0(x)|^2 - \frac{2}{\alpha+1}\,|u_0(x)|^{\alpha+1}\right) dx,$$

then $E(u_0^+) > 0$, $E(u_0^-) < 0$ and $E(u_0^0) = 0$.

6.6 [W3] Consider the following case of the Gagliardo–Nirenberg inequality (3.14):

$$\|f\|_{\alpha+1} \leq c\|\nabla f\|_2^{\theta}\, \|f\|_2^{1-\theta}, \tag{6.63}$$

$$\theta = n\left(\frac{1}{2} - \frac{1}{\alpha+1}\right) \quad \text{and} \quad 2 < \alpha+1 < \frac{2n}{n-2}.$$

For $f \in H^1(\mathbb{R}^n)$, define

$$J_{\alpha,n}(f) = \frac{\|\nabla f\|_2^{\theta(\alpha+1)}\, \|f\|_2^{(1-\theta)(\alpha+1)}}{\|f\|_{\alpha+1}^{\alpha+1}}, \tag{6.64}$$

$c^*_{\alpha,n}$ and $\mu_{\alpha,n}$ as:

$$c^*_{\alpha,n}{}^{-(\alpha+1)} = \mu_{\alpha,n} = \operatorname*{Inf}_{\substack{f\in H^1(\mathbb{R}^n)\\ f\neq 0}} J_{\alpha,n}(f),$$

i.e., $c^*_{\alpha,n} = \mu_{\alpha,n}^{-1/(\alpha+1)}$ is the optimal constant in (6.63).

(i) Assuming that $\mu_{\alpha,n}$ is attained, i.e., there exists $f^* \in H^1(\mathbb{R}^n)$ such that $J_{\alpha,n}(f^*) = \mu_{\alpha,n}$, with $f^* \geq 0$, prove that for any $\lambda, \nu > 0$

$$J_{\alpha,n}(\lambda\, f^*(\nu\, \cdot)) = \mu_{\alpha,n}. \tag{6.65}$$

(ii) Choosing λ_0, ν_0 in (6.65) such that $g(x) = \lambda_0\, f^*(\nu_0\, x)$ satisfies $\|g\|_2 = \|\nabla g\|_2 = 1$, i.e. $\mu_{\alpha,n} = \|g\|_{\alpha+1}^{-(\alpha+1)}$, and using that $D\, J_{\alpha,n}(g) \equiv 0$ (i.e., $\frac{d}{d\epsilon}\, J_{\alpha,n}(g+\epsilon h)|_{\epsilon=0} = 0, \ \ \forall h \in H^1(\mathbb{R}^n)$) prove that

$$-\theta\, \Delta g + (1-\theta)\, g - \mu_{\alpha,n}\, g^{\alpha} \equiv 0, \quad g \geq 0.$$

(iii) Prove that the function $g_{\beta,\rho}(x) = \beta\, g(\rho\, x)$, $\beta, \rho > 0$ satisfies

$$-\frac{\theta}{\rho^2}\, \Delta g_{\beta,\rho} + (1-\theta) g_{\beta,\rho} - \mu_{\alpha,n}\, \beta^{\alpha-1} g_{\beta,\rho} \equiv 0. \tag{6.66}$$

(iv) Choose β_0, ρ_0 in (6.66) such that g_{β_0,ρ_0} solves (5.8) with $\omega \equiv 1$, i.e.,

$$-\Delta\varphi + \varphi - |\varphi|^{\alpha-1}\varphi = 0, \quad \varphi \geq 0,$$

to prove that

$$\mu_{\alpha,n} = \frac{(1-\theta)^{1+\frac{n}{4}(\alpha-1)}}{\theta^{\frac{n}{4}(\alpha-1)}}\, \frac{1}{\|g_{\beta_0,\rho_0}\|_2^{\alpha-1}} = \frac{(1-\theta)^{1+\frac{n}{4}(\alpha-1)}}{\theta^{\frac{n}{4}(\alpha-1)}}\, \frac{1}{\|\varphi\|_2^{\alpha-1}},$$

so,

$$c^*_{\alpha,n} = (\mu_{\alpha,n})^{-1/(\alpha+1)} = \frac{\theta^{\frac{n(\alpha-1)}{4(\alpha+1)}}}{(1-\theta)^{\frac{4+n(\alpha-1)}{4(\alpha+1)}}}\, \|\varphi\|_2^{(\alpha-1)/(\alpha+1)}.$$

(v) Verify that f^* is a "rescaling" of $\varphi(\cdot)$.

6.7 Consider the IVP (6.1) in the focussing case ($\lambda = 1$) and for the L^2-critical power ($\alpha = 1 + 4/n$). Assume that the local H^1-solution provided by Theorem 5.4 blows up in finite time, i.e., there exists $T^* > 0$ such that

$$\lim_{t\uparrow T^*} \|\nabla u(t)\|_2 = \infty. \tag{6.67}$$

(i) Prove that for any $s \in (0, 1]$ it follows that

$$\liminf_{t\uparrow T^*} \|D^s u(t)\|_2 = \infty.$$

(ii) Prove that for any $p \in (2, \infty]$ it follows that

$$\liminf_{t\uparrow T^*} \|u(t)\|_p = \infty.$$

(iii) If instead of $\alpha = 1 + 4/n$ one considers a nonlinearity $\alpha \in (1 + 4/n, 1 + 4/(n-2))$ and assume that (6.67) holds, for which values of s (i) holds, and for which values of p (ii) holds.

6.8 [GHW] Consider the one-dimensional cubic focussing NLS with a δ-potential:

$$i\,\partial_t v + \partial_x^2 v + \mu\,\delta(\cdot)\,v + |v|^2 v = 0, \quad \mu \in \mathbb{R}. \tag{6.68}$$

(i) Prove that if $\mu = 0$, the equation (6.68) has a one-parameter family of standing wave solutions of the form:

$$v_\omega(x,t) = e^{i\omega t}\,\sqrt{\omega}\,\phi(\sqrt{\omega}x), \quad \omega > 0,$$

with $\phi(x) = \operatorname{sech}(x)$ (positive, even, and radially decreasing) being the ground state.

(ii) Prove that for $\mu \neq 0$ formally

$$v_{\omega,\mu}(x,t) = e^{i\omega t}\sqrt{\omega}\,\varphi_\mu(\sqrt{\omega}x)$$

is a standing wave solution of (6.68), if

$$\varphi(x) = \varphi_{\mu,\omega}(x) = \sqrt{\omega}\,\varphi_\mu(\sqrt{\omega}x)$$

satisfies the elliptic equation

$$-\omega\,\varphi + \varphi'' + |\varphi|^2\varphi + \mu\,\delta(\cdot)\,\varphi = 0, \tag{6.69}$$

with $\varphi \in H^1(\mathbb{R}) \cap H^2(\mathbb{R}\backslash\{0\})$.

(iii) Prove that if $\omega \leq \mu^2/4$, then (6.69) has no even positive radially decreasing nonnull L^2-solution.

(iv) Prove that if $\omega > \mu^2/4$ and $\mu > 0$, then (6.69) has an even, positive, radially decreasing solution of the form:

$$\varphi_{\mu,\omega}(x) = \sqrt{\omega}\operatorname{sech}\left(\sqrt{\omega}|x| + \tanh^{-1}\left(\frac{\mu}{2\sqrt{\omega}}\right)\right). \tag{6.70}$$

(v) Show that if the radially decreasing requirement is removed, then for $\omega > \mu^2/4$ and $\mu < 0$, the formula (6.70) describes another set of solutions of (6.69).
Hint: show that if φ solves (6.69), then

$$\varphi'(0^+) - \varphi'(0^-) + \mu\,\varphi(0) = 0. \tag{6.71}$$

Use that for $x > 0$, one should have

$$\varphi(x) = \sqrt{\omega}\,\mathrm{sech}(\sqrt{\omega}(x + x_0)), \quad x_0 > 0,$$

and a similar argument for $x < 0$. Combine this and (6.71) to obtain the equation:

$$\mu = 2\sqrt{\omega}\,\tanh\,(\sqrt{\omega}x_0),$$

which yields the desired result.

Chapter 7
Korteweg–de Vries Equation

In this chapter, we study the local well-posedness (LWP) for the initial value problem (IVP):

$$\begin{cases} \partial_t v + \partial_x^3 v + v^k \partial_x v = 0, \\ v(x,0) = v_0(x), \end{cases} \tag{7.1}$$

$x, t \in \mathbb{R}$, $k \in \mathbb{Z}^+$. The family of equations above is called the *k-generalized Korteweg–de Vries (k-gKdV) equation*. The case $k = 1$ is known as the Korteweg–de Vries (KdV) equation and is the most famous of the family. It was first derived as a model for unidirectional propagation of nonlinear dispersive long waves [KdV] but it also has been considered in different contexts, namely in its relation with inverse scattering (see Chapter 9, Section 9.6 for a brief introduction to it), in plasma physics, and in algebraic geometry (see [Mu] and references therein). The case $k = 2$ is called the *modified Korteweg–de Vries (mKdV) equation*. Like the KdV equation, it models propagation of weak nonlinear dispersive waves and it also can be solved via inverse scattering, i.e., this is a completely integrable system. There is an important relationship between these two equations given by the Miura transformation [Mu1]. More precisely, if we assume u to be a solution of the mKdV equation, then

$$v = i\sqrt{6}\,\partial_x u + u^2 \tag{7.2}$$

is a solution for the KdV equation. This relation was first used to obtain the inverse scattering results for both equations. Below we return to this transformation when we discuss global and ill-posedness results.

The KdV and mKdV equations have an infinite number of conserved quantities (see [MGK]). For $k > 2$, that is not the case. However, real solutions to the k-gKdV equation have the following conserved quantities: total mass

$$I_1(v) = \int_{-\infty}^{\infty} v(x,t)\,dx = \int_{-\infty}^{\infty} v_0(x)\,dx, \tag{7.3}$$

F. Linares, G. Ponce, *Introduction to Nonlinear Dispersive Equations*, Universitext, DOI 10.1007/978-1-4939-2181-2_7

the L^2-norm

$$I_2(v) = \int_{-\infty}^{\infty} v^2(x,t)\,dx = \int_{-\infty}^{\infty} v_0^2(x)\,dx, \tag{7.4}$$

and the energy

$$I_3(v) = \int_{-\infty}^{\infty} \left((\partial_x v)^2 - c_k\, v^{k+2}\right)(x,t)\,dx = \int_{-\infty}^{\infty} \left((v_0')^2 - c_k\, v_0^{k+2}\right)(x)\,dx, \tag{7.5}$$

where $c_k = 2\{(k+1)(k+2)\}^{-1}$.

The k-gKdV equation admits solitary wave solutions having strong decay at infinity. These solutions are given by $v_{c,k}(x,t) = \phi_{c,k}(x-ct)$, $c > 0$ (c is the propagation speed) where

$$\phi_{c,k}(x) = \left\{\frac{(k+1)(k+2)}{2} c\, \text{sech}^2\left(\frac{k}{2}\sqrt{c}\, x\right)\right\}^{1/k} \tag{7.6}$$

are the unique (up to translation) positive solutions decaying at infinity of

$$-c\varphi + \varphi'' + \frac{1}{k+1}\varphi^{k+1} = 0. \tag{7.7}$$

To motivate the local results that we describe in this chapter, we first note that if v solves (7.1), then, for $\lambda > 0$, so does $v_\lambda(x,t) = \lambda^{2/k} v(\lambda x, \lambda^3 t)$, with data $v_\lambda(x,0) = \lambda^{2/k} v(\lambda x, 0)$.

Note that

$$\|v_\lambda(\cdot,0)\|_{\dot{H}^s} = \|D_x^s\, v_\lambda(\cdot,0)\|_2 = \lambda^{2/k+s-1/2}\|v(\cdot,0)\|_{\dot{H}^s}. \tag{7.8}$$

This suggests that the optimal s, for the power k, is $s = s_k = 1/2 - 2/k$. Thus, $s_k \geq 0$ if and only if $k \geq 4$.

A simple computation shows that

$$\|\phi_{c,k}\|_{\dot{H}^{s_k}} = \|D^{s_k}\phi_{c,k}\|_2 = a_k, \text{ independent of } c, \tag{7.9}$$

and if $s \neq s_k$, $\|D^s\phi_{c,k}\|_2 \to 0$ as either $c \to 0$ or $+\infty$. Later on we illustrate the significance of this.

The best LWP results in Sobolev spaces $H^s(\mathbb{R})$ known for the k-gKdV equation can be summarized as follows:

k	Scaling	Result
1	$s = -\frac{3}{2}$	$s \geq -\frac{3}{4}$
2	$s = -\frac{1}{2}$	$s \geq \frac{1}{4}$
3	$s = -\frac{1}{6}$	$s \geq -\frac{1}{6}$
$k \geq 4$	$s = \frac{1}{2} - \frac{2}{k}$	$s \geq \frac{1}{2} - \frac{2}{k}$

In this chapter, the local results apply to both real and complex solutions. In Chapter 8, where we study global well-posedness for the k-gKdV, we will only consider real solutions since they satisfy the conservation laws (7.4) and (7.5).

Here we provide the proofs of the local results for the initial value problem (IVP) associated to the KdV ($k = 1$), mKdV ($k = 2$), and the L^2-critical gKdV ($k = 4$) equations.

The approach we follow for the last two equations is closely related to the previous one described for the nonlinear Schrödinger (NLS) equation. However, we shall remark that the situation faced here is more difficult to deal with due to the presence of derivatives on the nonlinearity that causes the so-called loss of derivatives. The idea is to analyze the special properties of solutions of the associated linear problem, such as smoothing effects like those of Strichartz (4.32) and Kato type (4.55), maximal function estimates combined with interpolated estimates. These along with some commutator estimates for fractional derivatives and the contraction mapping principle are the main ingredients in this method.

On the other hand, to establish LWP for the IVP associated to the KdV equation we use the function spaces $X_{s,b}$ introduced in the context of dispersive equations in [Bo1]. These functions spaces have a norm given in terms of the symbol of the associated linear operator (in this case $\partial_t + \partial_x^3$) and have been very useful in analysis of the interaction between the nonlinear and the dispersive effects. In this point, the so-called bilinear estimates play a main role to obtain sharp results.

In Section 7.3, we also list some results regarding the supercritical case ($k > 4$). There we use (7.6) and (7.9) mentioned above to illustrate ill-posedness results and thus the sharpness of the LWP results for the k-gKdV equation for $k \geq 4$.

7.1 Linear Properties

In this section, we establish a series of estimates for solutions of the linear initial value problem (IVP):

$$\begin{cases} \partial_t v + \partial_x^3 v = 0, \\ v(x,0) = v_0(x), \end{cases} \tag{7.10}$$

$x,\, t \in \mathbb{R}$. These estimates are useful to show sharp LWP results for the IVP (7.1) for $k = 2$ and $k \geq 4$.

We first recall that the solution of the IVP (7.10) is given by

$$v(x,t) = V(t)v_0(x) = S_t * v_0(x), \tag{7.11}$$

where

$$S_t(x) = \int_{-\infty}^{\infty} e^{2\pi i x\xi}\, e^{8\pi^3 i t\xi^3}\, d\xi = \frac{1}{\sqrt[3]{t}}\, S_1\Big(\frac{x}{\sqrt[3]{t}}\Big)$$

(see (1.30)).

Notice that $\{V(t)\}_{t=-\infty}^{\infty}$ defines a unitary group operator in $H^s(\mathbb{R})$ (see Proposition 4.2).

We begin by showing a sharpened version of the "local smoothing" effect found by Kato and Kruskov and Faminskii (see (4.55)) for solutions of the linear equation (7.10) and the inhomogeneous problem:

$$\begin{cases} \partial_t v + \partial_x^3 v = f, \\ v(x,0) = 0, \end{cases} \tag{7.12}$$

$x,\, t \in \mathbb{R}$.

Lemma 7.1. *The group* $\{V(t)\}_{t=-\infty}^{\infty}$ *satisfies*

$$\|\partial_x V(t) v_0\|_{L_x^\infty L_t^2} \leq c\, \|v_0\|_2, \tag{7.13}$$

$$\Big\|\partial_x^2 \int_0^t V(t-t')f(t')\,dt'\Big\|_{L_x^\infty L_t^2} \leq c\, \|f\|_{L_x^1 L_t^2}. \tag{7.14}$$

Remark 7.1. The proofs show that in (7.13) and (7.14), we can also have $D^{1+i\gamma}$, $D^{2+i\gamma}$, γ real.

Remark 7.2. To simplify the exposition from now on we omit the 2π factor in the Fourier transform. Thus, in particular we write

$$V(t)v_0(x) = \int_{-\infty}^{\infty} e^{i(x\xi + t\xi^3)} \widehat{v_0}(\xi)\,d\xi.$$

Proof. We only give the proof of (7.13), and refer to the proof of Theorem 4.4 estimate (4.28) for an argument similar to that needed to obtain (7.14).

The change of variables $\xi^3 = \eta$ shows that

$$\partial_x V(t) v_0(x) = \frac{1}{3} \int_{-\infty}^{\infty} e^{it\eta} e^{ix\eta^{1/3}} \eta^{-2/3+1/3} \widehat{v_0}(\eta^{1/3})\,d\eta.$$

Using Plancherel's identity (1.11) in the t variable, we get

$$\|\partial_x V(t) v_0\|_{L_t^2}^2 = \frac{1}{9} \int_{-\infty}^{\infty} \big|e^{ix\eta^{1/3}} \eta^{-2/3+1/3} \widehat{v_0}(\eta^{1/3})\big|^2\,d\eta$$

$$= c \int_{-\infty}^{\infty} |\widehat{v_0}(\xi)|^2\,d\xi,$$

using $\eta^{1/3} = \xi$. This proves (7.13). □

A consequence of (7.13) is

Corollary 7.1.

$$\|\partial_x \int_{-\infty}^{\infty} V(t-t')g(\cdot,t')\,dt'\|_{L^2} \le c\,\|g\|_{L^1_x L^2_t}. \tag{7.15}$$

Remark 7.3. The result in (7.15) is equivalent to (7.13) by duality. Note that this corollary implies

$$\sup_{t\in[-T,T]} \|\,\partial_x \int_0^t V(t-t')g(\cdot,t')\,dt'\|_{L^2} \le c\,\|g\|_{L^1_x L^2_t}. \tag{7.16}$$

The next lemma is useful to obtain maximal function estimates.

Lemma 7.2. *For any $x \in \mathbb{R}$,*

$$|I^t(x)| = \Big|\int_{-\infty}^{\infty} e^{i(x\xi+t\xi^3)} \frac{d\xi}{|\xi|^{1/2+i\gamma}}\Big| \le \frac{c\,(1+|\gamma|)}{|x|^{1/2}} \tag{7.17}$$

for γ real.

Proof. Since for $t=0$ the result is obvious (Exercise 1.14), we assume $t \neq 0$ and see that a dilation argument reduces the proof to show

$$|I^1(x)| \le \frac{c\,(1+|\gamma|)}{|x|^{1/2}}.$$

This can be done using a similar argument as the one in the proof of Proposition 1.6, taking into account the following sets:

$$\begin{aligned}
\Omega_1 &= \{\xi \in \mathbb{R} \;:\; |\xi| \le 2\},\\
\Omega_2 &= \{\xi \notin \Omega_1 \;:\; |3\xi^2 + x| \le |x|/2\},\\
\Omega_3 &= \mathbb{R} - (\Omega_1 \cup \Omega_2).
\end{aligned}$$

□

Next, we have maximal function estimates for solutions of (7.10).

Lemma 7.3.

$$\|\sup_{-\infty<t<\infty} |V(t)v_0|\,\|_{L^4_x} = \|V(t)v_0\|_{L^4_x L^\infty_t} \le c\,\|D_x^{1/4} v_0\|_2, \tag{7.18}$$

$$\Big\|D_x^{-1/2+i\gamma} \int_0^t V(t-t')f(t')\,dt'\Big\|_{L^4_x L^\infty_t} \le c_\gamma \|f\|_{L^{4/3}_x L^1_t}, \tag{7.19}$$

where γ is real.

Remark 7.4. The estimate (7.18) due to [KR] is sharp in the sense that for any $p \neq 4$ on the left-hand side require even in a finite time interval more than $1/4$ derivatives on the right-hand side of the inequality.

Proof. Showing estimate (7.18) is equivalent to proving

$$\|D_x^{-1/4+i\gamma}\, V(t)v_0\|_{L_x^4 L_t^\infty} \leq c_\gamma \|v_0\|_2. \tag{7.20}$$

Hence, we do so for $\gamma = 0$ and prove (7.20).

We see that a version of (7.19) implies (7.20) by a method that follows the proof of the Stein–Tomas L^2-restriction theorem for the Fourier transform. In fact, duality shows that (7.18) is equivalent to

$$\Big\| \int_{-\infty}^{\infty} D_x^{-1/4} V(t) g(\cdot,t)\, dt \Big\|_{L^2} \leq c\, \|g\|_{L_x^{4/3} L_t^1}.$$

Squaring the left-hand side of the inequality we obtain

$$\Big\| \int_{-\infty}^{\infty} D_x^{-1/4} V(t) g(\cdot,t)\, dt \Big\|_2^2 = \int\int g(x,t) \int_{-\infty}^{\infty} D_x^{-1/2}\, V(t-t')\overline{g(\cdot,t')}\, dt'\, dx\, dt,$$

so that (7.18) follows from

$$\Big\| \int_{-\infty}^{\infty} D_x^{-1/2}\, V(t-t') g(\cdot,t')\, dt' \Big\|_{L_x^4 L_t^\infty} \leq c\, \|g\|_{L_x^{4/3} L_t^1}. \tag{7.21}$$

Next, we observe that (7.17) shows that

$$\Big| \int_{-\infty}^{\infty} D_x^{-1/2}\, V(t-t') g(\cdot,t')\, dt' \Big| \leq \frac{c}{|x|^{1/2}} * \int_{-\infty}^{\infty} |g(\cdot,t')|\, dt'.$$

Thus, inequality (7.21) can be deduced from the Hardy–Littlewood–Sobolev theorem (Theorem 2.6), since $\dfrac{c}{|x|^{1/2}} * : L^{4/3} \to L^4$. The estimate (7.19) follows by the same argument. □

Lemma 7.4.

1. If $v_0 \in L^2(\mathbb{R})$, *then*

$$\|V(t)v_0\|_{L_x^5 L_t^{10}} \leq c\|v_0\|_2. \tag{7.22}$$

2. If $g \in L_x^{5/4} L_t^{10/9}$, *then*

$$\Big\| \int_0^t V(t-t') g(t')\, dt' \Big\|_{L_x^5 L_t^{10}} \leq c\|g\|_{L_x^{5/4} L_t^{10/9}}. \tag{7.23}$$

Proof. To prove (7.22), we consider the analytic family of operators

$$T_z v_0 = D_x^{-z/4} D_x^{(1-z)} V(t) v_0, \quad \text{with } z \in \mathbb{C},\ 0 \le \mathcal{R}e\, z \le 1.$$

When $z = i\gamma$,

$$T_{i\gamma} v_0 = \frac{\partial}{\partial x} V(t) D_x^{-i5\gamma/4} \mathsf{H} v_0,$$

where H denotes the Hilbert transform (see (1.18)). Hence, estimate (7.13) implies

$$\|T_{i\gamma}\, v_0\|_{L_x^\infty L_t^2} \le c\, \|v_0\|_2,$$

where we used that $\|D_x^{-i5\gamma/4} \mathsf{H} v_0\|_2 = \|v_0\|_2$.

On the other hand, setting $z = 1 + i\gamma$ we get

$$T_{1+i\gamma}\, v_0 = D_x^{-1/4}\, V(t) D_x^{-i\gamma 5/4}\, v_0.$$

Thus, estimate (7.18) yields

$$\|T_{1+i\gamma}\, v_0\|_{L_x^4 L_t^\infty} \le c\, \|v_0\|_2.$$

Hence, from Stein's analytic interpolation (Theorem 2.7), the estimate (7.22) follows by choosing $z = 4/5$.

The proof of part 2 uses a similar, but more delicate arguments (see Corollary 3.8 in [KPV4]). □

Lemma 7.5. *If $v_0 \in L^2(\mathbb{R})$, then*

$$\|D_x V(t) v_0\|_{L_x^{20} L_t^{5/2}} \le \|D_x^{1/4} v_0\|_2. \tag{7.24}$$

Proof. The result follows using the Stein interpolation theorem (see Theorem 2.7) and the estimates (7.13), i.e.,

$$\|D_x^{5/4} D_x^{i\gamma}\, V(t) v_0\|_{L_x^\infty L_t^2} \le c\, \|D_x^{1/4} v_0\|_2,$$

and (7.18), i.e.,

$$\|D_x^{i\gamma}\, V(t) v_0\|_{L_x^4 L_t^\infty} \le c\, \|D_x^{1/4} v_0\|_2,$$

with $\theta = 4/5$. □

Lemma 7.6 (Leibniz rule).

(i) *Let $\alpha \in (0,1)$. Let $p \in (1,\infty)$, $f = f(x)$, $g = g(x)$, then*

$$\|D^\alpha (fg) - f D^\alpha g\|_p \le c\, \|g\|_\infty \|D^\alpha f\|_p. \tag{7.25}$$

(ii) *Let* $\alpha \in (0,1)$, $\alpha_1, \alpha_2 \in [0,\alpha]$ *with* $\alpha = \alpha_1 + \alpha_2$. *Let* $p, q, p_1, p_2, q_2 \in (1,\infty)$, $q_1 \in (1,\infty]$ *be such that*

$$\frac{1}{p} = \frac{1}{p_1} + \frac{1}{p_2} \qquad \text{and} \qquad \frac{1}{q} = \frac{1}{q_1} + \frac{1}{q_2}.$$

Let $f = f(x,t)$ *and* $g = g(x,t)$. *Then,*

$$\|D_x^\alpha(fg) - f D_x^\alpha g - g D_x^\alpha f\|_{L_x^p L_T^q} \leq c\, \|D_x^{\alpha_1} f\|_{L_x^{p_1} L_T^{q_1}} \|D_x^{\alpha_2} g\|_{L_x^{p_2} L_T^{q_2}}. \tag{7.26}$$

Moreover, for $\alpha_1 = 0$ *the value* $q_1 = \infty$ *is allowed.*

Lemma 7.7 (Chain rule). *Let* $\alpha \in (0,1)$ *and* $p, q, p_1, p_2, q_2 \in (1,\infty)$, $q_1 \in (1,\infty]$ *be such that*

$$\frac{1}{p} = \frac{1}{p_1} + \frac{1}{p_2} \qquad \text{and} \qquad \frac{1}{q} = \frac{1}{q_1} + \frac{1}{q_2}.$$

Then,

$$\|D_x^\alpha F(f)\|_{L_x^p L_T^q} \leq c\, \|F'(f)\|_{L_x^{p_1} L_T^{q_1}}\, \|D_x^\alpha f\|_{L_x^{p_2} L_T^{q_2}}. \tag{7.27}$$

For the proof of Lemmas 7.6 and 7.7, we refer to [KPV4] (see also [CrW]).

The extra difficulty in obtaining these estimates comes from the fact that one needs to control derivatives in the space variable in a norm depending on the t variable first.

7.2 mKdV Equation

In this section, we establish the LWP theory for the IVP associated to the modified Korteweg–de Vries (mKdV) equation,

$$\begin{cases} \partial_t v + \partial_x^3 v + v^2\, \partial_x v = 0, \\ v(x,0) = v_0(x), \end{cases} \tag{7.28}$$

$x,\ t \in \mathbb{R}$. The idea of the proof is to use the linear estimates that we have obtained in the previous section plus a contraction mapping argument. As in the case of the nonlinear Schrödinger (NLS) equation, we employ the integral equation form of (7.28) for the same reason, i.e., it does not require differentiability of the solution.

Theorem 7.1. *Let* $s \geq 1/4$. *Then, for any* $v_0 \in H^s(\mathbb{R})$ *there exist* $T = T(\|D_x^{1/4} v_0\|_2) = c\, \|D_x^{1/4} v_0\|_2^{-4}$ *and a unique strong solution* $v(t)$ *of the IVP (7.28) such that*

$$v \in C([-T,T] : H^s(\mathbb{R})), \tag{7.29}$$

$$\|D_x^s\,\partial_x v\|_{L_x^\infty L_T^2} = \sup_{-\infty<x<\infty}\left(\int_{-T}^{T}|D_x^s\partial_x v(x,t)|^2\,dt\right)^{1/2} < \infty, \tag{7.30}$$

$$\|D_x^{s-1/4}\partial_x v\|_{L_x^{20}L_T^{5/2}} + \|D_x^s v\|_{L_x^5 L_T^{10}} < \infty, \tag{7.31}$$

and

$$\|v\|_{L_x^4 L_T^\infty} < \infty. \tag{7.32}$$

Moreover, there exists a neighborhood $\mathcal{V}$ of v_0 in $H^s(\mathbb{R})$ such that the map $\widetilde{v}_0 \mapsto \widetilde{v}(t)$ from $\mathcal{V}$ into the class defined by (7.29)–(7.32) is smooth.

Proof. We define

$$\mathcal{X}_T = \{v \in C([-T,T]:H^s(\mathbb{R})):\ \|\!|v|\!\|_T < \infty\}$$

and

$$\mathcal{X}_T^a = \{v \in C([-T,T]:H^s(\mathbb{R})):\ \|\!|v|\!\|_T \le a\},$$

where

$$\|\!|v|\!\|_T = \|v\|_{L_T^\infty H^s} + \|D_x^s\,\partial_x v\|_{L_x^\infty L_T^2} + \|D_x^{s-1/4}\partial_x v\|_{L_x^{20}L_T^{5/2}} + \|D_x^s v\|_{L_x^5 L_T^{10}} + \|v\|_{L_x^4 L_T^\infty}.$$

We shall prove that for appropriate values of a and T the operator

$$\Psi_{v_0}(v)(t) = \Psi(v)(t) = V(t)v_0 - \int_0^t V(t-t')(v^2\partial_x v)(t')\,dt' \tag{7.33}$$

defines a contraction map on $\mathcal{X}_T^a$.

We only consider the case $s = 1/4$. As the higher derivatives derivatives appear linearly in the norms (7.29)–(7.31), argument below also provide the proof in the general case $s > 1/4$.

Using the operator (7.33), group properties, and the Minkowsky and Cauchy–Schwarz inequalities it follows that

$$\begin{aligned}\|D_x^{1/4}\Psi(v)(t)\|_2 &\le c\,\|D_x^{1/4}v_0\|_2 + \int_0^t \|D_x^{1/4}(v^2\,\partial_x v)\|_2\,dt\\ &\le c\,\|D_x^{1/4}v_0\|_2 + c\,T^{1/2}\|D_x^{1/4}(v^2\,\partial_x v)\|_{L_x^2 L_T^2}.\end{aligned}$$

To estimate the last term we make use of the Leibniz rule for fractional derivatives (7.26) and the chain rule (7.27). Thus,

$$\|D_x^{1/4}(v^2\,\partial_x v)\|_{L_x^2 L_T^2}$$

$$
\begin{aligned}
&\leq \|v^2\|_{L^2_xL^\infty_T} \|D^{1/4}_x \partial_x v\|_{L^\infty_x L^2_T} + \|D^{1/4}_x(v^2)\|_{L^{20/9}_x L^{10}_T} \|\partial_x v\|_{L^{20}_x L^{5/2}_T} \\
&\leq \|v\|^2_{L^4_xL^\infty_T} \|D^{1/4}_x \partial_x v\|_{L^\infty_x L^2_T} + \|v\|_{L^4_xL^\infty_T} \|D^{1/4}_x v\|_{L^5_x L^{10}_T} \|\partial_x v\|_{L^{20}_x L^{5/2}_T} \qquad (7.34)\\
&\leq c\, |\!|\!| v |\!|\!|^3_T .
\end{aligned}
$$

Hence,

$$
\|D^{1/4}_x \Psi(v)(t)\|_2 \leq c\, \|D^{1/4}_x v_0\|_2 + c\, T^{1/2}\, |\!|\!| v |\!|\!|^3_T . \qquad (7.35)
$$

From Definition 7.33, Minkowski's inequality, group properties, the smoothing effect (7.13), and the Cauchy–Schwarz inequality it follows that

$$
\begin{aligned}
&\|D^{1/4}_x \partial_x \Psi(v)(t)\|_{L^\infty_x L^2_T} \\
&\quad \leq c\, \|D^{1/4}_x v_0\|_2 + \int_0^t \|\partial_x (V(t)V(-t')D^{1/4}_x(v^2\, \partial_x v))\|_{L^\infty_x L^2_T}\, dt' \qquad (7.36)\\
&\quad \leq c\, \|D^{1/4}_x v_0\|_2 + c\, T^{1/2} \|D^{1/4}_x (v^2\, \partial_x v)\|_{L^2_x L^2_T} .
\end{aligned}
$$

The maximal function norm of Ψ can be estimated applying Minkowski's inequality, group properties, (7.18), and the Cauchy–Schwarz inequality:

$$
\begin{aligned}
\|\Psi(v)(t)\|_{L^4_x L^\infty_T} &\leq c\, \|D^{1/4}_x v_0\|_2 + \int_0^t \|V(t)V(-t')(v^2\, \partial_x v)\|_{L^4_x L^\infty_T}\, dt' \\
&\leq c\, \|D^{1/4}_x v_0\|_2 + c\, T^{1/2} \|D^{1/4}_x (v^2\, \partial_x v)\|_{L^2_x L^2_T} .
\end{aligned} \qquad (7.37)
$$

A similar argument as the previous one, but using (7.24) instead of (7.18) yields

$$
\begin{aligned}
\|\partial_x \Psi(v)(t)\|_{L^{20}_x L^{5/2}_T} &\leq c\, \|D^{1/4}_x v_0\|_2 + \int_0^t \|\partial_x V(t)V(-t')(v^2\, \partial_x v)\|_{L^{20}_x L^{5/2}_T}\, dt' \\
&\leq c\, \|D^{1/4}_x v_0\|_2 + c\, T^{1/2} \|D^{1/4}_x (v^2\, \partial_x v)\|_{L^2_x L^2_T} .
\end{aligned} \qquad (7.38)
$$

Finally, the estimate (7.22) and the above argument gives us

$$
\begin{aligned}
\|D^{1/4}_x \Psi(v)(t)\|_{L^5_x L^{10}_T} &\leq c \|D^{1/4}_x v_0\|_2 + \int_0^t \|V(t)V(-t')D^{1/4}_x(v^2\, \partial_x v)\|_{L^5_x L^{10}_T}\, dt' \\
&\leq c\, \|D^{1/4}_x v_0\|_2 + c\, T^{1/2} \|D^{1/4}_x (v^2\, \partial_x v)\|_{L^2_x L^2_T} .
\end{aligned} \qquad (7.39)
$$

Hence, the argument in (7.34) applied in (7.36)–(7.39) yields

$$
|\!|\!| \Psi(v)(t) |\!|\!|_T \leq c\, \|v_0\|_{1/4,2} + c\, T^{1/2} |\!|\!| v |\!|\!|^3_T . \qquad (7.40)
$$

Choosing $a = 2c\|v_0\|_{1/4,2}$ and T such that

$$ca^2\, T^{1/2} < \frac{1}{2} \tag{7.41}$$

we obtain that $\Psi_{v_0} : \mathcal{X}_T^a \to \mathcal{X}_T^a$.

A similar argument shows that

$$|||\Psi(v) - \Psi(\tilde{v})|||_T \le cT^{1/2}(|||v|||_T^2 + |||\tilde{v}|||_T^2)\, |||v - \tilde{v}|||_T \le 2c\, T^{1/2} a^2\, |||v - \tilde{v}|||_T.$$

Then, the choice of a and T in (7.41) implies that Ψ is a contraction. Consequently, we have that there exists a unique $v \in \mathcal{X}_T^a$ with $\Psi_{v_0}(v) \equiv v$, i.e.,

$$v(t) = V(t)v_0 - \int_0^t V(t-t')(v^2\partial_x v)(t')\, dt'. \tag{7.42}$$

Using similar arguments as above, we also deduce that for $T_1 \in (0, T)$

$$|||\Psi_{v_0}(v) - \Psi_{\tilde{v}_0}(\tilde{v})||| \le c\, \|v_0 - \tilde{v}_0\|_{s,2} + c\, T_1^{1/2}(|||v|||^2 + |||\tilde{v}|||^2)\, |||v - \tilde{v}|||.$$

This shows that for $T_1 \in (0, T)$, the map $\tilde{v}_0 \mapsto \tilde{v}$ on a neighborhood $\mathcal{W}$ of v_0 depending on T_1 to $\mathcal{X}_T^a$ is Lipschitz. We notice that an argument as the one used in Corollary 5.6 allows one to prove that this map actually is smooth.

Hence, the solution $v(\cdot) \in \mathcal{X}_T^a$ of the integral equation (7.42) is a strong solution of the IVP (7.28). In particular, v satisfies the equation in (7.28) in the distribution sense.

Next, we extend the uniqueness result to the class $\mathcal{X}_T$. Suppose $w \in \mathcal{X}_{T_1}$ for small $T_1 \in (0, T)$ is a strong solution of the IVP (7.28). The argument used in (7.40) shows that for some $T_2 \in (0, T_1)$, $w \in \mathcal{X}_{T_2}^a$. Thus, (7.41) implies $w \equiv v$ in $\mathbb{R} \times [-T_2, T_2]$. By reapplying this argument, the result can be extended to the whole interval $[-T, T]$. This yields the uniqueness result in $\mathcal{X}_T$. □

7.3 Generalized KdV Equation

The local theory for the IVP (7.1) when $k \ge 4$ is discussed in this section. We will prove the local theory for the critical case $k = 4$. For the case $k > 4$, we give the statements of the LWP results without proofs and talk over the sharpness of these results.

We first consider the L^2-critical case (see 7.8), i.e.,

$$\begin{cases} \partial_t v + \partial_x^3 v + v^4\partial_x v = 0, \\ v(x,0) = v_0(x). \end{cases} \tag{7.43}$$

To show the LWP for (7.43), we follow a similar approach to the one applied for the mKdV equation.

Theorem 7.2. *There exists* $\delta > 0$ *such that for any* $v_0 \in L^2(\mathbb{R})$ *with*

$$\|v_0\|_2 < \delta,$$

there exists a unique strong solution $v(\cdot)$ *of the IVP (7.43) satisfying*

$$v \in C(\mathbb{R} : L^2(\mathbb{R})) \cap L^\infty(\mathbb{R} : L^2(\mathbb{R})), \tag{7.44}$$

$$\|\partial_x v\|_{L_x^\infty L_t^2} < \infty, \tag{7.45}$$

and

$$\|v\|_{L_x^5 L_t^{10}} < \infty. \tag{7.46}$$

Moreover, the map $v_0 \mapsto v(t)$ *from* $\{v_0 \in L^2(\mathbb{R}) : \|v_0\|_2 < \delta\}$ *into the class defined by (7.44)–(7.46) is smooth.*

Remark 7.5. Observe that this global L^2 result is valid for real or complex solutions. This is due to the homogeneity of the equation (scaling argument) and not to the L^2 conserved quantity.

Remark 7.6. It is expected that δ in the theorem be equal to the size of the solitary wave solution in the L^2-norm (7.6) with $k = 4$.

Proof of Theorem 7.2 We now define, for $v_0 \in L^2(\mathbb{R})$, $\|v_0\|_2 < \delta$,

$$\Psi(v)(t) = \Psi_{v_0}(v)(t) = V(t)v_0 - \int_0^t V(t - t')\, v^4 \partial_x v(t')\, dt'. \tag{7.47}$$

We shall show that there is $\delta > 0$ and $a > 0$ such that if $\|v_0\|_2 < \delta$, then

$$\Psi : \mathcal{X}_a \to \mathcal{X}_a$$

is a contraction map, where

$$\mathcal{X}_a = \{w \in C(\mathbb{R} : L^2(\mathbb{R})) : |||w||| \le a\}$$

and

$$|||v||| = \|\partial_x v\|_{L_x^\infty L_t^2} + \|v\|_{L_t^\infty L_x^2} + \|v\|_{L_x^5 L_t^{10}}.$$

From (7.47) and (7.15), we have

$$\begin{aligned} \|\Psi(v)(t)\|_2 &\le \|v_0\|_2 + c\, \|\partial_x \int_0^t V(t - t')\, v^5(t')\, dt'\|_2 \le \|v_0\|_2 + c\, \|v^5\|_{L_x^1 L_t^2} \\ &\le \|v_0\|_2 + c\, \|v\|^5_{L_x^5 L_t^{10}} \le \|v_0\|_2 + c|||v|||^5. \end{aligned} \tag{7.48}$$

Similarly, (7.22) and (7.23) lead to

$$
\begin{aligned}
\|\Psi(v)(t)\|_{L^5_x L^{10}_t} &\le c\|v_0\|_2 + c\,\Big\| \int_0^t V(t-t')\,\partial_x(v^5)(t')\,dt'\Big\|_{L^5_x L^{10}_t} \\
&\le c\,\|v_0\|_2 + c\,\|v^4\,\partial_x v\|_{L^{5/4}_x L^{10/9}_t} \\
&\le c\,\|v_0\|_2 + c\,\|v\|^4_{L^5_x L^{10}_t} \|\partial_x v\|_{L^\infty_x L^2_t} \\
&\le c\,\|v_0\|_2 + c|\!|\!|v|\!|\!|^5.
\end{aligned} \tag{7.49}
$$

Finally, we use (7.13) and (7.14) to have

$$
\begin{aligned}
\|\partial_x \Psi(v)(t)\|_{L^\infty_x L^2_t} &\le c\|v_0\|_2 + c\,\Big\|\partial_x^2 \int_0^t V(t-t')\,(v^5)(t')\,dt'\Big\|_{L^\infty_x L^2_t} \\
&\le c\,\|v_0\|_2 + c\,\|v^5\|_{L^1_x L^2_t} \\
&\le c\,\|v_0\|_2 + c|\!|\!|v|\!|\!|^5.
\end{aligned} \tag{7.50}
$$

Using Remark 7.3, it follows that $\Psi(v) \in C(\mathbb{R} : L^2(R))$. Thus, from (7.48) to (7.50) we obtain that

$$|\!|\!|\Psi(v)|\!|\!| \le c\|v_0\|_2 + c|\!|\!|v|\!|\!|^5.$$

Now, choosing δ such that

$$c(4c\delta)^4 < \frac{1}{2} \qquad \text{and} \qquad a \in (2c\delta,\, 3c\delta),$$

we conclude that $\Psi : \mathcal{X}_a \to \mathcal{X}_a$.

A similar argument leads to

$$|\!|\!|\Psi(v) - \Psi(\tilde{v})|\!|\!| \le c\,(|\!|\!|v|\!|\!|^4 + |\!|\!|\tilde{v}|\!|\!|^4)|\!|\!|v-\tilde{v}|\!|\!| \le 2ca^4|\!|\!|v-\tilde{v}|\!|\!| \le \frac{1}{2}|\!|\!|v-\tilde{v}|\!|\!|.$$

As the remainder of the proof follows the argument employed in Theorem 7.1, it is omitted. □

As a corollary of this result, we have the LWP for the L^2-critical case.

Theorem 7.3 (Critical case). *Let $k = 4$. Given any $v_0 \in L^2(\mathbb{R})$ there exist $T = T(v_0) > 0$ and a unique strong solution $v(\cdot)$ of the IVP (7.43) satisfying*

$$v \in C([-T,T] : L^2(\mathbb{R})), \tag{7.51}$$

$$\|v\|_{L^5_x L^{10}_T} < \infty, \tag{7.52}$$

and

$$\|\partial_x v\|_{L^\infty_x L^2_T} < \infty. \tag{7.53}$$

Given $T' \in (0,T)$, there exists a neighborhood $\mathcal{W}$ of v_0 in $L^2(\mathbb{R})$ such that the map $\tilde{v}_0 \mapsto \tilde{v}(t)$ from $\mathcal{W}$ into the class defined by (7.51)–(7.53), with T' instead of T is smooth.

If $v_0 \in H^s(\mathbb{R})$ with $s > 0$, the previous result extends to the class

$$v \in C([-T,T] : H^s(\mathbb{R}))$$

and

$$\|D_x^s\, \partial_x v\|_{L_x^\infty L_T^2} < \infty,$$

in the above time interval $[-T,T]$.

Remark 7.7. The norm we define to prove this result is as follows:

$$|||v||| = \|v - V(t)v_0\|_{L_T^\infty L_x^2} + \|\partial_x v\|_{L_x^\infty L_T^2} + \|v\|_{L_x^5 L_T^{10}},$$

which is "similar" to the L^2-critical case for the semilinear Schrödinger equation (see Theorem 5.3). Notice that in Theorem 7.3 the time of existence of the local solution depends on v_0 itself and not on its norm.

Next, we have the subcritical local existence result.

Theorem 7.4. *Let $s > 0$. Then, for any $v_0 \in H^s(\mathbb{R})$ there exist $T = T(\|v_0\|_{s,2})$ (with $T(\rho,s) \to \infty$ as $\rho \to 0$) and a unique strong solution $u(\cdot)$ of the IVP (7.43) satisfying*

$$v \in C([-T,T] : H^s(\mathbb{R})), \tag{7.54}$$

$$\|v\|_{L_x^5 L_T^{10}} + \|D_x^s\, v\|_{L_x^5 L_T^{10}} + \|D_t^{s/3}\, v\|_{L_x^5 L_T^{10}} < \infty, \tag{7.55}$$

and

$$\|\partial_x v\|_{L_x^\infty L_T^2} + \|D_x^s\, \partial_x v\|_{L_x^\infty L_T^2} + \|D_t^{s/3}\, \partial_x v\|_{L_x^\infty L_T^2} < \infty. \tag{7.56}$$

Given $T' \in (0,T)$, there exists a neighborhood $\mathcal{V}$ of v_0 in $H^s(\mathbb{R})$ such that the map $\tilde{v}_0 \to \tilde{v}(t)$ from $\mathcal{V}$ into the class defined by (7.54), (7.55), and (7.56) with T' instead of T is smooth.

Next, we consider the IVP (7.1) in the L^2-supercritical case, i.e., $k > 4$. The results are listed without proof.

Theorem 7.5. *Let $k > 4$ and $s_k = (k-4)/2k$. Then, there exists $\delta_k > 0$ such that for any $v_0 \in \dot{H}^{s_k}(\mathbb{R})$ with*

$$\|D_x^{s_k} v_0\|_2 \le \delta_k,$$

there exists a unique strong solution $v(\cdot)$ of the IVP (7.1) satisfying

$$v \in C(\mathbb{R} : \dot{H}^{s_k}(\mathbb{R})) \cap L^\infty(\mathbb{R} : \dot{H}^{s_k}(\mathbb{R})), \tag{7.57}$$

$$\|D_x^{s_k}\,\partial_x v\|_{L_x^\infty L_t^2} < \infty, \tag{7.58}$$

$$\|D_x^{s_k}\,v\|_{L_x^5 L_t^{10}} < \infty, \tag{7.59}$$

and

$$\left\| D_x^{1/10-2/5k}\, D_t^{3/10-6/5k}\, v \right\|_{L_x^{p_k} L_t^{q_k}} < \infty, \tag{7.60}$$

where

$$\frac{1}{p_k} = \frac{2}{5k} + \frac{1}{10} \quad \text{and} \quad \frac{1}{q_k} = \frac{3}{10} - \frac{4}{5k}.$$

Moreover, the map $v_0 \mapsto v(t)$ from

$$\mathcal{V} = \{v_0 \in \dot{H}^{s_k}(\mathbb{R}) \,:\, \|D_x^{s_k} v_0\|_2 \leq \delta_k\}$$

into the class defined by (7.57)–(7.60) is smooth.

Next, we have the result corresponding to any size data.

Theorem 7.6. *Let $k > 4$ and $s_k = (k-4)/2k$. Given $v_0 \in \dot{H}^{s_k}(\mathbb{R})$, there exist $T = T(v_0) > 0$ and a unique strong solution $v(\cdot)$ of the IVP (7.1) satisfying*

$$v \in C([-T, T] : \dot{H}^{s_k}(\mathbb{R})), \tag{7.61}$$

$$\|D_x^{s_k}\,\partial_x v\|_{L_x^\infty L_T^2} < \infty, \tag{7.62}$$

$$\|D_x^{s_k}\,v\|_{L_x^5 L_T^{10}} < \infty, \tag{7.63}$$

and

$$\left\| D_x^{1/10-2/5k}\, D_t^{3/10-6/5k}\, v \right\|_{L_x^{p_k} L_T^{q_k}} < \infty \tag{7.64}$$

with p_k and q_k as in (7.60).

Given $T' \in (0, T)$, there exists a neighborhood $\mathcal{W}$ of $v_0 \in \dot{H}^{s_k}(\mathbb{R})$ such that the map $\tilde{v}_0 \to \tilde{v}(t)$ from $\mathcal{W}$ into the class defined by (7.61)–(7.64) is smooth.

If $v_0 \in H^s(\mathbb{R})$ with $s \geq s_k$, the previous results extend to the class

$$v \in C([-T, T] \,:\, H^s(\mathbb{R}))$$

and

$$\|D_x^s\,\partial_x v\|_{L_x^\infty L_t^2} < \infty$$

in the above interval $[-T, T]$.

Corollary 7.2. *Let $k > 4$ and $s > s_k = (k-4)/2k$. Then, for any $v_0 \in H^s(\mathbb{R})$ there exist $T = T(\|v_0\|_{s,2}) > 0$ ($T(\rho; s) \to \infty$ as $\rho \to 0$) and a unique strong solution $v(\cdot)$ of the IVP (7.1) satisfying, in addition to (7.62)–(7.64):*

$$v \in C([-T, T] \,:\, H^s(\mathbb{R})), \tag{7.65}$$

$$\| D_x^s\, \partial_x v \|_{L_x^\infty L_T^2} + \| D_t^{s/3}\, \partial_x v \|_{L_x^\infty L_T^2} < \infty, \tag{7.66}$$

$$\| D_x^s\, v \|_{L_x^5 L_T^{10}} + \| D_t^{s/3}\, v \|_{L_x^5 L_T^{10}} < \infty, \tag{7.67}$$

and

$$\| D_x^{s/5}\, D_t^{3s/5}\, v \|_{L_x^{p_k} L_T^{q_k}} < \infty \tag{7.68}$$

with p_k *and* q_k *as in (7.60).*

Given $T' \in (0,T)$*, there exists a neighborhood* $\mathcal{W}$ *of* $v_0 \in \dot{H}^{s_k}(\mathbb{R})$ *such that the map* $\tilde{v}_0 \to \tilde{v}(t)$ *from* $\mathcal{W}$ *into the class defined by (7.65)–(7.68) is smooth.*

If $v_0 \in H^{s'}(\mathbb{R})$ *with* $s' > s$*, the previous results hold with* s' *instead of* s *in the same time interval* $[-T, T]$*.*

To conclude, we discuss the sharpness of the results described in this section.

In [BKPSV], it was proved that if the notion of well-posedness given in Chapter 5 is strengthened, then the IVP (7.1) is ill-posed for $k \geq 4$. More precisely, we have the following.

Theorem 7.7. *The IVP (7.1) with* $k \geq 4$ *is ill-posed in* $H^{s_k}(\mathbb{R})$ *with* $s_k = 1/2 - 2/k$ *in the sense that the time of existence and the continuous dependence cannot be expressed in terms of the size of the data in the* H^{s_k}*-norm.*

Proof. We only consider the case $k = 4$. We prove that if we assume $T = T(\|v_0\|_{L^2}) > 0$, then the part in the theorem regarding the continuous dependence of the solution upon the data fails. The proof below also establishes the second part of the theorem.

Consider the solitary wave solutions $\phi_{c,4}$ in (7.6) and $v_{c_k,4}(x,t)$ the solution corresponding to initial data $v_0(x) = \phi_{c_k,4}(x)$. We compare

$$\|\phi_{c_1,4} - \phi_{c_2,4}\|_2^2 \qquad \text{and} \qquad \|v_{c_1,4}(\cdot,t) - v_{c_2,4}(\cdot,t)\|_2^2$$

for $t \neq 0$. We show that one can choose c_1 and c_2 so that the first expression tends to 0 while the second one does not. Thus, well-posedness cannot hold for these data.

Let $a_4^2 = \int \phi_{c_j,4}^2$, $j = 1,2$, and note that

$$\|\phi_{c_1,4} - \phi_{c_2,4}\|_2^2 = \|\phi_{c_1,4}\|_2^2 + \|\phi_{c_2,4}\|_2^2 - 2\langle\, \phi_{c_1,4},\, \phi_{c_2,4} \rangle.$$

Writing $\varphi_4(x) = 3^{1/4}\,(\operatorname{sech}^2(2x))^{1/4}$, the inner product equals

$$c_1^{1/4} c_2^{1/4} \int_{-\infty}^{\infty} \varphi_4(\sqrt{c_1}x)\varphi_4(\sqrt{c_2}x)\, dx.$$

If $\sqrt{c_1}x = y$, we get

$$\Big(\frac{c_1}{c_2}\Big)^{1/4} \int_{-\infty}^{\infty} \varphi_4(y)\varphi_4\Big(\sqrt{\frac{c_1}{c_2}}y\Big)\, dy \to a_4^2 \quad \text{if} \quad \frac{c_1}{c_2} \to 1.$$

Thus,

$$\|\phi_{c_1,4} - \phi_{c_2,4}\|_2^2 \to 0. \tag{7.69}$$

Analyzing $\|v_{c_1,4}(\cdot,t) - v_{c_2,4}(\cdot,t)\|_2^2$ similarly, we obtain

$$a_4^2 + a_4^2 - \Big(\frac{c_1}{c_2}\Big)^{1/4} \int_{-\infty}^{\infty} \varphi_4(y - c_1^{3/2}t)\, \varphi_4\Big(\sqrt{\frac{c_1}{c_2}}\, y - c_2^{3/2}t\Big)\, dy$$

$$= a_4^2 + a_4^2 - \Big(\frac{c_1}{c_2}\Big)^{1/4} \int_{-\infty}^{\infty} \varphi_4(z)\, \varphi_4\Big(\sqrt{\frac{c_1}{c_2}}\, z - c_2^{1/2}(c_1 - c_2)t\Big)\, dz.$$

Choose now $c_1/c_2 \to 1$, but $c_2^{1/2}(c_1 - c_2) \to \infty$ (for instance, $c_1 = N+1$, $c_2 = N$, $N \in Z^+$). The rapid decay of φ_4 shows that the integral approaches 0. Thus,

$$\sup_{[0,T]} \|v_{c_1,4}(\cdot,t) - v_{c_2,4}(\cdot,t)\|_2^2 \to 2\,a_4^2 \quad \text{for any } T > 0 \tag{7.70}$$

as $c_1/c_2 \to 1$.

Finally, (7.8), (7.69), and (7.70) yield the result. □

7.4 KdV Equation

In this section, we establish the local theory of the IVP associated to the KdV equation, that is,

$$\begin{cases} \partial_t v + \partial_x^3 v + v\partial_x v = 0, \\ v(x,0) = v_0(x), \end{cases} \tag{7.71}$$

$x, t \in \mathbb{R}$. The method used here is quite different from the one illustrated for the NLS equation in Chapter 5 and in the previous three sections of this chapter for the mKdV and generalized KdV (gKdV) equations.

We start out by defining the function spaces introduced in the context of dispersive equations by Bourgain in [Bo1]:

Definition 7.1. For $s, b \in \mathbb{R}$ and $f \in \mathcal{S}'(\mathbb{R}^2)$, we say that $f \in X_{s,b}$, if

$$\|f\|_{X_{s,b}} = \Big(\int_{\mathbb{R}^2} (1 + |\tau - \xi^3|)^{2b} (1 + |\xi|)^{2s}\, |\widehat{f}(\xi,\tau)|^2\, d\xi d\tau\Big)^{1/2} < \infty, \tag{7.72}$$

where $\widehat{\ }$ denotes the Fourier transform in $\mathbb{R}^2$.

We solve (a variant) of the integral equation, namely

$$v(t) = \theta(t)V(t)v_0 + \theta(t) \int_0^t V(t - t')v\partial_x v(t')\, dt', \tag{7.73}$$

where $\theta \in C_0^\infty(\mathbb{R})$, $0 \le \theta \le 1$, $\theta \equiv 1$ near 0, supp $\theta \subseteq [-1,1]$ with $v_0 \in H^s(\mathbb{R})$ and $v \in X_{s,b}$.

Remark 7.8. Let $\widehat{J^s f}(\xi) = (1+|\xi|^2)^{s/2}\widehat{f}(\xi)$, and $\widehat{\Lambda^b f}(\tau) = (1+|\tau|^2)^{b/2}\widehat{f}(\tau)$, where $\widehat{\ }$ denotes the Fourier transform in one variable. Then,

$$\|f\|_{X_{s,b}} = \|\Lambda^b J^s V(-t)f\|_{L^2_\xi L^2_\tau}.$$

Corollary 7.3. *If $b > 1/2$,*

$$X_{s,b} \subset C((-\infty,\infty) : H^s(\mathbb{R})).$$

This is an easy consequence of Remark 7.8 and the usual Sobolev embedding theorem.

Let us set $\theta_\rho(t) = \theta(\rho^{-1}t)$, $\rho \in (0,1]$, where θ is as above.

Lemma 7.8. *For any $b > 1/2$ and $s \in \mathbb{R}$,*

$$\|\theta_\rho V(t)v_0\|_{X_{s,b}} \le c\,\rho^{(1-2b)/2}\,\|v_0\|_{s,2}. \tag{7.74}$$

Proof.

$$\theta_\rho(t)V(t)v_0 = \theta(\rho^{-1}t)\int_{-\infty}^{\infty}\int_{-\infty}^{\infty} e^{ix\xi}\,e^{it\tau}\,\delta(\tau-\xi^3)\widehat{v_0}\,d\xi\,d\tau,$$

so that $(\theta(\rho^{-1}t)V(t)v_0)^\wedge(\xi,\tau) = \rho\,\widehat{\theta}(\rho(\tau-\xi^3))\,\widehat{v_0}(\xi)$. Thus,

$$\begin{aligned}
&\|\theta_\rho(t)V(t)v_0\|^2_{X_{s,b}}\\
&= c\,\rho^2\int_{-\infty}^{\infty}\int_{-\infty}^{\infty} |\widehat{\theta}(\rho(\tau-\xi^3))|^2(1+|\tau-\xi^3|)^{2b}(1+|\xi|)^{2s}\,|\widehat{v_0}(\xi)|^2\,d\xi\,d\tau\\
&= c\int_{-\infty}^{\infty}(1+|\xi|)^{2s}|\widehat{v_0}(\xi)|^2\Big(\rho^2\int_{-\infty}^{\infty}|\widehat{\theta}(\rho(\tau-\xi^3))|^2(1+|\tau-\xi^3|)^{2b}\,d\tau\Big)\,d\xi.
\end{aligned}$$

Since $b > 1/2$ and $\rho \in (0,1)$, the inner integral can be estimated as follows:

$$\begin{aligned}
&\rho^2\int_{-\infty}^{\infty}|\widehat{\theta}(\rho(\tau-\xi^3))|^2(1+|\tau-\xi^3|)^{2b}\,d\tau\\
&\le c\rho^2\int_{-\infty}^{\infty}|\widehat{\theta}(\rho(\tau-\xi^3))|^2\,d\tau + c\rho^2\int_{-\infty}^{\infty}|\widehat{\theta}(\rho(\tau-\xi^3))|^2|\tau-\xi^3|^{2b}\,d\tau\\
&\le c\rho + c\rho^{1-2b} \le c\rho^{1-2b}.
\end{aligned}$$

This completes the proof of the lemma. □

Lemma 7.9. *For all $s \in \mathbb{R}$ and $1/2 < b \leq 1$,*

$$\|\theta_\rho v\|_{X_{s,b}} \leq c\,\rho^{(1-2b)/2}\,\|v\|_{X_{s,b}}. \tag{7.75}$$

Proof. $(\theta_\rho(t)v(x,t))^\wedge = \widehat{v} *_\tau (\rho\,\widehat{\theta}(\rho\,\cdot))$, so that by definition of the $X_{s,b}$-norm, the proof reduces to showing that, for $a \in \mathbb{R}$,

$$\int_{-\infty}^{\infty} |(\rho\widehat{\theta}(\rho\,\cdot)) * \widehat{v}(\tau)|^2\,(1+|\tau-a|)^{2b}\,d\tau \leq c\,\rho^{(1-2b)} \int_{-\infty}^{\infty} |\widehat{v}(\tau)|^2\,(1+|\tau-a|)^{2b}\,d\tau.$$

Since

$$\int_{-\infty}^{\infty} |\rho\,\widehat{\theta}(\rho\tau)|\,d\tau < \infty,$$

it follows that

$$\int_{-\infty}^{\infty} |(\rho\widehat{\theta}(\rho\,\cdot)) * \widehat{v}(\tau)|^2\,d\tau \leq c \int_{-\infty}^{\infty} |\widehat{v}(\tau)|^2\,d\tau.$$

We turn to

$$\int_{-\infty}^{\infty} |(\rho\widehat{\theta}(\rho\,\cdot)) * \hat{v}(\tau)|^2\,|\tau-a|^{2b}\,d\tau = \int_{-\infty}^{\infty} |D_t^b(e^{iat}\,v(t)\theta(\rho^{-1}t))|^2\,dt.$$

The Leibniz rule (7.25) yields

$$\|D_t^b(e^{iat}v\theta(\rho^{-1}\,\cdot)) - e^{iat}v\,D_t^b\theta(\rho^{-1}\,\cdot)\|_{L_t^2} \leq c\,\|D_t^b(e^{iat}v)\|_{L_t^2}\|\theta\|_{L_t^\infty}.$$

Note that $\|\theta\|_{L_t^\infty} \leq c$, and

$$\|D_t^b(e^{iat}v)\|_{L_t^2}^2 = \int_{-\infty}^{\infty} |\widehat{v}(\tau)|^2|\tau-a|^{2b}\,d\tau.$$

Thus, we only have to bound the term:

$$\int_{-\infty}^{\infty} |e^{iat}vD_t^b\theta(\rho^{-1}t)|^2\,dt.$$

But the Sobolev embedding theorem and the fact that $b > 1/2$ lead to

$$\Big(\int_{-\infty}^{\infty} |e^{iat}v\,D_t^b\theta(\rho^{-1}\,\cdot)|^2\,dt\Big)$$

$$
\begin{aligned}
&\le c\Big(\int_{-\infty}^{\infty} |e^{iat}\,v(t)|^2\,dt + \int_{-\infty}^{\infty} |D_t^b(e^{iat}v)|^2\,dt\Big)\,\|D_t^b\theta(\rho^{-1}\cdot)\|_{L_t^2}^2\\
&\le c\Big(\int_{-\infty}^{\infty} |\widehat{v}(\tau)|^2\,d\tau + \int_{-\infty}^{\infty} |\tau-a|^{2b}|\widehat{v}(\tau)|^2\,d\tau\Big)\,\|D_t^b\theta(\rho^{-1}\cdot)\|_{L_t^2}^2.
\end{aligned}
$$

By Plancherel's identity (1.11) and since $b > 1/2$ we have

$$
\|D_t^b\theta(\rho^{-1}\cdot)\|_{L_t^2}^2 = \int_{-\infty}^{\infty} |\tau|^{2b}\rho^2\,|\widehat{\theta}(\rho\tau)|^2\,d\tau \le c\,\rho^{(1-2b)}\,\|\theta\|_{H_t^1}^2.
$$

The proof of the lemma then follows. □

Lemma 7.10. *Let* $w(x,t) = \int_0^t V(t-t')h(t')\,dt'$. *If* $1/2 < b \le 1$; *then*

$$
\|\theta_\rho\, w\|_{X_{s,b}} \le c\,\rho^{(1-2b)/2}\,\|h\|_{X_{s,b-1}}. \tag{7.76}
$$

Proof. We write

$$
\begin{aligned}
&\theta_\rho(t)\int_0^t V(t-t')h(t')\,dt'\\
&= \theta_\rho(t)\int\!\!\int e^{ix\xi}\,\frac{e^{it\tau}-e^{it\xi^3}}{\tau-\xi^3}\,\widehat{h}(\xi,\tau)\,d\xi\,d\tau\\
&= \theta_\rho(t)\int\!\!\int e^{ix\xi}\,\frac{e^{it\tau}-e^{it\xi^3}}{\tau-\xi^3}\,\theta(\tau-\xi^3)\,\widehat{h}(\xi,\tau)\,d\xi\,d\tau \qquad (7.77)\\
&\quad+\theta_\rho(t)\int\!\!\int e^{ix\xi}\,\frac{e^{it\tau}-e^{it\xi^3}}{\tau-\xi^3}\,(1-\theta)(\tau-\xi^3)\,\widehat{h}(\xi,\tau)\,d\xi\,d\tau\\
&\equiv I + II\,.
\end{aligned}
$$

A Taylor expansion gives us

$$
I = \sum_{k=1}^{\infty}\frac{i^k}{k!}\,t^k\theta_\rho(t)\int_{-\infty}^{\infty} e^{ix\xi}\,e^{it\xi^3}\Big(\int_{-\infty}^{\infty}\widehat{h}(\xi,\tau)(\tau-\xi^3)^{k-1}\theta(\tau-\xi^3)\,d\tau\Big)\,d\xi. \tag{7.78}
$$

Let $t^k\theta_\rho(t) = \rho^k\,(t/\rho)^k\,\theta(\rho^{-1}t) = \varphi_k(t)$. Then,

$$
\rho^2\int_{-\infty}^{\infty} |\widehat{\varphi_k}(\rho\tau)|^2\,(1+|\tau|)^{2b}\,d\tau
$$

$$\leq c\rho^2\Big(\int |\widehat{\varphi}_k(\rho\tau)|^2\,d\tau + \int |\tau|^{2b}\,|\widehat{\varphi}_k(\rho\tau)|^2\,d\tau\Big)$$
$$\leq c\rho^{(1-2b)}(\|\varphi_k\|_{L_t^2}^2 + \|D_t^b\,\varphi_k\|_{L_t^2}^2) \leq c\rho^{(1-2b)}\,(1+k)^2.$$

Thus, by the proof of (7.74) and (7.78):

$$\|I\|_{X_{s,b}} \leq \sum_{k=1}^{\infty} \frac{1+k^2}{k!}\rho^k\,\rho^{(1-2b)}\,\Big\|\Big(\int_{-\infty}^{\infty} \hat{h}(\xi,\tau)(\tau-\xi^3)^{k-1}\theta(\tau-\xi^3)\,d\tau\Big)^{\vee}\Big\|_{s,2}.$$

But

$$\Big\|\Big(\int_{-\infty}^{\infty} \widehat{h}(\xi,\tau)(\tau-\xi^3)^{k-1}\theta(\tau-\xi^3)\,d\tau\Big)^{\vee}\Big\|_{s,2}^2$$
$$\leq \int_{-\infty}^{\infty} (1+|\xi|)^{2s}\Big(\int_{-\infty}^{\infty} |\hat{h}(\xi,\tau)(\tau-\xi^3)^{k-1}\theta(\tau-\xi^3)|\,d\tau\Big)^2\,d\xi$$
$$\leq \int_{-\infty}^{\infty} (1+|\xi|)^{2s}\Big(\int_{|\tau-\xi^3|<1} |\hat{h}(\xi,\tau)|\,d\tau\Big)^2\,d\xi$$
$$\leq \int_{-\infty}^{\infty} (1+|\xi|)^{2s}\Big(\int_{-\infty}^{\infty} \frac{|\widehat{h}(\xi,\tau)|}{(1+|\tau-\xi^3|)^{(1-b)}}\frac{1}{(1+|\tau-\xi^3|)^b}\,d\tau\Big)^2\,d\xi$$
$$\leq c\,\|h\|_{X_{s,b-1}}^2$$

since $b > 1/2$.

Next, we estimate II in (7.77). We rewrite it as $II = II_1 + II_2$, where

$$II_1 = -\,\theta_\rho(t)\int_{-\infty}^{\infty} e^{i(x\xi+t\xi^3)}\Big(\int_{-\infty}^{\infty} \frac{(1-\theta)(\tau-\xi^3)}{\tau-\xi^3}\,\widehat{h}(\xi,\tau)\,d\tau\Big)\,d\xi$$

$$II_2 = \theta_\rho(t)\int_{-\infty}^{\infty}\int_{-\infty}^{\infty} e^{i(x\xi+t\tau)}\,\frac{(1-\theta)(\tau-\xi^3)}{\tau-\xi^3}\,\widehat{h}(\xi,\tau)\,d\xi\,d\tau.$$

Using Lemma 7.8, the Cauchy–Schwarz inequality, and $b > 1/2$, we deduce

$$\|II_1\|_{X_{s,b}} \leq c\rho^{(1-2b)/2}\,\Big\|\Big(\int_{-\infty}^{\infty} \frac{(1-\theta)(\tau-\xi^3)}{\tau-\xi^3}\,\widehat{h}(\xi,\tau)\,d\tau\Big)^{\vee}\Big\|_{s,2}$$
$$\leq c\rho^{(1-2b)/2}\Big[\int_{-\infty}^{\infty} (1+|\xi|)^{2s}$$

$$\times \Big(\int\limits_{|\tau-\xi^3|\geq 1/2} \frac{1}{1+|\tau-\xi^3|} |\hat{h}(\xi,\tau)|\, d\tau \Big)^2 d\xi \Big]^{1/2}$$

$$\leq c\rho^{(1-2b)/2} \Big[\int\limits_{-\infty}^{\infty} (1+|\xi|)^{2s}$$

$$\times \Big(\int\limits_{|\tau-\xi^3|\geq 1/2} \frac{|\hat{h}(\xi,\tau)|}{(1+|\tau-\xi^3|)^{(1-b)}} \frac{1}{(1+|\tau-\xi^3|)^b} d\tau \Big)^2 d\xi \Big]^{1/2}$$

$$\leq c\rho^{(1-2b)/2} \|h\|_{X_{s,b-1}}.$$

Finally, by (7.75) and the definition of $X_{s,b-1}$,

$$\|II_2\|_{X_{s,b}} \leq c\,\rho^{(1-2b)/2} \Big\| \int\limits_{-\infty}^{\infty}\int\limits_{-\infty}^{\infty} e^{i(x\xi+t\tau)} \frac{(1-\theta)(\tau-\xi^3)}{\tau-\xi^3} \widehat{h}(\xi,\tau)\, d\xi\, d\tau \Big\|_{X_{s,b}}$$

$$\leq c\,\rho^{(1-2b)/2} \|h\|_{X_{s,b-1}}.$$

This completes the proof of the lemma. □

Lemma 7.11.

$$\|\theta_\rho(t) \int\limits_0^t V(t-t')h(t')\, dt'\|_{s,2} \leq c\rho^{(1-2b)/2} \|h\|_{X_{s,b-1}}. \tag{7.79}$$

Proof. A similar argument as the one used to show Lemma 7.10 yields (7.79). Thus, we omit it. □

Lemma 7.12. *Let $s \in \mathbb{R}$, $b', b \in (1/2, 7/8)$ with $b < b'$ and $\rho \in (0,1)$, then for $v \in X_{s,b'-1}$, we have*

$$\|\theta_\rho\, v\|_{X_{s,b-1}} \leq c\,\rho^{(b'-b)/8(1-b)} \|v\|_{X_{s,b'-1}}. \tag{7.80}$$

Proof. To prove (7.80), we use duality and prove the estimate

$$\|\theta_\rho\, v\|_{X_{-s,1-b'}} \leq c\,\rho^{(b'-b)/8(1-b)} \|v\|_{X_{-s,1-b}}. \tag{7.81}$$

This result follows by interpolation. To do so, we need to establish the next inequalities:

$$\|\theta_\rho\, v\|_{X_{-s,0}} \leq c\,\rho^{1/8} \|v\|_{X_{-s,1-b}} \tag{7.82}$$

and

$$\|\theta_\rho v\|_{X_{-s,1-b}} \leq c\, \|v\|_{X_{-s,1-b}}. \tag{7.83}$$

Combining Remark 7.8, the Hölder inequality, and the Sobolev inequality (Theorem 3.3), we have

$$\begin{aligned}\|\theta_\rho v\|_{X_{-s,0}} &= \|J^{-s}V(t)(\theta(\rho^{-1}\cdot)v)\|_{L^2_tL^2_x} = \|V(t)\theta(\rho^{-1}\cdot)\,J^{-s}v\|_{L^2_tL^2_x}\\ &= \|\theta(\rho^{-1}\cdot)V(t)\,J^{-s}v\|_{L^2_tL^2_x} \le c\rho^{1/8}\,\|V(t)\,J^{-s}v\|_{L^2_xL^{8/3}_t}\\ &\le c\rho^{1/8}\,\|V(t)\,J^{-s}v\|_{L^2_xH^{1/8}_t} = c\rho^{1/8}\,\|v\|_{X_{-s,1/8}}\\ &\le c\rho^{1/8}\,\|v\|_{X_{-s,1-b}},\end{aligned}$$

where we use that $1-b>1/8$. This shows (7.82).

On the other hand, to prove (7.83) we use a similar argument to the one applied in the proof of Lemma 7.9. Since $(\theta_\rho(t)\,v(x,t))^\wedge = \widehat{\theta}_\rho *_\tau \widehat{v}$, by the definition of the $X_{s,b}$-space, the proof reduces to showing that, for $a\in\mathbb{R}$,

$$\int_{-\infty}^{\infty}|\widehat{\theta}_\rho *_t \widehat{v}|^2(1+|\tau-a|)^{2(1-b)}d\tau \le c\int_{-\infty}^{\infty}|\widehat{v}|^2(1+|\tau-a|)^{2(1-b)}d\tau. \tag{7.84}$$

Since $\|\rho\,\widehat{\theta}(\rho\,\cdot)\|_{L^1_\tau}<\infty$, we have that

$$\int_{-\infty}^{\infty}|\widehat{\theta}_\rho *_t \widehat{v}|^2\,d\tau \le c\int_{-\infty}^{\infty}|\widehat{v}|^2\,d\tau.$$

Next, we estimate

$$\int_{-\infty}^{\infty}|\widehat{\theta}_\rho *_t \widehat{v}|^2\,|\tau-a|^{2(1-b)}\,d\tau = \int_{-\infty}^{\infty}|D_t^{1-b}(e^{iat}v(t)\,\theta(\rho^{-1}t))|^2\,dt.$$

Using the Leibniz rule (7.25) we have that

$$\|D_t^{1-b}(e^{iat}v\theta_\rho) - e^{iat}vD_t^{1-b}\theta_\rho\|_{L^2_t} \le c\,\|D_t^{1-b}(e^{iat}v)\|_{L^2_t}\|\theta_\rho\|_{L^\infty_t}. \tag{7.85}$$

The first term on the right-hand side of (7.85) can be estimated as follows. We first notice that $\|\theta_\rho\|_{L^\infty}<\infty$. Thus, Plancherel identity (1.11) gives us

$$\|D_t^{1-b}(e^{iat}v)\|_{L^2_t} = \Big(\int_{-\infty}^{\infty}|\widehat{v}(\tau)|^2|\tau-a|^{2(1-b)}\,d\tau\Big)^{1/2}. \tag{7.86}$$

To bound $\|e^{iat}v\,D_t^{1-b}\theta_\rho\|_{L^2_t}$, we use the Hölder inequality to obtain

$$\|e^{iat}v\,D_t^{1-b}\theta_\rho\|_{L^2_t} \le \|e^{iat}v\|_{L^{2p}_t}\|\,D_t^{1-b}\theta_\rho\|_{L^{2q}_t},$$

with $1/p+1/q=1$. Then, we choose p such that $1/2-1/2p=1-b$. Using the Sobolev inequality (Theorem 3.3), we have

$$\|e^{iat}v\|_{L^{2p}_t} \le \|e^{iat}v\|_{H^{1-b}_t} = c\Big(\int_{-\infty}^{\infty}(1+|\tau-a|)^{2(1-b)}|\widehat{v}(\tau)|^2\,d\tau\Big)^{1/2}. \tag{7.87}$$

Since the inverse Fourier transform is bounded from $L^{\frac{2q}{(2q-1)}}(\mathbb{R})$ into $L^{2q}(\mathbb{R})$, we have

$$
\begin{aligned}
\| D_t^{1-b}\theta_\rho\|_{L_t^{2q}} &\leq \left(\int_{-\infty}^{\infty} ||\tau|^{1-b}\rho\,\widehat{\theta}(\rho\tau)|^{\frac{2q}{2q-1}}\,d\tau \right)^{\frac{2q-1}{2q}} \\
&= \left(\int_{-\infty}^{\infty} ||\tau|^{1-b}\widehat{\theta}(\tau)|^{\frac{2q}{2q-1}}\,d\tau \right)^{\frac{2q-1}{2q}} < \infty.
\end{aligned}
\tag{7.88}
$$

Combining (7.87) and (7.88), we have

$$
\|e^{iat}v\,D_t^{1-b}\theta_\rho\|_{L_t^2} \leq c \int_{-\infty}^{\infty} (1+|\tau-a|)^{2(1-b)}|\widehat{v}(\tau)|^2\,d\tau. \tag{7.89}
$$

Thus, (7.86) and (7.89) yield (7.84).

The estimates (7.82) and (7.83) and interpolation yield the inequality (7.81). Thus, the lemma follows. □

The next estimate is the key argument to obtain the local result for the IVP (7.71). Notice that when we estimate the $X_{s,b}$-norm of the integral part in Lemma 7.10, we end up in the space $X_{s,b-1}$, we have lost "one derivative," so to apply a contraction mapping argument we need to have an estimate that takes the nonlinear part back to the space $X_{s,b}$.

Lemma 7.13.

1. *If $v \in X_{s,b}$, $s > -3/4$, there exists $b > 1/2$ such that $v\partial_x v \in X_{s,b-1}$ and*

$$
\|\partial_x(v^2)\|_{X_{s,b-1}} \leq c\,\|v\|_{X_{s,b}}^2.
$$

2. *Given $s \leq -3/4$ the estimate above fails for any b.*

We restate Lemma 7.13 in an equivalent form:

For $v \in X_{s,b}$, let $f(\xi,\tau) = \hat{v}(\xi,\tau)(1+|\xi|)^s(1+|\tau-\xi^3|)^b$, so that $\|v\|_{X_{s,b}} = \|f\|_2$. In terms of f we can express $v\partial_x v$ in the following way:

$$
\widehat{\partial_x(v^2)}(\xi,\tau) = i\xi \int_{-\infty}^{\infty}\int_{-\infty}^{\infty} f(\xi_1,\tau_1)\,f(\xi-\xi_1,\tau-\tau_1)
$$

$$
\times \frac{d\xi_1\,d\tau_1}{(1+|\xi_1|)^s(1+|\tau_1-\xi_1^3|)^b(1+|\xi-\xi_1|)^s(1+|(\tau-\tau_1)-(\xi-\xi_1)^3|)^b}.
$$

Thus, if we let

$$
\begin{aligned}
B(f,f,s,b) = &\frac{(1+|\xi|)^s}{(1+|\tau-\xi^3|^{1-b}}|\xi| \\
&\times \int_{\mathbb{R}^2} K(\xi,\xi_1,\tau,\tau_1)f(\xi_1,\tau_1)f(\xi-\xi_1,\tau-\tau_1)\,d\xi_1\,d\tau_1,
\end{aligned}
\tag{7.90}
$$

where

$$K(\xi, \xi_1, \tau, \tau_1) = \frac{(1 + |\xi_1|)^{-s}(1 + |\xi - \xi_1|)^{-s}}{(1 + |\tau_1 - \xi_1^3|)^b(1 + |(\tau - \tau_1) - (\xi - \xi_1)^3|)^b},$$

Lemma 7.13 is equivalent to proving the next result for the bilinear operator $B(\cdot, \cdot)$ defined in (7.90).

Lemma 7.14.

1. If $s > -3/4$, *then*

$$\|B(f, f, s, b)\|_2 \le c \, \|f\|_2^2. \tag{7.91}$$

2. If $s \le -3/4$, *the above estimate fails for each b.*

We prove (7.91) in detail for $s = 0$. For this purpose, we need some lemmas. The first one is regarding some elementary inequalities.

Lemma 7.15. *If* $b > 1/2$, *there exists* $c > 0$ *such that*

$$\int_{-\infty}^{\infty} \frac{dx}{(1 + |x - \alpha|)^{2b}(1 + |x - \beta|)^{2b}} \le \frac{c}{(1 + |\alpha - \beta|)^{2b}}, \tag{7.92}$$

$$\int_{-\infty}^{\infty} \frac{dx}{(1 + |x|)^{2b}\sqrt{|a - x|}} \le \frac{c}{(1 + |a|)^{1/2}}. \tag{7.93}$$

Lemma 7.16. *Let*

$$\begin{aligned} G(\xi, \tau) = & \frac{|\xi|}{(1 + |\tau - \xi^3|)^{1-b}} \\ & \times \left(\int_{-\infty}^{\infty}\int_{-\infty}^{\infty} \frac{d\xi_1 d\tau_1}{(1 + |\tau_1 - \xi_1^3|)^{2b}(1 + |\tau - \tau_1 - (\xi - \xi_1)^3|)^{2b}} \right)^{1/2}. \end{aligned} \tag{7.94}$$

If $1/2 < b \le 3/4$, *then*

$$|G(\xi, \tau)| \le c.$$

Proof. Let us set $\alpha = \xi_1^3$ and $\beta = \tau - (\xi - \xi_1)^3$ in (7.94). Then by (7.92), we have

$$\begin{aligned} & \int_{-\infty}^{\infty}\int_{-\infty}^{\infty} \frac{d\xi_1 d\tau_1}{(1 + |\tau_1 - \xi_1^3|)^{2b}(1 + |\tau - \tau_1 - (\xi - \xi_1)^3|)^{2b}} \\ & \le \int_{-\infty}^{\infty} \frac{d\xi_1}{(1 + |\tau - (\xi - \xi_1)^3 - \xi_1^3|)^{2b}}. \end{aligned}$$

Next, we use the change of variable

$$\mu = \tau - (\xi - \xi_1)^3 - \xi_1^3 = \tau - \xi^3 + 3\xi\xi_1(\xi - \xi_1), \qquad d\mu = 3\xi(\xi - 2\xi_1)\, d\xi_1,$$

and

$$\xi_1 = \frac{1}{2}\left\{\xi \pm \sqrt{\frac{4\tau - \xi^3 - 4\mu}{3\xi}}\right\}.$$

Thus,

$$|\xi(\xi - 2\xi_1)| \simeq \sqrt{|\xi|}\,\sqrt{4\tau - \xi^3 - 4\mu}$$

and

$$d\xi_1 \simeq \frac{d\mu}{\sqrt{|\xi|}\,\sqrt{4\tau - \xi^3 - 4\mu}}.$$

Substituting this on the right-hand side of the previous inequality and using (7.93), we obtain

$$\frac{1}{\sqrt{|\xi|}} \int_{-\infty}^{\infty} \frac{d\mu}{(1 + |\mu|)^{2b}\sqrt{4\tau - \xi^3 - 4\mu}} \leq \frac{1}{\sqrt{|\xi|}} \frac{c}{(1 + |4\tau - \xi^3|)^{1/2}}.$$

Hence, using the hypotheses we conclude that

$$\begin{aligned}|G(\xi, \tau)| &\leq \frac{|\xi|}{(1 + |\tau - \xi^3|)^{1-b}} \frac{1}{|\xi|^{1/4}} \frac{c}{(1 + |4\tau - \xi^3|)^{1/4}}\\ &\leq \frac{c\,|\xi|^{3/4}}{(1 + |\tau - \xi^3|)^{1-b}(1 + |4\tau - \xi^3|)^{1/4}} \leq c.\end{aligned}$$

Thus, the lemma follows. □

Proof of Lemma 7.14 We will prove 1. in the case $s = 0$. Definition (7.90), the Cauchy–Schwarz inequality, Lemma 7.16, Fubinni's theorem, and Young's inequality yield

$$\begin{aligned}\|B(f, f, 0, b)\|_{L^2_\tau L^2_\xi} &= \Big\| \frac{|\xi|}{1 + |\tau - \xi^3|^{1-b}}\\ &\times \int_{-\infty}^{\infty}\int_{-\infty}^{\infty} \frac{f(\xi_1, \tau_1)\, f(\xi - \xi_1, \tau - \tau_1)\, d\xi_1\, d\tau_1}{(1 + |\tau_1 - \xi_1^3|)^b (1 + |(\tau - \tau_1) - (\xi - \xi_1)^3|)^b} \Big\|_{L^2_\tau L^2_\xi}\\ &\leq \Big\| \frac{|\xi|}{(1 + |\tau - \xi^3|)^{1-b}}\\ &\times \Big(\int_{-\infty}^{\infty}\int_{-\infty}^{\infty} \frac{d\xi_1 d\tau_1}{(1 + |\tau_1 - \xi_1^3|)^{2b}(1 + |(\tau - \tau_1) - (\xi - \xi_1)^3|)^{2b}} \Big)^{1/2} \Big\|_{L^\infty_\tau L^\infty_\xi}\\ &\times \Big\| \Big(\int_{-\infty}^{\infty}\int_{-\infty}^{\infty} |f(\xi_1, \tau_1)|^2 |f(\xi - \xi_1, \tau - \tau_1)|^2\, d\xi_1 d\tau_1 \Big)^{1/2} \Big\|_{L^2_\tau L^2_\xi}\end{aligned}$$

$$\leq c\,\|f\|^2_{L^2_\tau L^2_\xi}.$$

This proves the lemma. □

As a corollary, we have the next result.

Corollary 7.4. *For $s > -3/4$ and $b \in (1/2, 3/4]$ and $b' \in (1/2, b]$ we have that*

$$\|B(f,f)\|_{X_{s,b-1}} \leq c\,\|f\|^2_{X_{s,b'}}. \tag{7.95}$$

Now, we are in position to establish the LWP result for the IVP (7.71). More precisely, we have the following.

Theorem 7.8. *Let $s \in (-\,3/4, 0]$. Then, there exists $b \in (1/2, 1)$ such that for any $v_0 \in H^s(\mathbb{R})$ there exist $T = T(\|v_0\|_{s,2})$ with ($T(\rho) \to \infty$ as $\rho \to 0$) and a unique solution $v(t)$ of the IVP (7.71) in the time interval $[-T, T]$ satisfying*

$$v \in C([-T,T] : H^s(\mathbb{R})), \tag{7.96}$$

$$v \in X_{s,b} \subset L^p_{x,\text{loc}}(\mathbb{R} : L^2_t(\mathbb{R})) \quad \text{for } 1 \leq p \leq \infty, \tag{7.97}$$

$$\partial_x(v^2) \in X_{s,b-1} \tag{7.98}$$

and

$$\partial_t v \in X_{s-3,b-1}. \tag{7.99}$$

Moreover, for any $T' \in (0, T)$ there exists $R = R(T') > 0$ such that the map $\tilde{v}_0 \mapsto \tilde{v}(t)$ from $\{\tilde{v}_0 \in H^s(\mathbb{R}) : \|v_0 - \tilde{v}_0\|_{s,2} < R\}$ into the class defined by (7.96)–(7.99) with T' instead of T is smooth.

In addition, if $v_0 \in H^{s'}(\mathbb{R})$ with $s' > s$, the previous results hold with s' instead of s in the same time interval $[-T, T]$.

Proof. We define

$$\mathcal{X}_a = \{v \in X_{s,b} \;:\; \|v\|_{X_{s,b}} < a\}. \tag{7.100}$$

For $v_0 \in H^s(\mathbb{R})$, $s > -3/4$, we define the operator:

$$\Psi_{v_0}(v) = \Psi(v) = \theta_1(t)V(t)v_0 - \frac{\theta_1(t)}{2}\int_0^t V(t-t')\,\theta_\rho(t')\,\partial_x(v^2(t'))\,dt'. \tag{7.101}$$

We see that $\Psi(\cdot)$ defines a contraction on $\mathcal{X}_a$.

Let $\beta = (b-b')/8(1-b')$. By using (7.74), (7.75), (7.76), and (7.91) in Lemma 7.13 we deduce that

$$\begin{aligned}\|\Psi(v)\|_{X_{s,b}} &\leq c\|v_0\|_{s,2} + c\|\theta_\rho(t)\partial_x v^2(\cdot,t))\|_{X_{s,b-1}}\\ &\leq c\|v_0\|_{s,2} + c\,\rho^\beta\,\|\partial_x v^2(\cdot,t))\|_{X_{s,b'-1}}\end{aligned}$$

$$\leq c\|v_0\|_{s,2} + c\,\rho^\beta \|v\|^2_{X_{s,b}}$$
$$\leq c\|v_0\|_{s,2} + c\,\rho^\beta\, a^2. \tag{7.102}$$

Setting $a = 2c\|v_0\|_{H^s}$ and ρ such that

$$c\,\rho^\beta\, a < \frac{1}{2}$$

we have

$$\|\Psi(v)\|_{X_{s,b}} \leq a.$$

A similar argument shows that for $v, \tilde{v} \in \mathcal{X}_a$

$$\begin{aligned}\|\Psi(v) - \Psi(\tilde{v})\|_{X_{s,b}} &= \frac{1}{2}\left\|\theta_1(t)\int_0^t V(t-t')\theta_\rho(t')\,\partial_x(v^2 - \tilde{v}^2)(t')\,dt'\right\|_{X_{s,b}}\\ &\leq c\rho^\beta\,\|v+\tilde{v}\|_{X_{s,b}}\,\|v-\tilde{v}\|_{X_{s,b}}\\ &\leq 2\,c\rho^\beta\, a\,\|v-\tilde{v}\|_{X_{s,b}}\\ &\leq \frac{1}{2}\,\|v-\tilde{v}\|_{X_{s,b}}.\end{aligned}$$

Therefore, $\Psi(\cdot)$ is a contraction from $\mathcal{X}_a$ into itself and we obtain a unique fixed point that solves the equation for $T < \rho$, i.e.,

$$v(t) = \theta_1(t)V(t)v_0 - \frac{\theta_1(t)}{2}\int_0^t V(t-t')\theta_\rho(t')\,\partial_x(v^2(t'))\,dt'. \tag{7.103}$$

The additional regularity

$$v \in C([0,T] : H^s(\mathbb{R}))$$

is proved as follows:

Using the integral equation (7.103), Lemmas 7.11, and 7.12, for $0 \leq \tilde{t} < t \leq 1$ and $t - \tilde{t} \leq \Delta t$ it follows that

$$\begin{aligned}\|v(t) - v(\tilde{t})\|_{s,2} &\leq \|V(t-\tilde{t})v(\tilde{t}) - v(\tilde{t})\|_{s,2}\\ &\quad + c\left\|\int_{\tilde{t}}^t V(t-t')\theta^2\Big(\frac{t'-\tilde{t}}{\Delta t}\Big)\,\partial_x(v^2(t'))\,dt'\right\|_{s,2}\\ &\leq \|V(t-\tilde{t})v(\tilde{t}) - v(\tilde{t})\|_{s,2} + c\,\Big\|\theta\Big(\frac{\cdot-\tilde{t}}{\Delta t}\Big)\partial_x\, v^2\Big\|_{X_{s,b-1}} \qquad (7.104)\\ &\leq \|V(t-\tilde{t})v(\tilde{t}) - v(\tilde{t})\|_{s,2} + c\,(\Delta t)^{\frac{(b-b')}{8(1-b')}}\,\|\partial_x\, v^2\|_{X_{s,b'-1}}\\ &\leq \|V(t-\tilde{t})v(\tilde{t}) - v(\tilde{t})\|_{s,2} + c\,(\Delta t)^{\frac{(b-b')}{8(1-b')}}\,\|v\|^2_{X_{s,b'}} = o(1)\end{aligned}$$

as $\Delta t \to 0$. This yields the persistence property. □

7.5 Comments

The well-posedness of the k-gKdV equation has been studied extensively for many years. Improving results in [BS], [BSc], [ST], it was shown in [K5] that the IVP (7.10) is locally well-posed in $H^s(\mathbb{R})$, $s > 3/2$. However, as Kato remarked in [K2], "In fact, local well-posedness has almost nothing to do with the special structure of the KdV equation." In other words, the local result in $H^s(\mathbb{R})$, $s > 3/2$, does not distinguish the powers k and works for any skew-symmetric operator (instead of ∂_x^3) or omitting it (hyperbolic case).

For the study of the stability of solitary wave solutions of the k-gKdV equation, it was important to have LWP in Sobolev spaces $H^s(\mathbb{R})$ with $s \leq 1$, specially for the case $k = 3$.

For the KdV equation the L^2 LWP was established by Bourgain [Bo1]. The proof given here was taken from [KPV6], where LWP was obtained for data in $H^s(\mathbb{R})$, $s > -3/4$ (Lemmas 7.14 and 7.15 were proved in [KPV6] and by Nakanishi, Takaoka and Tsutsumi [NTT1] in the case $s = -3/4$).

Extensions of the bilinear estimates (Lemma 7.14 (1)) were first obtained by Colliander, Staffilani and Takaoka [CST], motivated by the study of global well-posedness below the L^2-norm for the KdV equation. A further extension was given by Tao [To3].

The well-posedness in the limiting case $s = -\frac{3}{4}$ was established in [Ki1] [Gu] using a Besov-like generalization of the $X_{s,b}$ spaces with $(s, b) = (-3/4, 1/2)$ in the low frequency (see also [BTo]).

It is interesting to compare this LWP result for the KdV with those for the viscous Burgers' equation:

$$\begin{cases} \partial_t u = \partial_x^2 u + u\partial_x u, \\ u(x,0) = u_0(x) \in H^s(\mathbb{R}). \end{cases} \tag{7.105}$$

In [Dx], Dix showed that (7.105) is locally well-posed in $H^s(\mathbb{R})$, $s \geq -1/2$ (scaling $s = -1/2$) and uniqueness fails for $s < -1/2$ (a construction based in the Cole–Hopf transformation (see Exercise 9.18). Therefore, from the LWP point of view the KdV equation is better than the viscous Burgers' equation.

The proof of the LWP result for the mKdV was taken from [KPV4]. The estimate (7.13) is the sharp version of the Kato smoothing effect. It was already commented on at the end of Chapter 4 (see (4.54)–(4.67)) which was used to obtain weak L^2 solutions for the KdV equation.

The estimate (7.18) regarding the continuity of the maximal function associated to the group $V(t)$, i.e., $\sup\limits_{t\in[0,T]} |V(t)v_0|$, is due to Kenig and Ruiz [KR] and was obtained in their study of the problem mentioned in Chapter 4 (see (4.57)).

It was shown in [KPV5] that the result $s \geq 1/4$ is optimal, i.e., the map data-solution $v_0 \mapsto v(t)$ cannot be uniformly continuous in $H^s(\mathbb{R})$ for $s < 1/4$. The proof of this assertion follows a close argument to the one provided in Chapter 5 for the cubic (focusing) NLS equation in one dimension. There it was constructed a two-parameter family of solutions for the cubic (focusing) NLS by combining the Galilean and the scaling invariance of the solutions. However, the mKdV equation is not Galilean invariant. So to overcome this, one first considers the complex version of the mKdV equation, namely,

$$\partial_t w + \partial_x^3 w + |w|^2 \partial_x w = 0 \tag{7.106}$$

(see [GO], [KSC]), which has a set of solutions that is Galilean invariant. In fact, we have the two-parameter family

$$w_{N,\omega}(x,t) = \sqrt{3}\, e^{-it(3N\omega^2 - N^3)} e^{ixN} \varphi_\omega(x - t\omega^2 + 3tN^2), \tag{7.107}$$

where φ solves (5.8) (i.e., $-\varphi + \varphi'' + \varphi^3 = 0$ so $\varphi(x) = \sqrt{2}\,\mathrm{sech}(x)$); and $\varphi_\omega(x) = \omega\varphi(\omega x)$ (notice that $\sqrt{3}\,\varphi_\omega(x - t\omega^2)$ solves 7.106). With the two-parameter family, we follow an argument similar to the one employed in Theorem 5.12 to obtain the result for equation (7.106). To pass to the mKdV equation, one uses a special solution called a "breather," see [Wa],

$$\begin{aligned} v_{N,\omega}(x,t) &= -2\sqrt{6}\,\omega\,\mathrm{sech}(\omega x + \gamma t) \\ &\times \left(\frac{\cos(Nx + \delta t) - (\omega/N)\sin(Nx + \delta t)\tanh(\omega x + \gamma t)}{1 + (\omega/N)^2 \sin^2(Nx + \delta t)\,\mathrm{sech}(\omega x + \gamma t)} \right) \end{aligned} \tag{7.108}$$

with $\delta = N(N^2 - 3\omega^2)$ and $\gamma = \omega(3N^2 - \omega^2)$ and observe that for $\omega/N \ll 1$,

$$\begin{aligned} v_{N,\omega}(x,t) \approx &- 2\sqrt{6}\,\cos(Nx + N(N^2 - 3\omega^2)t) \\ &\times \omega\,\mathrm{sech}(\omega x + \omega(3N^2 - \omega^2)t), \end{aligned}$$

which is basically a multiple of the real part of (7.107).

As it was remarked above, Bourgain introduced the spaces $X_{s,b}$ in the context of dispersive equations. Previously they were used by Rauch and Reed [RuR] and Beals [Bs] in their respective studies of propagation of singularities for solutions of semilinear wave equation.

In the same spirit that [KTa2], [CrCT2] and [CrHoT] a priori estimates were established for solutions of the modified KdV below the Sobolev index 1/4 which guarantees the well-posedness. More precisely, it was shown in [CrHoT] that solutions of the IVP associated to the mKdV satisfies for $s \in (-1/4, -1/8)$,

$$\sup_{[0,T]} \|u(t)\|_{s,2} \le c\,(T;\ \|u_0\|_{s,2}).$$

This result does not give control on the difference of two solutions (uniqueness).

In a similar regard, in [BuKo] an a priori estimate in $H^{-1}(\mathbb{R})$ for smooth solutions of the KdV equation was obtained. More precisely, in [BuKo] the following result was established: if $v \in C([0,\infty) : H^s(\mathbb{R}))$, with $s \ge -3/4$, is a solution of the IVP with $k = 1$, then

$$\|v(\cdot,t)\|_{-1,2} \le c\,(\|v_0\|_{-1,2} + \|v_0\|_{-1,2}^3), \quad \text{for any } t \ge 0.$$

This allows to construct global H^{-1}-weak solutions of the associated IVP. In [Mo2], it was shown that the map data-solution associated to the IVP for the KdV equation cannot be continuously extended in $H^s(\mathbb{R})$ for $s < -1$.

In [FaPa], modified proofs of Theorems 7.4, 7.5, and 7.6 and Corollary 7.2 were obtained which simplify the argument by not using norm involving time derivatives D_t of the unknown function.

The LWP result for the case $k = 3$ for data in $H^s(\mathbb{R})$, $s > -1/6$, was obtained by Grünrock [Gr2]. The key tool to prove that result was the following bilinear estimate for solutions of the linear problem. More precisely,

$$\|D_x(V(t)f \cdot V(t)g)\|_{L^2_x L^2_t} \leq c\,\|f\|_2\|g\|_2.$$

Tao [To6] extended Grünrock's result to the critical case by showing LWP for data in $\dot{H}^{-1/6}(\mathbb{R})$. From this result, it follows readily the global one for small data due to the criticality of the space. Thus, the case $k = 3$ exhibits similar properties to the case $k \geq 4$; see Theorems 7.2 and 7.3 and the Remarks 7.15 and 7.16.

The results for $k \geq 4$ were taken from [KPV4] (Theorems 7.2–7.6). Their sharpness was established in [BKPSV] (Theorem 7.7).

In [GrV], LWP was established in the spaces $\widehat{H}^s_r(\mathbb{R})$ (see (5.75)) for the parameters $r \in (1,2)$ and $s \geq 1/2 - 1/2r$.

Results concerning the smoothing effects of solutions of (7.10) due to special decay of the data were first given by Cohen [Co1] and Cohen and Kappeler [CoK] for the KdV equation for step data using the inverse scattering theory.

In [K2], Kato studied the IVP (7.1) in weighted Sobolev spaces and showed that if

$$v_0 \in \mathcal{F}^s_{2k} = H^s(\mathbb{R}) \cap L^2(|x|^{2k}\,dx), \quad k \in \mathbb{Z}^+, \quad s \geq 2k, \tag{7.109}$$

then the local solution describes a continuous curve on $\mathcal{F}^s_{2k}$ as far as it exists. In particular, the solution flow preserves the Schwartz class $\mathcal{S}(\mathbb{R})$. Roughly, this is due to the fact that the operators $L = \partial_t + \partial_x^3$ and $\Gamma = x - 3t\,\partial_x^2$ commute. The extension of this result to solutions of (7.1) with data:

$$v_0 \in \mathcal{F}^s_{2l} \text{ with } s \geq \max\{2l;\ s_k\},\ l > 0, \tag{7.110}$$

$s_k = -3/4$ if $k = 1$, $s_k = 1/2$ if $k = 2$, and $s_k = 1/2 - 2/k$ was established in [Nh] and [FLP4]. In [ILP1], it was shown that this persistence result in $\mathcal{F}^s_l$ with $s \geq l > 0$ is optimal. More precisely, if $u \in C([-T,T] : H^s(\mathbb{R})) \cap \ldots$, with $s \geq \max\{s_k;\ 0\}$, is a solution of the IVP (7.1) and there exist $t_1, t_2 \in [-T,T]$, $t_1 \neq t_2$ such that $|x|^\alpha u(\cdot, t_j) \in L^2(\mathbb{R})$ for $2\alpha > s$, then $u \in C([-T,T] : H^{2\alpha}(\mathbb{R}))$. In other words, persistence in $L^2(|x|^{2\alpha}\,dx)$ (decay) can only hold for solutions in $C([-T,T] : H^{2\alpha}(\mathbb{R}))$.

Also in [K2], Kato proved the following result for the KdV equation (which also holds for solutions of (7.1) with $k \in \mathbb{Z}^+$): if $v \in C([0,T] : H^2(\mathbb{R})) \cap \ldots$ is a solution of (7.1) with $v_0 \in H^2(\mathbb{R}) \cap L^2(e^{2\beta x}dx)$, $\beta > 0$, then

$$e^{\beta x} v \in C([0,T] : L^2(\mathbb{R})) \cap C((0,T) : H^\infty(\mathbb{R})). \tag{7.111}$$

Formally, one has that the semigroup $\{V(t) = e^{-t\partial_x^3} : t \geq 0\}$ in $L^2(e^{\beta x}dx)$ is equivalent to $\{e^{-t(\partial_x-\beta)^3} : t \geq 0\}$ in $L^2(\mathbb{R})$, i.e., if

$$\partial_t u + \partial_x^3 u = 0,$$

then $v(x,t) = e^{\beta x}u(x,t)$ satisfies

$$\partial_t v + (\partial_x - \beta)^3 v = \partial_t v + \partial_x^3 v - 3\beta\partial_x^2 v + 3\beta^2\partial_x v - \beta^3 v = 0, \tag{7.112}$$

which explain this "parabolic effect." In [KF], Kruskov and Faminskii obtained another version of this result by considering weights of the form $x_+^{\alpha} = x^{\alpha}\,\chi_{(0,\infty)}(x)$.

Tarama [Ta2] showed that solutions of the KdV equation with real-valued initial data $v_0(x) \in L^2(\mathbb{R})$ satisfying the condition:

$$\int_{-\infty}^{\infty} e^{\delta|x|^{1/2}}|v_0(x)|^2 dx < \infty$$

with some positive constant δ, become analytic with respect to the variable x for all $t > 0$. The proof of this theorem is based on the inverse scattering method (see Section 9.6), which transforms the KdV equation into a linear dispersive equation for which the analyticity smoothing effect can be established through the analytic properties of the fundamental solutions. However, for higher powers a similar result is unknown even for the integrable case $k = 2$, i.e., for the mKdV.

In [GT], Ginibre and Tsutsumi proved for the mKdV that if $v_0 \in L^2(\mathbb{R})$ and $x_+^{1/8}\,v_0 \in L^2(\mathbb{R})$, then the uniqueness holds (observe that decay corresponds to $1/4$ derivatives via the operator Γ above which is the sharp LWP). In the KdV case, this result improves by a factor of 2 the one obtained in [KF].

In (7.111), we have seen that persistence property holds in $L^2(w(x)\,dx)$, $w(x) = e^{\beta x}$, $\beta > 0$ for $t > 0$. The following unique continuation result found in [EKPV3] gives an upper bound on the weight $w(x)$ for which this property remains: there exists $c_k > 0$ such that if

$$v_1, v_2 \in C([0,1] : H^3(\mathbb{R}) \cap L^2(|x|^2\,dx))$$

such that

$$(v_1 - v_2)(\cdot,0),\ (v_1 - v_2)(\cdot,1) \in L^2(e^{c_k\,x_+^{3/2}}\,dx), \quad x_+ = \max\{x;0\}. \tag{7.113}$$

Then,

$$v_1 \equiv v_2 \text{ on } \mathbb{R} \times [0,1].$$

Notice that taking $v_2 \equiv 0$ it follows that persistence in $L^2(w(x)\,dx)$ with $w(x) = e^{c_k\,x_+^{3/2}}$, c_k arbitrarily large cannot hold in the interval $[0,1]$ for a nonnull solution.

In [ILP1], it was shown that (7.113) is optimal. More precisely:

If $v_0 \in H^1(\mathbb{R}) \cap L^2(e^{a_0 x_+^{3/2}}\,dx)$, $a_0 > 0$, then $v = v(x,t)$ the solution of the IVP (7.1) defined in the time interval $[0,T]$ satisfies

$$\sup_{0\le t\le T} \int_{-\infty}^{\infty} e^{a(t) x_+^{3/2}} |v(x,t)|^2\,dx \le c = c(a_0; \|v_0\|_{1,2}; \|e^{\frac{a_0}{2} x_+^{3/2}} v_0\|_2; T) \tag{7.114}$$

with

$$a(t) = \frac{a_0}{(1+27a_0^2 t/4)^{1/2}}. \tag{7.115}$$

In other words, the initial decay of $v_0 \in L^2(e^{a_0 x_+^{3/2}}\,dx)$ remains in time but with a constant $a(t)$ decreasing in t as in (7.115).

Concerning the propagation of asymmetric regularity in solutions of the IVP (7.1), one has the following result established in [ILP2]: Let $v \in C([-T,T] : H^{3/4^+}(\mathbb{R})) \cap \ldots$ be a solution of the IVP (7.1). If for some $l \ge 1$

$$\int_0^{\infty} |\partial_x^l v_0(x)|^2\,dx < \infty, \tag{7.116}$$

then for any $\varepsilon > 0$ and any $b > 0$

$$\sup_{0<t<T} \int_{x>\varepsilon-bt} |\partial_x^l v(x,t)|^2\,dx < \infty. \tag{7.117}$$

Roughly speaking, this tells us that the regularity in the right-hand side of the data v_0 propagates with infinite speed to its left as time evolves.

Next, we consider the LWP for the periodic boundary value problem associated to the k-gKdV.

For the case $k=1$, the KdV equation, local well-posedness was proven in $H^s(\mathbb{T})$, $s \ge -1/2$ by Kenig, Ponce and Vega [KPV6] (improving an earlier result of Bourgain [Bo1] for $s \ge 0$). The proofs are based on the modified version of the $X_{s,b}$ spaces, i.e., the spaces $Y_{s,b}$, which are the completion under the norm:

$$\|f\|_{Y_{s,b}} = \left(\sum_{m\neq 0} \int_{-\infty}^{\infty} (1+|\tau - m^3|)^{2b}\,|m|^{2s} |\widehat{f}(m,\tau)|^2\,d\tau \right)^{1/2} \tag{7.118}$$

of the space $\mathcal{Y}$ defined as the function space of all f such that

(i) $f : \mathbb{T}\times\mathbb{R} \to \mathbb{C}$,
(ii) $f(x,\cdot) \in \mathcal{S}$ for each $x \in \mathbb{T}$,
(iii) $x \mapsto f(x,\cdot)$ is C^∞,
(iv) $\widehat{f}(0,\tau) = 0$ for all $\tau \in \mathbb{R}$.

Bourgain [Bo8] also showed that below $-1/2$ (for $s < -1/2$) the smoothness of the map data-solution fails. We recall that the smoothness of this map is a by-product of the contraction principle. So this type of result in particular shows that the iteration process by itself does not provide a result in $H^s(\mathbb{T})$, $s < -1/2$. In this regard, using the inverse scattering method Kappeler and Topalov [KpTo] showed that the solution flow of the KdV extends continuously to $H^{-1}(\mathbb{T})$.

For the mKdV equation ($k = 2$), LWP was established in [KPV6] in $H^s(\mathbb{T})$, $s \geq 1/2$. This was proven to be sharp in [CrCT1]. By requiring the dependence of solutions on the initial data be just continuous and considering real solutions, Takaoka and Tsutsumi [TTs] were able to lower the Sobolev index $s > 1/2$ to $s > 3/8$. One of the main new ideas in their approach was the modification of the Bourgain norm (7.118) by

$$\|f\|_{Z_{s,b}} = \Big(\sum_{m\neq 0}\int_{-\infty}^{\infty} (1+|\tau - m^3 - m|\widehat{u_0}(m)|^2|)^{2b}\,|m|^{2s}\,|\widehat{f}(m,\tau)|^2\,d\tau\Big)^{1/2},$$

where u_0 is the considered initial data. Notice that the definition of the norm $\|\cdot\|_{Z_{s,b}}$ depends on the initial data. In [NTT2], Nakanishi, Takaoka and Tsutsumi extended this LWP result for the mKdV to $H^s(\mathbb{T})$, for $s > 1/3$ (and under some additional hypotheses on the data to $H^s(\mathbb{T})$, $s > 1/4$).

For $k \geq 3$, the best LWP result is in $H^s(\mathbb{T})$, $s \geq 1/2$ (see [CKSTT4]).

7.6 Exercises

7.1 ([BSa2]) Let $A_i(x)$ be defined as in (1.32):

(i) Prove that

$$A_i''(z) - z\,A_i(z) = 0.$$

(ii) Defining

$$v(x,t) = \frac{1}{\sqrt[6]{t}}A_i^2\Big(\frac{1}{2^{2/3}}\frac{x}{\sqrt[3]{3t}}\Big),$$

prove that

$$\partial_t v + \partial_x^3 v = 0, \qquad x \in \mathbb{R},\ t > 0.$$

(iii) Using (1.37), show that for any $t > 0$,

$$\|v(\cdot,t)\|_p < \infty \text{ if and only if } p > 2.$$

(iv) Show that for any $p > 2$,

$$\lim_{t\downarrow 0}\|v(\cdot,t)\|_p = +\infty.$$

7.2 Denote by $(k, \pm 1)$ the equation:

$$\partial_t v + \partial_x^3 v \pm v^k \partial_x v = 0, \quad x, t \in \mathbb{R}, \;\; k \in \mathbb{Z}^+,$$

(the cases $(k, -1)$ correspond to the generalized defocusing KdV equation).

(i) Prove that if $v(x,t)$ is a solution of $(k, \pm)$, then

(a) $v(-x, -t)$ also solves $(k, \pm)$
(b) $-v(x,t)$ solves $(k, (-1)^k)$
(c) $\lambda^{2/k}\, v(\lambda x, \lambda^3 t)$ solves $(k, \pm)$ for $\lambda > 0$

(ii) Prove that for $\lambda > 0$

$$v_\lambda(x,t) = \lambda\sqrt{6}\tanh\left(\lambda(x + 2\lambda^2 t)\right) \text{ (kink solution)}$$

is a solution of $(2, -1)$.

(iii) Prove that if $v(\cdot,\cdot)$ is a solution of $(1, \pm 1)$, then for $h \in \mathbb{R}$, $v(x \pm h\,t, t) \pm h$ also solves $(1, \pm 1)$ (Galilean invariance).

(iv) Let v be a solution of $(2, -1)$. Show that the function (Miura transformation) $w = \sqrt{6}\,\partial_x v + v^2$ solves $(1, -1)$.

(v) Let v be a solution of $(2, 1)$. Show that $w = i\sqrt{6}\,\partial_x v + v^2$ solves $(1, 1)$.

7.3 Consider the IVP associated to the KdV equation (7.71):

(i) Prove that $v(x,t) = \dfrac{x}{1+t}$ is solution of (7.71) with $v(x, 0) = x$ for any time $t > 0$.

(ii) Prove that $v(x,t) = \dfrac{-x}{1-t}$ solves (7.71) with $v(x,0) = -x$, but it blows up in finite time.

(iii) Prove that parts (i) and (ii) also hold for the inviscid and viscous Burgers' equation ((3.46) and (7.105), respectively) and for the Benjamin–Ono equation (9.9).

7.4 (Soliton) Let $u(x,t) = \phi_{c,k}(x - ct) = \phi(x - ct)$ be solution of

$$\partial_t u + \partial_x^3 u + u^k \partial_x u = 0$$

with strong decay at infinity.

(i) Show that

(a) $-c\phi' + \phi''' + \phi^k \phi' = 0.$

(b) $-c\phi + \phi'' + \frac{\phi^{k+1}}{k+1} = 0.$

(c) $-c\frac{\phi^2}{2} + \frac{(\phi')^2}{2} + \frac{\phi^{k+2}}{(k+1)(k+2)} = 0,$
(integrating this equation one gets the explicit solution (7.6)).

(ii) Starting in (b) define $x = \phi$ and $y = \phi'$.

(a) Show that the second order ODE can be written as the Hamiltonian system:

$$\begin{cases} \dfrac{dx}{dt} = \partial_y H \\ \dfrac{dy}{dt} = -\partial_x H, \end{cases}$$

where

$$H = \frac{y^2}{2} - c\frac{x^2}{2} + \frac{x^{k+2}}{(k+1)(k+2)}.$$

(b) Using the decay condition at infinity prove that the level set $\{H(x,y) = 0\}$ represents the traveling wave.

(c) Prove that $\phi > 0$, symmetric and $\|\phi\|_\infty = [\frac{c}{2}\,(k+1)(k+2)]^{1/k}$.

7.5 ([Za]) Show that $v(x,t) = \frac{1}{\sqrt{6}}\Big(c - \dfrac{4c}{4c^2(x-6c^2t)^2+1}\Big)$ solves the mKdV equation (7.28).

7.6 (Critical KdV) Show that if $u_0 \in \dot{H}^{-3/2}(\mathbb{R}) \cap \mathcal{S}(\mathbb{R})$, then the solution of the KdV equation (7.71) $u(\cdot,t) \notin \dot{H}^{-3/2}(\mathbb{R})$ for all $t \neq 0$.

7.7 Using a formal scaling argument, obtain the estimate of the life span of the local solutions as a function of the size of the initial data given in Theorem 7.1, i.e., $T(\|D_x^{1/4}v_0\|_2) = c\,\|D^{1/4}v_0\|_2^{-4}$.

7.8 (Two-soliton solution of the KdV) Given the solution of the KdV:

$$v(x,t) = 72\,\frac{3 + 4\cosh(2(x-4t)) + \cosh(4(x-16t))}{[3\cosh(x-28t) + \cosh(3(x-12t))]^2}$$

show that for $\xi_1 = x - 16t$ fixed

$$v(x,t) \sim 48\,\text{sech}^2\Big(2\xi_1 \mp \frac{\log 3}{2}\Big) \qquad \text{as } t \to \pm\infty;$$

show that for $\xi_2 = x - 4t$ fixed

$$v(x,t) \sim 12\,\text{sech}^2\Big(2\xi_2 \pm \frac{\log 3}{2}\Big) \qquad \text{as } t \to \pm\infty.$$

Conclude that

$$v(x,t) \sim 48\,\text{sech}^2\Big(2\xi_1 \mp \frac{\log 3}{2}\Big) + 12\,\text{sech}^2\Big(2\xi_2 \pm \frac{\log 3}{2}\Big) \quad \text{as } t \to \pm\infty.$$

7.9 ([KPV6])

(i) Assuming that the following inequality holds for $s \in (-3/4, -1/2)$ and $b = b(s) \in (1/2, 1)$

$$\Big|\int\!\!\int\!\!\int\!\!\int \frac{|\xi|\,h(\xi,\tau)}{(1+|\tau-\xi^3|)^{1-b}\,(1+|\xi|)^{-s}}\,\frac{(1+|\xi_1|)^{-s}\,f(\xi_1,\tau_1)}{(1+|\tau_1-\xi_1^3|)^{b}}$$

$$\frac{(1+|\xi-\xi_1|)^{-s}\, g(\xi-\xi_1,\tau-\tau_1)}{(1+|\tau-\tau_1-(\xi-\xi_1)^3|)^b}\, d\tau_1 d\xi_1 d\tau d\xi \Big| \quad (7.119)$$
$$\le c\, \|h\|_{L^2_\xi L^2_\tau} \|f\|_{L^2_\xi L^2_\tau} \|g\|_{L^2_\xi L^2_\tau},$$

prove Corollary 7.4 with $b' = b$. Sketch the LWP result for the IVP associated to the KdV equation (7.71) in $H^s(\mathbb{R})$, $s \in (-3/4, -1/2)$.

(ii) Prove that if either $|\xi_1| \le 1$ or $|\xi - \xi_1| \le 1$, then

$$(1+|\xi_1|)^{-s}(1+|\xi-\xi_1|)^{-s} \le c\,(1+|\xi|)^{-s},$$

so the proof of (7.119) in this domain reduces to the estimate (7.94).

(iii) Show by symmetry that to prove (7.119) it suffices to consider

$$|\tau-\tau_1-(\xi-\xi_1)^3| \le |\tau-\xi_1^3|.$$

(iv) Combine (ii) and (iii) to show that in order to obtain Corollary 7.4 with $b' = b$ it suffices to establish the following inequalities:

$$\sup_{\xi,\tau} \frac{|\xi|}{(1+|\tau-\xi^3|)^{1-b}(1+|\xi|)^{-s}} \times \left(\iint_A \frac{|\xi_1(\xi-\xi_1)|^{-2s}\; d\tau_1 d\xi_1}{(1+|\tau_1-\xi_1^3|)^{2b}(1+|\tau-\tau_1-(\xi-\xi_1)^3|)^{2b}} \right)^{1/2} < c, \quad (7.120)$$

with $A = A(\xi,\tau)$ as:

$$A = \{(\xi_1,\tau_1) \in \mathbb{R}^2 :\; |\xi_1|,\ |\xi-\xi_1| \ge 1,\ |\tau-\tau_1-(\xi-\xi_1)^3| \le |\tau_1-\xi_1^3| \le |\tau-\xi^3|\}$$

and

$$\sup_{\xi_1,\tau_1} \frac{1}{(1+|\tau_1-\xi_1^3|)^b} \times \left(\iint_B \frac{|\xi|^{2(1+s)}|\xi\xi_1(\xi-\xi_1)|^{-2s}(1+|\xi|)^{2s}\; d\tau d\xi}{(1+|\tau-\xi^3|)^{2(1-b)}(1+|\tau-\tau_1-(\xi-\xi_1)^3|)^{2b}} \right)^{1/2} < c, \quad (7.121)$$

with $B = B(\xi,\tau)$ as:

$$B = \left\{ \begin{array}{l} (\xi_1,\tau_1) \in \mathbb{R}^2 :\; |\xi_1|,\ |\xi-\xi_1| \ge 1,\ \ |\tau-\tau_1-(\xi-\xi_1)^3| \\ \qquad\qquad \le |\tau_1-\xi_1^3|, |\tau-\xi^3| \le |\tau_1-\xi_1^3|. \end{array} \right\}.$$

(v) Following the argument given in Lemma 7.16 prove the inequality (7.120) (for the proof of (7.121) we refer the reader to [KPV6]).

7.10 Assuming $b > 1/2$ prove the following inequalities:

(i)
$$\|g\|_{L^8(\mathbb{R}^2)} \le c\,\|g\|_{X_{0,b}},$$

(ii)
$$\|D_x^{1/6} g\|_{L^6(\mathbb{R}^2)} \le c\,\|g\|_{X_{0,b}},$$

(iii)
$$\|\partial_x g\|_{L_t^\infty L_x^2} \le c\,\|g\|_{X_{0,b}},$$

(iv)
$$\|g\|_{L_x^\infty L_t^2} \le c\,\|g\|_{X_{0,b}}.$$

7.11 Let w be a solution of the IVP:

$$\begin{cases} (\partial_t + a\partial_x^3 + ib\partial_x^2)w = F, \\ w(x,0) = w_0(x). \end{cases} \tag{7.122}$$

Show that

$$z(x,t) = e^{i\frac{b^3}{27a^2}t}\, e^{i\frac{b}{3a}x}\, w\big(x + \frac{b^2}{3a}t, t\big)$$

solves the IVP

$$\begin{cases} \partial_t z + a\partial_x^3 z = \widetilde{F}, \\ z(x,0) = z_0(x). \end{cases} \tag{7.123}$$

where

$$z_0(x) = e^{i\frac{b}{3a}x}\, w_0(x) \quad \text{and} \quad \widetilde{F}(x,t) = e^{i\frac{b^3}{27a^2}t}\, e^{i\frac{b}{3a}x}\, F\big(x + \frac{b^2}{3a}t, t\big). \tag{7.124}$$

7.12 ([ILP2]) Consider the linear IVP (7.10) with $v_0 \in L^2(\mathbb{R})$. Prove that if for some $k \in \mathbb{Z}^+$,

$$v_0|_{(0,\infty)} \in H^k((0,\infty)),$$

then the corresponding solution $v(x,t)$ satisfies that for any $b > 0$,

$$v(\cdot,t)|_{(-b,\infty)} \in H^k((-b,\infty)), \quad \text{for each } t > 0$$

and

$$v(\cdot,t)|_{(-\infty,b)} \in H^k((-\infty,b)) \quad \text{for each } t < 0.$$

Hint: Let $\eta \in C^\infty(\mathbb{R})$, $\eta' \ge 0$, $\eta(x) = 0$ for $x \le 0$ and $\eta(x) = 1$ for $x \ge 1$. Following Kato's argument in [K2] one easily gets the (formal) identity

$$\begin{aligned} \frac{1}{2}\frac{d}{dt}\int v^2(x,t) \quad \eta(x)\,dx + \frac{3}{2}\int (\partial_x v)^2(x,t)\,\eta'(x)\,dx \\ -\frac{1}{2}\int v^2(x,t)\,\eta^{(3)}(x)\,dx = 0. \end{aligned} \tag{7.125}$$

Modify $\eta(\cdot)$ in each step $j = 1, 2, \ldots, k$ ($\eta_j(\cdot)$) and consider $\eta_j(x + ct)$ with $c > 0$ (for $t > 0$) to obtain the appropriate version of (7.125).

7.13 ([ILP2]) Consider the linear IVP (7.10) with $v_0 \in L^2(\mathbb{R})$. Prove that if for some $m \in \mathbb{Z}^+$,

$$x_+^{m/2} v_0 \in L^2(\mathbb{R}),$$

then for any $t > 0$,

$$x_+^{m/2} v(\cdot, t) \in L^2(\mathbb{R})$$

and for any $b > 0$ and $t > 0$

$$\int_b^\infty (\partial_x^m u(x,t))^2 \, dx < \infty.$$

Hint: Modify the argument in the hint of Exercise 7.12.

7.14 ([ILP2]) Consider the linear IVP (7.10) with $v_0 \in L^2(\mathbb{R})$. Prove that if for some $m \in \mathbb{Z}^+$ and $t_1, t_2 \in \mathbb{R}$ with $t_1 < t_2$ the corresponding solution $v(x,t)$ satisfies

$$x_+^{m/2} v(\cdot, t_1),\; x_-^{m/2} v(\cdot, t_2) \in L^2(\mathbb{R}),$$

then $v_0 \in H^m(\mathbb{R})$.
Hint: Use Exercise 7.13.

7.15 (i) Consider the IVP for the 1-D heat equation

$$\begin{cases} \partial_t u - \partial_x^2 u = 0, \\ u(x,0) = u_0(x), \end{cases} \tag{7.126}$$

$x \in \mathbb{R}$, $t > 0$. Prove that if $u_0 \in L^2(\mathbb{R})$, then for each t positive the solution $u(\cdot, t) = e^{t\partial_x^2} u_0$ of (7.126) has an analytic extension to $\mathbb{C}$. Moreover, if $z = x + iy$, then

$$|u(z,t)| \le \frac{e^{y^2/4t}}{(4\pi t)^{1/4}} \|u_0\|_2.$$

(ii) Consider the linear IVP (7.10). Prove that if $v_0 \in L^2(\mathbb{R}) \cap L^2(e^{\beta x} dx)$, $\beta > 0$, then for each t positive the solution $v(\cdot, t)$ has an analytic extension to $\mathbb{C}$.
Hint: Combine the result in part (i) and the formula (7.112).

Chapter 8
Asymptotic Behavior of Solutions for the k-gKdV Equations

This chapter is concerned with the longtime behavior of solutions to the initial value problem (IVP) associated to the k-generalized Korteweg-de Vries equations,

$$\begin{cases} \partial_t v + \partial_x^3 v + v^k \partial_x v = 0, \\ v(x,0) = v_0(x), \end{cases} \tag{8.1}$$

$x,\, t \in \mathbb{R},\ k \in \mathbb{Z}^+$.

We shall restrict ourselves to consider only real solutions of (8.1). In this case, the following quantities are preserved by the solution flow:

$$I_1 = \int_{-\infty}^{\infty} v(x,t)\,dx, \tag{8.2}$$

$$I_2 = \int_{-\infty}^{\infty} v^2(x,t)\,dx, \tag{8.3}$$

$$E(v_0) = I_3 = \int_{-\infty}^{\infty} \Big((\partial_x v)^2 - \frac{2}{(k+1)(k+2)}\, v^{k+2}\Big)(x,t)\,dx. \tag{8.4}$$

Combining them with the local existence theory in Chapter 7, we shall see that for $k = 1, 3$ the IVP (8.1) with $v_0 \in L^2(\mathbb{R})$ has a unique globally bounded solution. For the case $k = 2$, the same holds in $H^1(\mathbb{R})$.

In fact, we shall see in Section 8.1 that a quite stronger set of results has been established in the case $k = 1, 2, 3$.

In Section 8.2, we treat the L^2-critical case $k = 4$. A major set of results due to Martel and Merle as well as extensions and further analysis due to Martel, Merle, and Raphael is discussed. In particular, they have settled a long-standing open problem by proving the finite time blowup of some local H^1-solutions. Similar results for $k \geq 5$ remain as open problems.

In Section 8.3, we add some further comments.

F. Linares, G. Ponce, *Introduction to Nonlinear Dispersive Equations*,
Universitext, DOI 10.1007/978-1-4939-2181-2_8

8.1 Cases k = 1,2,3

We observe that if $v(t)$ is a local real H^1-solution of (8.1), combining Gagliardo–Nirenberg (3.14) and (8.3)–(8.4) gives

$$\begin{aligned} E(v_0) &= \int_{-\infty}^{\infty} \Big[(\partial_x v)^2 - \frac{2}{(k+1)(k+2)}\, v^{k+2}\Big](x,t)\,dx \\ &\geq \|\partial_x v(t)\|_2^2 - \frac{2}{(k+1)(k+2)} \|v(t)\|_{k+2}^{k+2} \\ &\geq \|\partial_x v(t)\|_2^2 - c_k\, \|\partial_x v(t)\|_2^{k/2} \|v(t)\|_2^{2+k/2} \\ &\geq \|\partial_x v(t)\|_2^2 - c_k\, \|\partial_x v(t)\|_2^{k/2} \|v_0\|_2^{2+k/2}. \end{aligned} \tag{8.5}$$

Hence, using the notation $\eta = \eta(t) = \|\partial_x v(t)\|_2$ it follows that

$$E(v_0) \geq \eta^2 - c_k \|v_0\|_2^{2+k/2}\, \eta^{k/2}. \tag{8.6}$$

So for $k < 4$, one obtains the a priori estimate:

$$\eta(t) \leq M(\|v_0\|_2; k). \tag{8.7}$$

In this sense as well as in the scaling argument for the L^2-norm (see (7.8)), the case $k = 4$ is critical.

Thus for $k = 2$, (8.7) allows us to reapply the local existence theory (local well-posedness in $H^s(\mathbb{R})$, $s \geq 1/4$) for data $v_0 \in H^1(\mathbb{R})$.

Theorem 8.1. *For $v_0 \in H^1(\mathbb{R})$ real valued the corresponding local solution of the initial value problem (IVP)* (8.1) *with $k = 2$ given by Theorem* 7.1 *extends in the same class to any time interval with*

$$v \in C(\mathbb{R} : H^1(\mathbb{R})) \cap L^\infty(\mathbb{R} : H^1(\mathbb{R})). \tag{8.8}$$

Moreover, if $v_0 \in H^s(\mathbb{R})$, $s > 1$, then

$$v \in C(\mathbb{R} : H^s(\mathbb{R})). \tag{8.9}$$

In the cases $k = 1, 3$, the local well-posedness was established in $H^s(\mathbb{R})$ for $s \geq -3/4$ and $s > -1/6$, respectively (see Theorem 7.8 and [Gr2]). These cases are subcritical, so the interval of existence in each case $[0, T]$ satisfies that $T = T(\|u_0\|_{s,2}) > 0$. Therefore, if $v_0 \in L^2(\mathbb{R})$ by I_2 (see (8.3)), we can reapply this local theory to obtain the following global result.

Theorem 8.2. *For $v_0 \in L^2(\mathbb{R})$ real the corresponding local solution of the IVP (8.1) with $k = 1$ or 3 extends in the same class to any time interval with*

$$v \in C(\mathbb{R} : L^2(\mathbb{R})) \cap L^\infty(\mathbb{R} : L^2(\mathbb{R})). \tag{8.10}$$

Moreover, if $v_0 \in H^s(\mathbb{R})$, $s > 0$, *then*

$$v \in C(\mathbb{R} : H^s(\mathbb{R})). \tag{8.11}$$

In the cases $k = 1, 2$, due to the form of the infinite conservation laws, one can replace (8.9) and (8.11) by $v \in C(\mathbb{R} : H^s(\mathbb{R})) \cap L^\infty(\mathbb{R} : H^s(\mathbb{R}))$ if $s \in \mathbb{Z}^+$.

These local and global results present the following questions:

Question 1. What happens with the longtime behavior of the solution corresponding to data $v_0 \in H^s(\mathbb{R})$ with $s \in [-3/4, 0)$, $[1/4, 1)$, and $(-1/6, 0)$ in the cases $k = 1, 2$, and 3, respectively?

The first result in this direction was given by Bourgain [Bo5] in his study of the critical two-dimensional nonlinear Schrödinger (NLS) equation. He set up a general argument to obtain global solutions corresponding to data whose regularity is below that required if one is using the conservation law.

To illustrate his approach, we take the mKdV equation, $k = 2$ in (8.1), with $v_0 \in H^s(\mathbb{R})$, $s \in [1/4, 1)$ (see [FLP2]).

First, one splits the data according to the frequency (low–high). For N large to be determined one considers

$$v_0 = (\chi_{\{|\xi| \le N\}} \widehat{v_0})^\vee + (\chi_{\{|\xi| > N\}} \widehat{v_0})^\vee = v_{1,0} + v_{2,0}. \tag{8.12}$$

Thus, $v_{1,0} \in H^\infty(\mathbb{R})$, with $E(v_{1,0}) + \|v_{1,0}\|_{1,2} \le c\, N^{1-s}$ and $\|v_{2,0}\|_{r,2} \le c\, N^{r-s}$ for $r \in [1/4, s)$.

One solves the mKdV with data $v_{1,0}$ as in Theorem 7.1, so the corresponding solution $v_1(t)$ satisfies

$$\|v_1(t)\|_{1,2} \le c\, N^{1-s}, \quad t \in [0, \Delta T], \quad \Delta T \simeq \|D_x^{1/4} v_{1,0}\|_2^{-4}, \tag{8.13}$$

and $v_2(t)$ satisfies the error equation (using that $v = v_1 + v_2$):

$$\partial_t v_2 + \partial_x^3 v_2 + v^2 \partial_x v - v_1^2 \partial_x v_1 = 0, \qquad t \in [0, \Delta T], \tag{8.14}$$

with data $v_{2,0}$ (small) in $H^r(\mathbb{R})$ for $r \in [0, s)$. The interval $[0, \Delta T]$ is given by the local well-posedness theory. Using that

$$\begin{aligned} v_2(t) &= V(t) v_{2,0} + \int_0^t V(t - t')[(v_1 + v_2)^2\, \partial_x (v_1 + v_2) - v_1^2 \partial_x v_1](t')\, dt' \\ &= V(t) v_{2,0} + z(t), \end{aligned}$$

one observes that $z(t)$ is smoother than $V(t)v_{2,0}$ (see Exercise 8.1 and comments there). Indeed, it belongs to $H^1(\mathbb{R})$ with a "good" estimate for its norm. Define

$$\begin{cases} v_{1,0}(\Delta T) = v_1(\Delta T) + z(\Delta T), \\ v_{2,0}(\Delta T) = V(\Delta T) v_{2,0} \end{cases}$$

and repeat the argument in $[\Delta T, 2\Delta T]$.

Briefly, to reach the time T^* we apply it $T^*/\Delta T$ times. If one proves that

$$E\left(v_{0,1} + \sum_{j=1}^{T^*/\Delta T} z(j\Delta T)\right) + \sum_{j=1}^{T^*/\Delta T} \|z(j\Delta T)\|_{1,2} \leq c\, N^{1-s}, \tag{8.15}$$

then all the previous estimates are uniform and one can extend the solution to $[0, T^*]$. It is in (8.15) where the restriction on s appears.

By introducing the I-method (see [KT2]) in this context Colliander, Keel, Staffilani, Takaoka and Tao [CKSTT4], [CKSTT5], [CKSTT6] have improved most of the results obtained by the above argument. By defining

$$If(x) = I_{N,s} f(x) = (m(\xi)\widehat{f}\,)^{\vee}, \tag{8.16}$$

where $m(\xi)$ is a smooth and monotone function given by

$$m(\xi) = \begin{cases} 1, & |\xi| \leq N, \\ N^{-s}|\xi|^{s}, & |\xi| > 2\,N, \end{cases} \tag{8.17}$$

with N to be determined and $s < 0$, they obtain a series of "almost conserved quantities."

By using the "cancellations" in the multilinear form working directly with the equation, in this case the KdV, they show that

$$\sup_{t\in[0,T]} \|Iv(t)\|_2 \leq \|Iv(0)\|_2 + cN^{-\beta}\, \|Iv(0)\|_2^3, \tag{8.18}$$

for some small $\beta > 0$. So if N is large, the increment in $\|Iv(t)\|_2$ is controlled. In particular for the IVP (8.1) they have shown the following.

Theorem 8.3 ([CKSTT5]).

1. *The local real solutions of the IVP* (8.1) *with* $k = 1$ *corresponding to data* $v_0 \in H^s(\mathbb{R})$, $s > -3/4$, *extend to any time interval* $[0, T^*]$.
2. *The local real solutions of the IVP* (8.1) *with* $k = 2$ *corresponding to data* $v_0 \in H^s(\mathbb{R})$, $s > 1/4$, *extend to any time interval* $[0, T^*]$.

In [Gu], Guo and [Ki1] Kishimoto have extended these global results to the limiting cases $s = -3/4$ and $s = 1/4$ for the KdV and the mKdV equations, respectively.

For the sake of completeness, we explain how the first step of this method works for the IVP associated to the KdV equation (8.1) ($k = 1$).

The material described below was essentially taken from the lecture notes given by Staffilani at IMPC (see [Sta3]).

One first notices that the operator defined in (8.16) is the identity operator on low frequencies $\{\xi\ :\ |\xi| < N\}$ and simply an integral operator in high frequencies. In general, it commutes with differential operators and maps $H^s(\mathbb{R})$ to $L^2(\mathbb{R})$.

As we mentioned before, the goal is to establish an estimate as the one in (8.18). To do so, we first use the fundamental theorem of calculus, the equation and integration by parts to get

$$\begin{aligned}
\|Iv(t)\|_2^2 &= \|Iv(0)\|_2^2 + \int_0^t \frac{d}{ds}(Iv(s), Iv(s))\,ds \\
&= \|Iv(0)\|_2^2 + 2\int_0^t \left(\frac{d}{ds}Iv(s), Iv(s)\right)\,ds \\
&= \|Iv(0)\|_2^2 + 2\int_0^t (I(-v_{xxx} - vv_x), Iv(s))\,ds \qquad (8.19)\\
&= \|Iv(0)\|_2^2 + 2\int_0^t (I(-vv_x), Iv(s))\,ds \\
&= \|Iv(0)\|_2^2 + R(t),
\end{aligned}$$

where

$$R(t) = \int_0^t \int_{\mathbb{R}} \partial_x(-Iv^2)\,Iv\,dxds \tag{8.20}$$

is an error term. Hence,

$$\|Iv(t)\|_2^2 = \|Iv_0\|_2^2 + R(t). \tag{8.21}$$

We shall show then that locally in time $R(t)$ is small. This can be achieved using local well-posedness estimates. Since one introduces the operator I in this analysis, a well-posedness result involving it has to be proved. A similar argument as the one given in the proof of Theorem 7.8 and the bilinear estimates (7.91) obtained by Kenig, Ponce and Vega [KPV6] provide us the local well-posedness result. More precisely:

Theorem 8.4. *For any $v_0 \in H^s(\mathbb{R})$, $s > -3/4$, the IVP* (8.1), *$k = 1$, is locally well-posed in the Banach space $I^{-1}L^2 = \{\phi \in H^s(\mathbb{R})\}$ furnished with the norm $\|I\phi\|_{L^2}$, with time existence satisfying*

$$T \gtrsim (\|Iv_0\|_2)^{-\alpha}, \qquad \alpha > 0. \tag{8.22}$$

Moreover,

$$\|\theta(\cdot/T)Iv\|_{X_{0,b}} \leq C\|Iv_0\|_2, \tag{8.23}$$

where θ was defined in (7.73).

The proof of this theorem follows by using the same procedure to establish Theorem 7.8 once one has the bilinear estimate:

$$\|\partial_x I(uv)\|_{X_{0,-\frac{1}{2}+}} \le c\,\|Iu\|_{X_{0,\frac{1}{2}+}}\|Iv\|_{X_{0,\frac{1}{2}+}}. \tag{8.24}$$

To prove the bilinear estimate (8.24), one applies the usual bilinear estimate (7.91) due to Kenig, Ponce and Vega [KPV6] combined with the following extra smoothing bilinear estimate whose proof is given in [CKSTT5].

Proposition 8.1. *The bilinear estimate:*

$$\|\partial_x\{IuIv - I(uv)\}\|_{X_{0,-\frac{1}{2}-}} \le cN^{-\frac{3}{4}+}\,\|Iu\|_{X_{0,\frac{1}{2}+}}\|Iv\|_{X_{0,\frac{1}{2}+}} \tag{8.25}$$

holds.

Proof. Just to give a flavor of the proof we consider the case when u is localized in a very small frequency ($|\xi| \ll 1$) and v localized in a very large one ($|\xi| \gg N$). One notices that in this situation

$$|(\widehat{I(uv) - IuIv})(\xi)| = \int\limits_{\xi=\xi_1+\xi_2} |m(\xi) - m(\xi_2)||\widehat{u}(\xi_1)||\widehat{v}(\xi_2)|.$$

Since m is smooth, the mean value theorem yields

$$|(\widehat{I(uv) - IuIv})(\xi)| \le \int\limits_{\xi=\xi_1+\xi_2} |m'(\bar{\xi}_2)||\widehat{u}(\xi_1)||\widehat{v}(\xi_2)|,$$

where $|\bar{\xi}_2| \sim |\xi_2| \gg N$. Moreover, it is easy to check that $m'(\bar{\xi}_2) \lesssim N^{-1}m(\xi_2)$. Thus,

$$\|\partial_x(I(uv) - IuIv)\|_{X_{0,-1/2+}} \le N^{-1}\|\partial_x(I(u)I(v))\|_{X_{0,-1/2+}}. \tag{8.26}$$

In this point, one uses the bilinear estimate (8.24) to get (8.25). For the estimates involving intermediate size frequencies the best gain that one can obtain is $N^{-3/4}$.□

Next we will obtain the so-called almost conserved quantity from (8.21). Note that the cancellation property

$$\int_0^t\int_{-\infty}^{\infty} \partial_x(Iu)^2 Iu\,dx\,dt = 0 \tag{8.27}$$

holds. In what follows this identity play an important role.

Using (8.27) one can write $R(t)$ as:

$$R(t) = \int_0^t\int_{-\infty}^{\infty} \partial_x\{(Iv)^2 - I(v^2)\}Iv\,dx\,ds. \tag{8.28}$$

The Plancherel identity and the Cauchy–Schwarz inequality yield

$$|R(t)| \leq c\, \|\partial_x\{(Iv)^2 - I(v^2)\}\|_{X_{0,-\frac{1}{2}-}}\, \|Iv\|_{X_{0,\frac{1}{2}+}}. \tag{8.29}$$

Now, using (8.29) and Proposition 8.1 the identity (8.21) gives the almost conservation law,

$$\|Iv(t)\|_2^2 \leq \|Iv(0)\|_2^2 + c\, N^{-\frac{3}{4}+}\, \|Iv\|^3_{X_{0,\frac{1}{2}+}}. \tag{8.30}$$

From (8.30), it is clear that the contribution of the error term $R(t)$ is very small for large N and therefore one can use (8.30) in the iteration process to extend the local solution.

Now, we are in position to prove the following global well-posedness result.

Theorem 8.5. *The IVP* (8.1), $k = 1$, *is globally well-posed in* $H^s(\mathbb{R})$ *for all* $s > -3/10$.

Proof. It is enough to show that the IVP (8.1) can be extended to $[0, T]$ for arbitrary $T > 0$. To make the analysis easy, one uses the scaling (7.8) mentioned in Chapter 7. More precisely, if v solves the IVP (8.1), $k = 1$, with initial data v_0, then for $1 > \lambda > 0$ so does v_λ, where $v_\lambda(x,t) = \lambda^2 v(\lambda x, \lambda^3 t)$, with initial data $v_0^\lambda(x) = \lambda^2 v_0(\lambda x)$. Observe that v exists in $[0, T]$ if and only if v_λ exists in $[0, \lambda^{-3} T]$. So we are interested to extend v_λ in $[0, \lambda^{-3} T]$.

An easy calculation shows that

$$\|Iv_0^\lambda\|_2 \leq c\, \lambda^{\frac{3}{2}+s} N^{-s}\, \|v_0\|_{s,2}, \tag{8.31}$$

where $N = N(T)$ is chosen later, but now we pick $\lambda = \lambda(N)$ by demanding

$$c\, \lambda^{\frac{3}{2}+s} N^{-s}\, \|v_0\|_{s,2} = \sqrt{\frac{\varepsilon_0}{2}} \ll 1. \tag{8.32}$$

From (8.32) one deduces that $\lambda \sim N^{\frac{2s}{3+2s}}$ and using (8.32) in (8.31) one gets

$$\|Iv_0^\lambda\|_2^2 \leq \frac{\varepsilon_0}{2} \ll 1. \tag{8.33}$$

Therefore, if we choose ε_0 arbitrarily small, then from Theorem 8.4 we see that IVP (8.1), $k = 1$, is well-posed for all $t \in [0, 1]$.

Now, using the almost-conserved quantity (8.30), the identity (8.33), and Theorem 8.4, one gets

$$\|Iv_\lambda(1)\|_2^2 \leq \frac{\varepsilon_0}{2} + c\, N^{-\frac{3}{4}+}\left[3\frac{\varepsilon_0}{2}\left(\frac{\varepsilon_0}{2}\right)^{1/2}\right] \leq \varepsilon_0 + c\, N^{-\frac{3}{4}+}\varepsilon_0. \tag{8.34}$$

So, one can iterate this process $c^{-1}N^{\frac{3}{4}-}$ times before doubling $\|Iv_\lambda(t)\|_2^2$. Hence, one can extend the solution in the time interval $[0, c^{-1}N^{\frac{3}{4}-}]$ by taking $c^{-1}N^{\frac{3}{4}-}$

times steps of size $O(1)$. As one is interested to define the solution in the time interval $[0, \lambda^{-3} T]$, one chooses $N = N(T)$ such that $c^{-1} N^{\frac{3}{4}-} \geq \lambda^{-3} T$. That is,

$$N^{\frac{3}{4}-} \geq c \frac{T}{\lambda^3} \sim T N^{\frac{-6s}{3+2s}}.$$

Therefore, for large N, the existence interval is arbitrarily large if we choose s such that $s > -3/10$. This completes the proof of the theorem. □

Question 2. For these global solutions whose regularity is below or between those given by the conservation law, one can ask for upper and lower bounds for the growth of the H^s-norm.

Theorem 8.1 provides some upper bound. In the case $k = 2$, where infinitely many conservation laws are available, one has the upper bound

$$\sup_{t \in [0,T]} \|v(t)\|_{s,2} \leq c\, T^{\theta(s)}, \quad \theta(s) = \min\{s - [s], [s+1] - s\} \tag{8.35}$$

(see [Fo], [Sta1]). A similar result for the case $k = 1$ is unknown as well as any lower bound estimate of the growth of the H^s-norm of the solutions.

For the case $k = 3$, the best-known global result for large H^s-data is due to [GPS] for $s > -1/42$. We recall that $s_3 = -1/6$ and the results in [To6] included global well-posedness for small data in $\dot{H}^{-1/6}(\mathbb{R})$.

8.2 Case $k = 4$

In this section, we shall first attempt to describe some of the main results in a series of works by Martel and Merle. Among other conclusions, they proved that blowup in finite time occurs in some H^1 local solutions of the IVP (8.1) with $k = 4$. Later, we shall add some further analysis with a more precise description of the dynamics of this blow-up result given by Martel, Merle, and Raphael.

For convenience sake we shall follow their notation, so we rewrite the equation in (8.1) with $k = 4$ in divergence form to get

$$\begin{cases} \partial_t u + \partial_x(\partial_x^2 u + u^5) = 0, \\ u(x,0) = u_0(x), \end{cases} \tag{8.36}$$

i.e., $v(x,t) = \sqrt[4]{5}\, u(x,t)$. In this setting, the conservation law E (or I_3) becomes

$$E(u_0) = \int_{-\infty}^{\infty} \Big[(\partial_x u)^2 - \frac{2}{6} u^6 \Big](x,t)\, dx. \tag{8.37}$$

We shall recall that the "traveling wave" $\varphi(x) = 3^{\frac{1}{4}} \operatorname{sech}^{\frac{1}{2}}(2x)$ satisfies

$$\varphi'' + \varphi^5 = \varphi \tag{8.38}$$

and $E(\varphi) = 0$.

In [W3], Weinstein (see Exercise 6.6) obtained the following sharp version of a Gagliardo–Nirenberg inequality,

$$\text{for all } w \in H^1(\mathbb{R}), \qquad \frac{1}{6}\int w^6\,dx \le \frac{1}{2}\left(\frac{\int w^2}{\int \varphi^2}\right)^2 \int (\partial_x w)^2\,dx. \tag{8.39}$$

Thus, if $u_0 \in H^1(\mathbb{R})$ with $\|u_0\|_2 < \|\varphi\|_2$, one has

$$\frac{1}{2}\left(1 - \frac{\int u_0^2}{\int \varphi^2}\right)^2 \int (\partial_x u)^2(x,t)\,dx \le E(u_0) \quad \text{for all} \quad t \in \mathbb{R}. \tag{8.40}$$

This a priori estimate together with I_2 ($\|u(t)\|_2 = \|u_0\|_2$) allows one to extend the local solution of (8.36) globally in time.

Notice that based on homogeneity, Theorem 7.2 guarantees the existence of global solutions for $u_0 \in L^2(\mathbb{R})$ with $\|u_0\|_2 < \delta$ sufficiently small. From these results, it is reasonable to conjecture that $\delta = \|\varphi\|_2$ (see the comments at the end of this chapter).

Also from the proof of Theorem 7.4 with $u_0 \in H^s(\mathbb{R})$, $s \in (0,1]$, and using an idea in [CzW4] one has that if there exists $T^* \in (0,\infty)$ such that

$$\lim_{t \uparrow T^*} \|u(t)\|_{s,2} = \infty \qquad \text{for } s \in [0,1), \tag{8.41}$$

then

$$\|u(t)\|_{s,2} \ge c\,(T^* - t)^{-s/3}, \tag{8.42}$$

and by [W], [Me5] there exist c_0, $R_0 > 0$ both depending on $\|u_0\|_2$ such that

$$\liminf_{t \uparrow T^*} \int_{|x - x(t)| \le R_0 (T^* - t)^{1/3}} |u(x,t)|^2\,dx \ge c_0\,, \tag{8.43}$$

for some function $x(t)$.

The next result by Martel and Merle [MM3] tells us that any global H^1 solution of (8.36) that at $t = 0$ is close to a traveling wave and does not disperse has to be precisely the traveling wave.

Theorem 8.6 ([MM3] of Liouville's type). *Let* $u_0 \in H^1(\mathbb{R})$ *and let*

$$\|u_0 - \varphi\|_{1,2} = \alpha. \tag{8.44}$$

Suppose that the corresponding H^1 *solution* $u = u(x,t)$ *of* (8.36) *satisfies:*

(i) There exist $c_1, c_2 > 0$ *such that*

$$c_1 \le \|u(t)\|_{1,2} \le c_2 \qquad \textit{for all } t \in \mathbb{R}. \tag{8.45}$$

(ii) There exists $x(t)$ such that for every $\varepsilon > 0$ there exists $R_0 > 0$ so that

$$\inf_{x(t)\in\mathbb{R}} \int_{|x-x(t)|>R_0} u^2(x,t)\,dx \le \varepsilon \quad \text{for all } t \in \mathbb{R}. \tag{8.46}$$

Then, there exists $\alpha_0 > 0$ such that for $\alpha \in (0,\alpha_0)$ in (8.44) *one has*

$$u(x,t) = \lambda_0^{1/2}\,\varphi(\lambda_0(x-x_0) - \lambda_0^3 t) \tag{8.47}$$

for some $\lambda_0 \in \mathbb{R}^+$ and $x_0 \in \mathbb{R}$.

The proof of this theorem is quite interesting.

First, the problem is renormalized by properly fixing the "center of mass" $x(t)$ and the "scaling" $\lambda(t)$, which is possible due to the invariance up to translations and dilations of the equation. Next, the authors establish a uniform-in-time exponential decay in the x-variable by using (8.46). Once this exponential decay is available they reduce the problem in studying which solutions of the associated linearized equation have such decay. They show that the solutions should have nontrivial projection on the singular spectrum of the linearized problem. But this possibility is withdrawn by using the choice of the parameters $x(t)$, $\lambda(t)$. So the solution of the linearized problem has to be the trivial one.

The next theorem complements the result in Theorem 8.6.

Theorem 8.7 ([MM5]). *Under the hypotheses* (8.44) *and* (8.45) *in* Theorem 8.6 *there exists α_1 such that if $\alpha \in (0,\alpha_1)$, then there exist $\lambda(t)$, $x(t)$ such that*

$$\lambda^{1/2}(t)\,u(\lambda(t)(x-x(t)),t) = \varphi(x) + u_R(x,t), \tag{8.48}$$

where

$$u_R(t) \underset{\text{(weakly)}}{\rightharpoonup} 0 \quad \text{in } H^1 \text{ as } t \uparrow \infty. \tag{8.49}$$

In fact, one has that

$$\lambda(t) \in (\lambda_1, \lambda_2) \quad \text{for all } t \text{ and } \quad x(t) \uparrow \infty \quad \text{as} \quad t \uparrow \infty. \tag{8.50}$$

In [MM1], Martel and Merle studied the stability of the traveling wave solution of the IVP (8.1) with $k = 4$.

We recall that it was shown in [Be1] and [BSS] that for the IVP (8.1) with $k = 1,2,3$, the corresponding traveling waves were stable and in [BSS] that for $k \ge 5$ they were unstable. Also, we recall that for the IVP (8.36), we have that φ satisfies

$$E(\varphi) = \int \left((\varphi')^2 - \frac{2}{6}\varphi^6\right) dx = 0$$

and using (8.38) that

$$DE(\varphi)\phi = \frac{d}{d\eta}E(\varphi+\eta\phi)\big|_{\eta=0} = 2\int (\varphi'\,\phi' - \varphi^5\,\phi)\,dx$$

$$= -2\int (\varphi'' + \varphi^5)\phi\, dx = -2\int \varphi\phi\, dx = \langle -2\varphi, \phi\rangle.$$

So,

$$DE(\varphi) = -2\varphi.$$

Let $\varepsilon \in H^1(\mathbb{R})$ with $\|\varepsilon\|_{1,2} \ll 1$; thus, $E(\varphi + \varepsilon) \sim \langle -2\varphi, \varepsilon\rangle$.

The next result establishes the instability of the traveling wave in this critical case $k = 4$ in (8.1) (see also (8.36)).

Theorem 8.8 ([MM1]). *There exist α_0, a_0, b_0, $c_0 > 0$ such that if $u_0 = \varphi + \varepsilon$ with*

$$\varepsilon \in H^1(\mathbb{R}), \qquad \|\varepsilon\|_{1,2} < a_0, \qquad x\varepsilon^2 \in L^1(\mathbb{R}), \tag{8.51}$$

$$|\varepsilon(x)| < b_0(1+x)^{-2}, \qquad \textit{for all } x > 0 \tag{8.52}$$

and

$$0 < \int \varepsilon\varphi\, dx < c_0 \int \varphi^2\, dx, \tag{8.53}$$

then there exists $t_0 = t_0(u_0)$ such that

$$\inf_{y\in\mathbb{R}} \|u(\cdot, t_0) - \varphi(\cdot - y)\|_{1,2} \geq \alpha_0. \tag{8.54}$$

In fact, they show that (8.54) holds in $L^2(\mathbb{R})$. Observe that taking $\varepsilon_n = n^{-1}\varphi$ for n large enough, ε_n satisfies the hypotheses (8.51)–(8.53). Similarly, if $\varepsilon = a\varphi + \varepsilon_0$ with $x\varepsilon^2 \in L^1$, $(1+x)^2\,|\varepsilon_0(x)| \leq c_0$ for all $x \geq 0$ with $\|\varepsilon_0\|_{1,2} \leq b_0\sqrt{a_0}$, then ε also satisfies (8.51)–(8.53).

In [Me4], Merle proved the existence of blow-up solutions of (8.36) in finite or infinite time.

Theorem 8.9 ([Me4]). *There exists $\alpha_0 > 0$ such that if $u_0 \in H^1(\mathbb{R})$ with*

$$E(u_0) < 0 \qquad \textit{and} \qquad \int \varphi^2 < \int u^2 < \int \varphi^2 + \alpha_0, \tag{8.55}$$

then the corresponding solution $u(t)$ of (8.36) blows up in the H^1-norm in finite or infinite time.

Observe that since $E(\varphi) = 0$ and $DE(\varphi) = -2\varphi$ there is a large class of data u_0 satisfying (8.55) whose corresponding solution blows up.

In [MM5], the authors showed that any blowup solution close to the traveling wave φ behaves asymptotically like it up to rescaling and translation, i.e., for some C^1 functions $x(t)$, $\lambda(t)$,

$$\pm\lambda^{1/2}(t)\, u(\lambda(t)x + x(t), t) \rightharpoonup \varphi \quad \text{in } H^1(\mathbb{R}) \text{ as } t \uparrow T, \quad T \leq \infty.$$

(See [ABLS] for a related result.)

As a consequence they established that the blowup at finite time must occur at a rate that in particular excludes the possibility of blowup at the self-similar rate:

$$u(x,t) \sim \frac{1}{(T-t)^{1/6}}\, h\left[\frac{x-x(t)}{(T-t)^{1/3}}\right]$$

since they establish that in this case (finite blow up time T)

$$\lim_{t\uparrow T}\, (T-t)^{1/3}\, \|\partial_x u(\cdot,t)\|_2 = \infty.$$

Based on these works, Martel and Merle were able to show the blowup in finite time [MM4] for solutions corresponding to data u_0 with negative energy ($E(u_0) < 0$), L^2-norm close to that of the solitary wave, see (8.55), and with sufficient decay at the right, i.e., there exists $\theta > 0$ such that for all $x_0 > 0$

$$\int_{x\geq x_0} u_0^2(x)\, dx \leq \frac{\theta}{x_0^6}. \tag{8.56}$$

Theorem 8.10 ([MM4]). *Under the hypotheses* (8.55) *and* (8.56) *the corresponding solutions of the IVP* (8.36) *blowup in finite time* $T < \infty$, *i.e.,*

$$\lim_{t\uparrow T}\, \|\partial_x u(t)\|_2 = \infty. \tag{8.57}$$

Moreover, let $t_n \uparrow T$ *be the sequence defined as:*

$$\|\partial_x u(\cdot,t_n)\|_2 = 2^n\, \|\partial_x \varphi\|_2 \tag{8.58}$$

with

$$\|\partial_x u(\cdot,t)\|_2 > 2^n\, \|\partial_x \varphi\|_2, \quad t \in (t_n, T).$$

Then there exists $n_0 = n(u_0)$ *such that for all* $n \geq n_0$,

$$\|\partial_x u(\cdot,t_n)\|_2 \leq \frac{c_0}{|E(u_0)|\,(T-t_n)}, \tag{8.59}$$

where $c_0 = 4(\int \varphi)^2\, \|\partial_x \varphi\|_2$.

The proof of this theorem used the results in the previous ones together with some elliptic and oscillatory integral-type estimates.

Finally, we have their following result regarding the nonexistence of minimal mass blow up solutions.

Theorem 8.11 ([MM6]). *Let* $u_0 \in H^1(\mathbb{R})$ *be such that*

$$\|u_0\|_2 = \|\varphi\|_2.$$

Assume that for some $c > 0$ and $\theta > 3$

$$\int_{x>x_0} u_0^2(x)\,dx \leq \frac{c}{x_0^{\theta}} \qquad \text{for all } x_0 > 0.$$

Then the corresponding solution $u(t)$ of the IVP (8.36) does not blowup in $H^1(\mathbb{R})$ either in finite or in infinite time.

We recall that for $u_0 \in H^1(\mathbb{R})$ with $\|u_0\|_2 < \|\varphi\|_2$ global existence is known (see 8.40). Also that for the NLS with critical power there exists a unique (up to the invariants of the equation) blow-up solution with minimal mass, i.e., a blow-up solution for

$$\begin{cases} i\partial_t u + \Delta u + |u|^{4/n}u = 0, \\ u(x,0) = u_0(x), \end{cases}$$

$\alpha = 1 + 4/n$, and $\|u_0\|_2 = \|\varphi\|_2$, where φ is a solution of (7.10) (see [Me3]).

The blow-up problem for the local solutions of the IVP (8.36) has been revised in the sequence of works [MMR1], [MMR2], [MMR3]. In these papers, a more concise description of the results in Theorems 8.9, 8.10, and 8.11 was established.

By defining the L^2-tubular neighborhood of the soliton manifold:

$$\mathcal{V}_{\alpha^*} = \{u \in H^1(\mathbb{R}) : \inf_{\lambda_0>0,\, x_0\in\mathbb{R}} \left\|u - \frac{1}{\lambda_0}\varphi((\,\cdot\, - x_0)/\lambda_0)\right\|_2 < \alpha^*\},$$

and the set of data:

$$\mathcal{A}_{\alpha_0} = \{u_0 = \varphi + \varepsilon_0 \,:\, \|\varepsilon_0\|_2 < \alpha_0, \; \int_{y>0} y^{10}\varepsilon_0(y)dy < 1\}$$

with $0 < \alpha_0 << \alpha^* < 1$ and φ the soliton (8.38), it was obtained in [MMR1] the following blow-up scenario near the soliton in $\mathcal{A}_{\alpha_0}$.

Theorem 8.12 ([MMR1]). *There exist universal constants α_0, α^* with $0 < \alpha_0 << \alpha^* < 1$ such that if $u_0 \in \mathcal{A}_{\alpha_0}$, with $E(u_0) \leq 0$ and $u_0 \not\equiv \varphi$, then the corresponding solution $u(t)$ blows up in finite time T and for $t \in [0,T)$ $u(t) \in \mathcal{V}_{\alpha^*}$. Moreover, there exists $l_0 = l_0(u_0) > 0$ such that*

$$\|\partial_x u(t)\|_2 \sim \frac{\|\varphi'\|_2}{l_0(T-t)}, \text{ as } t \uparrow T,$$

and $\exists\ \lambda(t)$, $x(t)$, $u^* \in H^1$, $u^* \not\equiv 0$ such that

$$u(x,t) - \frac{1}{\lambda^{1/2}(t)}\varphi((x - x_0)/\lambda(t)) \to u^* \text{ in } L^2 \text{as } t \uparrow T,$$

with

$$\lambda(t) \sim l_0(T-t) \quad \text{and} \quad x(t) \sim \frac{1}{l_0^2(T-t)}.$$

Also, there exists $\rho_0 = \rho_0(u_0) > 0$ such that if $v_0 \in \mathcal{A}_{\alpha_0}$ with $\|u_0 - v_0\|_{1,2} < \rho_0$, then the corresponding solution $v(t)$ blows up in finite time $T(v_0)$ in the manner described above.

Notice that $x(t) \to \infty$ as $t \uparrow T$. The next result found in [MMR1] gives a picture of the dynamic of the solution flow in $\mathcal{A}_{\alpha_0}$.

Theorem 8.13 ([MMR1]). *There exist universal constants* α_0, α^* *with* $0 < \alpha_0 \ll \alpha^* < 1$ *such that if* $u_0 \in \mathcal{A}_{\alpha_0}$, *then one of the following three possibilities occurs:*

(i) $\exists\, t^* \in [0, T)$ *such that* $u(t^*) \notin \mathcal{V}_{\alpha^*}$.
(ii) The solution $u(t)$ *blows up in finite time in the regime of the previous theorem.*
(iii) The solution is global, for all t, $u(t) \in \mathcal{V}_{\alpha^*}$ *and there exist* $\lambda_\infty > 0$, $x(t)$ *such that*

$$\lambda_\infty^{1/2} u(\lambda_\infty(\cdot + x(t)), t) \to \varphi \quad \text{in} \quad H^1_{loc} \quad \text{as} \quad t \uparrow \infty$$

$$x(t) \sim \frac{t}{\lambda_\infty} \quad \text{and} \quad |\lambda_\infty - 1| = o(1) \quad \text{as} \quad \alpha_0 \downarrow 0.$$

Thus, the set of data found in (i) and (ii) are open. Also, results in [MMR3] delineates the exit scenario (i) in Theorem 8.13 and the existence and uniqueness of the minimal mass blowup element.

The following result shows that the decay assumption in the definition of $\mathcal{A}_{\alpha_0}$ is essential in the above theorems. More precisely, H^1-data with slower right decay may produce "exotic" blow-up rates.

Theorem 8.14 ([MMR3]).

(i) $\forall\, \gamma > 11/13\ \exists\, u \in C([0, T) : H^1(\mathbb{R}))$ *solution of the IVP* (8.36) *which blows up at* $t = T$ *with*

$$\|\partial_x u(t)\|_2 \sim \frac{1}{(T - t)^\gamma}.$$

(ii) $\exists\, u \in C([0, \infty) : H^1(\mathbb{R}))$ *solution of the IVP* (8.36) *such that*

$$\|\partial_x u(t)\|_2 \sim e^t.$$

(iii) $\forall \gamma > 0\ \exists\, u \in C([0, \infty) : H^1(\mathbb{R}))$ *solution of the IVP* (8.36) *such that*

$$\|\partial_x u(t)\|_2 \sim t^\gamma.$$

The possibility of continua blow up rates were first observed in [KST] for solutions of the $H^1(\mathbb{R}^3)$-critical semilinear wave equations.

8.3 Comments

The global solution for the IVP (8.1) with $k \geq 5$ with small data $v_0 \in H^1(\mathbb{R})$ follows by the argument used in (8.5). This tells us that

$$E(v_0) \geq \|\partial_x v(t)\|_2^2 - c_k\, \|\partial_x v(t)\|_2^{k/2}\, \|v_0\|_2^{2+2/k}. \tag{8.60}$$

Since at $t = 0$ we have

$$E(v_0) \geq \|\partial_x v_0\|_2^2 - c_k \|\partial_x v_0\|_2^{k/2} \|v_0\|_2^{2+2/k}$$

then for $\|v_0\|_2 + \|\partial_x v_0\|_2 \ll 1$ one has $E(v_0) > 0$, which inserted into (8.60) provides an a priori estimate for $\|\partial_x v(t)\|_2$ through an argument similar to the one in (6.11). This combined with I_2 gives an a priori estimate for $\|v(t)\|_{1,2}$.

More precisely, in [FaLP] Farah, Linares and Pastor following some arguments in [HR1] proved

Theorem 8.15. *Let $u_0 \in H^1(\mathbb{R})$. Let $k > 4$ and $s_k = (k-4)/2k$. Suppose that*

$$E(u_0)^{s_k} I_2(u_0)^{1-s_k} < E(Q)^{s_k} I_2(Q)^{1-s_k}, \quad E(u_0) \geq 0. \tag{8.61}$$

If

$$\|\partial_x u_0\|_2^{s_k} \|u_0\|_2^{1-s_k} < \|\partial_x Q\|_2^{s_k} \|Q\|_2^{1-s_k}, \tag{8.62}$$

then for any t as long as the solution exists,

$$\|\partial_x u(t)\|_2^{s_k} \|u_0\|_2^{1-s_k} = \|\partial_x u(t)\|_2^{s_k} \|u(t)\|_2^{1-s_k} < \|\partial_x Q\|_2^{s_k} \|Q\|_2^{1-s_k}, \tag{8.63}$$

where $Q(x) = (k+1)^{1/k} \phi_{1,k}(x)$ and $\phi_{1,k}$ is unique positive even solution of the equation (7.7).

This in turn implies that H^1 solutions exist globally in time.

We also recall that in the case $k \geq 5$ global well-posedness based on the homogeneity (scaling argument) was established in Theorem 7.5 for small data in $\dot{H}^{s_k}(\mathbb{R})$, $s_k = 1/2 - 2/k$.

The problem of describing the long time behavior of solutions to the generalized KdV equation corresponding to "small" data was studied by Hayashi and Naumkin [HN1], [HN2].

In [HN1], they answered the following question: what is the smallest power ρ which guarantees that "small" solutions of the generalized KdV equation:

$$\partial_t u + \partial_x^3 u + |u|^{\rho-1} \partial_x u = 0, \quad \rho > 1, \tag{8.64}$$

behave as the solutions of the associated linear problem (7.21) and scatter? They showed that if $\rho > 3$ and the data u_0 satisfies that

$$\|(1+x^2)^{1/2} \Lambda u_0\|_2 \leq \varepsilon \quad \text{(for some } \varepsilon \text{ fixed} \ll 1), \tag{8.65}$$

then the corresponding solution $u(\cdot, t)$ of (8.64) satisfies that for any $t > 0$,

$$\|u(\cdot, t)\|_p \leq c\,(1+t)^{-1/3\, p'}, \quad p \in (4, \infty], \ \frac{1}{p} + \frac{1}{p'} = 1. \tag{8.66}$$

Moreover, there exists $u_+ \in L^2(\mathbb{R})$ such that for $t > 0$

$$\|u(\cdot, t) - V(t)u_+\|_2 \leq c\, t^{-(\rho-3)/3}, \tag{8.67}$$

(see the notations in (3.1) and (7.22)).

In [HN2], they proved that the above result is optimal by establishing that "small" solutions of the mKdV ($\rho = 3$ in (8.64)), although satisfy (8.65), they do not hold (8.66). (The description of their asymptotic behavior involves the self-similar solutions ($= \frac{1}{\sqrt[3]{t}}\,\omega(\frac{x}{\sqrt[3]{t}})$) of the mKdV).

Consider the periodic boundary value problem:

$$\begin{cases} \partial_t v + \partial_x^3 v + v^k \partial_x v = 0, \\ v(x,0) = v_0(x) \in H^s(\mathbb{T}), \end{cases} \tag{8.68}$$

$t \in \mathbb{R}$, $x \in \mathbb{T}$, $k \in Z^+$. Global well-posedness for (8.68) with $k = 1, 2, 3$ has been established in $H^s(\mathbb{T})$ with $s \geq -1/2$, $s \geq 1/2$, $s > 5/6$, respectively, by Colliander, Keel, Staffilani, Takaoka and Tao [CKSTT4], [CKSTT5].

For $k \geq 4$ the best results are due to Staffilani [Sta2] ($s \geq 1$ with a smallness condition on the $\|v_0\|_2$ norm).

In the same regard for the IVP (8.36), global well-posedness was obtained in $H^s(\mathbb{R})$ with $s > 6/13$ (see [MSWX]) for data satisfying $\|u_0\|_2 < \|\varphi\|_2$ (improving previous results in [FLP1] ($s > 3/4$) and in [Fa] ($s > 3/5$)). As it was mentioned this result should hold in L^2, i.e., if $u_0 \in L^2(\mathbb{R})$ and $\|u_0\|_2 < \|\varphi\|_2$, then the local solution extends globally or $\delta = \|\varphi\|_2$ in Theorem 7.2 with φ as in (8.38).

Next, we shall briefly comment on stability for the solitary wave solutions (7.6) for the k-generalized Korteweg–de Vries (k-gKdV) equation. In [Be1] and [Bn2], the stability of the solitary wave solution for the KdV equation was established. The stability is understood in the following sense: Given $\varepsilon > 0$, there exists $\delta > 0$ such that if $\|v_0 - \phi_{c,1}\|_{1,2} < \delta$, then for all $t \in \mathbb{R}$, there is $x(t)$ such that

$$\|v(\cdot + x(t), t) - \phi_{c,1}(\cdot)\|_{1,2} < \varepsilon, \tag{8.69}$$

this is known as orbital stability.

For the k-gKdV, it was proved in [BSS] that for $k < 4$ (subcritical case) the solitary waves are stable, and for $k > 4$ they are unstable (see also [GSS]). Martel and Merle [MM1] have shown the instability of the solitary waves in the critical case $k = 4$. Regarding asymptotic stability of the solitary waves $\phi_{c,k}$, Pego and Weinstein [PW] obtained results for the cases $k = 1$ and $k = 2$ for data decaying exponentially as $x \to \infty$. In [MM2], the following assertion was proved: Given c_0 there exists a δ_{0,c_0} such that for $\|v_0 - \phi_{c_0,k}\|_{1,2} \leq \delta_{0,c_0}$ there exist c_∞ a constant and $x(t)$ a real function so that

$$v(x + x(t), t) \rightharpoonup \phi_{c_\infty,k} \quad \text{in} \quad H^1 \quad \text{as } t \to \infty$$

for $k = 1, 2, 3$, i.e., the subcritical case.

The results listed above were obtained in the H^1-norm. Merle and Vega [MV] have shown the stability and asymptotic stability for the solitary wave solutions of the KdV equation in the L^2-norm. More precisely, in [MV] the following result was proved (see also [MiT]): Let $c_0 > \sigma > 0$. Then, there exist $\hat{c}, \delta > 0$ such that if $u(x,t)$ is a solution of the IVP (8.1) with $k = 1$ (KdV) such that

$$u(x,0) = \phi_{c_0,1}(x) + v_0(x), \quad \text{with} \quad \|v_0\|_2 < \delta,$$

then there exist $c_+ > 0$ and $x : [0,\infty) \mapsto \mathbb{R}$ a C^1 function such that

$$\sup_{t>0} \|u(\cdot,t) - \phi_{c_0,1}(\cdot - x(t))\|_2 \le \hat{c}\, \|v_0\|_2^{1/2}$$

$$c_+ = \lim_{t\to\infty} x(t)$$

$$|c_+ - c_0| + \sup_{t\ge 0}|\dot{x}(t) - c_0| \le c\, \|v_0\|_2$$

and

$$\lim_{t\to\infty} \int_{x>\sigma t} |u(x,t) - \phi_{c_+,1}(x - x(t))|^2\, dx = 0.$$

In [KM], the stability of the traveling wave solution for the quartic KdV, i.e., $k = 3$ in (8.1), was studied in the critical space $\dot{H}^{-1/6}(\mathbb{R})$.

In [W4], Weinstein deduced the following variational characterization of the traveling wave $\phi_{c,k}$ in (7.6):

If $u(x,t)$ is a solution of (8.1) with $k = 1,2,3,4$ such that

$$I_3(u(t)) = I_3(\phi_{c,k}) \quad \text{and} \quad I_2(u(t)) = I_2(\phi_{c,k}) \quad \text{for some} \quad c > 0. \tag{8.70}$$

Then, $u(x,t) = \phi_{c,k}(x - x_0 - ct)$ for some $x_0 \in \mathbb{R}$.

In particular, this implies (see Exercise 8.4) that if $u = u(x,t)$ is a solution of (8.1) with $k = 1,2,3,4$ such that

$$\lim_{t\to\infty} \operatorname*{Inf}_{y\in\mathbb{R}} \|u(\cdot,t) - \phi_{c,k}(\cdot - y)\|_{1,2} = 0, \tag{8.71}$$

then $u(x,t) = \phi_{c,k}(x - x_0 - ct)$ for some $x_0 \in \mathbb{R}$.

Based on the previous works [DM], [DRu], concerning related results for the NLS in [Cb] it was shown that (8.71) fails for $n \ge 5$. More precisely, it was proved the existence of a one parameter family of special solutions of (8.1) with $k \ge 5$ $\{U^A(x,t)\}_{A\in\mathbb{R}}$ such that

$$\lim_{t\to\infty} \operatorname*{Inf}_{y\in\mathbb{R}} \|U^A(\cdot,t) - \phi_{c,k}(\cdot - y)\|_{1,2} = 0.$$

Moreover, if $u = u(x,t)$ is a global solution of (8.1) with $k \ge 5$ such that

$$\lim_{t\to\infty} \operatorname*{Inf}_{y\in\mathbb{R}} \|u(\cdot,t) - \phi_{c,k}(\cdot - y)\|_{1,2} = 0,$$

then

$$u(x,t) = U^A(x - x_0, t), \quad t \ge t_0, \quad \text{for some} \quad A, t_0, x_0 \in \mathbb{R}.$$

In [AlMn1], it was established that the breather solutions of the mKdV equation (7.108) are orbitally stable in the H^1 topology.

In the introduction to this chapter we mentioned the fact that the KdV and mKdV equations can be solved via the inverse scattering method. Now, we would like to describe some interesting applications deduced from this method. The first one regards the construction of explicit solutions called N-solitons. These solutions generalize the solitary wave solutions or "solitons" (7.6) ($k = 1,2$) (see [Lb], [Hi1], [Sc]). In particular, they describe the interaction between several solitons with different speeds. In addition, the N-soliton solutions decompose exactly as a sum of N solitons as $t \to +\infty$. In other words, for any given $0 < c_1 < c_2 < \cdots < c_N$, $x_1, \ldots, x_N$, there exists an explicit N-soliton solution $v(t)$ such that

$$\left\| v(t) - \sum_{j=1}^{N} \phi_{c_j,k}(\cdot - x_j - c_j t) \right\|_{1,2} \to 0 \quad \text{as} \quad t \to +\infty. \tag{8.72}$$

Another interesting result obtained in [ES] for the case $k = 1$ is the following: Any sufficiently smooth and decaying solution v of (7.1) splits into two parts as $t \to \infty$, i.e.,

$$v(x,t) = v_d(x,t) + v_c(x,t),$$

where v_d is an N-soliton solution and $v_c(x,t) \to 0$ uniformly for $x > 0$ as $t \to +\infty$. (see also [Sc]).

Concerning the stability of N-solitons in the sense given in (8.69) for the solitary waves, Martel, Merle and Tsai [MMT] obtained for powers $k = 1,2$ (integrable cases) and $k = 3$ (nonintegrable) the following result:

Theorem 8.16. *Let $0 < c_1 < \cdots < c_N$ and $k = 1,2,3$. There exists γ_0, A, L_0, $\alpha_0 > 0$ such that the following is satisfied. Assume that there exist $L > L_0$, $\alpha < \alpha_0$, and $x_1^0 < \cdots < x_N^0$ such that*

$$\| v(0) - \sum_{j=1}^{N} \phi_{c_j,k}\left(\cdot - x_j^0\right) \|_{1,2} \le \alpha, \quad \text{with} \quad x_j^0 > x_{j-1}^0 + L$$

for all $j = 2, \ldots, N$. Then there exist $x_1(t), \ldots, x_N(t) \in \mathbb{R}$ such that for all $t \geq 0$,

$$\left\| v(t) - \sum_{j=1}^{N} \phi_{c_j,k}(x - x_j(t)) \right\|_{1,2} \le A(\alpha + e^{-\gamma_0 L}).$$

The above result tells us that if $v(0)$ is close in the H^1-norm to the sum of N-solitons whose speeds are ordered (so they do not interact for $t > 0$) and whose centers are far apart, then the corresponding solution $v(t)$ remains close in H^1-norm to a translated sum of N-solitons for all $t > 0$.

Using the ideas in [MV] and [MMT], Alejo, Muñoz and Vega [AlMnVe] were able to establish the L^2-stability of the N-solitons solutions.

In [Ma], the following existence and uniqueness result of an asymptotic N-soliton-like solution was established for the subcritical $k = 1,2,3$ and critical case $k = 4$ in (8.1).

Theorem 8.17. *Let $N \in \mathbb{Z}^+$, $0 < c_1 < c_2 < \cdots < c_N$, $x_1, \ldots, x_N \in \mathbb{R}$. There exists a unique $v \in C([T_0,\infty) : H^1(\mathbb{R}))$ for some $T_0 > 0$ solution of the equation in* (8.1) *with $k = 1, 2, 3,$ or 4 such that* (8.72) *holds. Moreover, there exist $A, \gamma > 0$ such that*

$$\|v(t) - \sum_{j=1}^{N} \phi_{c_j,k}(-x_j - c_j t)\|_{1,2} \leq A\, e^{-\gamma t}.$$

Notice that Theorem 8.17 extends the estimate (8.72) to the nonintegrable cases $k = 3, 4$.

In [FePaUl], Fermi, Pasta and Ulam and latter in [ZaKr] Zabusky and Kruskal presented numerical evidences describing the remarkable phenomena of the soliton collision. They illustrated the *elastic* character of the collision of two solitons (elastic: the collision preserves the shape of the solitons). So the unique consequence of the collision is a shift translation on each soliton.

For the equation:

$$\partial_t u + \partial_x^3 u + \partial_x(f(u)) = 0,$$

it was established in [Mu] that the collision between two solitons is not elastic in general, except for the KdV equation, for the mKdV equation and for the Gardner equation ($f(u) = u^2 - \mu u^3$) all completely integrable systems. This work was preceded by [MM8], where for the case $f(u) = u^4$ with two solitons of different masses it was shown that the collision is inelastic by proving the nonexistence of a pure two-soliton solution. More precisely, if the solution $u(x,t)$ satisfies that

$$u(x,t) = \phi_{c_1,3}(x - c_1 t) + \phi_{c_2,3}(x - c_2 t) + \eta(x,t), \quad \text{as } t \downarrow -\infty$$

(see 7.6) with

$$\|\eta(t)\|_{1,2} \ll \|\phi_{c_2,3}\|_{1,2} \ll \|\phi_{c_1,3}\|_{1,2},$$

then for $t \gg 1$

$$u(x,t) = \phi_{c_1(t),3}(x - y_1(t)) + \phi_{c_2(t),3}(x - y_2(t)) + \eta(x,t)$$

with

$$\|\eta(t)\|_{1,2} \ll \|\phi_{c_2(t),3}\|_{1,2},$$

and

$$c_1(t) \to c_1^+, \;\; c_2(t) \to c_2^+ \quad \text{as} \quad t \uparrow \infty.$$

In the case where $u(x,t)$ is a pure two-soliton as $t \downarrow -\infty$, one has that

$$c_1^+ > c_1, \;\; c_2^+ < c_2, \quad \lim_{t\to\infty} \|\eta(t)\|_{1,2} > 0.$$

In the case of the modified KdV equation and the Gardner equation, it is an interesting problem to characterize the initial data which precede to the formation of these special solution solitons or "breathers" (see 7.108). Using the *inverse scattering method* (IST) this question was studied in [SaYa].

Also, it is interesting to describe the interaction between these solutions traveling in opposite directions. In this regard, one has the results concerning the generalized Gardner equation found in [CGD]:

$$\partial_t u + \alpha\, u \partial_x u + \beta\, u^2\, \partial_x u + \delta\, \partial_x^3 u = 0, \quad \alpha,\, \beta,\, \delta \in \mathbb{R}. \tag{8.73}$$

It has been proved that this equation is integrable and also arises in the study of wave propagation (see [GKM]). Notice that in (8.73) the interaction between the dispersion and the nonlinearity cubic and quadratic should be considered. It was shown in [CGD] that (8.73) possesses breather solutions and solitons traveling in both directions when $\beta,\, \delta > 0$. Also based on the Hirota method of constructing multisoliton solution to integrable models (see [Hi2]) explicit expressions describing the interaction of these solutions were deduced. It was proved that these solutions retain their shape after the interaction, except for a phase shift, and numerical simulations were presented to confirm this fact.

In the same regard, one has the special solutions of the modified KdV equation:

$$\partial_t v + \partial_x^3 v + v^2 \partial_x v = 0,$$

given by solitons (described in 7.6) traveling to the right and the breathers (see (7.108)), which travel to the left if $3\, N^2 > \omega^2$. The description of the interaction of these solutions is largely open.

Regarding the "soliton resolution conjecture": any "reasonable" solution of the k-gKdV (7.1) will eventually resolve into a radiation-dispersive wave moving to the right plus a finite number of traveling waves moving to the left. Notice that the breather solution of the mKdV equation (7.108) contradicts this statement. In [ES], Eckhaus and Schuur were able to prove this conjecture for the KdV equation ($k = 1$ in (7.1)). Their proof uses the inverse scattering theory and is based on the relation between properties of the datum $u_0 = q$ and the reflected coefficient $b(k)$ (see (9.59)–(9.64)). More precisely, they proved that if u_0 and its derivatives up to order fourth have an appropriate algebraic decay as $|x| \to \infty$, then $b(k) = b \in C^r(\mathbb{R})$ and

$$\partial^m b(k) = O(|k|^{-5}) \quad \text{as} \quad |k| \to \infty,$$

for $m = 0, 1, \ldots, r$. We recall that in the cases when $b(k) \equiv 0$ the solution is the sum of N solitons with N being the number of discrete eigenvalues in (9.62).

8.4 Exercises

8.1 Consider the IVP (8.36) with a real-valued datum $u_0 \in H^1(\mathbb{R})$ such that $\|u_0\|_2 < \|\varphi\|_2$ with φ as in (8.38). As it was shown in this case, $u \in C(\mathbb{R} : H^1(\mathbb{R}))$ is the global solution of the problem.

(i) Prove that for any time interval $(t_0, t_0 + \Delta T)$ with $\Delta T > 0$,

$$\sum_{j=0}^{1}\big(\|\partial_x^j u\|_{L_x^5 L_t^{10}((t_0,t_0+\Delta T))} + \|\partial_x^{j+1} u\|_{L_x^\infty L_t^2((t_0,t_0+\Delta T))}\big) \le c\,(\|u_0\|_{1,2};\ \Delta T). \tag{8.74}$$

Hint: Use Theorem 7.4, and the conservation laws I_2 and I_3 in (7.4) and (7.5). Notice that in this case $s = 1$ one can take ∂_x instead of D_x in (7.55) and (8.38).

(ii) Prove that for any time interval $(t_0, t_0 + \Delta T)$

$$\|u\|_{L_x^4 L_t^\infty((t_0,t_0+\Delta T))} \le c\,(\|u_0\|_{1,2}; \Delta T). \tag{8.75}$$

Hint: Use Lemma 7.3 and the integral equation:

$$u(t) = V(t)u_0 - \int_0^t V(t-t')\,\partial_x(u^5)(t')\,dt' = V(t)u_0 + z(t). \tag{8.76}$$

(iii) Prove that $z(\cdot)$ in (8.76) satisfies

$$z \in C(\mathbb{R} : H^2(\mathbb{R})). \tag{8.77}$$

Hint: First observe that to obtain (8.77) it suffices to show that $\partial_x^2 z \in C(\mathbb{R} : L^2(\mathbb{R}))$. Use (7.16) to reduce the problem to bound $\|\partial_x^2(u^5)\|_{L_x^1 L_t^2((t_0,t_0+\Delta T))}$ with $\Delta T \ll 1$. Now combine parts (i) and (ii) to get the desired result.

Remark 8.1. Roughly speaking, Exercise 8.1 illustrates a general principle, i.e., if $v \in C([0,T] : H^{\hat{s}}(\mathbb{R}))$ is a solution of the k-gKdV (7.10) with $\hat{s} > s_{0,k}$, where $s_{0,k}$ is the smallest Sobolev exponent, where local well-posedness can be established (i.e., $s_{0,1} = -3/4$, $s_{0,2} = 1/4, \ldots$), then the integral term in $z_k(t)$,

$$v(t) = V(t)v_0 - \int_0^t V(t-t')\,v^k\partial_x v(t')\,dt' = V(t)v_0 + z_k(t)$$

is more regular in the $H^s(\mathbb{R})$ scale than both $v(t)$ and the linear part $V(t)v_0$.

8.2 Consider the linear IVP (7.10). Prove:

(i) If $v_0 \in L^2(\mathbb{R}) \cap L^2(|x|^2dx)$, then $V(t)v_0 \in C^1(\mathbb{R})$ for $t \neq 0$.
Hint: Use the commutative property of the operators $\Gamma = x + 3t\partial_x^2$ and $L = \partial_t + \partial_x^3$.

(ii) Given $\varepsilon > 0$ and the set:

$$A_N = \{(x_j, t_j) : j = 1, \ldots, N\} \subset \mathbb{R}^2,$$

there exists $v_0 \in H^1(\mathbb{R})$ (real valued) with $\|v_0\|_2 < \varepsilon$ such that

(a) If $t \notin \{t_1, \ldots, t_N\}$, then $V(t)v_0 \in C^1(\mathbb{R})$.
(b) If $t = t_j$, then $V(t_j)v_0 \in C^1(\mathbb{R}) - \{x_k : (x_k, t_k) \in A_n\}$, and $\partial_x V(t_j)v_0(x_k)$ does not exist if $(x_k, t_j) \in A_N$.

(iii) If $u = u(x,t)$ is the solution of the IVP (8.36) with data $u(x,0) = v_0(x)$ as in part (ii) with $\varepsilon = \|\varphi\|_2$, then (a) and (b) hold for $u(x,t)$.
Hint: Use Exercise 8.1

Remark 8.2. This is a particular case of the so-called dispersive blow up, studied by Bona and Saut [BSa1], [BSa2].

8.3 Let $v \in C(\mathbb{R} : H^2(\mathbb{R}))$ be a solution of the KdV equation.

(i) Prove that for $t \in \mathbb{R}$,

$$\begin{aligned} I_4(v)(t) &= \int_{-\infty}^{\infty} \left[\frac{9}{5}(\partial_x^2 v)^2 - 3u(\partial_x v)^2 + \frac{1}{4}v^4 \right](x,t)\,dx \\ &= I_4(v)(0) = I_4(v_0). \end{aligned} \tag{8.78}$$

(ii) Prove that there exists $c > 0$ such that

$$\sup_{t\in\mathbb{R}} \|v(t)\|_{2,2} \le c\,\|v_0\|_{2,2}.$$

Hint: Combine (8.78) and I_2, I_3 in (7.4) and (7.5).

(iii) If $\tilde{v} \in C(\mathbb{R} : H^1(\mathbb{R}))$ is solution of the IVP associated to the KdV equation and $\tilde{v}_0 \in H^{1+\delta}(\mathbb{R})$, prove that $\tilde{v} \in C(\mathbb{R} : H^{1+\delta}(\mathbb{R}))$ and deduce an upper bound for

$$\Phi(T) = \sup_{0\le t\le T} \|\tilde{v}(t)\|_{1+\delta,2}$$

in terms of T and $\|\tilde{v}_0\|_{1+\delta,2}$ (for the case of the mKdV, see (8.35)).

8.4 (i) Using the notation in (7.6), prove that

$$\frac{d}{dc}\|\phi_{c,k}\|_2 \quad \begin{cases} > 0, & \text{if} \quad k = 1,2,3, \\ = 0, & \text{if} \quad k = 4, \\ < 0, & \text{if} \quad k = 5,\ldots. \end{cases}$$

(ii) Using the notation in (7.6)–(7.6), prove that

$$I_3(\phi_{c,k}) = \frac{k-4}{2(k+1)(k+2)} \int_{-\infty}^{\infty} \phi_{c,k}^{k+2}(x)\,dx.$$

Thus,

$$I_3(\phi_{c,k}) \quad \begin{cases} < 0, & \text{if} \quad k = 1,2,3, \\ = 0, & \text{if} \quad k = 4, \\ > 0, & \text{if} \quad k = 5,\ldots. \end{cases}$$

Hint: Combine the equation (7.7) and the identity (5.83).

8.5 [W2] Defining the functional $B : H^1(\mathbb{R}) \mapsto \mathbb{R}$ as:

$$B(v) = I_3(v) + c\, I_2(v). \tag{8.79}$$

(i) Prove that B is differentiable, and

$$DB(v)w = \frac{d}{d\varepsilon} B(v+\varepsilon w)|_{\varepsilon=0} = 2 \int_{-\infty}^{\infty} \left(-\partial_x^2 v + c\, v - \frac{v^{k+1}}{k+1}\right) w\, dx,$$

if $v \in H^2(\mathbb{R})$ and $w \in H^1(\mathbb{R})$.

(ii) Prove that $\phi_{c,k}$ is a critical point of B, i.e. $DB(\phi_{c,k}) \equiv 0$.

(iii) Prove that $DB(\cdot)$ is differentiable, and

$$D^2B(v)(h,w) = \frac{d}{d\varepsilon} DB(v+\varepsilon h)w|_{\varepsilon=0} = 2 \int_{-\infty}^{\infty} \left(-\partial_x^2\, h - v^k h + ch\right) w\, dx$$

if $v \in H^1(\mathbb{R})$ and $h,\, w \in H^2(\mathbb{R})$.

(iv) Using the notation:

$$\mathcal{L}_{\phi_{c,k}} f(x) = -\frac{d^2}{dx^2} f(x) - \phi_{c,k}^k(x) f(x) + c f(x),$$

show that

(a) $D^2B(\phi_{c,k})(h,w) = 2 \int_{-\infty}^{\infty} \mathcal{L}_{\phi_{c,k}} h\, w\, dx = 2 \int_{-\infty}^{\infty} h\, \mathcal{L}_{\phi_{c,k}} w\, dx.$

(b) $\mathcal{L}_{\phi_{c,k}} \phi_{c,k}^{\frac{k+2}{2}} = -c^{\frac{k(k+4)}{4}}\, \phi_{c,k}^{\frac{k+2}{2}}.$

(c) $\mathcal{L}_{\phi_{c,k}} \phi'_{c,k} = 0.$

(d) $\mathcal{L}_{\phi_{c,k}} \left(-\frac{d}{dc}\phi_{c,k}\right) = \phi_{c,k}.$

8.6 Using the notation in (7.6) prove that $h : (0,\infty) \mapsto \mathbb{R}$ defined as:

$$h(c) = I_3(\phi_{c,k}) + c\, I_2(\phi_{c,k})$$

is strictly convex if and only if $k = 1, 2, 3$.

Hint: Use Exercises 8.4(i) and 8.5(ii).

8.7 Assuming the characterization of the traveling wave described in (8.70) for $k = 1, 2, 3, 4$ prove property (8.71).

8.8 Let $u \in C([0,T] : H^4(\mathbb{R}) \cap L^2(|x|^2\, dx))$ be a real solution of the k-gKdV equation (8.1).

(i) Prove the identity:

$$\frac{d}{dt} \int x\, u^2(x,t)\, dx = -3 \left[I_3(u_0) + \frac{4-2k}{3(k+1)(k+2)} \int u^{k+2}\, dx \right].$$

(ii) Prove that if $k = 2$ (mKdV) and $a_0 \in \mathbb{R}$ such that

$$\int (x - a_0)\, u_0^2(x)\, dx = 0,$$

then

$$\int (x - a(t))\, u_0^2(x)\, dx = 0,$$

with $a(t) = a_0 + t\dfrac{3\, I_3(u_0)}{I_2(u_0)}$.

8.9 Let $u \in C([0,T] : H^1(\mathbb{R}))$ be a solution of the k-gKdV equation (8.1). Prove that if $|x|\, u(0), |x|\, u(1) \in L^2(\mathbb{R})$, then $u \in C([0,T] : H^2(\mathbb{R}))$.

8.10 Consider the equation (8.73) with the parameters $\alpha = \delta = 1$, $\beta = -\gamma$, i.e.,

$$\partial_t u + u\partial_x u - \gamma\, u^2\, \partial_x u + \partial_x^3 u = 0, \quad \gamma \in \mathbb{R}. \tag{8.80}$$

(i) Prove that for $\gamma > 0$ the equation (8.80) has traveling wave solutions of the form $q_{c,\gamma}(x - ct)$ with

$$q_{c,\gamma}(x) = \frac{6c}{1 + \rho\, \cosh(\sqrt{c}x)}, \quad \rho = (1 - 6c\gamma)^{1/2}, \ \ c \in (0, 1/6\gamma). \tag{8.81}$$

(ii) Prove that if $u \in C([0,T] : H^4(\mathbb{R}))$ is a solution of the equation (8.80), then

$$v = v(x,t) = u - \sqrt{6\gamma}\, \partial_x u - \gamma\, u^2 \in C([0,T] : H^3(\mathbb{R}))$$

satisfies the KdV equation.

(iii) Prove that if $\gamma \downarrow 0$, then

$$q_{c,\gamma}(x) \to \phi_{c,1}(x) = 3c\, \mathrm{sech}^2\left(\frac{\sqrt{c}x}{2}\right),$$

the soliton solution of the KdV equation (7.6).

8.11 Let $u \in C([0,T^*) : H^1(\mathbb{R})) \cap \ldots$ be a local solution of the IVP (8.1) with $k = 4$ (L^2-critical case). Assume that

$$\lim_{t \uparrow T^*} \|\partial_x u(t)\|_2 = \infty.$$

(i) Prove that for any $s \in (0,1]$,

$$\liminf_{t \uparrow T^*} \|D_x^s u(t)\|_2 = \infty.$$

(ii) Prove that for any $p \in (2,\infty]$,

$$\liminf_{t \uparrow T^*} \|u(t)\|_p = \infty.$$

Chapter 9
Other Nonlinear Dispersive Models

In this chapter, we will discuss local and global well-posedness for some nonlinear dispersive models arising in different physical situations. Our goal is to present some relevant results associated to the equations to be contemplated here and it is by no means an exhaustive study of each of them. In Section 9.1 we will treat the Davey–Stewartson systems. The Ishimori equations will be considered in Section 9.2. The Kadomtsev–Petviashvili (KP) equations will be discussed in Section 9.3. The Benjamin–Ono equation will be studied in Section 9.4 and in Section 9.5 we will be examine the Zakharov systems. Finally, in Section 9.6 we will briefly review the inverse scattering method for the KdV equation and well-posedness results regarding higher order KdV equations.

9.1 Davey–Stewartson Systems

The cubic nonlinear Schrödinger equation

$$i\partial_t u + \partial_x^2 u = \pm |u|^2 u, \quad x, t \in \mathbb{R},$$

among other phenomena models the propagation of wave packets in the theory of water waves. It is also a complete integrable system. The corresponding bi-dimensional model is called the Davey–Stewartson system, which is given by the nonlinear system of partial differential equations,

$$\begin{cases} i\partial_t u + c_0\partial_x^2 u + \partial_y^2 u = c_1|u|^2 u + c_2 u\partial_x\varphi, \\ \partial_x^2\varphi + c_3\partial_y^2\varphi = \partial_x(|u|^2), \end{cases} \tag{9.1}$$

$x, y \in \mathbb{R},\ t > 0$, where $u = u(x, y, t)$ is a complex-valued function, $\varphi = \varphi(x, y, t)$ is a real-valued function, and c_0, c_3 are real parameters and c_1, c_2 are complex parameters. It was first derived by Davey and Stewartson in [DS] in the case $c_3 > 0$. When capillary effects are important, Djordjevic and Redekopp [DR] showed that c_3 can be negative (see also Benney and Roskes [BnR]). Independently, Ablowitz and Haberman [AH] obtained a particular form of (9.1) as an example of a completely integrable model generalizing the two-dimensional nonlinear Schrödinger

F. Linares, G. Ponce, *Introduction to Nonlinear Dispersive Equations*, Universitext, DOI 10.1007/978-1-4939-2181-2_9

equation. In the context of inverse scattering theory the system above with parameters $(c_0, c_1, c_2, c_3) = (1, -1, -2, -1)$, $(-1, -2, 1, 1)$, and $(-1, 2, -1, 1)$ are known as DSI, DSII defocusing, and DSII focusing, respectively. For these particular cases several results regarding the existence of solitons and the Cauchy problem have been established by inverse scattering techniques (see [AnF], [BC1], [FS1], [Su1]). For instance, in [FS1] Fokas and Sung proved that for initial data in the Schwartz class $\mathcal{S}(\mathbb{R}^2)$ and boundary data $\partial_x \varphi_1(x,t)$ and $\partial_y \varphi_2(y,t)$ in the Schwartz class in the spatial variable and continuous in t, (9.1) has a unique global solution in time t which, for each t belongs to the Schwartz class in the spatial variable. The same result was obtained in [BC1] for the DSII defocusing.

Ghigladia and Saut [GS] classified the system as elliptic-elliptic, hyperbolic-elliptic, elliptic-hyperbolic, and hyperbolic-hyperbolic according to the signs of the parameters (c_0, c_3), i.e., $(+,+)$, $(-,+)$, $(+,-)$, and $(-,-)$, respectively.

Solutions of (9.1) satisfy the following two conservation laws:

$$M(u_0) = \int\limits_{\mathbb{R}^2} |u(x,y,t)|^2 \, dxdy,$$

$$\begin{aligned} E(u_0) = &\int\limits_{\mathbb{R}^2} (c_0 \, |\partial_x u(x,y,t)|^2 + |\partial_y u(x,y,t)|^2) \, dxdy \\ &+ \frac{1}{2} \int\limits_{\mathbb{R}^2} (c_1 \, |u(x,y,t)|^4 + c_2 \, (\partial_x \varphi)^2(x,y,t) + c_3 \, (\partial_y \varphi)^2(x,y,t)) \, dxdy. \end{aligned}$$

The elliptic-elliptic and hyperbolic-elliptic cases were considered by Ghigladia and Saut [GS]. In these cases they reduced the system (9.1) to the nonlinear cubic Schrödinger equation with a nonlocal nonlinear term, i.e.,

$$i\,\partial_t u + c_0 \partial_x^2 u + \partial_y^2 u = c_1 |u|^2 \, u + A(u),$$

where $A(u) = (\Delta^{-1} \, \partial_x^2 |u|^2) u$. They showed local well-posedness for data in $L^2(\mathbb{R}^2)$, $H^1(\mathbb{R}^2)$, and $H^2(\mathbb{R}^2)$ using Strichartz estimates (see 4.23) and the continuity properties of the operator $\Delta^{-1}\partial_x^2$. They also established global well-posedness and blow up results for the elliptic–elliptic case (see also [SG]). Ozawa in [Oz] found exact blow up solutions in the hyperbolic–elliptic case (see Exercise 9.6).

For the elliptic–hyperbolic and hyperbolic–hyperbolic cases the Strichartz estimates by itself does not provide the desired result. To explain this we will consider without loss of generality $c_0 = \pm 1$ and $c_3 = -1$. So using a rotation in the xy-plane and assuming that φ satisfies the radiation condition

$$\lim_{y \to \infty} \varphi(x,y,t) = \varphi_1(x,t) \quad \text{and} \quad \lim_{x \to \infty} \varphi(x,y,t) = \varphi_2(y,t),$$

for some given functions φ_1, φ_2, then the system (9.1) can be written as

$$\begin{cases} i\partial_t u + Hu = d_1\, |u|^2 u + d_2\, u \int\limits_y^\infty \partial_x(|u(x,y',t)|^2)\, dy' \\ \qquad\qquad + d_3\, u \int\limits_x^\infty \partial_y(|u(x',y,t)|^2)\, dx' + d_4 u \partial_x \varphi_1 + d_5 u \partial_y \varphi_2, \\ u(x,y,0) = u_0(x,y), \end{cases} \tag{9.2}$$

where $H = \Delta$ in the elliptic–hyperbolic case and $H = 2\partial_x\partial_y$ in the hyperbolic-hyperbolic case. The difficulty of these problems comes from the fact that the nonlinear terms contain derivatives of the unknown function and that the terms

$$\int\limits_y^\infty \partial_x(|u(x,y',t)|^2)\, dy' \quad \text{and} \quad \int\limits_x^\infty \partial_y(|u(x',y,t)|^2)\, dx'$$

do not decay as $|x| \to \infty$, $|y| \to \infty$, respectively.

To describe the results in these two cases we introduce the weighted Sobolev spaces $\mathcal{F}_l^m$ defined as follows:

$$\mathcal{F}_l^m = H^m(\mathbb{R}^2) \cap L^2(|x|^l\, dx).$$

First we look at the elliptic–hyperbolic case. In [LiPo] Linares and Ponce proved local well-posedness for the IVP (9.2) for sufficiently small data in $\mathcal{F}_{12}^m$, $m \geq 12$, $\varphi_1 = \varphi_2 \equiv 0$. They use the smoothing effect of Kato's type associated to the group $\{e^{iHt}\}$. Chihara [Ch1], using pseudo differential operators, obtained a local result for data in $u_0 \in H^m(\mathbb{R}^2)$ satisfying $\|u_0\|_2 \leq 1/(2\sqrt{\max\{d_1,d_2\}})$, $\varphi_1 = \varphi_2 \equiv 0$, with m sufficiently large. Hayashi in [H2] showed local well-posedness for small data in $\mathcal{F}_{2\,l}^m$, $m, l > 1$. The main tool for accomplishing this was the use of smoothing effects. In [HH2] Hayashi and Hirata proved that one can have local result in the usual Sobolev space $H^{5/2}(\mathbb{R}^2)$ for data with L^2-norm small. The latest updated result is due to Hayashi [H3], where he got local well-posedness for the IVP (9.2) for data of any size in $H^s(\mathbb{R}^2)$, $s \geq 2$. Global results were obtained by Hayashi and Hirota in [HH1] for small data in $\mathcal{F}_6^3$; see also [Ch1]. For analytic function spaces a global result for small data was established by Hayashi and Saut in [HS].

For the hyperbolic–hyperbolic case, using Kato's smoothing effect Linares and Ponce proved local well-posedness for small data in $\mathcal{F}_4^6$, $\varphi_1 = \varphi_2 = 0$ in [LiPo]. Hayashi [H2] showed local well-posedness for small data in $\mathcal{F}_{2\delta}^\delta$, $\delta > 1$. No local well-posedness results are known without restriction on the size of the data.

9.2 Ishimori Equation

In this section we comment on local and global well-posedness results for a two dimensional generalization of the Hesinberg equation, called the Ishimori equation which reads

$$\begin{cases} \partial_t S = S \wedge (\partial_x^2 S \pm \partial_y^2 S) + b(\partial_x \phi \partial_y S + \partial_y \phi \partial_x S), \\ \partial_x^2 \phi \mp \partial_y^2 \phi = \mp 2S \cdot (\partial_x S \wedge \partial_y S), \end{cases} \tag{9.3}$$

$x, y, t \in \mathbb{R}$, where $S(\cdot, t) : \mathbb{R}^2 \to \mathbb{R}^3$ with $\|S\| = 1$, $S \to (0, 0, 1)$ as $\|(x, y)\| \to \infty$, and $\wedge$ denotes the wedge product in $\mathbb{R}^3$.

This model was proposed by Ishimori in [Is1] as a two-dimensional generalization of the Heisenberg equation in ferromagnetism, which corresponds to the case $b = 0$ and signs $(-, +, +)$ in (9.3) and it was studied in [SSB].

For $b = 1$ the system (9.3) is completely integrable by inverse scattering (see [AH], [BC1], [KMa], [Su2], [ZK], and references therein).

Using the stereographic variable $u : \mathbb{R}^2 \mapsto \mathbb{C}$ one can get rid of the constraint $\|S\| = 1$. Thus, for

$$\begin{aligned} u &= \frac{S_1 + iS_2}{1 + S_3}, \\ S = (S_1, S_2, S_3) &= \frac{1}{1 + |u|^2}(u + \bar{u}, -i(u - \bar{u}), 1 - |u|^2), \end{aligned} \tag{9.4}$$

the IVP for (9.3) can be written as

$$\begin{cases} i\partial_t u + \partial_x^2 u + a\partial_y^2 u = 2u\dfrac{(\partial_x u^2 - \partial_y u^2)}{(1 + |u|^2)} - ib(\partial_x \phi \partial_y u - \partial_y \phi \partial_x u), \\ \partial_x^2 \phi + a' \partial_y^2 \phi = 8\,\mathcal{I}m\, \dfrac{(\partial_x u \partial_y u)}{(1 + |u|^2)^2}, \\ u(x, y, 0) = u_0(x, y), \end{cases} \tag{9.5}$$

with the condition $u(x, y, t) \to 0$ as $\|(x, y)\| \to \infty$, where $a, a' \in \mathbb{R} \setminus \{0\}$.

To discuss the local and global results we will distinguish two cases: case $(-, +)$, i.e., $a < 0$ in the first equation, and $a' > 0$ in the second equation in (9.5) and case $(+, -)$ with similar connotation.

The case $(-, +)$ was studied by Soyeur [Sy]. He obtained local well-posedness for the IVP (9.5) for small data in $H^m(\mathbb{R}^2)$, $m \geq 4$. Assuming additional regularity on the data he extended the local solution globally in $H^m(\mathbb{R}^2)$, $m \geq 6$. The argument used here does not extend to the case $(+, -)$.

The case $(+, -)$ was first studied by Hayashi and Saut [HS]. They considered the problem in a class of analytic functions obtaining local and global existence results for small analytic data. This approach allows them to overcome the loss of derivatives introduced by the nonlinearity.

Hayashi in [H4] removed the analyticity hypotheses used in [HS]. He established local well-posedness for the IVP (9.5) for small data in the weighted Sobolev $\mathcal{F}_8^4$.

In [KPV9] Kenig, Ponce and Vega established a local well-posedness result for data of arbitrary size in the space $\mathcal{F}_{2m}^s = H^s(\mathbb{R}^2) \cap L^2(|x|^{2m}\, dx)$, $s > m$. The method of proof follows closely the method explained in detail in the next chapter.

9.3 KP Equations

Here we shall discuss some well-posedness results for the Kadomtsev–Petviashvili (KP) equations. The KP equations are two-dimensional versions of the KdV equation. They arise in many physical contexts as models for the propagation of weakly nonlinear dispersive long waves, which are essentially one-directional, with weak transverse effects. For instance, in the plasma physics context these models were derived by Kadomtsev and Petviashvili [KP]. Meanwhile, in surface water wave theory, they were deduced by Ablowitz and Segur in [AS1]. It is also one of the classical prototype problems in the field of exactly solvable equations (see [AC] for a complete set of references on this subject).

The equation reads as follows.

$$\begin{cases} \partial_x(\partial_t u + \partial_x^3 u + u\partial_x u) \mp \partial_y^2 u = 0, \\ u(x,y,0) = u_0(x,y), \end{cases} \tag{9.6}$$

$x, y \in \mathbb{R},\ t > 0$. Under some conditions on the initial data, (9.6) can be written as

$$\begin{cases} \partial_t u + \partial_x^3 u + u\partial_x u \mp \partial_x^{-1}\partial_y^2 u = 0, \\ u(x,y,0) = u_0(x,y), \end{cases} \tag{9.7}$$

$x, y \in \mathbb{R},\ t > 0$. When the sign in front of $\partial_x^{-1}\partial_y^2$ in (9.7) is minus we refer to this equation as the KPI equation; otherwise we called it the KPII equation.

The results concerning well-posedness for KPI and KPII equations are quite different. We will first list the results regarding the KPII equation.

Bourgain [Bo10] showed local and global well-posedness for data in $H^s(\mathbb{R}^2)$, $s \geq 0$. The local result was obtained by the Fourier transform restriction method introduced by him to study nonlinear dispersive equations. In [Tz1] Tzvetkov obtained local results in anisotropic Sobolev spaces $H^{s_1,s_2}(\mathbb{R}^2)$ defined as

$$H^{s_1,s_2}(\mathbb{R}^2) = \{f \in \mathcal{S}'(\mathbb{R}^2) :$$
$$\|f\|^2_{H^{s_1,s_2}} = \int_{\mathbb{R}^2} (1+|\xi_1|)^{2s_1}(1+|\xi_2|)^{2s_2}\,|\widehat{f}(\xi_1,\xi_2)|^2\,d\xi_1\,d\xi_2 < \infty\},$$

with $s_1 > -1/4$, $s_2 \geq 0$. He combined the ideas in [Bo1] with bilinear estimates in [KPV6] and Strichartz estimates. Improvements of these results were obtained in [Tz2], [Tk2]. Independently, Isaza and Mejia [IM1] and Takaoka and Tzvetkov [TT] established local well-posedness for data in $H^{s_1,s_2}(\mathbb{R}^2)$ for $s_1 > -1/3$ and $s_2 \geq 0$. Global results are also obtained in [IM1], [Tk2] using Bourgain's method in [Bo5]. In [IM2] Isaza and Mejia using the I-method introduced by [CKSTT6] showed global well-posedness for data in $H^{s_1,s_2}(\mathbb{R}^2)$ for $s_1 > -1/14$ and $s_2 \geq 0$. In [HaHK] Hadac, Herr and Koch obtained local well-posedness in the critical space $\dot{H}^{-1/2,0}(\mathbb{R}^2)$ (see Exercise 9.14(i)). These solutions corresponding to small data are global and scattered. They also showed local well-posedness in the inhomogeneous case $H^{1/2,0}(\mathbb{R}^2)$ for arbitrary data.

The problem for the KPI equation is completely different. The techniques used in Bourgain [Bo1] do not work here due to the lack of symmetry of the symbol associated to the equation. In [IN] Iorio and Nunes proved local existence result using the Kato quasilinear theory for data in $H^s(\mathbb{R}^2)$, $s > 2$. Molinet, Saut and Tzvetkov [MST1] showed that the difficulty with respect to the symmetry of the symbol was not at all technical, by proving that a Picard's scheme cannot be applied to study local in well-posedness for that equation in standard Sobolev spaces. However, they obtained [MST2] using the conservation laws for the solution flow of the KPI equation and a compactness argument the global existence of solutions for (9.7).

In [Ke] Kenig showed local well-posedness in

$$Y_s = \{u \in L^2(\mathbb{R}^2) : \|u\|_2 + \|J_x^s u\|_2 + \|\partial_x^{-1}\partial_y u\|_2 < \infty\} \tag{9.8}$$

for the KPI equation, $s > 3/2$. Combining this local result with the results in [MST2] he established global well-posedness in the space

$$Z_0 = \{u \in L^2(\mathbb{R}^2) : \|u\|_2 + \|\partial_x^{-1}\partial_y u\|_2 + \|\partial_x^2 u\|_2 + \|\partial_x^{-2}\partial_y^2 u\|_2 < \infty\}.$$

In [CIKS] Colliander, Ionescu, Kenig and Staffilani obtained local well-posedness in the space $Y_1 \cap L^2(|y|\,dxdy)$.

In [IKT] Ionescu, Kenig and Tataru proved global well-posedness in the energy space Y_1, i.e. $u_0, \partial_x u_0, \partial_y \partial_x^{-1} u_0 \in L^2(\mathbb{R}^2)$.

Regarding the periodic setting there are some results by Bourgain [Bo10], Iorio and Nunes [IN], and Isaza, Mejía and Stallbohm [IMS].

9.4 BO Equation

$$\begin{cases} \partial_t u + \mathsf{H}\partial_x^2 u + u\partial_x u = 0, \\ u(x,0) = u_0(x), \end{cases} \tag{9.9}$$

$x \in \mathbb{R}$, $t > 0$, where H denotes the Hilbert transform (see Definition 1.7).

This integro-differential equation serves as a generic model for the study of weakly nonlinear long waves incorporating the lowest-order effects of nonlinearity and non-local dispersion. In particular, the propagation of internal waves in stratified fluids of great depth is described by the BO equation (see [Be2], [On]) and turns out to be important in other physical situations as well (see [DaR], [Is2], [MK]). Among noticeable properties of this equation we can mention that it defines a Hamiltonian system, can be solved by an analogue of the inverse scattering method (see [AF]), admits (multi)soliton solutions (see [Ca]), and satisfies infinitely many conserved quantities (see [Ca]).

Regarding the IVP associated to the BO equation, local and global results have been obtained by various authors. Iorio [Io1] showed local well-posedness for data in $H^s(\mathbb{R})$, $s > 3/2$, and making use of the conserved quantities he extended globally the result in $H^s(\mathbb{R})$, $s \geq 2$. He also studied the problem in weighted Sobolev spaces. In [Po], Ponce extended the local result for data in $H^{3/2}(\mathbb{R})$ and the global result for any solution in $H^s(\mathbb{R})$, $s \geq 3/2$. The argument of proof combines parabolic regularization, smoothing properties, and energy estimates. In [MST3], Molinet, Saut and Tzvetkov showed that the Picard iteration process cannot be carry out to prove local results for the BO equation in $H^s(\mathbb{R})$ for any $s \in \mathbb{R}$. Koch and Tzvetkov [KTz] established a local result for data in $H^s(\mathbb{R})$, $s > 5/4$, improving the one given in [Po]. In [KeKo] Kenig and Köenig refined the argument in [KTz] to obtain $s > 9/8$. The main idea is the use of the Strichartz estimates to control one derivative of the solution. More precisely, through energy estimates and Kato–Ponce commutator estimates (3.16) a smooth solution of the BO equation satisfies

$$\|D^s u\|_{L^\infty_T L^2_x} \leq \|u(0)\|_{s,2} \exp\Big(c \int_0^T \|\partial_x u(t)\|_{L^\infty}\, dt\Big). \tag{9.10}$$

Then the Strichartz estimates allow them to establish the existence of a constant c such that

$$\int_0^1 \|\partial_x u(t)\|_{L^\infty}\, dt \leq c \tag{9.11}$$

whenever $u_0 \in H^s(\mathbb{R})$, $s > 5/4$. Thus, a combination of (9.10) and (9.11) and a standard compactness argument yields the result.

Tao in [To4] showed that the IVP associated to the BO equation is globally well-posed in $H^1(\mathbb{R})$. The new tool introduced by him was the following gauge transformation

$$w = P_+(e^{-iF}u), \quad F = F(u) = \int_{-\infty}^x u(y,t)\, dy, \tag{9.12}$$

where $\widehat{P_+ f}(\xi) = \chi_{[0,\infty)}(\xi)\widehat{f}(\xi)$. This is a variant of the Cole–Hopf transformation for viscous Burgers' equation (see Exercise 9.18), which in this setting allows one to remove most of the worst terms involving the derivative.

Tao's gauge transformation idea was further developed by Burq and Planchon [BuPl] to carry the local well-posedness to $H^s(\mathbb{R})$, for $s > 1/4$ and by Ionescu and Kenig [IK1] to extend it to $H^s(\mathbb{R})$, $s \geq 0$. We refer to [MoPi] for further discussion of the latter results.

In the periodic setting, Molinet [Mo1] has shown global well-posedness for data in $L^2(\mathbb{T})$.

In the previous chapters we discussed some decay and smoothness properties for solutions of the NLS and k-gKdV equations and their relationship. In particular, for initial data in the Schwartz class $\mathcal{S}$, the corresponding solutions (for smooth nonlinearity) also belong to this class in their life span. The decay of the data is reflected in the decay of the corresponding solutions of the associated IVP (persistence). Solutions of the BO equation do not share this property, not even mild persistence properties regarding the decay hold. To illustrate this unusual character of solutions of the BO equation we shall recall the following spaces:

$$\mathcal{F}_r^s = H^s(\mathbb{R}) \cap L^2(|x|^r\,dx),$$

and

$$\dot{\mathcal{F}}_r^s = \Big\{ f \in \mathcal{F}_r^s : \int f(x)\,dx = \widehat{f}(0) = 0 \Big\}.$$

The following result is due to Iório [Io2].

Theorem 9.1. *Let $u \in C([0,T] : H^2(\mathbb{R}))$, $T > 0$, be the solution of the IVP (9.9).*

(i) If $u_0 \in \mathcal{F}_{2j}^2$, $j = 1,2$. Then

$$u \in C([0,T] : \mathcal{F}_{2j}^2), \quad j = 1,2.$$

(ii) If $u_0 \in \mathcal{F}_6^3$ and $\int u_0(x)\,dx = 0$. Then

$$u \in C([0,T] : \dot{\mathcal{F}}_6^3).$$

(iii) If $u \in C([0,T] : \dot{\mathcal{F}}_8^4)$. Then $u \equiv 0$.

In [Io3] Iorio strengthened the result in Theorem 9.1(iii) by proving that if at three different times a solution of the BO equation satisfies $u(\cdot,t_j) \in \mathcal{F}_8^4$, $j = 1,2,3$, then $u \equiv 0$. In [FoPo] Iorio's result was extended to non-integer values. In particular it was shown that if $u(\cdot,t_j) \in \mathcal{F}_7^{7/2}$, $j = 1,2,3$, then $u \equiv 0$, and that for every $\epsilon > 0$ if $u_0 \in \dot{\mathcal{F}}_{7-\epsilon}^{7/2}$, then the corresponding solution satisfies $u \in C([0,T] : \dot{\mathcal{F}}_{7-\epsilon}^{7/2})$. In [FLP3] it was shown that the uniqueness result of Iorio mentioned above involving a condition a three different times is necessary. More precisely, it was proved that there exist non zero solutions of the BO equation $u \in C([0,T] : \dot{\mathcal{F}}_6^3)$ such that $u(\cdot,t_j) \in \mathcal{F}_8^4$, $j = 1,2$.

Notice that the above results are mainly a consequence of the lack of smoothness of the symbol $\sigma(\xi) = \xi|\xi|$ modeling the dispersion.

For the sake of completeness we shall explain the parabolic regularization method or artificial viscosity method for the case of the BO equation. This method which is quite general will be used in the next chapter.

The goal is to establish the following local well posedness result for the IVP (9.9) associated to the BO equation.

Theorem 9.2. *Let $s > 3/2$. Given any $u_0 \in H^s(\mathbb{R})$ there exist $T(\|u_0\|_{s,2}) > 0$ and a unique solution u of the IVP (9.9) such that*

$$u \in C([0,T] : H^s(\mathbb{R})) \cap C^1((0,T) : H^{s-2}(\mathbb{R})). \tag{9.13}$$

Moreover, the map data → solution from $H^s(\mathbb{R})$ to $C([0,T] : H^s(\mathbb{R}))$ is locally well defined and continuous.

In addition, if $u_0 \in H^{s'}(\mathbb{R})$ with $s' > s$, then

$$u \in C([0,T] : H^{s'}(\mathbb{R})) \cap C^1((0,T) : H^{s'-2}(\mathbb{R})).$$

To simplify the exposition we shall sketch the details in the case $s = 2$. It will be clear from our proof below and the calculus of inequalities in Chapter 3 how to obtain the general result $s > 3/2$.

We consider the IVP associated to the viscous BO equation

$$\begin{cases} \partial_t u + \mathsf{H}\partial_x^2 u + u\partial_x u = \gamma\partial_x^2 u, \\ u(x,0) = u_0(x), \end{cases} \tag{9.14}$$

$t > 0, x \in \mathbb{R}, \gamma \in (0,1)$.

Step 1 *A priori* estimate for solutions (9.14)

Assume that $u^\gamma \in C([0,T^*] : H^2(\mathbb{R})) \cap C^\infty((0,T^*) : H^\infty(\mathbb{R}))$ is a solution of the IVP (9.14), then the standard energy estimate (see 3.12 and 3.13) show that

$$\frac{d}{dt}\|u^\gamma(t)\|_{2,2}^2 + \gamma\|\partial_x^3 u^\gamma(t)\|_2^2 \leq c\|\partial_x u^\gamma(t)\|_\infty \|u^\gamma(t)\|_{2,2}^2. \tag{9.15}$$

Thus, from Sobolev Embedding (Theorem 3.2)

$$\frac{d}{dt}\|u^\gamma(t)\|_{2,2} \leq c\|u^\gamma(t)\|_{2,2}^2, \tag{9.16}$$

where c here and below will denote a constant whose value may change from line to line but it is independent of the data and the parameters in (9.9) and (9.14) (and later in (9.32)). From (9.16) one has that

$$\|u^\gamma(t)\|_{2,2} \leq \frac{\|u_0\|_{2,2}}{1 - ct\|u_0\|_{2,2}}, \tag{9.17}$$

therefore taking T such that

$$cT\|u_0\|_{2,2} = 1/2 \tag{9.18}$$

it follows that

$$\sup_{[0,T]} \|u^\gamma(t)\|_{2,2} \leq 2\|u_0\|_{2,2}, \qquad \forall \gamma \in (0,1). \tag{9.19}$$

Now integrating in the t variable in (9.15), using Sobolev embedding and (9.18)–(9.19) one gets that

$$\gamma\int_0^T \|\partial_x^3 u^\gamma(t)\|_2^2 dt \leq \|u_0\|_{2,2}^2 + c\int_0^T (2\|u_0\|_{2,2})^3 dt \tag{9.20}$$

$$\leq \|u_0\|_{2,2}^2 + 8cT\|u_0\|_{2,2}^3 \leq c\|u_0\|_{2,2}^2.$$

To complete this step we observe that if $u_0 \in H^{s'}(\mathbb{R})$ with $s' > 2$, then as in (9.15) it follows that

$$\frac{d}{dt}\|u^\gamma(t)\|_{s',2} \leq c\|u^\gamma(t)\|_{2,2}\|u^\gamma(t)\|_{s',2}.$$

Hence,

$$\sup_{[0,T]} \|u^\gamma(t)\|_{s',2} \leq \|u_0\|_{s',2}\, e^{cT\|u_0\|_{2,2}} = K\,\|u_0\|_{s',2}, \qquad K = K(\|u_0\|_{2,2}),$$

i.e. higher derivatives of the solution are also bounded by the data in the same time interval $[0, T]$.

Step 2 Existence of solutions to the IVP (9.14).

We consider the semigroup $\{U^\gamma(t) : t \geq 0\}$ defined as

$$U^\gamma(t)f(x) = \left(e^{4\pi^2 i|\xi|\xi t} e^{-\gamma 4\pi^2 \xi^2 t}\widehat{f}\right)^\vee(x).$$

It is easy to see that for any $t \geq 0$

$$(a) \quad \|U^\gamma(t)f\|_2 \leq \|f\|_2,$$
$$(b) \quad \|\partial_x U^\gamma(t)f\|_2 \leq \frac{c}{(\gamma t)^{1/2}}\|f\|_2. \tag{9.21}$$

The solution of the IVP (9.14) is a fixed point of the operator $\Psi = \Psi_{\gamma,u_0}$ with

$$\Psi(v)(t) = U^\gamma(t)u_0 - \int_0^t U^\gamma(t-t')v\partial_x v(t')dt', \tag{9.22}$$

defined on

$$\Omega_{\widehat{T},r} = \{v : \mathbb{R}\times[0,\widehat{T}] \to \mathbb{R} : v \in C([0,\widehat{T}] : H^2(\mathbb{R})),\ \sup_{[0,\widehat{T}]}\|v(t)\|_{2,2} \leq r\}, \tag{9.23}$$

with $\widehat{T}$ and $r > 0$ to be chosen. From (9.21) it follows that

$$\sup_{[0,T_\gamma]}\|\Psi(v)(t)\|_{2,2} \leq c\|u_0\|_{2,2} + \frac{cT_\gamma^{1/2}}{\gamma^{1/2}}\sup_{[0,T_\gamma]}\|v(t)\|_{2,2}^2,$$

and

$$\sup_{[0,T_\gamma]}\|(\Psi(v)-\Psi(\tilde{v}))(t)\|_{2,2}$$
$$\leq \frac{cT_\gamma^{1/2}}{\gamma^{1/2}}\sup_{[0,T_\gamma]}(\|v(t)\|_{2,2} + \|\tilde{v}(t)\|_{2,2})\sup_{[0,T_\gamma]}\|(v-\tilde{v})(t)\|_{2,2}.$$

Hence, choosing r and T_γ as

$$r = 2c\|u_0\|_{2,2} \quad \text{and} \quad \frac{cT_\gamma^{1/2} r}{\gamma^{1/2}} = \frac{2c^2 T_\gamma^{1/2}\|u_0\|_{2,2}}{\gamma^{1/2}} = \frac{1}{4}, \tag{9.24}$$

it follows that the operator Ψ defines a contraction in $\Omega_{T_\gamma,r}$, and so for any $\gamma > 0$ the IVP (9.14) has a unique solution

$$u^\gamma \in C([0,T_\gamma] : H^2(\mathbb{R})) \cap C^\infty(\mathbb{R} \times (0,T_\gamma)), \qquad \text{with} \qquad T_\gamma \sim \gamma. \tag{9.25}$$

Now using the *a priori* estimate (step–1) we can reapply the above local existence argument (which only depends on the size of the initial data, see 9.24) to extend for each $\gamma \in (0,1)$ the solution u^γ in the class (9.25) to the whole time interval $[0,T]$ with T as in (9.18). Moreover, we have that

$$\sup_{\gamma>0} \big(\sup_{[0,T]} \|u^\gamma(t)\|_{2,2}^2 + \gamma \int_0^T \|\partial_x^3 u^\gamma(t)\|_2^2 dt \big) \le c\|u_0\|_{2,2}^2. \tag{9.26}$$

Step 3 Convergence of the u^γ's as $\gamma \downarrow 0$.

For $1 > \gamma > \gamma' > 0$ we define

$$\omega(t) = \omega^{\gamma,\gamma'}(t) = u^\gamma(t) - u^{\gamma'}(t), \tag{9.27}$$

which satisfies the equation

$$\partial_t \omega + \mathsf{H}\partial_x^2 \omega + \omega \partial_x u^\gamma + u^{\gamma'} \partial_x \omega = \gamma' \partial_x^2 \omega + (\gamma - \gamma')\partial_x^2 u^\gamma, \qquad t \in [0,T], \tag{9.28}$$

with data $\omega(x,0) = 0$. Using energy estimates it follows that

$$\frac{d}{dt}\|\omega(t)\|_2 \le c(\|\partial_x u^\gamma(t)\|_\infty + \|\partial_x u^{\gamma'}(t)\|_\infty)\|\omega(t)\|_2 + (\gamma - \gamma')\|\partial_x^2 u^\gamma(t)\|_2.$$

Hence, from (9.26) one has that

$$\begin{aligned} \sup_{[0,T]} \|(u^\gamma - u^{\gamma'})(t)\|_2 &= \sup_{[0,T]} \|\omega(t)\|_2 \\ &\le (\gamma - \gamma') \int_0^T \|\partial_x^2 u^\gamma(t)\|_2 dt \, \exp\{\int_0^T (\|\partial_x u^\gamma\|_\infty + \|\partial_x u^{\gamma'}\|_\infty) dt\} \qquad (9.29) \\ &\le 2(\gamma - \gamma')T\|u_0\|_{2,2} \cdot e^{4cT\|u_0\|_{2,2}} \le (\gamma - \gamma')K, \end{aligned}$$

$K = K(\|u_0\|_{2,2})$ which shows that the u^γ's converge as $\gamma \downarrow 0$ in $C([0,T] : L^2(\mathbb{R}))$. Moreover, combining (9.26) and interpolation the u^γ's converge in $C([0,T] : H^{2-\mu}(\mathbb{R}))$ for any $\mu > 0$ to a limit function $\widetilde{u}(x,t)$

$$\widetilde{u} \in C([0,T] : H^{2-\mu}(\mathbb{R})) \cap L^\infty([0,T] : H^2(\mathbb{R})), \qquad \forall \mu > 0 \tag{9.30}$$

(using a weak compactness argument) satisfying that

$$\sup_{[0,T]} \|\widetilde{u}(t)\|_{2,2} \le c\|u_0\|_{2,2}. \tag{9.31}$$

To complete this step we observe that if $u_0 \in H^{s'}(\mathbb{R})$ with $s' > 2$, then the u^{γ}'s converge as $\gamma \downarrow 0$ in $C([0,T] : H^{s'-\mu}(\mathbb{R}))$, $\mu > 0$ and by taking limit with $s' - \mu \ge 2$ one has that

$$\widetilde{u} \in C([0,T] : H^{s'-\mu}(\mathbb{R})) \cap L^{\infty}([0,T] : H^{s'}(\mathbb{R})) \qquad \forall\, \mu > 0$$

is a solution of the IVP (9.9).

Step 4 Persistence property: $u \in C([0,T] : H^s(\mathbb{R}))$.

We need some preliminary estimates. Let $\rho \in C_0^{\infty}(\mathbb{R})$ be such that

$$\rho(x) \ge 0 \ \ \forall x \in \mathbb{R}, \qquad \int \rho(x)dx = 1, \qquad \int x^k\, \rho(x)\, dx = 0, \ \ k = 1, \dots, m, \tag{9.32}$$

for some $m \in \mathbb{Z}^+$. Denote

$$\rho_\epsilon(x) = \frac{1}{\epsilon}\rho\Big(\frac{x}{\epsilon}\Big), \qquad \epsilon > 0.$$

Proposition 9.1. *Let $r > 0$ and $f \in H^r(\mathbb{R})$. Then*

$$\begin{aligned} &(a) \quad \|\rho_\epsilon * f\|_{r+\alpha,2} \le c\epsilon^{-\alpha}\|f\|_{r,2}, && \forall \alpha > 0,\\ &(b) \quad \|f - \rho_\epsilon * f\|_{r-\beta,2} \le c\epsilon^{\beta}\|f\|_{r,2}, && \forall \beta \in [0,r]. \end{aligned} \tag{9.33}$$

Moreover,

$$\begin{aligned} &(a) \quad \|\rho_\epsilon * f\|_{r+\alpha,2} = O(\epsilon^{-\alpha}) \qquad as \qquad \epsilon \downarrow 0 \quad \forall \alpha > 0,\\ &(b) \quad \|f - \rho_\epsilon * f\|_{r-\beta,2} = o(\epsilon^{\beta}) \qquad as \quad \epsilon \downarrow 0 \quad \forall \beta \in [0,r]. \end{aligned} \tag{9.34}$$

Proof. The proof of part (a) in (9.33) and (9.34) is immediate so we only consider part (b). We shall restrict ourselves to prove the case $r = 1$ and $\beta = 1$. The proof for the case where $r,\ \beta \in \mathbb{Z}^+$ is similar to the argument below. The general case follows by interpolation between the previous cases.

By hypothesis on $\rho(\cdot)$ one has

$$\begin{aligned} f(x) - \rho_\epsilon * f(x) &= \int \rho_\epsilon(y)(f(x) - f(x-y))\,dy\\ &= \int \rho_\epsilon(y)\Big(- \int_0^1 \frac{d}{dt} f(x-ty)dt\Big)dy = \int_0^1 \int \rho_\epsilon(y) f'(x-ty)y\,dydt\\ &= \int_0^1 \int \rho_\epsilon(y)(f'(x-ty) - f'(x))y\,dydt. \end{aligned}$$

Hence,

$$\|f - \rho_\epsilon * f\|_2 \le \epsilon \int_0^1 \int \frac{|y|}{\epsilon} \rho_\epsilon(y) \|f'(\cdot - ty) - f'(\cdot)\|_2 \, dy dt.$$

Since $f \in H^1(\mathbb{R})$ a density argument shows that

$$\lim_{\delta \downarrow 0} \sup_{|y| \le \delta, \, |t| \le 1} \|f'(\cdot - ty) - f'(\cdot)\|_2 = 0.$$

This together with the fact that for any $\delta > 0$ fixed

$$\lim_{\epsilon \downarrow 0} \int_{|y| \ge \delta} \frac{|y|}{\epsilon} \rho_\epsilon(y) \, dy = \lim_{\epsilon \downarrow 0} \int_{|x| \ge \delta/\epsilon} |x| \rho(x) \, dx = 0$$

yields the desired result.

Next, we turn to the proof of step 3. We consider the IVP

$$\begin{cases} \partial_t u + \mathsf{H}\partial_x^2 u + u \partial_x u = 0, \\ u(x,0) = u_0^\epsilon(x) = \rho_\epsilon * u_0(x), \end{cases} \tag{9.35}$$

$t > 0$, $x \in \mathbb{R}$.

Since the data in (9.35) $u_0^\epsilon \in H^\infty(\mathbb{R})$ the argument in steps 1 and 2 shows for any $\epsilon > 0$ the IVP (9.32) has a solution

$$u^\epsilon \in C([0,T] : H^\infty(\mathbb{R})),$$

with T as in (9.18), i.e.

$$cT \|u_0^\epsilon\|_{2,2} = cT \|u_0\|_{2,2} = 1/2$$

satisfying that

$$\sup_{\epsilon > 0} \sup_{[0,T]} \|u^\epsilon(t)\|_{2,2} \le c \|u_0\|_{2,2} \tag{9.36}$$

with c independent of ε. Also by (9.33) one has that

$$\sup_{[0,T]} \|u^\epsilon(t)\|_{l,2} = O(\epsilon^{-l+2}) \quad \text{as} \quad \epsilon \downarrow 0, \quad \forall l > 2. \tag{9.37}$$

Next, for $\epsilon > \epsilon' > 0$ we define

$$v(t) = v^{\epsilon,\epsilon'}(t) = (u^\epsilon - u^{\epsilon'})(t),$$

which solves the IVP

$$\begin{cases} \partial_t v + \mathsf{H}\partial_x^2 v + v \partial_x u^\epsilon + u^{\epsilon'} \partial_x v = 0, \\ v(x,0) = u_0^\epsilon(x) - u_0^{\epsilon'}(x) = (\rho_\epsilon * u_0 - \rho_{\epsilon'} * u_0)(x) = v_0(x), \end{cases} \tag{9.38}$$

$t \in [0,T]$. Using energy estimates it follows that

$$\frac{d}{dt}\|v(t)\|_2 \le c(\|\partial_x u^{\epsilon}(t)\|_\infty + \|\partial_x u^{\epsilon'}(t)\|_\infty)\|v(t)\|_2, \tag{9.39}$$

which combined with (9.36) and Proposition 9.1 lead to

$$\sup_{[0,T]} \|v(t)\|_2 \le c\|v_0\|_2\, e^{2cT\|u_0\|_{2,2}} \le c\epsilon^2\, K, \qquad K = K(\|u_0\|_{2,2}). \tag{9.40}$$

In the same manner one has

$$\begin{aligned}
\frac{d}{dt}\|\partial_x^2 v(t)\|_2 &\le c(\|\partial_x u^{\epsilon}(t)\|_\infty + \|\partial_x u^{\epsilon'}(t)\|_\infty)\|\partial_x^2 v(t)\|_2 \\
&\quad + c(\|\partial_x^2 u^{\epsilon}(t)\|_2 + \|\partial_x^2 u^{\epsilon'}(t)\|_2)\|\partial_x v(t)\|_\infty \\
&\quad + \|\partial_x^3 u^{\epsilon}\|_2\, \|v(t)\|_\infty \\
&\equiv E_1(t) + E_2(t) + E_3(t).
\end{aligned} \tag{9.41}$$

Gronwall's inequality will be applied to (9.41) after estimating E_j, $j = 1,2,3$. The estimate for E_1 follows from (9.36) and Sobolev Embedding Theorem. Using the Gagliardo–Nirenberg inequality (see (3.13)), (9.36) and (9.40) the contribution of E_2 can be bounded as

$$\sup_{[0,T]} E_2(t) \le c\|u_0\|_{2,2} \sup_{[0,T]} (\|v(t)\|_2^{1/4}\|\partial_x^2 v(t)\|_2^{3/4}) \le c\epsilon^{1/2}K, \qquad K = K(\|u_0\|_{2,2}).$$

Similarly, using (9.37) and (9.40) one controls the contribution of the term E_3 in (9.41)

$$\begin{aligned}
\sup_{[0,T]} E_3(t) &\le \sup_{[0,T]} (\|\partial_x^3 u^{\epsilon}\|_2 \|v(t)\|_2^{3/4}\|\partial_x^2 v(t)\|_2^{1/4}) \\
&\le c\epsilon^{-1}\epsilon^{3/2}K = c\epsilon^{1/2}K, \quad K = K(\|u_0\|_{2,2}).
\end{aligned}$$

Hence, collecting the above information, using Gronwall's inequality and (9.41), and adding the result to (9.40) we conclude that

$$\sup_{[0,T]} \|v(t)\|_{2,2} = \sup_{[0,T]} \|(u^{\epsilon} - u^{\epsilon'})(t)\|_{2,2} = o(1) \quad \text{as} \quad \epsilon \downarrow 0. \tag{9.42}$$

Thus,

$$u^{\epsilon} \to u \quad \text{in} \quad C([0,T] : H^2(\mathbb{R})), \quad \text{as} \quad \epsilon \downarrow 0,$$

with $u(\cdot)$ solving the IVP (9.9) where the equation holds in $C([0,T] : L^2(\mathbb{R}))$. The uniqueness of the solution $u = u(x,t)$ in the class $C([0,T] : H^2(\mathbb{R}))$ follows by using the argument in (9.39) and (9.40). One can show that our solution u agrees with the function $\widetilde{u}$ found in the step 3, see (9.30).

Step 5 (from [BS]) Proof of the continuous dependence of the solution u upon the data u_0.

We shall show that

$$\forall \lambda > 0 \; \exists \delta > 0 \left[\|u_0 - z_0\|_{2,2} < \delta \;\Rightarrow\; \sup_{[0,T/2]} \|(u-z)(t)\|_{2,2} < \lambda \right], \tag{9.43}$$

where u, z represent the solutions of the IVP (9.9) with data $u_0, z_0 \in H^2(\mathbb{R})$) respectively. Without loss of generality we assume $u_0 \neq 0$.

In (9.43) we take the time interval $[0, T/2]$ with T as in (9.18) to guarantee that if $\|u_0 - z_0\|_{2,2} < \delta$, then the solution $z(t)$ is defined in the time interval $[0, T/2]$.

For $1 > \epsilon > \epsilon' > 0$ we define

$$w(t) = w^{\epsilon,\epsilon'}(t) = (u^{\epsilon} - z^{\epsilon'})(t), \tag{9.44}$$

where u^{ϵ}, $z^{\epsilon'}$ are the solutions of the IVP (9.35) with data $u_0^{\epsilon} = \rho_{\epsilon} * u_0$, $z_0^{\epsilon'} = \rho_{\epsilon'} * z_0$ respectively. Thus, taking $\delta_1 = \|u_0\|_{2,2}/2$ from the above results one has that

$$\sup_{[0,T/2]} \|u^{\epsilon}(t)\|_{2,2} + \sup_{[0,T/2]} \|z^{\epsilon'}(t)\|_{2,2} \leq c\|u_0\|_{2,2} + 2c\|u_0\|_{2,2}. \tag{9.45}$$

Since $w(t)$ satisfies the IVP

$$\begin{cases} \partial_t w + \mathsf{H}\partial_x^2 w + w\partial_x u^{\epsilon} + z^{\epsilon'}\partial_x w = 0, \\ w(x,0) = u_0^{\epsilon}(x) - z_0^{\epsilon'}(x) = (\rho_{\epsilon} * u_0 - \rho_{\epsilon'} * z_0)(x) = v_0(x), \end{cases} \tag{9.46}$$

$t \in [0, T/2]$, one has (combining (9.45) and a familiar argument) that

$$\begin{aligned} \sup_{[0,T/2]} \|w(t)\|_2 &\leq \|u_0^{\epsilon} - z_0^{\epsilon'}\|_2 \, K \\ &\leq K(\|u_0 - u_0^{\epsilon}\|_2 + \|z - z_0^{\epsilon'}\|_2 + \delta) \leq K(\epsilon^2 + \delta), \end{aligned} \tag{9.47}$$

$K = K(\|u_0\|_{2,2})$ and

$$\begin{aligned} \frac{d}{dt}\|\partial_x^2 w(t)\|_2 &\leq c(\|\partial_x u^{\epsilon}(t)\|_{\infty} + \|\partial_x z^{\epsilon'}(t)\|_{\infty})\|\partial_x^2 w(t)\|_2 \\ &\quad + c(\|\partial_x^2 u^{\epsilon}(t)\|_2 + \|\partial_x^2 z^{\epsilon'}(t)\|_2)\|\partial_x w(t)\|_{\infty} \\ &\quad + \|\partial_x^3 u^{\epsilon}\|_2 \, \|w(t)\|_{\infty} \\ &\equiv G_1(t) + G_2(t) + G_3(t). \end{aligned} \tag{9.48}$$

First using (9.45) and Sobolev embedding one gets that

$$G_1(t) \leq c\|u_0\|_{2,2}\|\partial_x^2 w(t)\|_2, \qquad \forall t \in [0, T/2].$$

Next, combining (9.45) and (9.47) one gets the bound

$$\begin{aligned} \sup_{[0,T/2]} G_2(t) &\leq c\|u_0\|_{2,2} \sup_{[0,T/2]} (\|w(t)\|_2^{1/4}\|\partial_x^2 w(t)\|_2^{3/4}) \\ &\leq c\|u_0\|_{2,2}^{7/4} K(\epsilon^2 + \delta)^{1/4} \leq K(\epsilon^2 + \delta)^{1/4} \leq cK(\epsilon^{1/2} + \delta^{1/4}), \end{aligned}$$

$K = K(\|u_0\|_{2,2})$. Finally, from (9.36), (9.37), and (9.47) the term G_3 in (9.48) can bounded as

$$\begin{aligned}\sup_{[0,T/2]} G_3(t) &= \sup_{[0,T/2]} (\|\partial_x^3 u^\epsilon(t)\|_2 \|w(t)\|_\infty)\\ &\le cK\epsilon^{-1} \sup_{[0,T/2]} (\|w(t)\|_2^{3/4} \|\partial_x^2 w(t)\|_2^{1/4})\\ &\le cK\epsilon^{-1}(\epsilon^2+\delta)^{3/4} \le cK'(\epsilon^{1/2}+\epsilon^{-1}\delta^{3/4}).\end{aligned}$$

Combining the estimates for $G_j, j = 1,2,3$, Proposition 9.1, and Gronwall's inequality and adding the result to (9.47) it follows that

$$\begin{aligned}\sup_{[0,T/2]} \|w(t)\|_{2,2} &= \sup_{[0,T/2]} \|(u^\epsilon - z^\epsilon)(t)\|_{2,2}\\ &\le K(\|u_0^\epsilon - z_0^\epsilon\|_{2,2} + T(\epsilon^{1/2}+\delta^{1/4}+\epsilon^{-1}\delta^{3/4})) \qquad (9.49)\\ &\le K(\|u_0^\epsilon - u_0\|_{2,2} + \|z_0^{\epsilon'} - z_0\|_{2,2} + \delta^{1/4} + \epsilon^{1/2} + \epsilon^{-1}\delta^{3/4})\end{aligned}$$

assuming $\delta < 1$. Therefore collecting these results one concludes that

$$\begin{aligned}\sup_{[0,T/2]} \|(u-z)(t)\|_{2,2} &\le \sup_{[0,T/2]} (\|u-u^\epsilon\|_{2,2} + \|z-z^{\epsilon'}\|_{2,2} + \|u^\epsilon - z^{\epsilon'}\|_{2,2})\\ &\le o(1)_\epsilon + o(1)_{\epsilon'}\\ &\quad + K(\|u_0^\epsilon - u_0\|_{2,2} + \|z_0^{\epsilon'} - z_0\|_{2,2} + \delta^{1/4} + \epsilon^{1/2} + \epsilon^{-1}\delta^{3/4})\\ &\le o(1)_\epsilon + o(1)_{\epsilon'} + K(\delta^{1/4} + \epsilon^{-1}\delta^{3/4}).\end{aligned}$$

So we fixed ε small enough such that

$$o(1)_\epsilon \le \lambda/3,$$

then we take $\delta < \min\{1; \|u\|_{2,2}/2\}$ such that

$$K(\delta^{1/4} + \epsilon^{-1}\delta^{3/4}) < \lambda/3,$$

and finally for each $z_0 \in H^2(\mathbb{R})$ such that $\|z_0 - u_0\|_{2,2} < \delta$ we take $\epsilon' = \epsilon'(z_0) > 0$, $\epsilon' \in (0,\epsilon)$ such that

$$o(1)_{\epsilon'} \le \lambda/3,$$

to conclude the proof of Theorem 9.2.

We observe that the only fact used on the operator $\mathsf{H}\partial_x^2$ describing the dispersive relation in the BO equation was that it is skew-symmetric.

Next, consider the IVP associated to the generalized BO equation, that is,

$$\begin{cases}\partial_t u + \mathsf{H}\partial_x^2 u + u^k \partial_x u = 0,\\ u(x,0) = u_0(x),\end{cases} \qquad (9.50)$$

$x \in \mathbb{R}, t > 0, k \in \mathbb{Z}^+, k \geq 2$.

In addition to preserve the L^2-norm, solutions of the IVP (9.50) leave invariant the quantity

$$E(u)(t) = \int_{-\infty}^{\infty} \left(|D_x^{1/2} u(x,t)|^2 - \frac{1}{(k+1)(k+2)} u(x,t)^{k+2} \right) dx. \tag{9.51}$$

These quantities will be useful for extending possible local results globally in the corresponding Sobolev spaces dictated for them.

We also notice that the scaling argument for the equation in (9.50) suggests well-posedness for the IVP in $H^s(\mathbb{R})$ for

$$s > s_k = \frac{1}{2} - \frac{1}{k}. \tag{9.52}$$

Using the oscillatory integral techniques described in Chapters 4 and 7 in [KPV11] local well-posedness for small data was established in Sobolev indices lower than the $3/2$ given by the energy method.

In [MR1] and [MR2] Molinet and Riboud improved the results in [KPV11]. In particular, they showed local well-posedness for small data in $H^s(\mathbb{R})$, $s > 1/3$ for $k = 3$, and $s > s_k$ for $k \geq 4$, and for data in $H^s(\mathbb{R})$ of arbitrary size, $s \geq 3/4$ for $k = 3$, $s > 1/2$ for $k = 4$, and $s \geq 1/2$ for $k \geq 5$. These results can be extended globally using the conserved quantities (9.51) whenever the local well-posedness is realized in $H^{1/2}(\mathbb{R})$. Kenig and Takaoka [KT] has obtained global well-posedness for (9.50) with $k = 2$ for $s \geq 1/2$. One of the main new tool used by these authors was a gauge transformation reminiscent of that introduced by Tao (see (9.12)). In [Ve] Vento established local well-posedness in the critical space $\dot{H}^{s_k}(\mathbb{R})$, $s_k = \frac{1}{2} - \frac{1}{k}$, (and its inhomogeneous version) for $k \geq 4$. It was also proved that for $k = 3$ local well-posedness holds in $H^s(\mathbb{R})$ for $s > 1/3$.

From the ill-posedness results obtained by Biagioni and Linares [BiL] the results in [Ve] for $k \geq 4$ should be optimal.

9.5 Zakharov System

In this section we will give a brief account of some results concerning local and global well-posedness for the Zakharov system,

$$\begin{cases} i\partial_t u + \Delta u = u\,v, \\ \lambda^{-2}\partial_t^2 v - \Delta v = \Delta(|u|^2), \\ u(x,0) = u_0(x), \;\; v(x,0) = v_0(x), \;\; \partial_t v(x,0) = v_1(x), \end{cases} \tag{9.53}$$

$x \in \mathbb{R}^n$, $t > 0$, where $u : \mathbb{R}^n \times [0,\infty) \mapsto \mathbb{C}^n$ and $v : \mathbb{R}^n \mapsto \mathbb{R}$.

This model was introduced by Zakharov [Zk] to describe the long wave Langmuir turbulence in a plasma. The function $u = u(x,t)$ represents the slowly varying envelope of the highly oscillatory electric field; $v = v(x,t)$ is the deviation of the ion density from the equilibrium; and λ is proportional to the ionic speed of sound. In the limit when $\lambda \to \infty$ the system (9.53) reduces formally to the cubic (focusing) nonlinear Schrödinger equation,

$$i\partial_t u_\infty + \Delta u_\infty = -|u_\infty|^2 u_\infty. \tag{9.54}$$

Solutions of this system satisfy the following conservation laws:

$$M(u_0) = \int_{\mathbb{R}^2} |u(x,t)|^2\,dx,$$

$$E(u_0, v_0, v_1) = \int_{\mathbb{R}^2} \Big(|\nabla u|^2 + v|u|^2 + \frac{v^2}{2} + \lambda^{-2}((-\Delta)^{-1/2}\partial_t v)^2\Big)(x,t)\,dx. \tag{9.55}$$

The Zakharov system has been studied by several authors. Sulem and Sulem [SS1] showed that for data

$$(u_0, v_0, v_1) \in H^s(\mathbb{R}^n) \times H^{s-1}(\mathbb{R}^n) \times (H^{s-2}(\mathbb{R}^n) \cap \dot{H}^{-1}(\mathbb{R}^n)) \tag{9.56}$$

with $s \geq 3$ and $1 \leq n \leq 3$, the IVP (9.53) has unique local solution

$$(u, v) \in L^\infty([0,T] : H^s(\mathbb{R}^n)) \times L^\infty([0,T] : H^{s-1}(\mathbb{R}^n)).$$

They also proved that in the case $n = 1$ these solutions can be extended globally in time. Later on in [AA2] Added and Added established the global existence for the solutions given in [SS1] in the case $n = 2$ corresponding to data u_0 with $\|u_0\|_2$ sufficiently small. Schochet and Weinstein [SWe] obtained a local existence and uniqueness results for data in (9.56) with time interval $[0,T]$ independent of the parameter λ. This allowed them to show that solutions (u^λ, v^λ) of (9.53) converge to a solution of (9.54) as $\lambda \to \infty$. For small amplitude solutions rates of this convergence were obtained in [AA1]. Latter Ozawa and Tsutsumi [OT3] found optimal rates of convergence of solutions of (9.53) to solutions of (9.54).

In [OT2] Ozawa and Tsutsumi obtained, for a fixed λ, unique local results for the IVP (9.53) for data $(u_0, v_0, v_1) \in H^2(\mathbb{R}^n) \times H^1(\mathbb{R}^n) \times L^2(\mathbb{R}^n)$ with $1 \leq n \leq 3$, removing the hypothesis $v_1 \in \dot{H}^{-1}$ in previous works (see (9.55)). Ozawa and Tsutsumi approach relies on the L^p-L^q estimates of Strichartz type.

Kenig, Ponce and Vega [KPV8] proved that an iteration scheme can be used directly to obtain small amplitude solutions. They showed that for $n \geq 1$, there exist $s > 0$, $m \in \mathbb{Z}^+$, and $\delta > 0$ such that for any data

$$(u_0, v_0, v_1) \in \mathcal{X}^{s,m} = H^s(\mathbb{R}^n) \cap H^{s_0}(|x|^m\,dx) \times H^{s-1/2}(\mathbb{R}^n) \times H^{s-3/2}(\mathbb{R}^n), \tag{9.57}$$

$s_0 = [(s+3)/2]$ (where $[r]$ denotes the largest integer $\leq r$) with $\|(u_0, v_0, v_1)\|_{\mathcal{X}^{s,m}} \leq \delta$, there exists a unique solution (u^λ, v^λ) in an interval of time $[0,T]$ independent of $\lambda \geq 1$. They also showed that under some additional hypotheses on v_0 and v_1,

$$\sup_{[0,T]} \|(u^\lambda - u_\infty)(t)\|_{H^{s_0}} = O(\lambda^{-1}) \qquad \text{as } \lambda \to \infty.$$

The main idea used in [KPV8] was to exploit the inhomogeneous n-dimensional version of Kato's smoothing effect (4.30) to overcome the loss of derivatives. This was complemented with maximal function estimates for the group $\{e^{it\Delta}\}$.

In [BoC] Bourgain and Colliander showed local well-posedness of IVP (9.53) in the energy space $(u_0, v_0, v_1) \in H^1(\mathbb{R}^n) \times L^2(\mathbb{R}^n) \times H^{-1}(\mathbb{R}^n)$, $n = 2, 3$, by extending the method developed in [Bo1]. Global well-posedness for small data was also established by combining local well-posedness and conservation laws, (see (9.55)).

Ginibre, Tsutsumi and Velo [GTV], using the Fourier restriction method introduced by Bourgain [Bo1], obtained a more complete set of results concerning local well-posedness. Their results are roughly as follows:

For data $(u_0, v_0, v_1) \in H^k(\mathbb{R}^n) \times H^l(\mathbb{R}^n) \times H^{l-1}(\mathbb{R}^n)$ the IVP is locally well-posed provided

(k, l)	Dimension
$-\frac{1}{2} < k - l \leq 1, \quad 2k \geq l + \frac{1}{2}$	$n = 1$
$l \geq 0, \quad 2k - (l + 1) \geq 0$	$n = 2, 3$
$l > \frac{n}{2} - 2, \quad 2k - (l + 1) > \frac{n}{2} - 2$	$n \geq 4$

The solutions satisfy

$$(u, v, \partial_t v) \in C([0, T] : H^k(\mathbb{R}^n) \times H^l(\mathbb{R}^n) \times H^{l-1}(\mathbb{R}^n)).$$

In [BHHT] for the two-dimensional case, Bejenaru, Herr, Holmer and Tataru obtained local well-posedness in the space $L^2(\mathbb{R}^2) \times H^{-1/2}(\mathbb{R}^2) \times H^{-3/2}(\mathbb{R}^2)$ and showed that this result should be optimal.

Regarding blow up results we shall mention the following. In the two-dimensional case Glangetas and Merle [GM] proved the existence of blow up solutions with radial symmetry and self-similar form:

$$u(x, t) = \frac{\omega}{(T - t)} e^{i\Phi(x,t)} P\big(\frac{\omega|x|}{T - t}\big),$$

$$v(x, t) = \big(\frac{\omega}{T - t}\big)^2 N\big(\frac{\omega|x|}{T - t}\big),$$

where $\omega \in \mathbb{R}$ and

$$\Phi(x, t) = \frac{\omega^2}{(T - t)} - \frac{|x|^2}{4(T - t)}.$$

They also showed that concentration happens in L^2 (see (6.4)). In [Me5] Merle found rates for the blow up. In [Me6] he also obtained some extensions of the blow up results.

In the one-dimensional case a global result below the energy space has been proved by Pecher [P2].

The corresponding IVP (9.53) in the periodic setting was treated by Bourgain in [Bo9] and Takaoka in [Tk1].

To end this section we comment on the results obtained by Colin and Métivier [CM] and Linares, Ponce and Saut [LPS] regarding the local theory concerning a system deduced by Zakharov where the Schrödinger linear part has a degenerate Laplacian. In [CM] it was established that the periodic boundary value problem is ill-posed. However, the use of some smoothing properties in [LPS] allow the authors to prove local well-posedness in spaces defined via those regularizing properties. This example illustrates the difference between the nonperiodic and periodic setting.

9.6 Higher Order KdV Equations

In 1967 Gardner, Greene, Kruskal and Miura [GGKM] discovered the remarkable fact that the spectrum of the Sturm–Liouville (or stationary Schrödinger) equation

$$L_q(y) = y'' - q(x)\,y = \frac{d^2y}{dx^2} - q(x)y = \lambda\, y, \quad -\infty < x < \infty, \tag{9.58}$$

does not change when the potential $q(x)$ evolves accordingly to the KdV equation, i.e., if $u(x,t)$ solves the IVP

$$\begin{cases} \partial_t u + \partial_x^3 u + u\partial_x u = 0, \\ u(x,0) = q(x), \end{cases} \tag{9.59}$$

$x,t \in \mathbb{R}$, with $q(\cdot)$ in an appropriate class, then

$$\text{spectrum of} L_q = \sigma(L_q) = \sigma(L_{u(\cdot,t)}) \quad \text{for any } t \in \mathbb{R}. \tag{9.60}$$

This principle allowed them to use results from (direct and inverse) spectral theory to solve the IVP (9.59) through a succession of linear computations. This procedure is called the *inverse scattering method* (ISM) as it was mentioned in previous chapters.

More precisely, to guarantee the validity of the process we will describe next, one assumes that $q(x)$ satisfies the decay assumption

$$\int_{-\infty}^{\infty} (1 + |x|^2)\, |q(x)|^2\, dx < \infty \quad \text{(no optimal condition)}. \tag{9.61}$$

The scattering data for the problem (9.58) is the spectral information needed to reconstruct the potential $q(x)$.

First, one has the spectrum $\sigma(L_q)$ where by (9.61)

$$\sigma(L_q) = (-\infty, 0] \cup \{k_j^2\}_{j=1}^N, \quad N \in \mathbb{Z}^+ \cup \{0\}, \tag{9.62}$$

where $(-\infty, 0]$ is the continuous spectrum and $\lambda_j = -k_j^2$, $k_j > 0$, $j = 1, \ldots, N$, are the eigenvalues corresponding to eigenfunctions $\{\psi_j\}_{j=1}^N \subseteq L^2(\mathbb{R})$ normalized, i.e., $\|\psi_j\|_2 = 1$, $j = 1, \ldots, N$. Thus from (9.58) and (9.61)

$$\psi_j(x) \sim c_j\, e^{-k_j x} \quad \text{as } x \uparrow \infty, \ j = 1, \ldots, N. \tag{9.63}$$

The $\{c_j\}_{j=1}^N$ are called the "normalizing coefficients."

For $\lambda < 0$ the generalized eigenfunctions can be written as $(k = \sqrt{-\lambda})$

$$\psi(x) \sim \begin{cases} e^{-ikx} + b(k)\, e^{ikx}, & x \to +\infty \\ a(k)\, e^{-ikx}, & x \to -\infty, \end{cases} \tag{9.64}$$

where $a(k)$ and $b(k)$ are called the *transmitted* and the *reflected coefficients*, respectively, extended to $k \in \mathbb{R}$.

The scattering data are given by the spectrum, the normalizing coefficients, and the reflected coefficients

$$\{\sigma(L_q);\ \{c_j\}_{j=1}^N;\ \{b(k) : \ k \in \mathbb{R}\}\}. \tag{9.65}$$

This information permits one to recover the potential $q(x)$ as follows: Define

$$F(x) = \sum_{j=1}^{N} c_j^2\, e^{-k_j x} + \frac{1}{2\pi} \int_{-\infty}^{\infty} b(k)\, e^{ikx}\, dk, \tag{9.66}$$

and let $K(x, z)$ be the solution of the Marchenko (Fredholm integral) equation

$$K(x,z) + F(x+z) + \int_{-x}^{\infty} K(x,x')\, F(x'+z)\, dx' = 0. \tag{9.67}$$

Then the potential is obtained via the formula

$$q(x) = \frac{1}{3} \frac{d}{dx} K(x,z)\big|_{z=x}. \tag{9.68}$$

Assuming now that the potential $q(x)$ evolves accordingly to (9.59), one can show (see [AS2], [DJ] for details of this discussion) that the scattering data change in time, the spectrum as (9.60) and the normalized and reflected coefficients as

$$\begin{cases} c_j(t) = c(0)\, e^{4k_j^3 t} = c_j\, e^{4k_j^3 t}, \\ b(k;t) = b(k;0)\, e^{8ik^3 t} = b(k)\, e^{8ik^3 t}. \end{cases} \tag{9.69}$$

Hence we know

$$\{\sigma(L_{u(\cdot,t)});\ \{c_j(t)\}_{j=1}^N;\ \{b(k;t) : \ k \geq 0\}\}, \tag{9.70}$$

the scattering data for

$$L_{u(\cdot,t)}(y) = y'' - u(\cdot,t)\, y = \lambda\, y,$$

which allows one to recover the potential $u(\cdot, t)$, i.e., the solution of the IVP (9.58) associated to the KdV.

In [Lx2] Lax generalized this principle by finding a class of evolution equations for which the operators

$$L_{u(\cdot,t)} = \frac{d^2}{dx^2} - u(x,t) \tag{9.71}$$

are unitary equivalent whenever $u(\cdot, t)$ is a solution of an equation in this class. One must find a family of unitary operators $\{U(t)\}_{t=-\infty}^{\infty}$ such that

$$U^*(t)\, L_{u(\cdot,t)}\, U(t) = L_{u(\cdot,0)}. \tag{9.72}$$

This family should satisfy an equation of the form

$$\frac{d}{dt}U(t) = B(t)U(t) \tag{9.73}$$

for some $B(t)$ skew-symmetric operator. Combining (9.72) and (9.73) one sees that

$$\frac{d}{dt}L_{u(\cdot,t)} = B(t)\, L_{u(\cdot,t)} - L_{u(\cdot,t)}\, B(t) \equiv [B(t); L_{u(\cdot,t)}]. \tag{9.74}$$

Choosing $B = B_0 = \dfrac{d}{dx}$ one gets

$$\frac{d}{dt}L_{u(\cdot,t)} = \partial_t u = \Big[\frac{d}{dx}; L_{u(\cdot,t)}\Big] = -\partial_x u, \tag{9.75}$$

i.e.,

$$\partial_t u + \partial_x u = 0,$$

whose solution $u(x,t) = u_0(x-t) = q(x-t)$ clearly leaves the spectrum of $L_{u(\cdot,t)}$ in (9.71) independently of t.

The choice

$$B_1 = \alpha\frac{d^2}{dx^2} + \beta\Big(u\,\frac{d}{dx} + \frac{d}{dx}(u\ \cdot)\Big) \tag{9.76}$$

with appropriate values of the constants α and β gives

$$[B_1(t);\ L_{u(\cdot,t)}] = -\partial_x^3 u - u\,\partial_x u. \tag{9.77}$$

Hence, (9.74) becomes the KdV equation.

In general, one has

$$B_j = \alpha_j \frac{d^{2j+1}}{dx^{2j+1}} + \sum_{k=0}^{j-1} [\beta_{jk}(u) \frac{d^{2k+1}}{dx^{2k+1}} + \frac{d^{2k+1}}{dx^{2k+1}} (\beta_{kj}(u) \cdot)] \tag{9.78}$$

with $\beta_{jk}(u)$ selected such that $[B_j;\ L_{u(\cdot,t)}]$ has order zero.

Thus for $B_2(u)$ one obtains (up to rescaling)

$$\partial_t u - \partial_x^5 u + 10\, u\, \partial_x^3 u + 20\, \partial_x u\, \partial_x^2 u - 30\, u^2\, \partial_x u = 0. \tag{9.79}$$

This class can also be described using the conservation laws satisfied by solutions of the KdV [Lx2]

$$F_0(u) = 3 \int u\, dx; \quad F_1(u) = \frac{1}{2} \int u^2\, dx; \quad F_2(u) = \int (\frac{u^3}{6} - \frac{(\partial_x u)^2}{2})\, dx; \quad \dots. \tag{9.80}$$

The gradient of these functionals ($\partial F_j = G_j$) are

$$G_0(u) = 3, \quad G_1(u) = u, \quad G_2(u) = \frac{1}{2} u^2 + \partial_x^2 u, \quad \dots, \tag{9.81}$$

which are related by the formula

$$H\, G_j = \partial G_{j+1}, \quad j = 0, 1, \dots, \tag{9.82}$$

where

$$H = \frac{d^3}{dx^3} + \frac{2}{3}\, u \frac{d}{dx} + \frac{1}{3} \frac{du}{dx},$$

and

$$\partial_t u + \frac{d}{dx} G_{j+1} = \partial_t u + [B_j; L_{u(\cdot,t)}] = 0, \quad j = 0, 1, \dots, \tag{9.83}$$

which is called the jth equation in the KdV hierarchy.

So (9.79) is the second equation in the KdV hierarchy. Related versions of this equation appear as a higher order approximations in the study of water wave problems for long, small amplitude waves over shallow horizontal bottom (see for instance [Ol], [Bn1] and references therein). In 1972 Zhakarov and Shabat [ZS] showed that the ISM used for the KdV and its hierarchy can be extended to other relevant physical equations. More precisely, they proved that the cubic one-dimensional defocusing Schrödinger equation

$$i\partial_t u = \partial_x^2 u + \lambda |u|^2 u, \quad \lambda > 0,$$

can be solved by considering an appropriate linear scattering problem and its inverse.

The local and global well-posedness of the IVP and PBVP associated to equation (9.79) was established in [St2]. Also the PBVP for the whole KdV hierarchy was studied in [Sch].

Here we restrict ourselves to consider the IVP for the KdV hierarchy in (9.83).

In a more general framework consider the initial value problem

$$\begin{cases} \partial_t u + \partial_x^{2j+1} u + P(u,\, \partial_x u, \dots,\, \partial_x^{2j} u) = 0, \\ u(x,0) = u_0(x), \end{cases} \tag{9.84}$$

$x, t \in \mathbb{R}$, $j \in \mathbb{Z}^+$, where $u = u(x,t)$ is real-(or complex-)valued function and

$$P : \mathbb{R}^{2j+1} \mapsto \mathbb{R} \quad (\text{or } \; P : \mathbb{C}^{2j+1} \mapsto \mathbb{C})$$

is a polynomial having no constant or linear terms, i.e.,

$$P(z) = \sum_{|\alpha|=\ell_0}^{\ell_1} a_\alpha\, z^\alpha \quad \text{with } \; \ell_0 \geq 2 \tag{9.85}$$

and $z = (z_1, \dots, z_{2j+1})$.

In [KPV13] local well-posedness of the IVP (9.84) in

$$\mathcal{F}^s_m = H^s(\mathbb{R}) \cap L^2(|x|^m\, dx)$$

was established. The proof combines the fact that the results in [HO] extend to diagonal systems and a change of dependent variable, which allows us to write the equation in (9.84) (after a few differentiations with respect to the x-variable) as a diagonal system

$$\partial_t \omega^k + \partial_x^{2j+1} \omega^k + Q_k(\omega^1, \dots,\, \omega^m,\, \partial_x \omega^1, \dots,\, \partial_x^{2j-1} \omega^m) = 0 \tag{9.86}$$

for $k = 1, \dots, m = m(j)$ where the nonlinear terms Q_k are independent of the highest derivatives, i.e., those of order $2j$. In this case some modifications are needed since the Q_k introduced by the change of variable involve nonlocal operators.

More precisely, in [KPV13] the following two results were proven:

Theorem 9.3. *Let $P(\cdot)$ be a polynomial of the type described in (9.85). Then there exist $s, m \in \mathbb{Z}^+$ such that for any $u_0 \in \mathcal{F}^s_m = H^s(\mathbb{R}) \cap L^2(|x|^m\, dx)$ there exist $T = T(\|u_0\|_{\mathcal{F}^s_m}) > 0$ (with $T(\rho) \to \infty$ as $\rho \to 0$) and a unique solution $u(\cdot)$ of the IVP (9.84) satisfying*

$$u \in C([0,T] : \mathcal{F}^s_m), \tag{9.87}$$

$$\sup_x \int_0^T |\partial_x^{s+j} u(x,t)|^2\, dt < \infty \tag{9.88}$$

and

$$\int_{-\infty}^{\infty} \sup_{[0,T]} |\partial_x^r u(x,t)|\, dx < \infty, \quad r = 0, \ldots, [\frac{s+j}{2}]. \tag{9.89}$$

If $u_0 \in \mathcal{F}_m^{s_0}$ with $s_0 > s$ the results above hold with s_0 instead of s in the same time interval $[0,T]$.

Moreover, for any $T' \in (0,T)$ there exists a neighborhood U_{u_0} of u_0 in $\mathcal{F}_m^s$ such that the map $\widetilde{u}_0 \mapsto \widetilde{u}(t)$ from U_{u_0} into the class defined in (9.87)–(9.89), with T' instead of T, is smooth.

Theorem 9.4. *Let $P(\cdot)$ be a polynomial of the type described in (9.85) with $\ell_0 \geq 3$, or $P(z) = P(z_1, \ldots, z_{j+1})$ in (9.85). Then the results in* Theorem 9.3 *hold with $m = 0$ and L_x^2-norm instead of L_x^1-norm in (9.89).*

Theorem 9.4 tells us that the IVP for the equation

$$\partial_t u + \partial_x^3 u + (u^2 + (\partial_x u)^2)\, \partial_x^2 u = 0, \tag{9.90}$$

$x, t \in \mathbb{R}$ is locally well-posed in $H^s(\mathbb{R})$, $s \geq s_0$, with s_0 sufficiently large. Roughly speaking Theorems 9.3 and 9.4 establish conditions that guarantee that the local behavior of the solution of (9.84) is controlled by the linear part of the equation. Moreover, it shows that the dispersive structure of the equation is strong enough to overcome nonlinear terms of lower order with arbitrary sign as in (9.90).

However, for a specific model of the kind described in (9.84) the results in Theorems 9.3 and 9.4 can be improved by reducing the index s and m depending on the order $(2j+1)$ considered and the structure of the nonlinear term (see for example [Kw2] and [Ci] for some fifth order cases). In particular in [KePi] Kenig and Pilod have shown that the third equation in the KdV hierarchy (9.79) is globally well posed in the energy space $H^2(\mathbb{R})$.

As it was previously mentioned the existence and uniqueness (in $H^s(\mathbb{T})$) for the periodic boundary value problem (PBVP) for the KdV hierarchy was established in [Sch]. The argument relied heavily on the structure of the equations in the hierarchy. Thus one can ask if a general result can be established for the PBVP associated to the general equation in (9.84). The answer is not, in [Bo12] Bourgain showed that the PBVP for the equation

$$\partial_t u + \partial_x^3 u = u^2\, (\partial_x u)^2$$

is ill-posed in $H^s(\mathbb{T})$ for every $s \in \mathbb{R}$.

9.7 Exercises

9.1 Prove that the Benjamin–Ono equation

$$\partial_t u + \mathsf{H}\partial_x^2 u + u\partial_x u = 0$$

has a traveling wave solution (decaying at infinity) $\phi_c(x+t)$, $c > 0$, with

$$\phi(x) = -\frac{4}{1+x^2},$$

and

$$\phi_c(x+t) = c\phi(c(x+t)).$$

Notice that $\phi(x+ct)$ is negative and moves to the left, so $\varphi(x-ct) = -\phi(x-ct)$ is a traveling wave, positive and traveling to the right, of the equation

$$\partial_t v - \mathsf{H}\partial_x^2 v + v\partial_x v = 0.$$

Hint: Integrate the equation for ϕ to get a first order ODE. Take Fourier transform and use Exercise 3.3 to get the result.

9.2 (Camassa–Holm equation [CH]) Consider the equation

$$\partial_t u - \partial_t \partial_x^2 u + 3\, u\partial_x u = 2\partial_x u\, \partial_x^2 u + u\, \partial_x^3 u. \tag{9.91}$$

(i) Prove that (9.91) can be written in the formally equivalent form

$$\partial_t u + u\partial_x u + \frac{1}{2}\, \partial_x\, e^{-|x|} * \big(u^2 + \frac{(\partial_x u)^2}{2}\big) = 0. \tag{9.92}$$

(ii) Prove that for any $c > 0$ the equation (9.91) has the nonsmooth traveling wave (peakon)

$$\varphi(x-ct) = c\, e^{-|x-ct|}.$$

Hint: (i) Use Exercise 3.4.
(ii) Notice first that it suffices to consider the ODE for φ with $c = 1$. Prove that

$$\big(e^{-|\cdot|} * e^{-2|\cdot|}\big)(x) = \frac{4}{3}\, e^{-|x|} - \frac{2}{3}\, e^{-2|x|}.$$

Integrate the ODE and use that $(\varphi'(x))^2 = \varphi(x)^2$.

9.3 (Benjamin–Bona–Mahony equation [BBM]) Consider the equation

$$\partial_t u + \partial_x u + u\, \partial_x u - \partial_{xxt}^3 u = 0. \tag{9.93}$$

(i) Prove that (9.93) can be written in the following forms

$$\partial_t u - \frac{\operatorname{sgn}(x)}{2}\, e^{-|x|} * \Big(u + \frac{u^2}{2}\Big) = 0 \tag{9.94}$$

and

$$u(x,t) = u_0(x) + \int_0^t \int \frac{\operatorname{sgn}(y)}{2}\, e^{-|y|}\Big(u + \frac{u^2}{2}\Big)(x-y,\tau)\, dy d\tau. \tag{9.95}$$

(ii) [BTz] Prove that given $u_0 \in L^2(\mathbb{R})$ there exist $T = T(\|u_0\|_2) > 0$ and a unique solution $u \in C([0,T] : L^2(\mathbb{R}))$ solution of (9.95).
Hint: Given $u_0 \in L^2(\mathbb{R})$ consider the set

$$X_T(\delta) = \{v : \mathbb{R} \times [0,T] \mapsto \mathbb{R} \ : \ \sup_{0\leq t\leq T} \|v(t)\|_2 \leq 2\delta\}$$

with $\delta = \|u_0\|_2$. Show that

$$\Phi(v)(x,t) = u_0(x) + \int_0^t \int \frac{\text{sgn}(y)}{2} e^{-|y|}(u + \frac{u^2}{2})(x-y,\tau)\,dy\,d\tau$$

defines a contraction map in $X_T(\delta)$ if $T(1+\delta) \leq 1/2$. (This result is sharp, see [BTz]).

(iii) Prove that for any $b \in (0,1)$ (9.93) has a traveling wave solution $u_b = u_b(x,t)$ of the form

$$u_b(x,t) = \frac{3b^2}{1-b^2}\,\text{sech}^2\Big(\frac{b}{2}\big(x - \frac{t}{1-b^2}\big)\Big) \tag{9.96}$$

Hint:
(a) For $\alpha > 1$ look for solutions of (9.93) of the form $\eta(x - \alpha\, t)$.
(b) By rescaling the ODE for $\eta(\cdot)$ obtain a relation between $\eta(\cdot)$ and $\phi(\cdot)$ in (7.6) and (7.7) with $k = 2$.

9.4 Consider the sine-Gordon equation

$$\partial_t^2 u - \partial_x^2 u + \sin(u) = 0. \tag{9.97}$$

(i) Show that the function

$$v_\pm^\mu(x,t) = 4\tan^{-1}\left(c\, e^{(x-\mu t)/\sigma_\pm(\mu)}\right)$$

with $\mu \in (-1,1)$, $\sigma_\pm(\mu) = \pm\sqrt{1-\mu^2}$ and $c \in \mathbb{R}$ is a traveling wave solution of the sine-Gordon equation.

(ii) Show that for $\mu \in (0,1)$, $v_+^\mu(\cdot)$ (*kink solution*) satisfies:
(a) for each $t_0 \in \mathbb{R}$ fixed $v_+^\mu(\cdot, t_0)$ is increasing with

$$\lim_{x\downarrow -\infty} v_+^\mu(x,t_0) = 0 \quad \text{and} \quad \lim_{x\uparrow\infty} v_+^\mu(x,t_0) = 1.$$

(b) $v_+^\mu(x,t)$ moves to the right as t increases.

(iii) Show that for $\mu \in (-1,0)$, $v_-^\mu(\cdot)$ (*anti-kink solution*) satisfies
(a) for each $t_0 \in \mathbb{R}$ fixed $v_-^\mu(\cdot, t_0)$ is decreasing with

$$\lim_{x\downarrow -\infty} v_-^\mu(x,t_0) = -1 \quad \text{and} \quad \lim_{x\uparrow\infty} v_-^\mu(x,t_0) = 0.$$

(b) $v_+^\mu(x,t)$ moves to the left as t increases.

(iv) Verify that

$$w_{\mu}(x,t) = 4\tan^{-1}\big(\frac{\sigma_{+}(\mu)}{\mu}\frac{\sin(\mu t)}{\cosh(\sigma_{+}(\mu)x)}\big)$$

with $0 < |\mu| < 1$ is a (*stationary breather*) solution of the sine-Gordon equation which satisfies

$$\lim_{|x|\to\infty} w_{\mu}(x,t) = 0 \text{ uniformly in } t \in \mathbb{R}.$$

(v) Show that if $t' = (t - \alpha x)/\sigma_{+}(\alpha)$, $x' = (x - \alpha t)/\sigma_{+}(\alpha)$, $\alpha \in (-1, 1)$ (Lorentz transformation), then

$$\partial_t^2 u - \partial_x^2 u = \partial_{t'}^2 u - \partial_{x'}^2 u,$$

i.e. the sine-Gordon equation is invariant under the Lorentz transformation.

(vi) Combine (iv) and (v) to conclude that for $|\alpha|$, $|\mu| < 1$ the function

$$Z_{\mu,\alpha}(x,t) = 4\,\tan^{-1}\Big(\frac{\sigma_{+}(\mu)}{\mu}\frac{\sin(\mu(t-\alpha x)/\sigma_{+}(\alpha))}{\cosh(\sigma_{+}(x-\alpha t)/\sigma_{+}(\alpha))}\Big)$$

is a (*moving breather*) solution of the sine-Gordon equation.

9.5 (Compactons [RH]) Consider the quasilinear equation

$$\partial_t u + \partial_x^3(u^2) + \partial_x(u^2) = 0. \tag{9.98}$$

Show that the C^1-function of compact support

$$\phi(x-ct) = \begin{cases} \dfrac{4c}{3}\cos^2\big(\frac{x-ct}{4}\big), & |x-ct| \le 2\pi, \\ 0, & |x-ct| > 2\pi, \end{cases}$$

$c > 0$, is a traveling wave (classical) solution of (9.98).

9.6 (i) Show that the function

$$u(x,y,t) = 4i\,\frac{e^{(-(x+y+8t)-i(x+y))}}{(1+\exp(-2x-8t))(1+\exp(-2y-8t))+1}$$

is a solution of the elliptic-hyperbolic DS system (9.1) with $(c_0, c_1, c_2, c_3) = (1, 2, -1, -1)$ where φ satisfies the boundary conditions

$$\lim_{y\to\infty} \partial_x\varphi(x,y,t) = 4\,\mathrm{sech}^2(x+4t)$$

$$\lim_{x\to\infty} \partial_y\varphi(x,y,t) = 4\,\mathrm{sech}^2(y+4t).$$

(ii) [Oz] For the hyperbolic-elliptic DS system (9.1) with $(c_0, c_1, c_2, c_3) = (-1, 16, 8, 1)$ prove:

(a)
$$u(x,y,t) = e^{i(x^2-y^2)/4(1-t)} \frac{1-t}{(1-t)^2+x^2+y^2}$$
$$\varphi(x,y,t) = \frac{x}{2} \frac{1}{(1-t)^2+x^2+y^2}$$
is a solution of (9.1) with $u(x,y,0) \in L^2(\mathbb{R}^2) \cap L^\infty(\mathbb{R}^2)$.

(b) For any $t \in \mathbb{R}$, $u(\cdot,\cdot,t) \notin H^1(\mathbb{R}^2)$ and $(x^2+y^2)^{1/2}\, u(\cdot,\cdot,t) \notin L^2(\mathbb{R}^2)$ but for $s \in (0,1)$ and $t \in (-\infty,1)$, one has $u(\cdot,\cdot,t) \in H^s(\mathbb{R}^2)$ and $(x^2+y^2)^{s/2}\, u(\cdot,\cdot,t) \in L^2(\mathbb{R}^2)$.

(c)
$$\lim_{t\uparrow 1} \|u(\cdot,\cdot,t)\|_\infty = +\infty.$$

9.7 Show that the function
$$u(x,y,t) = \frac{4i\, \exp(-(x+y+4t) - i(x+y))}{(1+\exp(-2x-4t))(1+\exp(-2y-4t))+1}$$
is a solution of the Davey–Stewartson system (DSI)
$$\begin{cases} i\,\partial_t u + \frac{1}{2}(\partial_x^2 u + \partial_y^2 u) = |u|^2 u + u\partial_x\varphi, \\ \partial_x^2\varphi - \partial_y^2\varphi = -2\partial_x(|u|^2), \end{cases} \tag{9.99}$$
when ϕ satisfies the following boundary conditions
$$\lim_{y\to\infty} \partial_x\varphi(x,y,t) = -2\text{sech}^2(x+2t), \text{and } \lim_{x\to\infty} \partial_x\varphi(x,y,t) = -2\text{sech}^2(y+2t).$$
This solution is called dromion (see [FSa]).

9.8 Show that the Boussinesq equation
$$\partial_t^2 u - \partial_x^2 u - \partial_x^4 u + \partial_x^2(u^2) = 0$$
has traveling wave solutions of the form
$$u(x,t) = a\, \text{sech}^2(b(x-ct)),$$
with appropriate values of a, b for $c > 0$ and $c < 0$, i.e., the wave can propagate in any direction.

9.9 Consider the linear part of the Benjamin–Ono equation
$$Lu = \partial_t u + \mathsf{H}\partial_x^2 u = 0.$$
Defining
$$\Gamma = x - 2t\,\mathsf{H}\,\partial_x = x - 2t\, D_x.$$
Show that
$$[L;\ \Gamma] = [L;\ \Gamma^2] = 0,$$

$$[L;\ \Gamma^3]\phi = 0 \quad \text{if and only if} \quad \widehat{\phi}(0,t) = 0, \quad \text{for all} \quad t \in \mathbb{R},$$

and

$$[L;\ \Gamma^4] \neq 0.$$

9.10 Consider the IVP

$$\begin{cases} \partial_t v + \mathsf{H}\partial_x^2 v = 0, \\ v(x,0) = v_0(x) \end{cases}$$

and

$$Z_{s,r} = H^s(\mathbb{R}) \cap L^2(|x|^{2r}).$$

Let $k \in \mathbb{Z}^+$, prove that $v(\cdot,t) \in L^2(|x|^{2k}\,dx)$ for all $t > 0$ if and only if $v_0 \in Z_{k,k}$, $k = 1,2$ and for $k \geq 3$,

$$\int_{-\infty}^{\infty} x^j\, v_0(x)\,dx = 0, \quad j = 0,1,\dots,k-3.$$

9.11 Let H be the Hilbert transform. Prove that $\mathsf{H}:\ L^2(\langle x\rangle^{\theta}\,dx) \to L^2(\langle x\rangle^{\theta}\,dx)$ is continuous if and only if $\theta \in (-1/2, 1/2)$.

Remark 9.1. Compare this result with that of Exercise 3.19 in Chapter 3.

9.12 Let $u \in C([0,T] : H^1(\mathbb{R}) \cap L^2(|x|^2\,dx))$ be a "strong" solution of the Benjamin-Ono equation. Assuming that $\int u(x,0)\,dx = 0$, prove the identity

$$\frac{d}{dt}\int x\,(u^2 + (\mathsf{H}u)^2)(x,t)\,dx = 4\,\|D_x^{1/2}u(\cdot,t)\|_2^2.$$

9.13 (i) Show that the following quantities are conserved by the BO solution flow:

$$I_1(u) = \int_{\mathbb{R}} u(x,t)\,dx,$$

$$I_2(u) = \int_{\mathbb{R}} u^2(x,t)\,dx,$$

$$I_3(u) = \int_{\mathbb{R}} \Big(u\partial_x \mathsf{H}u + \frac{u^3}{3}\Big)(x,t)\,dx = \int_{\mathbb{R}} \Big(|D_x^{1/2}u|^2 + \frac{u^3}{3}\Big)(x,t)\,dx$$

and

$$I_4(u) = \int_{\mathbb{R}} \Big(2(\partial_x u)^2 + \frac{3}{2}\,u^2\mathsf{H}\partial_x u + \frac{u^4}{4}\Big)(x,t)\,dx.$$

(ii) Prove that a solution $u \in C([0,T] : H^3(\mathbb{R}) \cap L^2(|x|^4\,dx))$ of the BO equation satisfies

$$\text{(a)} \quad \int_{\mathbb{R}} x\,u(x,t)\,dx = \int_{\mathbb{R}} x\,u_0(x)\,dx + \frac{t}{2}\,\|u_0\|_2^2.$$

$$\text{(b)} \quad \int_{\mathbb{R}} x\,u^2(x,t)\,dx = \int_{\mathbb{R}} x\,u_0^2(x)\,dx + 2t\,I_3(u_0).$$

$$\text{(c)} \quad \int_{\mathbb{R}} x^2\,u(x,t)\,dx = \int_{\mathbb{R}} x^2\,u_0(x)\,dx + t\int_{\mathbb{R}} x\,u_0^2(x)\,dx + t^2 I_3(u_0).$$

9.14 Consider the KPI(−) and KPII(+) equations,

$$\partial_t u + u\partial_x u + \partial_x^3 u \mp \partial_x^{-1}\partial_y^2 u = 0. \tag{9.100}$$

Prove that if $u = u(x,y,t)$ is a solution then

(i) $u_\lambda(x,y,t) = \lambda^2\,u(\lambda x, \lambda^2 y, \lambda^5 t)$, $\lambda > 0$ (scaling) is also a solution.
(ii) $u_c(x,y,t) = u(x - cy \pm c^2 t, y \mp 2ct, t)$ (Galilean invariance) is also a solution of the KPI and KPII, respectively.

9.15 Show that

$$u(x,y,t) = 3c\,\text{sech}^2\Big(\frac{1}{2}(\sqrt{c}x + l\,y - \theta\,t)\Big), \quad \theta = c^{3/2} + lc^{-1/2}, \quad c > 0,\ l \in \mathbb{R},$$

is a solution of the Eqs. (9.100) with + sign, i.e. the KPII equation.
Hint: Use that $u_c(x,t) = 3c\,\text{sech}^2(\frac{\sqrt{c}}{2}(x - ct))$ is the soliton solution of the KdV equation with speed c.

9.16 Prove that the function

$$\phi_c(x - ct, y) = \frac{24c(3 - c(x-ct)^2 + c^2y^2)}{(3 + c(x-ct)^2 + c^2y^2)^2}$$

is a finite energy (ϕ_c, $\partial_x\phi_c$, $\partial_x^{-1}\partial_y\phi_c \in L^2(\mathbb{R}^2)$) solitary wave solution of the KPI.

9.17 Consider the linear IVP associated to (9.84)

$$\begin{cases} \partial_t w + \partial_x^{2j+1} w = 0, \\ w(x,0) = w_0(x), \end{cases} \tag{9.101}$$

$x,\ t \in \mathbb{R}$, $j = 0, 1, \ldots$. Denote by

$$w(x,t) = V_j(t)\,w_0(x) = e^{-t\,\partial_x^{2j+1}}\,w_0(x) \tag{9.102}$$

its solution.

(i) Prove that for any $j = 0, 1, \ldots$ there exists $c_j > 0$ such that for any $x, t \in \mathbb{R}$ one has that

$$c_j \int_{-\infty}^{\infty} |\partial_x^j w(x,s)|^2\, ds = \int_{-\infty}^{\infty} |w(y,t)|^2\, dy = \|w_0\|_2^2. \tag{9.103}$$

Hint: Follow the argument given in the proof of Lemma 7.1.

(ii) Prove that for any $j = 0, 1, \ldots$ there exists $c'_j > 0$ such that for any $x, t \in \mathbb{R}$,

$$\|\partial_x^{2j} \int_0^t V_j(t-t')\, F(\cdot,t')\, dt'\|_{L_t^2} \leq c'_j \|F\|_{L_x^1 L_t^2}. \tag{9.104}$$

Hint: Follow the argument given in the proof of Theorem 4.4 (estimate (4.28)).

(iii) Show that for any $k = 0, 1, \ldots, j$ there exists $c = c(k; j) > 0$ such that for any $x \in \mathbb{R}$,

$$\Big(\int_0^T |\partial_x^{j+k} \int_0^t V_j(t-t')\, F(\cdot,t')\, dt'|^2\, dt\Big)^{1/2}$$

$$\leq c\, T^{(j-k)/2j} \Big\| \int_0^T |F(\cdot,t)|^2\, dt \Big\|_{2j/(j+k)}. \tag{9.105}$$

Hint: Combine (i) and Minkowski's integral inequality to obtain (9.105) for $k = 0$. Interpolate between this result and (9.104).

(iv) Combining the identity (9.103) and the (unsharp) estimate

$$\| \sup_{0\leq t\leq T} |V_j(t)u_0| \|_2 \leq c\,(1+T)\|u_0\|_{2j+1,2} \tag{9.106}$$

to prove Theorem 9.4 with $s \gg 1$ and $m = 0$ (no weight) in the case where

$$P = P(u, \ldots, \partial_x^j u)$$

with $P(\cdot)$ as in (9.85) (nonlinear terms at least quadratic) using a fixed point argument.

Hint: Consider the integral equation equivalent form of the corresponding IVP

$$\Phi(u)(t) = V_j(t)u_0 + \int_0^t V_j(t-t')\, P(u, \ldots, \partial_x^j u)(\cdot,t')\, dt'. \tag{9.107}$$

(v) Combining (9.103), (9.104) and (9.106) prove Theorem 9.4 with $s \gg 1$ and $m = 0$ (no weight) in the case where

$$P = P(u, \dots, \partial_x^{2j-1} u) \tag{9.108}$$

with $P : \mathbb{R}^{2j} \to \mathbb{R}$ being a polynomial such that

$$P(x) = \sum_{|\alpha|=3}^{l_1} a_\alpha x^\alpha. \tag{9.109}$$

i.e. the nonlinear terms are at least cubic.

Note: Parts (iv) and (v) show that the IVP (9.84) can be solved via contraction principle in $H^s(\mathbb{R})$ (without weight) with $s \gg 1$ if either

(i) $P = P(u, \dots, \partial_x^j u)$ with P as in (9.85), i.e., the nonlinear terms are at least quadratic or
(ii) $P = P(u, \dots, \partial_x^{2j-1} u)$ with P as in (9.108) and (9.109).

This is optimal. More precisely, it was established in [Pd] that the IVP (9.84) cannot be solved in $H^s(\mathbb{R})$ for any $s > 0$ with an argument based solely on the contraction principle if $P = \partial_x^{2j-1}(u^2)$.

9.18 (Cole–Hopf transformation) Let $w = w(x,t)$ be a positive C^3-solution of the heat equation

$$\partial_t w = \partial_x^2 w, \tag{9.110}$$

(i) $x \in \mathbb{R},\ t > 0$. Prove that $u(x,t) = -2\,\partial_x(\ln w(x,t))$ satisfies the viscous Burgers' equation (7.105).

(ii) Prove that if $u = u(x,t)$ is a C^2-solution of the viscous Burgers' equation (7.105) with $u \in L^\infty(\mathbb{R}^+ : L^1(\mathbb{R}))$, then $w(x,t) = \exp\left(-\frac{1}{2}\int_{-\infty}^{x} u(s,t)\,ds\right)$ is a positive solution of the heat equation (9.110).

9.19 (i) [BM] Consider the logarithmic Schrödinger equation

$$i\partial_t u + \Delta u + u \ln |u|^2 = 0, x \in \mathbb{R}^n, t \in \mathbb{R}. \tag{9.111}$$

(a) Prove that $\varphi(x) = \pi^{-n/4} e^{-|x|^2/2}$ satisfies the elliptic equation

$$-\Delta\varphi - \varphi \ln \varphi^2 = n(1 + \ln\sqrt{\pi})\varphi. \tag{9.112}$$

(b) Prove that

$$u(x,t) = e^{-i\,n(1+\ln\sqrt{\pi})}\varphi(x)$$

is a (standing wave) solution of (9.111).

(ii) [JP] Consider the logarithmic Korteweg-de Vries equation

$$\partial_t u + \partial_x^3 u + \partial_x (u \ln |u|) = 0, \tag{9.113}$$

$x, t \in \mathbb{R}$.
Prove that for any $c, \alpha \in \mathbb{R}$

$$u(x,t) = e^c \sqrt{e}\, e^{-(x-ct-\alpha)^2/4}$$

is a (Gaussian solitary wave) solution of (9.113).

9.20 [X] Consider the Burgers-Korteweg-de Vries equation

$$\partial_t u + u\partial_x u - \beta\, \partial_x^2 u + \alpha \partial_x^3 u = 0, \tag{9.114}$$

$\alpha,\ \beta > 0, x \in \mathbb{R},\ t > 0$.

(i) Prove that for $\beta = 0$ the equation (9.114) has a traveling wave solution of the form $u_c(x,t) = \phi_c(x - ct)$, $c > 0$, with

$$\phi_c(x) = 3c\, \text{sech}^2\Big(\frac{\sqrt{c}}{2\sqrt{\alpha}} x\Big).$$

(ii) Prove that for $\alpha = 0$ the equation (9.114) has a traveling wave solution of the form

$$u_c(x,t) = c - \varphi_c(x - ct), \quad c > 0,$$

with

$$\varphi_c(x) = \beta c\ \tanh\left(\frac{c}{2} x\right).$$

(iii) Check that if $\alpha,\ \beta > 0$, then the equation (9.114) has traveling wave solutions of the form

$$u_c(x,t) = \frac{1}{2} c\, \text{sech}^2\Big(\frac{1}{\sqrt{24\alpha}} \sqrt{c}(x \mp ct)\Big) - c\ \tanh\Big(\frac{1}{\sqrt{24\alpha}} \sqrt{c}(x \mp ct)\Big) \pm c$$

with $c = \frac{6\beta^2}{25\alpha} > 0$.

Chapter 10
General Quasilinear Schrödinger Equation

10.1 The General Quasilinear Schrödinger Equation

In this chapter, we shall study the local solvability of the initial value problem (IVP) associated with the general quasilinear Schrödinger equation:

$$\begin{cases} \partial_t u = i a_{jk}(x,t,u,\bar{u},\nabla_x u,\nabla_x \bar{u})\,\partial^2_{x_j x_k} u \\ \qquad\quad +b_{jk}(x,t,u,\bar{u},\nabla_x u,\nabla_x \bar{u})\,\partial^2_{x_j x_k} \bar{u} \\ \qquad\quad +\overrightarrow{b_1}(x,t,u,\bar{u},\nabla_x u,\nabla_x \bar{u})\cdot\nabla_x u \\ \qquad\quad +\overrightarrow{b_2}(x,t,u,\bar{u},\nabla_x u,\nabla_x \bar{u})\cdot\nabla_x \bar{u} \\ \qquad\quad +c_1(x,t,u,\bar{u})\,u + c_2(x,t,u,\bar{u})\,\bar{u} + f(x,t), \\ u(x,0) = u_0(x) \end{cases} \tag{10.1}$$

(using summation convention).

One may think of this equation as a nonlinear Schrödinger equation, where the operator modeling the dispersion relation is nonisotropic and depends also on the unknown function, its conjugate, and its space gradient.

Equations of this form arise in several fields of physics (plasma, fluids, classical and quantum ferromagnetic, laser theory, etc.)

A well-studied model is

$$\partial_t u = i\Delta u - 2iu\,h'(|u|^2)\,\Delta h(|u|^2) + iu\,g(|u|^2), \tag{10.2}$$

where h and g are given functions, $n \geq 1$. When $n = 1,2,3$, Bouard, Hayashi and Saut [BHS] proved local well-posedness of the associated IVP in $H^6(\mathbb{R}^n)$, for small data. This was extended by Colin [Cl] to data of arbitrary size in $H^s(\mathbb{R}^n)$, $s \geq s(n)$ for all n.

Problems of this type also arise in Kähler geometry, where the "Schrödinger flow" is defined as follows:

Let (M,g) be a Riemannian manifold and (N,J,h) be a complete Kähler manifold with complex structure J and Kähler metric h. Then given

$$u_0 : M \mapsto N \tag{10.3}$$

F. Linares, G. Ponce, *Introduction to Nonlinear Dispersive Equations*,
Universitext, DOI 10.1007/978-1-4939-2181-2_10

one seeks for

$$u : M \times [0,T] \mapsto N \tag{10.4}$$

such that

$$\begin{cases} \partial_t u = J(u(x,t)) \cdot \tau(u(x,t)), \\ u(x,0) = u_0(x), \end{cases} \tag{10.5}$$

where $\tau(u)$ the tension field of u is given in local coordinates by

$$\tau^\alpha(u) = \Delta_g u^\alpha + g^{jk}\,\Gamma^\alpha_{\beta\gamma}(u)\,\frac{\partial u^\beta}{\partial x_j}\,\frac{\partial u^\gamma}{\partial x_k}, \tag{10.6}$$

where $\Gamma^\alpha_{\beta\gamma}$ represents the Christoffel symbol for the target manifold N. These systems have been studied in [DW], [CSU], [MG], [NSU], [NSVZ], among others. For the minimal regularity problem, i.e., to determinate the minimal Sobolev index that guaranties (local or global) well-posedness see [IK2], [BIK] and references therein.

Before considering nonlinear models, it is convenient to study the IVP for the linear equation involving first-order terms. More precisely, we review the results mentioned at the end of Chapter 4. This will be helpful in understanding the hypotheses and the arguments of the proof of the nonlinear result to be discussed later in this chapter.

Consider the linear IVP,

$$\begin{cases} \partial_t u = iAu + \overrightarrow{b}(x) \cdot \nabla u + d(x)u + f(x,t), \\ u(x,0) = u_0(x) \in L^2(\mathbb{R}^n), \end{cases} \tag{10.7}$$

$x \in \mathbb{R}^n$, $t \in \mathbb{R}$, with $A = \partial_{x_j}(a_{jk}(x)\partial_{x_k})$ a second-order elliptic operator, $\overrightarrow{b} = (b_1, \dots, b_n)$, $b_j : \mathbb{R} \mapsto \mathbb{C}$, $j = 1, 2, \dots, n$, and $f \in C(\mathbb{R} : L^2(\mathbb{R}^n))$. To simplify the exposition, assume $b_j \in C_0^\infty(\mathbb{R}^n)$ and $f \equiv 0$. Concerning the L^2-local well-posedness of (10.7) one has:

(i) If $b = b(x)$ is a real-valued function, the result follows by integrating by parts.

(ii) If $n \geq 1$, $a_{jk}(x) = \delta_{jk}$, i.e., $A = \Delta$ and $b_j(x) = i\,c_0$, $c_0 \in \mathbb{R}$ for some j, then problem (10.7) is ill-posed.

(iii) If $n = 1$ and $A = \partial_x^2$, define $v = \phi\, u$, with ϕ real-valued to be determined (ϕ and $1/\phi$ bounded) so

$$\begin{aligned} \partial_t v = i\,\partial_x^2 v + i\left(2\frac{\partial_x \phi}{\phi} + \mathcal{I}m\, b(x)\right)\partial_x v + \mathcal{R}e\, b(x)\partial_x v \\ + \text{ terms of order zero in} v. \end{aligned} \tag{10.8}$$

Then to eliminate the term which cannot be handled by integration by parts one takes

$$\ln \phi(x) = -\frac{1}{2}\int_0^x \mathcal{I}m\, b(s)\, ds. \tag{10.9}$$

In [Ta1], Takeuchi proved in case, $n = 1$ and $A = \partial_x^2$, and general $b(\cdot)$ in (10.7), that the condition

$$\sup_{l \in \mathbb{R}} \left| \int_0^l \mathcal{I}m\ b(s)\,ds \right| < \infty$$

is sufficient for the L^2-well-posedness of (10.7).

(iv) If $n \geq 2$, and $A = \Delta$ one can reapply the argument above to find $\phi = \phi(x, \widehat{\xi})$, $\widehat{\xi} \in \mathbb{S}^{n-1}$, which should solve the equation

$$2\frac{\nabla \phi}{\phi} + \mathcal{I}m\ \overrightarrow{b}(x) = 0. \tag{10.10}$$

Hence, if $\mu = \ln \phi$,

$$2\partial_{\widehat{\xi}}\mu = -\mathcal{I}m\ \overrightarrow{b}(x) \cdot \widehat{\xi}, \qquad \text{for all } \widehat{\xi} \in \mathbb{S}^{n-1}.$$

Thus,

$$\mu(x, \widehat{\xi}) = -\frac{1}{2} \int_{-\infty}^{0} \mathcal{I}m\ \overrightarrow{b}(x + s\,\widehat{\xi}) \cdot \widehat{\xi}\,ds, \quad \widehat{\xi} \in \mathbb{S}^{n-1}, \tag{10.11}$$

and

$$\phi(x, \widehat{\xi}) = e^{-\frac{1}{2}\int_{-\infty}^{0} \mathcal{I}m\ \overrightarrow{b}(x+s\,\widehat{\xi})\cdot\widehat{\xi}\,ds}, \quad \widehat{\xi} \in \mathbb{S}^{n-1}. \tag{10.12}$$

In [Mz], Mizohata showed that if $n \geq 1$ and $A = \Delta$, the condition

$$\sup_{\widehat{\xi} \in \mathbb{S}^{n-1}} \sup_{\substack{x \in \mathbb{R}^n \\ l \in \mathbb{R}}} \left| \int_0^l \mathcal{I}m\ b_j(x + s\,\widehat{\xi})\,ds \right| < \infty \tag{10.13}$$

is necessary for the L^2-local well-posedness (10.7). Notice that (10.11) is an integrability condition on the coefficients $\overrightarrow{b} = (b_1, \ldots, b_n)$ of the first-order term along the bicharacteristics.

(v) Consider now $n \geq 1$ and $A = \partial_{x_j}(a_{jk}(x)\,\partial_{x_k}\cdot\,)$ a general elliptic operator (see (3.25)). In this case, we apply an invertible pseudo-differential operator $C(x, D)$ with real symbol $c(x, \xi)$ to the equation in (10.7) to get

$$\begin{aligned} \partial_t C = i\,ACu + i\,[C; A]u + iC(\mathcal{I}m\ \overrightarrow{b}(x) \cdot \nabla u) + \mathcal{R}e\ \overrightarrow{b}(x) \cdot \nabla Cu \\ + \text{terms of order zero in } u \text{ and } Cu. \end{aligned} \tag{10.14}$$

To cancel the bad first-order term one solves the equation:

$$i[C; A] + i\,C(\mathcal{I}m\ \overrightarrow{b}(x) \cdot \nabla) = 0, \tag{10.15}$$

up to operators of order zero. So, using their symbols one has

$$\{c(x,\xi);a(x,\xi)\} + c(x,\xi)\mathcal{I}m\,\overrightarrow{b}(x)\cdot\xi = -H_a(c) + c(x,\xi)\mathcal{I}m\,\overrightarrow{b}(x)\cdot\xi = 0,$$

i.e., (see Lemma 3.1),

$$\frac{d}{ds}c(X(s,x,\xi),\Xi(s,x,\xi)) = \tag{10.16}$$
$$c(X(s,x,\xi),\Xi(s,x,\xi))\mathcal{I}m\,\overrightarrow{b}(X(s,x,\xi))\cdot\Xi(s,x,\xi).$$

Therefore,

$$c(x,\xi) = e^{\int_{-\infty}^{0}\mathcal{I}m\,\overrightarrow{b}(X(s,x,\xi))\cdot\Xi(s,x,\xi)\,ds},$$

where $s \to (X(s,x,\xi)), \Xi(s,x,\xi))$ is the bicharacteristic flow associated to the symbol of A (see (3.28)).

In [I], Ichinose extended the Mizohata condition (10.13) to the case of elliptic variable coefficients deducing that

$$\sup_{\widehat{\xi}\in\mathbb{S}^{n-1}}\ \sup_{\substack{x\in\mathbb{R}^n\\ l\in\mathbb{R}}}\left|\int_0^l \mathcal{I}m\ b_j(X(s,x,\widehat{\xi})\cdot\Xi_j(s,x,\widehat{\xi})\,ds\right| < \infty \tag{10.17}$$

is a necessary condition for the L^2-well-posedness of (10.7).

Notice that the notion of nontrapping for the bicharacteristic flow associated to the symbol of A is essential for (10.17) to hold even for $b_j \in C_0^\infty(\mathbb{R}^n)$. Also, asymptotic flatness conditions in the coefficients $a_{jk}(x)$ (see for instance (4.66)) guarantee an appropriate behavior at infinity of the bicharacteristic flow.

Returning to the nonlinear problem consider the case of the Schrödinger equation, with the constant coefficients semilinear case, i.e.,

$$\partial_t u = i\Delta u + f(u,\bar{u},\nabla_x u,\nabla_x\bar{u}), \qquad x\in\mathbb{R}^n. \tag{10.18}$$

If f is smooth, integration by parts yields the estimate

$$\left|\sum_{|\alpha\leq s}\int_{\mathbb{R}^n}\partial_x^\alpha f(u,\bar{u},\nabla_x u,\nabla_x\bar{u})\partial_x^\alpha u\,dx\right| \leq c\,(1+\|u\|_{s,2}^{\rho})\,\|u\|_{s,2}^2, \tag{10.19}$$

for any $u\in H^{s+1}(\mathbb{R}^n)$, $s > n/2+1$, and $\rho = \rho(f)\in\mathbb{Z}^+$, then energy estimates lead to the desired local well-posedness.

Another technique used to overcome the "loss of derivatives" introduced by the nonlinearity f in (10.18) involving $\nabla_x u$ relies on an analytic function approach (see [H5]).

Local well-posedness for small data and general smooth function $f : \mathbb{C}^{2n+2} \mapsto \mathbb{C}$ was established by Kenig, Ponce and Vega [KPV3]. In [KPV3], the authors consider the integral equation associated to (10.18):

$$u(t) = e^{it\Delta}u_0 + \int_0^t e^{i(t-t')\Delta} f(u, \bar{u}, \nabla_x u, \nabla_x \bar{u})(t')\, dt' \tag{10.20}$$

and use the inhomogeneous version of the local smoothing effect (see (4.30)) of the group $\{e^{it\Delta}\}_{t=-\infty}^{\infty}$, i.e.,

$$\Big\| \nabla_x \int_0^t e^{i(t-t')\Delta} g(t')\, dt' \Big\|_{\ell_\alpha^\infty(L^2(Q_\alpha \times [0,T]))} \leq c\, \|g\|_{\ell_\alpha^1(L^2(Q_\alpha\times[0,T]))}, \tag{10.21}$$

(where the $\{Q_\alpha\}_\alpha$ is the family of unit cubes with disjoint interiors such that $\bigcup_\alpha Q_\alpha = \mathbb{R}^n$), to overcome the "loss of derivatives" introduced by the nonlinearity $f(\cdot)$ in (10.18), which depends up to first-order derivatives of the unknown. Briefly, one needs to estimate u in the $\ell_\alpha^\infty(L^2(Q_\alpha \times [0,T]))$-norm, which cannot be made "small" by taking $T \to 0$, so it is here where the conditions on the size of the data appear.

The smallness assumption on the data was removed by Hayashi and Ozawa [HO] in the one-dimensional case ($n = 1$). To do so they introduced a change of variables. To illustrate their argument, let us consider the equation:

$$\partial_t u = i\partial_x^2 u + u\partial_x u + u\partial_x \bar{u}. \tag{10.22}$$

When performing standard energy estimates, one sees that the "bad" term in (10.22) is $u\,\partial_x u$, i.e., the one involving $\partial_x u$. This term cannot be handled by integration by parts except when it has a real coefficient, for instance, $|u|^2\,\partial_x u$. Hence, the idea is to eliminate it. First, take the derivatives of (10.22) up to order 3, and use the notation $\partial_x^j u = u_{j+1}$ to rewrite equation (10.22) as the system:

$$\begin{cases} \partial_t u_1 = i\partial_x^2 u_1 + u_1\, u_2 + u_1 \bar{u}_2, \\ \partial_t u_2 = i\partial_x^2 u_2 + u_2\, u_2 + u_1\, u_3 + u_2\bar{u}_2 + u_1\bar{u}_3, \\ \partial_t u_3 = i\partial_x^2 u_3 + 3u_2 u_3 + u_1\, u_4 + u_3\bar{u}_2 + 2u_2\bar{u}_3 + u_1\bar{u}_4, \\ \partial_t u_4 = i\partial_x^2 u_4 + u_1\partial_x u_4 + u_1\partial_x\bar{u}_4 + Q(u_1, \bar{u}_1, \dots, u_4, \bar{u}_4). \end{cases} \tag{10.23}$$

The first three equations in (10.23) are semilinear as well as the term $Q(\cdot)$ in the fourth one. One then considers "$u_4\,\phi$" instead of "u_3" with ϕ to be determined.

So we substitute u_4 by $\phi^{-1}\,(u_4\,\phi)$ except in the main part of the fourth equation, i.e.,

$$\partial_t u_4 = i\partial_x^2 u_4 + u_1\,\partial_x u_4 + u_1\,\partial_x \bar{u}_4. \tag{10.24}$$

Here, multiplying by ϕ we rewrite (10.24) as

$$\partial_t(u_4\,\phi) - u_4\partial_t\phi = i\partial_x^2(u_4\,\phi) - 2i\,\partial_x u_4\,\partial_x\phi - i u_4\,\partial_x^2\phi + u_1\,\phi\,\partial_x u_4 \tag{10.25}$$

$$+\overline{\phi}^{-1}\phi\, u_1\, \partial_x(\overline{u_4\,\phi}) + \phi u_1(\overline{u_4\,\phi})\partial_x\overline{\phi}^{-1}.$$

We now choose ϕ to eliminate the terms involving $\partial_x\, u_4$, i.e.,

$$-2i\,\partial_x u_4\,\partial_x\phi + u_1\,\phi\,\partial_x u_4 = 0 \qquad \text{or} \qquad -2i\,\partial_x\phi + u_1\,\phi = 0, \tag{10.26}$$

that is,

$$\phi(x,t) = \exp\Big(-\frac{i}{2}\int_0^x u_1(\theta,t)\,d\theta\Big). \tag{10.27}$$

In the new variables $(u_1, u_2, u_3, u_4\,\phi)$ for the system (10.23), the standard energy estimates can be performed to obtain the desired local existence and uniqueness result.

Later, Chihara [Ch2] removed the smallness assumption on the data in any dimension. The change of variables used in [HO] in higher dimensions leads to an "exotic" class of pseudodifferential operators (ψ.d.o.) studied by Craig, Kappeler and Strauss [CKS].

Consider the symbol in (10.11), i.e.,

$$\mu(x,\xi) = -\frac{1}{2}\int_{-\infty}^{0} \mathcal{I}m\ \vec{b}\Big(x + s\,\frac{\xi}{|\xi|}\Big)\cdot\frac{\xi}{|\xi|}\,ds \tag{10.28}$$

with $\xi \in \mathbb{R}^n$, and $\vec{b} = (b_1,\dots,b_n),\ b_j \in C_0^\infty(\mathbb{R}^n)$. One has that for $|\xi| \geq 1$

$$\Big|\partial_x^\alpha\partial_\xi^\beta\mu(x,\xi)\Big| \leq c_{\alpha,\beta}\,\langle x\rangle^{|\alpha|}|\xi|^{-|\beta|} \quad \forall\alpha,\beta \in (\mathbb{Z}^+)^n, \tag{10.29}$$

where $\langle x\rangle^2 = 1 + |x|^2$.

Roughly speaking, the function space for the local well-posedness was $H^s(\mathbb{R}^n)$, $s > s(n)$ in the case where f is at least cubic, and where it was $\mathcal{F}_n^s = H^s(\mathbb{R}^n)\cap L^2(|x|^n\,dx),\ s \geq s(n)$ when f is just quadratic. This is a clear necessary condition in the light of the integrability (10.12).

In [KPV3], Kenig, Ponce and Vega showed that this local result can be proved by a Picard iteration, so the mapping data solution, $u_0 \mapsto u$, is not only continuous but also analytic. A crucial step in this proof was to establish a "local smoothing" effect (see 4.23) for solutions of (10.18), i.e., if $u_0 \in H^{s_0}(\mathbb{R}^n)$, then

$$\int_0^T\!\!\int \frac{1}{\langle x\rangle^2}\,|\Lambda^{s_0+1/2}u(x,t)|^2\,dx\,dt < \infty, \tag{10.30}$$

where $\langle x\rangle^2 = (1+|x|^2)^{1/2}$ and $\Lambda^s = (I-\Delta)^{s/2}$ is the operator with symbol $\langle\xi\rangle^s$.

This might seem like a technical device but Molinet, Saut and Tzvetkov [MST3] showed that for the IVP:

$$\begin{cases} \partial_t u = i\,\partial_x^2 u + u\partial_x u,\\ u(x,0) = u_0(x) \end{cases} \tag{10.31}$$

the map data-solution, $u_0 \mapsto u$, cannot be C^2 at $u_0 \equiv 0$ for u_0 in any Sobolev space $H^s(\mathbb{R})$. Hence, in order to use Picard iteration, the weights are needed.

Returning to our IVP (10.1), we have that in the one-dimensional case ($n = 1$) Poppenberg [Pp] established local well-posedness for coefficients independent of (x,t) under the following conditions:

Ellipticity. $a(\cdot)$ is real-valued and for $|(z_1, z_2, z_3, z_4)| < R$, there exists $\lambda(R) > 0$ such that

$$a(z_1, z_2, z_3, z_4) - |b(z_1, z_2, z_3, z_4)| \geq \lambda(R). \tag{10.32}$$

Degree of nonlinearity:

$$\begin{cases} \partial_z a(0,0,0,0) = \partial_z b(0,0,0,0) = 0, \\ b_1,\ b_2 \ \text{ vanishing quadratically at } \ (0,0,0,0). \end{cases} \tag{10.33}$$

Poppenberg showed local well-posedness in $H^\infty(\mathbb{R}) = \bigcap_{s \geq 0} H^s(\mathbb{R})$. His proof is based on the Nash–Moser techniques.

In [LmPo], Lim and Ponce showed, in the (x,t)-dependent setting, that under Poppenberg's hypotheses one has local well-posedness in $H^s(\mathbb{R})$, $s \geq s_0$, s_0 large enough, and if b_1, b_2 vanish linearly at $(0,0,0,0)$ and $\partial_z a(0,0,0,0) \neq 0$ or $\partial_z b\,(0,0,0,0) \neq 0$ in the weighted space $\mathcal{F}^s_m = H^s(\mathbb{R}) \cap L^2(|x|^m\,dx)$.

To clarify the elliptic condition notice that when $b \equiv 0$, this is the usual condition and in general, it says that $\partial_x^2 u$ "dominates" $\partial_x^2 \bar{u}$. This is certainly needed, as Exercise 4.14 shows.

Moreover, if the nontrapping condition fails dramatically, i.e., all orbits are periodic, ill-posedness in semilinear problems occurs, as Chihara [Ch3] has shown. He proved that for the IVP,

$$\begin{cases} \partial_t u = i\,\Delta u + \operatorname{div}\,(\overrightarrow{G}\,(u)), \\ u(x,0) = u_0(x), \end{cases} \tag{10.34}$$

$x \in \mathbb{T}^n$, $t \in [0,T]$, where $\overrightarrow{G} = (G_1, \ldots, G_n) \neq 0$, G_j holomorphic, is ill-posed on any Sobolev space $H^s(\mathbb{T}^n)$.

Now we turn to the positive results in [KPV10] concerning the local well-posedness of the IVP (10.1). To simplify the exposition, we shall consider only the case $b_{jk} \equiv 0$.

We shall assume the following:

(H1) *Ellipticity*. Given $M > 0$ there exists $\gamma_M > 0$ such that

$$\langle a_{jk}(x,t,\overrightarrow{z}\,)\xi, \xi\rangle \geq \gamma_M \quad \forall \xi \in \mathbb{R}^n, \ \text{for all } \overrightarrow{z} \in \mathbb{C}^{2n+2} \tag{10.35}$$

with $|\overrightarrow{z}\,| \leq M$.

(H2) *Asymptotic flatness*. There exists $c > 0$ such that for any $(x,t) \in \mathbb{R}^n \times \mathbb{R}$,

$$|\partial_{x_l} a_{jk}(x,t,\overrightarrow{0}\,)| + |\partial^2_{x_l x_r} a_{jk}(x,t,\overrightarrow{0}\,)| \leq \frac{c}{\langle x\rangle^2}, \tag{10.36}$$

where $l = 0, 1, \ldots, n$, $r = 1, \ldots, n$ with $\partial_{x_0} = \partial_t$.

(H3) *Growth of the first-order coefficients.* There exist $c, c_1 > 0$ such that for any $x \in \mathbb{R}^n$ and $(x,t) \in \mathbb{R}^n \times \mathbb{R}$,

$$|\overrightarrow{b_m}(x, 0, \overrightarrow{0})| \leq \frac{c_1}{\langle x \rangle^2}, \qquad |\partial_t \overrightarrow{b_m}(x, 0, \overrightarrow{0})| \leq \frac{c}{\langle x \rangle^2}, \qquad m = 1, 2. \tag{10.37}$$

(H4) *Regularity.* For any $N \in \mathbb{N}$ and $M > 0$ the coefficients $a_{jk}, \overrightarrow{b_1}, \overrightarrow{b_2}, c_1, c_2$ are in

$$C_b^N(\mathbb{R}^n \times \mathbb{R} \times (|\overrightarrow{z}| \leq M)).$$

(H5) *Nontrapping condition.* The data $u_0 \in H^s(\mathbb{R}^n)$, $s > n/2 + 2$, are such that the Hamiltonian flow $H_{h(u_0)}$ associated to the symbol

$$h(u_0) = h_{u_0}(x, \xi) = -a_{jk}(x, 0, u_0, \bar{u}_0, \nabla u_0, \nabla \bar{u}_0)\xi_j \xi_k \tag{10.38}$$

is nontrapping.

The main result in this chapter is the following theorem:

Theorem 10.1. *Under the hypotheses (H1) – (H4) there exists $N = N(n) \in \mathbb{Z}^+$ such that for any $u_0 \in H^s(\mathbb{R}^n)$ with*

$$\langle x \rangle^2 \, \partial_x^\alpha \, u_0 \in L^2(\mathbb{R}^n), \qquad |\alpha| \leq s_1,$$

and

$$f \in L^1(\mathbb{R} : H^s(\mathbb{R}^n)) \quad \text{and} \quad \langle x \rangle^2 \partial_x^\alpha f \in L^1(\mathbb{R} : L^2(\mathbb{R}^n)), \quad |\alpha| \leq s_1,$$

where $s, s_1 \in \mathbb{Z}^+$ with $s_1 \geq n/2 + 7$, $s = \max\{s_1 + 4, N + n + 3\}$ and u_0 satisfying the hypothesis (H5). There exists $T_0 > 0$ depending only on

$$\begin{aligned} &\|u_0\|_{s,2} + \sum_{|\alpha| \leq s} \|\langle x \rangle^2 \, \partial_x^\alpha \, u_0\|_2 \\ &\quad + \int_{-\infty}^{\infty} \|f(t)\|_{s,2} dt + \sum_{|\alpha| \leq s_1} \int_{-\infty}^{\infty} \|\langle x \rangle^2 \, \partial_x^\alpha \, f(t)\|_2 dt \equiv \lambda, \end{aligned} \tag{10.39}$$

so that the IVP (10.1) is locally well-posed in $[0, T_0)$ with the solution:

$$u \in C([0, T_0] : H^s(\mathbb{R}^n)), \quad \langle x \rangle^2 \, \partial_x^\alpha \, u \in C([0, T_0] : L^2(\mathbb{R}^n))$$

for $|\alpha| \leq s_1$.

Remark 10.1.

(i) When $n = 1$, the ellipticity hypothesis (H1) implies the nontrapping one (H5).

(ii) One can also prove that the solution possesses the "local smoothing" effect

$$\Lambda^{s+1/2}u \in L^2(\mathbb{R}^n \times [0, T_0] : \langle x\rangle^{-2}\, dx\, dt).$$

(iii) In the above statements, $\langle x\rangle^2$ can be replaced by $\langle x\rangle^{1+\epsilon}$, $\epsilon > 0$.
(iv) Koch and Tataru [KTa1] have noticed that the map data-solution, $u_0 \mapsto u$, is not C^2, hence the result in Theorem 10.1 cannot be established by using only Picard iteration.
(v) The proof of this theorem is based in the so-called artificial viscosity method, which was explained in details in Chapter 9 (Section 9.4).
(vi) The proof sketched below only uses classical pseudodifferential operators.

To apply the "artificial viscosity method," we first consider the IVP:

$$\begin{cases} \partial_t u = -\epsilon\, \Delta^2 u + i a_{jk}(x,t)\, \partial^2_{x_j x_k} u + \overrightarrow{b_1}(x,t)\cdot \nabla u + \overrightarrow{b_2}(x,t)\cdot \nabla \bar{u} \\ \qquad\quad + c_1(x,t)u + c_2(x,t)\bar{u} + f(x,t), \\ u(x,0) = u_0(x) \end{cases} \tag{10.40}$$

under the following assumptions:

(H_l1) *Ellipticity*. $A(x,t) = (a_{jk}(x,t))^n_{j,k=1}$ is a real symmetric matrix and there exists $\gamma \in (0,1)$ such that for any $\xi \in \mathbb{R}^n$ and $(x,t) \in \mathbb{R}^n \times [0,\infty)$,

$$\gamma |\xi|^2 \le \langle A(x,t)\xi, \xi\rangle \le \gamma^{-1}|\xi|^2. \tag{10.41}$$

(H_l2) *Asymptotic flatness*. There exists $c > 0$ such that for any $(x,t) \in \mathbb{R}^n \times [0,\infty)$,

$$|\partial_{x_l}\, a_{jk}(x,t)| + |\partial^2_{x_l x_r}\, a_{jk}(x,t)| \le \frac{c}{\langle x\rangle^2} \tag{10.42}$$

with $l = 0, 1, \dots, n$, $r = 1, \dots, n$, and $\partial_{x_0} = \partial_t$.

(H_l3) *Growth of the first-order coefficients*. There exists $c > 0$ such that

$$|\mathcal{I}m\, \overrightarrow{b_1}(x,0)| + |\mathcal{I}m\, \partial_t \overrightarrow{b_1}(x,t)| \le \frac{c}{\langle x\rangle^2} \tag{10.43}$$

for all $(x,t) \in \mathbb{R}^n \times [0,\infty)$.

(H_l4) *Regularity*. The coefficients a_{jk}, b_{1j}, b_{2j}, c_1, c_2 are in $C^N_b(\mathbb{R}^n \times [0,\infty))$ with $\overrightarrow{b_l} = (b_{l1}, \dots, b_{ln})$, $l = 1, 2$, for $N = N(n)$ sufficiently large.

(H_l5) *Nontrapping condition*. Let $A_0(x) = A(x,0) = (a_{jk}(x,0))^n_{j,k=1}$,

$$h(x,\xi) = -a_{jk}(x,0)\xi_j \xi_k, \tag{10.44}$$

and H_h be the corresponding Hamiltonian flow. Then H_h is nontrapping.

The following a priori estimate for solutions of the linear IVP (10.40) is the key in the proof of the nonlinear result for the IVP (10.1), Theorem 10.1.

Lemma 10.1. *Under the hypotheses $(H_l 1)$–$(H_l 5)$ above there exist $N = N(n)$, c_0 and $T_0 > 0$ (depending both c_0, T_0 on the nontrapping condition $(H_l 5)$ and on the coefficients at $t = 0$), so that for any $T \in (0, T_0)$ and any $\epsilon \in (0,1)$ we have that the solution of (10.40) satisfies*

$$\begin{aligned}\sup_{0\le t\le T} \| u(t)\|_2 &+ \left(\int_0^T\!\!\int \langle x\rangle^{-2}|\Lambda^{1/2}u|^2\,dx\,dt\right)^{1/2}\\ &\le c_0\left[\|u_0\|_2 + \left(\int_0^T\!\!\int \langle x\rangle^{2}|\Lambda^{-1/2} f(x,t)|^2\,dx\,dt\right)^{1/2}\right],\end{aligned}\tag{10.45}$$

and

$$\begin{aligned}\sup_{0\le t\le T} \| u(t)\|_2 &+ \left(\int_0^T\!\!\int \langle x\rangle^{-2}|\Lambda^{1/2}u|^2\,dx\,dt\right)^{1/2}\\ &\le c_0\left[\|u_0\|_2 + \left(\int_0^T\!\!\int |f(x,t)|^2\,dx\,dt\right)^{1/2}\right].\end{aligned}\tag{10.46}$$

In fact, the constant c_0 depends only on the nontrapping condition for $h(x,\xi)$ $(H_l 5)$, on the bounds at $t = 0$ of $\langle x\rangle^2\, \overrightarrow{b_j}(x,0)$, $j = 1,2$, and on size estimates for the coefficients and their derivatives at $t = 0$. Thus, in the nonlinear case, c_0 depends only on the data u_0. Assuming the result in Lemma 10.1, we shall prove Theorem 10.1.

We introduce the notations ($v = v(x,t)$, $u = u(x,t)$) for

$$\begin{aligned}\mathcal{L}(v)u = {}& i a_{jk}(x,t,v,\bar v,\nabla v,\nabla\bar v)\,\partial^2_{x_j x_k} u\\ &+ \overrightarrow{b_1}(x,t,v,\bar v,\nabla v,\nabla\bar v)\cdot\nabla u + \overrightarrow{b_2}(x,t,v,\bar v,\nabla v,\nabla\bar v)\cdot\nabla\bar u\\ &+ c_1(x,t,v,\bar v)\,u + c_2(x,t,v,\bar v)\,\bar u,\end{aligned}\tag{10.47}$$

$$\begin{aligned}X_{T,M_0} = \big\{& v : \mathbb{R}^n\times[0,T]\to\mathbb{C}\mid v\in C([0,T]:H^s(\mathbb{R}^n)),\\ &\langle x\rangle^2\,\partial_x^\alpha v\in C([0,T]:L^2(\mathbb{R}^n)),\ |\alpha|\le s_1,\ \ v(x,0)=u_0(x)\big\},\end{aligned}\tag{10.48}$$

with the norm

$$|\!|\!| v |\!|\!|_T = \sup_{[0,T]}\|v(t)\|_{s,2} + \sum_{|\alpha|\le s_1}\sup_{[0,T]}\|\langle x\rangle^2\,\partial_x^\alpha v(t)\|_2 \le M_0. \tag{10.49}$$

For $u \in X_{T,M_0}$, we study the linear IVP:

$$\begin{cases} \partial_t v = -\epsilon\, \Delta^2 v + \mathcal{L}(u)v + f(x,t), & \epsilon \in (0,1) \\ v(x,0) = u_0(x), \end{cases} \tag{10.50}$$

and its integral equation version

$$v(t) = e^{-\epsilon t\,\Delta^2}\, u_0 + \int_0^t e^{-\epsilon(t-t')\Delta^2}\,(\mathcal{L}(u)v + f)(t')\,dt'. \tag{10.51}$$

One defines the operator $\Phi(u)(t)$ as the right-hand side of (10.51). Using that

$$\|e^{-\epsilon t\,\Delta^2} g\|_2 \le \|g\|_2 \qquad \text{and} \qquad \|\Delta\, e^{-\epsilon t\,\Delta^2} g\|_2 \le \frac{1}{\epsilon^{1/2} t^{1/2}}\, \|g\|_2,$$

it is easy to check that the operator $\Phi(\cdot)$ is a contraction on X_{T_ϵ,M_0} with $T_\epsilon = O(\epsilon)$. One needs standard commutator identities to estimate the weighted norms in $X_{T,M}$. Thus, there exists $u^\epsilon \in X_{T_\epsilon,M_0}$ (the fixed point of Φ) solution of the IVP:

$$\begin{cases} \partial_t u = -\epsilon\, \Delta^2 u + \mathcal{L}(u)u + f(x,t), & \epsilon \in (0,1), \\ u(x,0) = u_0(x), \end{cases} \tag{10.52}$$

on the time interval $[0, T_\epsilon]$.

Now we will use Lemma 10.1 to extend all solutions $\{u^\epsilon\}_{\epsilon\in(0,1)}$ to the time interval $[0, T_0]$ with T_0 independent of $\epsilon \in (0,1)$, with $|||u^\epsilon|||_{T_0}$ uniformly bounded.

The first step is to show that if $|||u^\epsilon|||_T \le M_0 = 20\, c_0\, \lambda$ (see (10.39)), the coefficients of the linear equation for $\Lambda^{2m} u^\epsilon = (I - \Delta)^m u^\epsilon$, $2m \le s$, and $x_l^2\, \Lambda^{2m} u$ with $2m \le s_1$ (assuming s, s_1 even integers) can be written so that the constants in $(\text{H}_l 1)$–$(\text{H}_l 5)$ are uniform for all these equations in a time interval $[0, \widetilde{T}]$ independent of ϵ.

The equations for $\Lambda^{2m} u^\epsilon$ are obtained by applying the operator Λ^{2m} to the equation (10.52), which can be written as

$$\begin{aligned} \partial_t \Lambda^{2m} u = &- \epsilon\, \Delta^2 \Lambda^{2m} u + i\,\mathcal{L}_{2m}(u)\Lambda^{2m} u \\ &+ f_{2m}(x,t,(\partial^\beta u)_{|\beta|\le 2m-1},(\partial^\beta \bar{u})_{|\beta|\le 2m-1}) + \Lambda^{2m} f(x,t), \end{aligned} \tag{10.53}$$

where

$$\begin{aligned} \mathcal{L}_{2m}(u)v = &\, i a_{jk}(x,t,u,\bar{u},\nabla u,\nabla \bar{u})\partial^2_{x_j x_k} v \\ &+ b_{2m,2,j}(x,t,(\partial^\beta u)_{|\beta|\le 2m-1},(\partial^\beta \bar{u})_{|\beta|\le 2m-1})\, R_j \partial_{x_j} v \\ &+ b_{2m,2,j}(x,t,(\partial^\beta u)_{|\beta|\le 2m-1},(\partial^\beta \bar{u})_{|\beta|\le 2m-1})\, \widetilde{R}_j \partial_{x_j} \bar{v} \\ &+ c_{1,2m}(x,t,(\partial^\beta u)_{|\beta|\le 2m-1},(\partial^\beta \bar{u})_{|\beta|\le 2m-1})\, R_{2m,1}\, v \\ &+ c_{2,2m}(x,t,(\partial^\beta u)_{|\beta|\le 2m-1},(\partial^\beta \bar{u})_{|\beta|\le 2m-1})\, R_{2m,2}\, \bar{v}, \end{aligned} \tag{10.54}$$

where R_j, $\widetilde{R}_j$, $R_{2m,1}$, $R_{2m,2}$ are ψ.d.o. of order zero.

We observe that the principal part of $\mathcal{L}_{2m}(u)$ is independent of m. Moreover, the first-order coefficients $b_{2m,1,j}$, $b_{2m,2,j}$ depend on $2m$ as a multiplicative constant, and on the original coefficients a_{jk}, $\overrightarrow{b_1}$, $\overrightarrow{b_2}$ and their first derivatives and they verify the asymptotic flatness assumptions (H_l2). The term $f_{2m}(\cdot)$ involves derivatives that have been previously estimated in $L^\infty_T L^2_x$, and so putting it on the right-hand side in the $L^1_T L^2_x$-norm they appear with a factor T in front.

Similar remarks hold for the equation for $x_l^2 \Lambda^{2m} u$ after using some simple commutator identities.

Collecting this information, we can also show that there exists a $Q(\cdot)$ increasing function such that, for any $\omega \in X_{T,M_0}$ with $T > 0$ solution of the IVP (10.52),

$$\sup_{[0,T]} \sum_{|\alpha| \leq s_1 - 4} \|\langle x\rangle^2 \partial_x^\alpha \partial_t \omega\|_2 \leq Q(M_0) \tag{10.55}$$

holds.

All these facts will allow us to apply Lemma 10.1 to get the a priori estimate

$$|||u^\epsilon||| \leq c_0(\lambda + \widetilde{T} R(M_0)) \leq M_0/4 \tag{10.56}$$

for $\widetilde{T}$ small, but uniform in ϵ, where $R(\cdot)$ is a fixed increasing function. Thus, we can reapply the local existence theorem (originally on $[0, T_\epsilon]$) to extend the local solution u^ϵ to the time interval $[0, \widetilde{T}]$, with

$$|||u^\epsilon||| \leq M_0 = 20 c_0 \lambda. \tag{10.57}$$

Once (10.57) has been established (as in the viscosity method for the BO explained in Chapter 9), we consider the equation for the difference $u^\epsilon - u^{\epsilon'}$, $\epsilon > \epsilon' > 0$, and reapply the argument to obtain the existence as $\epsilon \to 0$ and the uniqueness of the solution. The continuous dependence is based on Bona–Smith regularization argument [BS] (see Step 5 in the proof of Theorem 9.2).

Now we turn our attention to the proof of Lemma 10.1. One of the main ingredients in the proof is the following lemma due to Doi [Do1].

Lemma 10.2 [Do1]. *Assume that h in (10.38) verifies the assumptions (H_l5) (nontrapping), (H_l4) (regularity) and (H_l2) (asymptotic flatness). Then, there exists a real-valued zeroth-order classical symbol $p \in S^0$ (see (3.18)) whose seminorm is bounded in terms of the "nontrapping character" of h, the ellipticity constant γ in (H_l1), and the bound for the smoothness norm at $t = 0$, c_1, and a constant $\beta \in (0, 1)$ (with the same dependence) such that*

$$H_h p = \{h, p\} \geq \beta \frac{|\xi|}{\langle x\rangle^2} - \frac{1}{\beta} \tag{10.58}$$

for all $(x, \xi) \in \mathbb{R}^n \times \mathbb{R}$.

Remark 10.2. The seminorm bounds for the symbol p and the constant β above are the quantitative way in which the "nontrapping" character of h enters into the proof.

We recall that

$$H_h\, p = \{h,\, p\} = \partial_{\xi_j} h\, \partial_{x_j} p - \partial_{x_j} h\, \partial_{\xi_j} p. \tag{10.59}$$

Observe that, if $\tilde{h}$ is only "approximately nontrapping" and we use the p corresponding to H_a, for $H_{\tilde{h}}$, we get a lower bound of $H_{\tilde{h}}\, p$ by $\beta|\xi|/2\langle x\rangle^2 - 2/\beta$.

To apply Doi's lemma, we need the sharp Garding inequality (see [Ho2]).

Lemma 10.3 (Sharp Garding's inequality). *Let $q \in S^1$ be a classical symbol of order 1 such that $\mathcal{R}e\, q(x,\xi) \geq 0$ for $|\xi| \geq R$, then there exist $j_0 = j_0(n)$ and $c = c(n, R)$ such that*

$$\mathcal{R}e\, \langle \Psi_q f, f\rangle \geq -c\, \|q\|_{S^1}^{(j_0)}\, \|f\|, \tag{10.60}$$

where Ψ_q denotes the ψ.d.o. with symbol q, i.e.,

$$\Psi_q\, f(x) = \int e^{ix\xi} q(x,\xi) \widehat{f}(\xi)\, d\xi. \tag{10.61}$$

Assuming Lemmas 10.2 and 10.3 we shall divide the proof of Lemma 10.1 into several steps.

Step 1. Write equation (10.40) as a system. Using

$$\overrightarrow{w} = \begin{pmatrix} u \\ \bar{u} \end{pmatrix}, \qquad \overrightarrow{f} = \begin{pmatrix} f \\ \bar{f} \end{pmatrix}, \qquad \overrightarrow{w}_0 = \begin{pmatrix} u_0 \\ \bar{u}_0 \end{pmatrix},$$

one has the system

$$\begin{cases} \partial_t \overrightarrow{w} = -\epsilon \Delta^2 I \overrightarrow{w} + (iH + B + C)\overrightarrow{w} + \overrightarrow{f}, \\ \overrightarrow{w}(x,0) = \overrightarrow{w}_0(x), \end{cases}$$

where

$$H = \begin{pmatrix} \mathcal{L} & 0 \\ 0 & -\mathcal{L} \end{pmatrix}, \qquad C = \begin{pmatrix} c_1 & c_2 \\ \bar{c}_2 & \bar{c}_1 \end{pmatrix}, \tag{10.62}$$

$$B = \begin{pmatrix} \overrightarrow{b_1} \cdot \nabla & \overrightarrow{b_2} \cdot \nabla \\ \overline{\overrightarrow{b_2}} \cdot \nabla & \overline{\overrightarrow{b_1}} \cdot \nabla \end{pmatrix} = \begin{pmatrix} B_{11} & B_{12} \\ B_{21} & B_{22} \end{pmatrix}, \tag{10.63}$$

and $\mathcal{L} = i a_{jk}(x,t) \partial^2_{x_j x_k}$.

Step 2. Diagonalization of the first-order terms. (To simplify the exposition take $\epsilon = 0$).

Notice that $\mathcal{L}$ is elliptic, with ellipticity constant $\gamma/2$ for $t \in [0,T]$ for T sufficiently small since

$$a_{jk}(x,t)\xi_j\xi_k = a_{jk}(x,0)\xi_j\xi_k + [a_{jk}(x,t) - a_{jk}(x,0)]\xi_j\xi_k \geq \gamma|\xi|^2 - c\,T|\xi|^2 \tag{10.64}$$

(by using the bound of $\partial_t a_{jk}(x,t)$ in (H_l2)).

This type of argument shall be used repeatedly.

Next we write

$$B = B_{diag} + B_{anti} = \begin{pmatrix} B_{11} & 0 \\ 0 & B_{22} \end{pmatrix} + \begin{pmatrix} 0 & B_{12} \\ B_{21} & 0 \end{pmatrix}. \tag{10.65}$$

Our goal is to eliminate B_{anti}. To do this we set

$$\Lambda = I - S, \quad \text{with} \quad S = \begin{pmatrix} 0 & S_{12} \\ S_{21} & 0 \end{pmatrix}$$

where S_{12}, S_{21} are ψ.d.o. of order -1 to be determined.

We want to write the system in the new variable

$$\vec{z} = \Lambda\vec{w} \tag{10.66}$$

for an appropriate choice of S, so that B_{anti} is eliminated.

We will use that S is a matrix of ψ.d.o. of order -1, to have that Λ is invertible in L^2 and so the estimates on $\vec{z}$ are equivalent to the estimates on $\vec{w}$.

We calculate

$$\begin{aligned}
&\begin{pmatrix} \mathcal{L} & 0 \\ 0 & -\mathcal{L} \end{pmatrix}\Lambda - \Lambda\begin{pmatrix} \mathcal{L} & 0 \\ 0 & -\mathcal{L} \end{pmatrix} \\
&\quad = -\begin{pmatrix} \mathcal{L} & 0 \\ 0 & -\mathcal{L} \end{pmatrix}\begin{pmatrix} 0 & S_{12} \\ S_{21} & 0 \end{pmatrix} + \begin{pmatrix} 0 & S_{12} \\ S_{21} & 0 \end{pmatrix}\begin{pmatrix} \mathcal{L} & 0 \\ 0 & -\mathcal{L} \end{pmatrix} \\
&\quad = \begin{pmatrix} 0 & -\mathcal{L}S_{12} - S_{12}\,\mathcal{L} \\ \mathcal{L}S_{21} + S_{21}\,\mathcal{L} & 0 \end{pmatrix}.
\end{aligned}$$

Since

$$|h(x,\xi)| = |a_{jk}(x,t)\xi_j\xi_k| \geq \gamma|\xi|^2 \quad \text{for} \quad |\xi| \geq R \tag{10.67}$$

uniformly in t, choosing $\varphi \in C_0^\infty(\mathbb{R}^n)$ with $\varphi(y) = 1$ if $|y| \leq 1$ and $\varphi(y) = 0$ if $|y| \geq 2$ we define

$$\widetilde{h}(x,t,\xi) = (h(x,\xi))^{-1}(1 - \varphi(\xi/R)) \tag{10.68}$$

and $\widetilde{\mathcal{L}} = \Psi_{\widetilde{h}}$, i.e., the ψ.d.o. of order -2 with symbol $\widetilde{h}$. Notice that

$$\widetilde{\mathcal{L}}\,\mathcal{L} = I + \Psi_{\ell_1} \tag{10.69}$$

with $\ell_1 \in S^{-1}$ (uniformly in t). Define

$$S_{12} = \frac{1}{2}\, i B_{12}\, \widetilde{\mathcal{L}}, \qquad S_{21} = -\frac{1}{2} i\, B_{21}\, \widetilde{\mathcal{L}} \tag{10.70}$$

and

$$S = \begin{pmatrix} 0 & S_{12} \\ S_{21} & 0 \end{pmatrix}. \tag{10.71}$$

Notice that the entries of S are ψ.d.o of order -1, whose S^0 seminorms tend to zero as $R \uparrow \infty$ (see (10.68)). Thus, for R large enough Λ is invertible in $H^s(\langle x \rangle^2\, dx)$, $H^s(\langle x \rangle^{-2}\, dx)$, and $H^s(\mathbb{R}^n)$ with operator norm in the interval $(1/2, 2)$. Also if Λ^{-1} denotes the inverse of Λ, the entries of Λ^{-1} are ψ.d.o. of order zero.

Finally, from our construction

$$\begin{cases} -\mathcal{L}\, S_{12} - S_{12}\, \mathcal{L} = -B_{12} + \text{ order } 0, \\ \ \ \mathcal{L}\, S_{21} - S_{21}\, \mathcal{L} = -B_{21} + \text{ order } 0. \end{cases} \tag{10.72}$$

We then observe that

$$\begin{aligned} \Lambda B_{diag} &= (I - S)\, B_{diag} = B_{diag} - S\, B_{diag} \\ &= B_{diag} \Lambda + B_{diag}\, S - S\, B_{diag} \\ &= B_{diag} \Lambda + [(B_{diag}\, S - S\, B_{diag}) \Lambda^{-1}]\, \Lambda, \end{aligned} \tag{10.73}$$

(notice that $[\,\cdot\,]$ is an operator of order zero).

Similarly,

$$\Lambda\, B_{anti} = B_{anti}\, \Lambda + C\, \Lambda, \tag{10.74}$$

by (10.70), (10.71), (10.72) with C a matrix of ψ.d.o. of order zero.

Thus, our system in $\vec{z} = \Lambda \vec{w}$ becomes

$$\begin{cases} \partial_t \vec{z} = i H \vec{z} + B_{diag}\, \vec{z} + C\, \vec{z} + \vec{g}, \\ \vec{z}(x, 0) = \vec{z_0}(x), \end{cases} \tag{10.75}$$

where $\vec{g} = \Lambda \vec{f}$, $\vec{z}_0 = \Lambda \vec{w}_0$, H, B_{diag} as before with $B_{11} = \vec{b_1} \cdot \nabla$, $B_{22} = \overline{\vec{b_1}} \cdot \nabla$ and C is a matrix of ψ.d.o. of order zero whose symbols have seminorm estimates controlled by c (not c_1).

Step 3. Energy estimates for a "gauged" system. We recall that $\mathcal{L} = a_{jk}(x,t)\,\partial^2_{x_j x_k}$ has symbol $a_{jk}(x,t)\,\xi_j\xi_k$, and our "nontrapping" assumption is on

$$h(x,\xi) = a_{jk}(x,0)\xi_j\xi_k. \tag{10.76}$$

Let $p \doteq S^0$ be the symbol associated to h as in Lemma 10.2 so that

$$H_h\, p = \{h, p\} \geq \beta \frac{|\xi|}{\langle x\rangle^2} - \frac{1}{\beta}.$$

Let

$$h_1(x,t,\xi) = a_{jk}(x,t)\xi_j\xi_k. \tag{10.77}$$

So

$$\begin{aligned} H_{h_1}\, p &= \frac{\partial h_1}{\partial \xi_l}\frac{\partial p}{\partial x_l} - \frac{\partial h_1}{\partial x_l}\frac{\partial p}{\partial \xi_l} \\ &= \frac{\partial h}{\partial \xi_l}\frac{\partial p}{\partial x_l} - \frac{\partial h}{\partial x_l}\frac{\partial p}{\partial \xi_l} + (a_{jk}(x,t) - a_{jk}(x,0))\frac{\partial}{\partial \xi_l}(\xi_j\xi_k)\frac{\partial p}{\partial x_l} \\ &- \Big(\frac{\partial}{\partial x_l}(a_{jk}(x,t) - a_{jk}(x,0))\Big)\xi_j\xi_k\frac{\partial p}{\partial \xi_l}. \end{aligned}$$

Thus, by "asymptotic flatness," assumption ($H_l 2$), we see that for small T

$$H_{h_1}\, p \geq \frac{\beta}{2}\frac{|\xi|}{\langle x\rangle^2} - \frac{2}{\beta}, \tag{10.78}$$

for the same p. We now consider the ψ.d.o. of order 0, Ψ_{r_1}, whose symbol is $e^{Mp(x,t)}$ for an M large to be determined depending only on c_1, and the "nontrapping character." Notice that the seminorms of r_1 depend only on c_1, and the nontrapping character: it is elliptic. The same holds for $\Psi_{r_1}^{-1}$ (modulo order -2 errors). Also

$$H_{h_1}\, r_1 = M(H_{h_1}\, p)\, r_1 \geq \Big\{M\,\frac{\beta}{2}\frac{|\xi|}{\langle x\rangle^2} - \frac{2M}{\beta}\Big\}\, r_1 \tag{10.79}$$

and that modulo 0th-order operators the symbol of $i\{\mathcal{L}\Psi_{r_1} - \Psi_{r_1}\mathcal{L}\} = H_{h_1}\, r_1$, for each t.

We also consider Ψ_{r_2}, whose symbol is $e^{-Mp(x,\xi)}$ so that symbol-wise we have

$$i\{\mathcal{L}\Psi_{r_2} - \Psi_{r_2}\mathcal{L}\} = H_{h_1} r_2 = -M(H_{h_1} p) r_2 \leq -\Big\{M\frac{\beta}{2}\frac{|\xi|}{\langle x\rangle^2} - \frac{2M}{\beta}\Big\}\, r_2.$$

Now we define

$$\vec{\alpha} = \begin{pmatrix} \Psi_{r_1} & 0 \\ 0 & \Psi_{r_2} \end{pmatrix} \vec{z},$$

and obtain a new system for $\overrightarrow{\alpha}$, in which for M chosen appropriately we will be able to perform energy estimates. For simplicity, let

$$\Psi = \begin{pmatrix} \Psi_{r_1} & 0 \\ 0 & \Psi_{r_2} \end{pmatrix}.$$

We want to obtain the system for $\overrightarrow{\alpha}$

$$i\{\Psi H - H\Psi\} = i \begin{pmatrix} \Psi_{r_1}\mathcal{L} - \mathcal{L}\Psi_{r_1} & 0 \\ 0 & \mathcal{L}\Psi_{r_2} - \Psi_{r_2}\mathcal{L} \end{pmatrix}$$

$$= \begin{pmatrix} \Psi_{H_{h_1} r_1} & 0 \\ 0 & \Psi_{H_{h_1} r_2} \end{pmatrix} + \text{ 0th order.}$$

Recalling that $-H_{h_1}r_1 = -M(H_{h_1}p)r_1$, $H_{h_1}\, r_2 = -M(H_{h_1}\, p)\, r_2$, and using that

$$\begin{pmatrix} \Psi_{H_{h_1} r_1} & 0 \\ 0 & \Psi_{H_{h_1} r_2} \end{pmatrix} = \begin{pmatrix} -M\Psi_{H_{h_1} p} & 0 \\ 0 & -M\Psi_{H_{h_1} p} \end{pmatrix} \begin{pmatrix} \Psi_{r_1} & 0 \\ 0 & \Psi_{r_2} \end{pmatrix} + \text{ 0th order}$$

we get

$$i\{\Psi H - H\Psi\} = \begin{pmatrix} -M\Psi_{H_{h_1} p} & 0 \\ 0 & -M\Psi_{H_{h_1} p} \end{pmatrix} \Psi + \text{ 0th order.}$$

Next,

$$\Psi\, B_{diag} = \begin{pmatrix} B_{11} & 0 \\ 0 & B_{22} \end{pmatrix} \Psi + \text{ 0th order}$$

and

$$\Psi\, C = (\Psi\, C \Psi^{-1})\Psi.$$

Thus, the system for $\overrightarrow{\alpha}$ is:

$$\begin{cases} \partial_t \overrightarrow{\alpha} = iH\overrightarrow{\alpha} + B_{diag}\, \overrightarrow{\alpha} - M \begin{pmatrix} \Psi_{H_{h_1} p} & 0 \\ 0 & \Psi_{H_{h_1} p} \end{pmatrix} \overrightarrow{\alpha} + C\overrightarrow{\alpha} + \overrightarrow{F}, \\ \overrightarrow{\alpha}(x,0) = \overrightarrow{\alpha_0}(x), \end{cases}$$

where $\overrightarrow{\alpha_0} = \Psi \overrightarrow{z_0}$ and $\overrightarrow{F} = \Psi \overrightarrow{g}$. It suffices to find M (see definition of Ψ_{r_1}) depending only on c_1 and the nontrapping character so that we can estimate $\overrightarrow{\alpha}$. To do this we consider

$$\langle \overrightarrow{\alpha}, \overrightarrow{\beta} \rangle = \int \alpha_1 \bar{\beta}_1 + \alpha_2 \bar{\beta}_2$$

and calculate

$$\begin{aligned}
\frac{d}{dt} \langle \overrightarrow{\alpha}, \overrightarrow{\alpha} \rangle &= i\big[\langle H \overrightarrow{\alpha}, \overrightarrow{\alpha} \rangle - \langle \overrightarrow{\alpha}, H \overrightarrow{\alpha} \rangle\big] + \langle B_{diag} \overrightarrow{\alpha}, \overrightarrow{\alpha} \rangle + \langle \overrightarrow{\alpha}, B_{diag} \overrightarrow{\alpha} \rangle \\
&\quad - M\Big[\Big\langle \begin{pmatrix} \Psi_{H_{h_1} p} & 0 \\ 0 & \Psi_{H_{h_1} p} \end{pmatrix} \overrightarrow{\alpha}, \overrightarrow{\alpha} \Big\rangle + \Big\langle \overrightarrow{\alpha}, \begin{pmatrix} \Psi_{H_{h_1} p} & 0 \\ 0 & \Psi_{H_{h_1} p} \end{pmatrix} \overrightarrow{\alpha} \Big\rangle\Big] \\
&\quad + \langle C \overrightarrow{\alpha}, \overrightarrow{\alpha} \rangle + \langle \overrightarrow{\alpha}, C \overrightarrow{\alpha} \rangle + \langle \overrightarrow{F}, \overrightarrow{\alpha} \rangle + \langle \overrightarrow{\alpha}, \overrightarrow{F} \rangle \\
&= i\big[\langle H \overrightarrow{\alpha}, \overrightarrow{\alpha} \rangle - \langle \overrightarrow{\alpha}, H \overrightarrow{\alpha} \rangle\big] + 2\mathcal{R}e\, \langle B_{diag} \overrightarrow{\alpha}, \overrightarrow{\alpha} \rangle \\
&\quad - 2M\mathcal{R}e \Big\langle \begin{pmatrix} \Psi_{H_{h_1} p} & 0 \\ 0 & \Psi_{H_{h_1} p} \end{pmatrix} \overrightarrow{\alpha}, \overrightarrow{\alpha} \Big\rangle + 2\mathcal{R}e\, \langle C \overrightarrow{\alpha}, \overrightarrow{\alpha} \rangle \\
&\quad + 2\mathcal{R}e\, \langle \overrightarrow{F}, \overrightarrow{\alpha} \rangle.
\end{aligned}$$

We recall that

$$H = \begin{pmatrix} \mathcal{L} & 0 \\ 0 & -\mathcal{L} \end{pmatrix},$$

with

$$\mathcal{L} = a_{jk}(x,t) \partial^2_{x_j x_k} = \partial_{x_j}(a_{jk}(x,t) \partial_{x_k}) - \partial_{x_j} a(x,t) \partial_{x_k} = \mathcal{L}_0 - \overrightarrow{b_3}(x,t) \cdot \nabla.$$

So

$$\begin{aligned}
i\big[\langle H \overrightarrow{\alpha}, \overrightarrow{\alpha} \rangle - \langle \overrightarrow{\alpha}, H \overrightarrow{\alpha} \rangle\big] &= i[\langle H_0 \overrightarrow{\alpha}, \overrightarrow{\alpha} \rangle - \langle \overrightarrow{\alpha}, H_0 \overrightarrow{\alpha} \rangle] \\
&\quad + \langle i(\overrightarrow{b_3}(x,t) \cdot \nabla) \overrightarrow{\alpha}, \overrightarrow{\alpha} \rangle + \langle \overrightarrow{\alpha}, -i(\overrightarrow{b_3}(x,t) \cdot \nabla) \overrightarrow{\alpha} \rangle \\
&= [\langle H_0 \overrightarrow{\alpha}, \overrightarrow{\alpha} \rangle - \langle \overrightarrow{\alpha}, H_0 \overrightarrow{\alpha} \rangle] + 2\mathcal{R}e\, \langle B^1_{diag} \overrightarrow{\alpha}, \overrightarrow{\alpha} \rangle,
\end{aligned}$$

where

$$H_0 = \begin{pmatrix} \mathcal{L}_0 & 0 \\ 0 & -\mathcal{L}_0 \end{pmatrix}, \qquad B^1_{diag} = \begin{pmatrix} i\,\overrightarrow{b_3}(x,t) \cdot \nabla & 0 \\ 0 & -i\,\overrightarrow{b_3}(x,t) \cdot \nabla \end{pmatrix}.$$

Note that our asymptotic flatness assumption implies that

$$\widetilde{B}_{diag} = B_{diag} + B^1_{diag} = \begin{pmatrix} \widetilde{B}_{11} & 0 \\ 0 & \widetilde{B}_{11}, \end{pmatrix},$$

where the symbols of $\widetilde{B}_{jj}$, $l = 1, 2$ satisfy

$$|\partial_t \widetilde{B}_{jj}(x,t,\xi)| \leq c \frac{|\xi|}{\langle x \rangle^2},$$

and

$$|\widetilde{B}_{jj}(x,0,\xi)| \leq c_1 \frac{|\xi|}{\langle x \rangle^2}.$$

As a consequence, for t small (depending on c) we have that

$$|\widetilde{B}_{jj}(x,t,\xi)| \leq 2c_1 \frac{|\xi|}{\langle x \rangle^2},$$

and

$$\begin{aligned}
\frac{d}{dt}\langle \overrightarrow{\alpha}, \overrightarrow{\alpha} \rangle &= i[\langle H_0 \overrightarrow{\alpha}, \overrightarrow{\alpha} \rangle - \langle \overrightarrow{\alpha}, H_0 \overrightarrow{\alpha} \rangle] \\
&\quad + \mathcal{R}e\, \langle \widetilde{B}_{diag} \overrightarrow{\alpha}, \overrightarrow{\alpha} \rangle - 2M\, \mathcal{R}e \left\langle \begin{pmatrix} \Psi_{H_{h_1} p} & 0 \\ 0 & \Psi_{H_{h_1} p} \end{pmatrix} \overrightarrow{\alpha}, \overrightarrow{\alpha} \right\rangle \\
&\quad + 2\mathcal{R}e\, \langle C \overrightarrow{\alpha}, \overrightarrow{\alpha} \rangle + 2\mathcal{R}e\, \langle \overrightarrow{F}, \overrightarrow{\alpha} \rangle.
\end{aligned} \tag{10.80}$$

Now it is easy to see that

$$\langle H_0 \overrightarrow{\alpha}, \overrightarrow{\alpha} \rangle - \langle \overrightarrow{\alpha}, H_0 \overrightarrow{\alpha} \rangle \equiv 0.$$

For the next two terms in (10.80), we get

$$\mathcal{R}e\, \langle \widetilde{B}_{11} \alpha_1, \alpha_1 \rangle - M\, \langle \Psi_{H_{h_1} p}\, \alpha_1, \alpha_2 \rangle \quad + \quad \text{similar terms in } \alpha_2.$$

We recall that

$$|\widetilde{B}_{11}(x,t,\xi)| \leq c_1 \frac{|\xi|}{\langle x \rangle^2}$$

and

$$H_{h_1}\, p \geq \frac{\beta}{2} \frac{|\xi|}{\langle x \rangle^2} - \frac{2}{\beta}.$$

We now choose M so large such that

$$M\, H_{h_1}\, p \pm \widetilde{B}_{11}(x,t,\xi) \leq \beta - \widetilde{\beta} \frac{\langle \xi \rangle}{\langle x \rangle^2},$$

then the sharp Garding inequality (Lemma 10.3) gives

$$\mathcal{R}e\, \langle \widetilde{B}_{11} \alpha_1, \alpha_1 \rangle - M\, \mathcal{R}e\, \langle \Psi_{H_{h_1} p}\, \alpha_1, \alpha_1 \rangle \leq c \|\alpha_1\|_2^2 - \langle \Psi_{\widetilde{\beta}\, \langle \xi \rangle / \langle x \rangle^2} \alpha_1, \alpha_1 \rangle.$$

Using that

$$\Psi_{\tilde{c}\langle\xi\rangle/\langle x\rangle^2} = \frac{1}{\langle x\rangle^2}\Psi_{\tilde{c}\langle\xi\rangle} + \Psi_{e_0},$$

with e_0 of order 0,

$$\Psi_{\tilde{c}\langle\xi\rangle} = \tilde{c}\Lambda^{1/2}\Lambda^{1/2},$$

and

$$\frac{1}{\langle x\rangle^2}\Lambda^{1/2}\Lambda^{1/2} = \Lambda^{1/2}\frac{1}{\langle x\rangle^2}\Lambda^{1/2} + \Psi_{e_0^1},$$

with e_0^1 of order 0, we see that

$$\langle \Psi_{\tilde{\beta}\langle\xi\rangle/\langle x\rangle^2}\alpha_1, \alpha_1\rangle = \tilde{\beta}\int \frac{|\Lambda^{1/2}\alpha_1|^2}{\langle x\rangle^2}(x,t)dx + O(\|\alpha_1\|_2).$$

So we pick $t_0 \in [0,T]$ such that

$$\|\overrightarrow{\alpha}(t_0)\|_2^2 \geq \frac{1}{2}\sup_{[0,T]}\|\overrightarrow{\alpha}(t)\|_2^2,$$

to get that

$$\begin{aligned}
\sup_{[0,T]}\|\overrightarrow{\alpha}(t)\|_2^2 + \tilde{\beta}\int_0^T\int \frac{|\Lambda^{1/2}\alpha_1|^2}{\langle x\rangle^2}(x,t)dx\,dt &\leq 2\int_0^{t_0}\frac{d}{dt}\langle\overrightarrow{\alpha},\overrightarrow{\alpha}\rangle\,dt + 2\|\overrightarrow{\alpha_0}\|_2^2 \\
&\leq c\int_0^{t_0}\|\overrightarrow{\alpha}\|_2^2\,dt + 2\int_0^{t_0}\|\overrightarrow{F}\|_2\|\overrightarrow{\alpha}\|_2\,dt + 2\|\overrightarrow{\alpha_0}\|_2^2 \\
&\leq CT\sup_{[0,T]}\|\overrightarrow{\alpha}(t)\|_2^2 + 2\sup_{[0,T]}\|\overrightarrow{\alpha}(t)\|_2^2\int_0^T\|\overrightarrow{F}\|\,dt + 2\|\overrightarrow{\alpha_0}\|_2^2,
\end{aligned}$$

which, upon choosing $CT < 1/2$ yields the desired estimate (10.46).

10.2 Comments

The main result in this chapter, Theorem 10.1, was obtained in [KPV10]. As it was seen, the proof is based on the artificial viscosity method (it cannot be established by a solely Picard argument; see [KT]) and uses only classical pseudodifferential operators. In this regard, the ellipticity assumption is crucial.

It was displayed in Chapter 9 that dispersive models of the form

$$\partial_t u = i(\partial_{x_1}^2 + \cdots + \partial_{x_k}^2 - \partial_{x_{k+1}}^2 - \cdots - \partial_{x_n}^2)u + f(u,\bar{u},\nabla_x u,\nabla_x\bar{u}) \qquad (10.81)$$

arise in the physical (for instance, wave propagation) and in the mathematical context, for example, related to higher-order models which can be solved by inverse scattering method.

In [KPV14], local well-posedness of the IVP associated to the equation (10.81) was obtained. The method of proof, among other arguments, employs pseudodifferential operators in the Calderón–Vaillancourt class. This approach does not seem to apply to the variable coefficient class

$$\begin{aligned} \partial_t u = i\,\partial_{x_k}(a_{jk}(x)\partial_{x_j}u) + (\overrightarrow{b_1}(x)\cdot\nabla)u + (\overrightarrow{b_2}(x)\cdot\nabla)\bar{u} \\ +c_1(x)u + c_2(x)u + f(u,\bar{u},\nabla_x u,\nabla_x\bar{u}), \end{aligned} \tag{10.82}$$

where $(a_{jk}(x))$ is a symmetric nondegenerate (invertible) matrix.

The local well-posedness of the IVP associated to equation (10.82) was studied in the massive work [KPRV1]. For that purpose, a new class of pseudodifferential operators was introduced, which takes into consideration the "geometry" of the nonelliptic operator. Under asymptotic flatness hypothesis of the coefficient $a_{jk}(x)$, $b_{1j}, b_{2j},\ k,j = 1,\dots,n$, and nontrapping assumptions of the bicharacteristic flow associated to the symbol $a_{jk}(x)\xi_k\xi_j$ it was proved in [KPV14] that the IVP for (10.82) is locally well-posed in weighted Sobolev spaces $\mathcal{F}^s_{2k} = H^s(\mathbb{R}^n)\cap L^2(|x|^{2k}\,dx)$ for large enough values of $s,k\in\mathbb{Z}^+$ $(s>k)$. The results in [KPRV1] were extended in [KPRV2] to the case where the coefficients $a_{jk}(\cdot)$, $b_{1j}(\cdot)$, $b_{2j}(\cdot)$ depend on $(x,t,u,\bar{u},\nabla_x u,\nabla_x\bar{u})$, $k,j=1,\dots,n$, and $c_1(\cdot)$, $c_2(\cdot)$ on $(x,t,u,\bar{u})$.

In [MMTa2] and [MMTa3], the problem of finding the minimal regularity assumptions required to guarantee local well-posedness was considered. In these works, the setting was restricted to the small data regime. In [MMTa2], for the quadratic case, a translation invariant space was used instead of the weighted one. One should remark that in the small data case the crucial step in the proof presented here, the use of the integrating factor, is not necessary. Similarly, in the small data case the nontrapping condition is not relevant and the proof follows by applying the contraction mapping principle which cannot be the case for data of arbitrary size (see [KT]).

10.3 Exercises

10.1 Fill out the details of the results discussed in (i)–(v) regarding the IVP (10.7).

10.2 Prove that $f = f(u,\bar{u},\nabla\bar{u})$, $n\ge 1$, and $f = \partial_x(|u|^2u)$, $n=1$, satisfy the inequality (10.19).

10.3 Assuming that $f(u,\bar{u},\nabla u,\nabla\bar{u})$ satisfies the inequality (10.19), sketch a local existence proof for the IVP:

$$\begin{cases} \partial_t u = i\epsilon\,\Delta u + f(u,\bar{u},\nabla u,\nabla\bar{u}), \\ u(x,0) = u_0(x)\in H^s(\mathbb{R}^n), \quad s > n/2+1, \end{cases} \tag{10.83}$$

for $\epsilon \geq 0$. This shows that under the hypothesis (10.19) the dispersion is not needed for a local theory.

10.4 Consider the IVP:

$$\begin{cases} \partial_t u = i\Delta u + P(u, \bar{u}, \nabla u, \nabla \bar{u}), \\ u(x,0) = u_0(x), \end{cases} \tag{10.84}$$

$x \in \mathbb{R}^n$, $t \in \mathbb{R}$, where $P : \mathbb{C}^n \mapsto \mathbb{C}$ is a polynomial such that

$$P(z) = \sum_{|\alpha|=3}^{N} a_\alpha \, z^\alpha.$$

Prove that there exists $\delta > 0$ (small) and $s \gg 1$ such that for any $u_0 \in H^s(\mathbb{R})$ with $\delta_0 \equiv \|u_0\|_{s,2} \leq \delta$ the IVP (10.84) has a unique solution:

$$u \in C([0,T] : H^s(\mathbb{R}^n)), \quad D^{s+1/2}u \in L^2_{\text{loc}}(\mathbb{R}^n \times [0,T]) \tag{10.85}$$

(with $T = T(\delta_0) \uparrow \infty$ as $\delta_0 \downarrow 0$) which can be obtained by a fixed-point argument. Hint: Consider the equivalent integral equation form of the IVP (10.84) and prove that the operator

$$\Phi_{u_0}(u)(t) = e^{it\Delta}u_0 + \int_0^t e^{i(t-t')\Delta} P(u, \bar{u}, \nabla u, \nabla \bar{u})(\cdot, t')\,dt'$$

has a unique fixed point in an appropriate space

$$X_T^s \hookrightarrow C([0,T] : H^s(\mathbb{R}^n))$$

by using the estimates in Theorems 4.2 and 4.3.

10.5 (i) Prove that the symbol in (10.28) for $|\xi| \geq 1$ satisfies the estimate (10.29).
(ii) Prove that in addition, for $|\xi| \geq 1$ one has

$$|(x_j \partial_{x_j})^\gamma \, \partial_x^\alpha \partial_\xi^\beta \, \mu(x,\xi)| \leq c_{\alpha\beta\gamma} \, \langle x \rangle^{|\alpha|} \, |\xi|^{-|\beta|}, \tag{10.86}$$

$\gamma \in \mathbb{Z}^+, \alpha, \beta \in (\mathbb{Z}^+)^n$, where $\langle x \rangle = (1 + |x|^2)^{1/2}$.

Appendix A
Proof of Theorem 2.8

Definition A.1. For $k \in \mathbb{Z}$ let $\mathcal{Q}_k$ be the collection of cubes in $\mathbb{R}^n$ which are congruent to $[0, 2^{-k})^n$ and whose vertices lie on the lattice $(2^{-k}\mathbb{Z})^n$.

The cubes in

$$\mathcal{Q}^* = \bigcup_{k \in \mathbb{Z}} \mathcal{Q}_k \tag{A.1}$$

are called the *dyadic* cubes.

A.4 Proof of Theorem 2.8

As it was mentioned after the statement of Theorem 2.8 it suffices to show that the operator T_m is of weak type (1,1), that is, there exists $c_1 > 0$ such that for every $f \in L^1(\mathbb{R}^n)$

$$\sup_{\alpha>0} \alpha \, |\{x \in \mathbb{R}^n \ : \ |T_m f(x)| > \alpha\}| \leq c_1 \, \|f\|_1. \tag{A.2}$$

To establish (A.2) we need the Calderón–Zygmund decomposition of L^1-functions.

Lemma A.1 (Calderón–Zygmund lemma). *Let $f \in L^1(\mathbb{R}^n)$. For any $\alpha > 0$, f can be decomposed as*

$$f = g + b = g + \sum_{j=1}^{\infty} b_j \tag{A.3}$$

such that

$$|g(x)| \leq 2^n \alpha \ \ a.e \ \ x \in \mathbb{R}^n, \tag{A.4}$$

$$b_j \text{ supported in } \overline{Q}_j, \ Q_j \text{ a dyadic cube with } \int_{Q_j} b_j \, dx = 0, \tag{A.5}$$

F. Linares, G. Ponce, *Introduction to Nonlinear Dispersive Equations*,
Universitext, DOI 10.1007/978-1-4939-2181-2

$$\textit{the } Q_j's \textit{ are disjoint,} \quad \sum_{j=1}^{\infty} |Q_j| \leq \alpha^{-1} \, \|f\|_1, \tag{A.6}$$

and

$$\|g\|_1 + \sum_{j=1}^{\infty} \|b_j\|_1 \leq 6 \, \|f\|_1. \tag{A.7}$$

Proof. Assume $f \geq 0$ (otherwise $f = f^+ - f^-$ and decompose each part). Since $f \in L^1(\mathbb{R}^n)$ there exists l such that $|Q|^{-1} \int_Q f \, dy < \alpha$ for any cube of side length l.

Fix $k_0 \in \mathbb{Z}$ such that

$$2^{k_0 n} \, \|f\|_1 < \alpha.$$

We start with the family of cubes in $\mathcal{Q}_{k_0}$. Let Q^0 be one of them. Divide each side of Q^0 in two to get 2^n new dyadic cubes of side length $2^{-(k_0+1)}$. Let Q^1 be such a cube; there are two possibilities:

$$\text{(a)} \; \frac{1}{|Q^1|} \int_{Q^1} f \, dy < \alpha \quad \text{or} \quad \text{(b)} \; \frac{1}{|Q^1|} \int_{Q^1} f \, dy \geq \alpha.$$

In case (b) one stops the subdivision, noticing that

$$\alpha \leq \frac{1}{|Q^1|} \int_{Q^1} f \, dy \leq \frac{2^n}{|Q^0|} \int_{Q^0} f \, dy \leq 2^n \alpha, \tag{A.8}$$

and collecting it in a sequence Q_j.

In case (a) the subdivision process continues. Thus, if $x \notin \bigcup_j Q_j$ it follows from the Lebesgue differentiation theorem (Exercise 2.6 (ii)) that

$$f(x) \leq \alpha \text{ a.e. } x \in \mathbb{R}^n \setminus \bigcup_j Q_j. \tag{A.9}$$

Finally, we define

$$g(x) = \begin{cases} \dfrac{1}{|Q_j|} \displaystyle\int_{Q_j} f \, dy & \text{if } x \in Q_j, \\ \\ f(x) & \text{if } x \notin Q_j, \end{cases} \tag{A.10}$$

and

$$b_j(x) = (f(x) - g(x)) \chi_{Q_j}(x), \quad j \in \mathbb{Z}^+, \tag{A.11}$$

which yields the result. □

We shall denote by Q_j^* the cube having the same center as Q_j and twice its side length as

$$\Omega = \underset{j}{\cup} Q_j \text{ and } \Omega^* = \underset{j}{\cup} Q_j^* \tag{A.12}$$

with

$$|\Omega^*| \leq \sum_j |Q_j^*| = 2^n \sum_j |Q_j|. \tag{A.13}$$

Proof of inequality (A.2). First we notice that using Calderón–Zygmund lemma

$$\begin{aligned} &|\{x \in \mathbb{R}^n \ : \ |T_m f(x)| > \alpha\}| \\ &\leq |\{x \in \mathbb{R}^n \ : \ |T_m g(x)| > \alpha/2\}| + |\{x \in \mathbb{R}^n \ : \ |T_m b(x)| > \alpha/2\}| \\ &\leq |\{x \in \mathbb{R}^n \ : \ |T_m g(x)| > \alpha/2\}| + |\{x \notin \Omega^* \ : \ |T_m b(x)| > \alpha\}| + |\Omega^*| \\ &= E_1 + E_2 + E_3. \end{aligned} \tag{A.14}$$

From (A.13) and (A.6) in Calderón–Zygmung lemma we have that

$$E_3 = |\Omega^*| \leq 2^n \sum_j |Q_j| \leq 2^n \alpha^{-1} \|f\|_1. \tag{A.15}$$

Tchebychev's inequality and (A.4) in the Calderón–Zygmund lemma yield

$$\begin{aligned} E_1 &= |\{x \in \mathbb{R}^n \ : \ |T_m g(x)| > \alpha/2\}| \leq c\left(\frac{\|T_m g\|_2}{\alpha/2}\right)^2 \leq c\,\frac{\|g\|_2^2}{\alpha^2} \\ &\leq \frac{c}{\alpha^2}\|g\|_1\,\|g\|_\infty \leq \frac{c}{\alpha}\|g\|_1 \leq \frac{c}{\alpha}\|f\|_1. \end{aligned} \tag{A.16}$$

Hence, it remains to prove that

$$E_2 = |\{x \notin \Omega^* \ : \ |T_m b(x)| > \alpha/2\}| \leq c\alpha^{-1}\,\|f\|_1. \tag{A.17}$$

It will suffice to show that

$$\int_{x \notin Q_j^*} |T_m b_j(x)|\, dx \leq c\|b_j\|_1, \qquad j \in Z^+. \tag{A.18}$$

To establish (A.18) we follow the argument in [BeL].

Let $\varphi \in C_0^\infty(\{\xi \ : \ |\xi| < 2\})$, such that $\varphi(\xi) = 1$ for $|\xi| \leq 1$. Let $\beta(\xi) = \varphi(\xi) - \varphi(2\xi)$. Thus

$$\sum_{l=-\infty}^{\infty} \beta(2^{-l}\xi) = 1 \quad \text{for } \xi \neq 0. \tag{A.19}$$

If $m_l(\xi) = \beta(\xi)\, m(2^l \xi)$, then by hypothesis (2.18)

$$\int |(1-\Delta)^{s/2} m_l(\xi)|^2 \, d\xi \, < c. \tag{A.20}$$

Thus, by Plancherel's identity using the notation $K_l(x) = \widehat{m}_l(x)$, one gets that

$$\int (1+|x|^2)^s |K_l(x)|^2 \, dx \, < c, \tag{A.21}$$

which, combined with the Cauchy–Schwarz inequality yields the estimate

$$\int\limits_{\{x:\, \max\limits_m |x_m|>R\}} |K_l(x)| \, dx \, < c\, R^{n/2-s}, \tag{A.22}$$

which is a good estimate for $R >> 1$.

Reapplying the estimates (A.20) and (A.21) for $\xi_k\, m_l(\xi)$ instead of $m_l(\xi)$ one finds that

$$\int |\nabla K_l(x)| \, dx \, < c. \tag{A.23}$$

Consequently, it follows that

$$\int |K_l(x+y) - K_l(x)| \, dx \, < c|y|. \tag{A.24}$$

We observe that as a temperate distribution,

$$K(x) = \sum_{l=-\infty}^{\infty} 2^{nl} K_l(2^l x) = \sum_{l=-\infty}^{\infty} \widehat{m}_l(2^{-l} x). \tag{A.25}$$

Assume that Q_j is a cube of side R centered at the origin. From (A.22) one has that

$$\begin{aligned} \int\limits_{x \notin Q_j^*} |2^{nl} K_l(2^l \cdot) * b_j| \, dx &\leq \int\limits_{Q_j} \int\limits_{x\notin Q_j^*} |2^{nl} K_l(2^l (x-y))||b_j(y)| \, dx dy \\ &\leq \|b_j\|_1 \int\limits_{\{x:\, \max\limits_m |x_m| \geq 2^l R\}} |K_l(x)| \, dx \\ &\leq c\, (2^l R)^{n/2-s} \, \|b_j\|_1. \end{aligned} \tag{A.26}$$

Now using that $\int\limits_{Q_j} b_j dy = 0$ it follows that

$$\int\limits_{x\notin Q_j^*} 2^{nl} \int\limits_{y \in Q_j} K_l(2^l (x-y)) b_j(y) \, dy dx \tag{A.27}$$

$$= \int_{x \notin Q_j^*} 2^{nl} \int_{y \in Q_j} \left(K_l(2^l(x-y)) - K_l(2^l x)\right) b_j(y)\, dy dx.$$

Therefore, (A.24) yields

$$\int_{x \notin Q_j^*} |2^{nl} K_l(2^{nl} \cdot) * b_j| \, dx$$

$$\le \int_{y \in Q_j} \int_{x \notin Q_j^*} 2^{nl} |K_l(2^l(x-y)) - K_l(2^l x)||b_j(y)| \, dx dy \tag{A.28}$$

$$\le c\,(2^l R) \|b_j\|_1.$$

Adding in l in (A.26) for $2^l R > 1$ and in (A.28) for $2^l R \le 1$ one gets that

$$\int_{x \notin Q_j^*} |T_m b_j(x)| \, dx \le c \, \|b_j\|_1, \tag{A.29}$$

which completes the proof. □

Appendix B
Proof of Lemma 4.2

B.1 Proof of Lemma 4.2

Let

$$\Omega = \{(x, y) \in [0, 1] \times [0, 1] \mid x < y\} = \underset{j}{\cup} Q_j, \tag{B.1}$$

and

$$\mathcal{A} \equiv \{Q_j\}_j \tag{B.2}$$

where the Q_j's are disjoint dyadic cubes (see Definition A.1) such that if $\bar{Q}_j = \bar{I}_j \times \bar{J}_j$, ($I_j, J_j$ intervals), then

(i) $\#\,(\bar{Q}_j \cap \{(x, x) \mid x \in [0, 1]\}) = 1$.
(ii) $\#\,\{Q_j \subseteq \Omega \mid \text{length side of } Q_j = 2^{-k}\} = 2^{k-1}, \; k \in \mathbb{Z}^+$.

Without loss of generality assume $\|f\|_r = 1$ and define

$$F(t) = \int_{-\infty}^{t} |f(s)|^r \, ds, \tag{B.3}$$

so $F : \mathbb{R} \to [0, 1]$ is a nondecreasing continuous function.

Notice that if $s < t$, then either

$$F(s) < F(t)$$

or

$$f \equiv 0, \quad \text{a.e. in}[s, t].$$

For $I = [a, b] \subseteq [0, 1]$ one has that

$$F^{-1}([a, b]) = [A, B] \quad \text{with} \quad F(A) = a \quad \text{and} \quad F(B) = b. \tag{B.4}$$

F. Linares, G. Ponce, *Introduction to Nonlinear Dispersive Equations*,
Universitext, DOI 10.1007/978-1-4939-2181-2

Hence

$$\int_A^B |f(s)|^r \, ds = F(B) - F(A) = b - a, \tag{B.5}$$

and

$$\|f\|_{L^r(F^{-1}(I))} = |I|^{1/r}. \tag{B.6}$$

Defining

$$B(f,g) = \int_{-\infty}^{\infty}\int_{-\infty}^{\infty} K(t,s) f(s) g(t) \, ds dt \tag{B.7}$$

and

$$\tilde{B}(f,g) = \int\int_{s<t} K(t,s) f(s) g(t) \, ds dt \tag{B.8}$$

it will suffices to see that there exists $c > 0$ such that

$$|\tilde{B}(f,g)| \leq c \, \|f\|_r \|g\|_{l'}, \qquad \frac{1}{l} + \frac{1}{l'} = 1. \tag{B.9}$$

We take $\|f\|_r = \|g\|_{l'} = 1$, thus

$$\begin{aligned}
|\tilde{B}(f,g)| &= |\int\int_{s<t} K(t,s) f(s) g(t) \, ds dt| \\
&= \Big| \sum_{\substack{Q_j = I_j \times J_j \\ Q_j \in \mathcal{A}}} B(\chi_{F^{-1}(I_j)} f, \chi_{F^{-1}(J_j)} g) \Big| \\
&\leq \sum_{Q_j \in \mathcal{A}} c \, \|f\|_{L^r(F^{-1}(I_j))} \|g\|_{L^{l'}(F^{-1}(J_j))} \qquad \text{(B.10)} \\
&\leq c \sum_{k \in \mathbb{Z}^+} (2^{-k})^{1/r} \sum_{|J_j| = 2^{-k}} \|g\|_{L^{l'}(F^{-1}(J_j))} \\
&\leq c \sum_{k \in \mathbb{Z}^+} (2^{-k})^{1/r} \|g\|_{l'} \Big(\sum_{|J_j| = 2^{-k}} 1 \Big)^{1/l} \\
&\leq c \sum_{k \in \mathbb{Z}^+} (2^{-k})^{1/r} (2^{k-1})^{1/l}.
\end{aligned}$$

Since by hypotheses $-\frac{1}{r} + \frac{1}{l} < 0$, this finishes the proof. □

References

[AA1] H. Added and S. Added. Equations of Langmuir turbulence and nonlinear Schrödinger equation: smoothness and approximation, J. Funct. Anal. **79** (1988), 183–210.

[AA2] H. Added and S. Added. Existence globale de solutions fortes pour les équations de la turbulence de Langmuir en dimension 2, C. R. Acad. Sci. Paris Sér. I Math. **299** (1984), 551–554.

[ABLS] J. Angulo, J.L. Bona, F. Linares, and M. Scialom. Scaling, stability and singularities for nonlinear, dispersive wave equations: the critical case, Nonlinearity **15** (2002), 759–786.

[AC] M.J. Ablowitz and P.A. Clarkson. *Solitons, Nonlinear Evolution Equations and Inverse Scattering*, London Mathematical Society Lecture Note Series, **149**. Cambridge University Press, Cambridge, 1991.

[AF] M.J. Ablowitz and A.S. Fokas. The inverse scattering transform for the Benjamin-Ono equation-a pivot to multidimensional problems, Stud. Appl. Math. **68** (1983), 1–10.

[AH] M.J. Ablowitz and R. Haberman. Nonlinear evolution equations in two and three dimensions, Phys. Rev. Lett. **35** (1975), 1185–1188.

[AL] J. Albert and F. Linares. Stability and symmetry of solitary-wave solutions to systems modeling interactions of long waves, J. Math. Pures Appl. **79** (2000), 195–226.

[AlCa] T. Alazard and R. Carles. Loss of regularity for the super-critical Schrödinger equations, Math. Ann. **343** (2009), 397–420.

[AlMn1] M. Alejo and C. Muñoz. Nonlinear stability of the mKdV breather, Comm. Math. Phys. **324** (2013), 233–262.

[AlMn2] M. Alejo and C. Muñoz. Dynamics of complex-valued modified KdV solitons with applications to the stability of breathers, arXiv:1308.0998

[AlMnVe] M. Alejo, C. Muñoz and L. Vega. The Gardner equation and the L^2-stability of the N-soliton solution of the Korteweg-de Vries equation, Trans. A.M.S. **365** (2013), 195–212.

[AnF] D. Anker and N.C. Freeman. On the soliton solutions of the Davey-Stewartson equation for long waves, Proc. R. Soc. London, Ser. A **360** (1978), 529–540.

[AS] N. Aronszajn and K.T. Smith. Theory of Bessel potentials I, Ann. Inst. Fourier (Grenoble) 11 (1961), 385–475.

[AS1] M.J. Ablowitz and H. Segur. On the evolution of packets of water waves, J. Fluid Mech. **92** (1979), 691–715.

[AS2] M.J. Ablowitz and H. Segur. Solitons and the Inverse Scattering Transform, Studies in Applied Math. Philadelphia: SIAM, 1981.

[AvHe] J.E. Avron and I.W. Herbst. Spectral and scattering theory of Schrödinger operators related to Stark effect, Comm. Math. Phys. **52** (1977) 239–254.

[B] W. Beckner. Inequalities in Fourier Analysis, Ann. of Math. **102** (1975), 159–182.

F. Linares, G. Ponce, *Introduction to Nonlinear Dispersive Equations*,
Universitext, DOI 10.1007/978-1-4939-2181-2

[BBCH] J. Bennet, N. Bez, A. Carbery, and D. Hundertmark. Heat-flow monotonicity of Strichartz norms, Anal. PDE **2** (2009), 147–158.

[BBM] T.B. Benjamin, J.L. Bona, and J.J. Mahony. Model equation for the long waves in nonlinear dispersive systems, Phil. Trans. Royal Soc. London **272** (1972), 47–78.

[BC1] R. Beals and R.R. Coifman. The spectral problem for the Davey-Stewartson and Ishimori hierarchies, Proc. Conf. on Nonlinear Evolution Equations: Integrability and Spectral Methods, Machester, U. K., 1988.

[BC2] H. Berestycki and T. Cazenave. Instabilité des états stationnaires dans les équations de Schrödinger et de Klein-Gordon non linéaires, C.R. Acad. Sci. Paris Sér. I Math. **293** (1981), 489–492.

[BeDS] I. Bejenaru and D. Da Silva. Low regularity solutions for a 2D quadratic Schrödinger equation, Trans. A.M.S. **360** (2008), 5805–5830.

[Be1] T.B. Benjamin. The stability of solitary waves, Proc. Roy. Soc. London, Ser. A **328** (1972), 153–183.

[Be2] T.B. Benjamin. Internal waves of permanent form in fluids of great depth, J. Fluid Mech. **29** (1967), 559–592.

[BeL] J. Bergh and J. Löfström. *Interpolation Spaces. An Introduction*, Grundhlehren 223, Springer-Verlag, New York, 1976.

[Bc] M. Beceanu. A centre-stable manifold for the focussing cubic NLS in $\mathbb{R}^{1+3}$, Comm. Math. Phys. **280** (2008), 145–205.

[BGa] H. Brezis and T. Gallouet. Nonlinear Schrödinger evolution equations, Nonlinear Analysis, TMS **4** (1980), 667–681.

[BGK] H. Berestycki, T. Gallouët and O. Kavian. Équations de champs scalaires Euclidiens non linéaires dans le plan, C.R. Acad. Sci. Paris Sér. I Math. **297** (1983), 307–310.

[BGT1] N. Burq, P. Gerard and N. Tzvetkov. Two singular dynamics of the nonlinear Schrödinger equation on a plane domain, Geom. Funct. Anal. **13** (2003), 1–19.

[BGT2] N. Burq, P. Gerard and N. Tzvetkov. An instability property of the nonlinear Schrödinger equation on $\mathbb{S}^d$, Math. Res. Lett. **9** (2002), 323–335.

[BGT3] N. Burq, P. Gerard and N. Tzvetkov. Strichartz inequalities and the nonlinear Schrödinger equation on compact manifolds, Amer. J. Math. **126** (2004), 569–605.

[BHHT] I. Bejenaru, S. Herr, J. Holmer and D. Tataru. On the 2D Zakharov system with L^2-Schrödinger data, Nonlinearity **22** (2009), 1063–1089.

[BHS] A. de Bouard, N. Hayashi and J-C. Saut. Global existence of small solutions to a relativistic nonlinear Schrödinger equation, Comm. Math. Phys. **189** (1997), 73–105.

[BIK] I. Bejenaru, A.D. Ionescu and C.E. Kenig. Global existence and uniqueness of Schrödinger maps in dimension $d \geq 4$, Advances in Math. **215** (2007), 263–291.

[BiL] H.A. Biagioni and F. Linares. Ill-posedness for the derivative Schrödinger and generalized Benjamin-Ono equations, Trans. A.M.S. **353** (2001), 3649–3659.

[BKl] M. Ben-Artzi and S. Klainerman. Decay and regularity for the Schrödinger equation, J. Anal. Math. **58** (1992), 25–37.

[BKPSV] B. Birnir, C.E. Kenig, G. Ponce, N. Svanstedt, and L. Vega. On the ill-posedness of the IVP for the generalized Korteweg-de Vries and nonlinear Schrödinger equations, J. London Math. Soc. **53** (1996), 551–559.

[BLi] H. Berestycki and P.-L. Lions. Nonlinear scalar field equations, Arch. Rat. Mech. Anal. **82** (1983), 313–375.

[BLiP] H. Berestycki, P.-L. Lions and L.A. Peletier. An ODE approach to the existence of positive solutions for semilinear problems in $\mathbb{R}^N$, Indiana Univ. Math. J. **30** (1981), 141–157.

[BM] I. Bialynicki-Birula and J. Mycielski. *Nonlinear wave mechanics*, Ann. Phys. **100** (1976), 62–93.

[Bn1] D.J. Benney. A general theory for interactions between short and long waves, Stud. Appl. Math. **56** (1977), 81–94.

[Bn2] J. Bona. On the stability theory of solitary waves, Proc. Roy. Soc. London, Ser. A **344** (1975), 363–374.

[BnR] D.J. Benney and G.J. Roskes. Wave instabilities, Stud. Appl. Math. **48**, (1969), 377–385.

[Bo1] J. Bourgain. Fourier transform restriction phenomena for certain lattice subsets and applications to nonlinear evolution equation, Geom. Funct. Anal. **3** (1993) 107–156, 209–262.

[Bo2] J. Bourgain. *Global Solutions of Nonlinear Schrödinger Equations*, American Mathematical Society Colloquium Publications **46**, American Mathematical Society, Providence, RI, (1999).

[Bo3] J. Bourgain. A remark on Schrödinger operators, Israel J. Math. **77** (1992), 1–16.

[Bo4] J. Bourgain. Scattering in the energy space and below for the 3D NLS, J. Anal. Math. **75** (1998), 267–297.

[Bo5] J. Bourgain. Refinements of Strichartz' inequalities and applications to 2D-NLS with critical nonlinearity, Int. Math. Res. Notices **5** (1998), 253–283.

[Bo6] J. Bourgain. On growth in time of Sobolev norms of smooth solutions of nonlinear Schrödinger equations in $\mathbb{R}^d$, J. Analyse Math. **72** (1997), 299–310.

[Bo7] J. Bourgain. Global well-posedness of defocusing critical nonlinear Schrödinger equation in the radial case. Journal A.M.S. **12** (1999), 145–171.

[Bo8] J. Bourgain. Periodic Korteweg-de Vries equation with measures as initial data, Selecta Math. **3** (1997), 115–159.

[Bo9] J. Bourgain. On the Cauchy and invariant measure problem for the periodic Zakharov system, Duke Math. J. **76** (1994), 175–202.

[Bo10] J. Bourgain. On the Cauchy problem for the Kadomtsev-Petviashvili equation, Geom. Funct. Anal. **3** (1993), 315–341.

[Bo11] J. Bourgain. On the Schrödinger maximal function in higher dimensions, arXiv:1201.3342.

[Bo12] J. Bourgain. On the Cauchy problem for the periodic KdV type equations, J. Fourier Anal. Appl. Special issue (1995), 17–86.

[BoC] J. Bourgain and J. Colliander. On well-posedness of the Zakharov system, Inter. Math. Res. Notices **11** (1996), 515–546.

[BoLi] J. Bourgain and D. Li. On an endpoint Kato-Ponce inequality, to appear in Diff. and Int. Eqs. **27** (2014), 1037–1072.

[BoW] J. Bourgain and W. Wang. Construction of blow up solutions for the nonlinear Schrödinger equations with critical nonlinearity, Ann. Scuola Norm. Sup. Pisa Cl. Sci. **25** (1997), 197–215.

[Br] A.R. Brodsky. On the asymptotic behavior of solutions of the wave equations, Proc. A.M.S. **18** (1967), 207–208.

[BRV] J.A. Barcelo, A. Ruiz and L. Vega. Some dispersive estimates for Schrödinger equations with repulsive potentials, J. Funct. Anal. **236** (2006) 1–24.

[Bs] M. Beals. Self-spreading and strength of singularities for the solutions to semilinear wave equations, Ann. of Math. **118** (1983), 187–214.

[BS] J.L. Bona and R. Smith. The initial value problem for the Korteweg-de Vries equation, Roy. Soc. London **Ser A 278** (1978), 555–601.

[BSa1] J.L. Bona and J-C. Saut. Dispersive blowup of solutions of generalized Korteweg-de Vries equations. J. Diff. Eqs. **103** (1993), 3–57.

[BSa2] J.L. Bona and J-C. Saut. Dispersive blow-up II. Schrödinger-type equations, optical and oceanic rogue waves, Chin. Ann. Math. Ser. B **31** (2010), 793–818.

[BSc] J.L. Bona and R. Scott. Solutions of the Korteweg-de Vries equation in fractional order Sobolev spaces, Duke Math. J. **43** (1976), 87–99.

[BSS] J.L. Bona, P.E. Souganidis, and W. Strauss. Stability and instability of solitary waves of Korteweg-de Vries type, Proc. Roy. Soc. London, Ser. A **A411** (1987), 395–412.

[BTo] I. Bejenaru and T. Tao. Sharp well-posedness and ill-posedness for a quadratic nonlinear Schrödinger equation, J. Funct. Anal. **233** (2006), 228–259.

[BTz] J.L. Bona and N. Tzvetkov. Sharp well-posedness result for the BBM equation, Discrete Cont. Dyn. Syst. **23** (2009), 1241–1252.

[BuKo] T. Buckmaster and H. Koch. The Korteweg-de Vries equation at H^{-1} regularity, arxiv 1112.4657

[BuPl] N. Burq and F. Planchon. On well-possedness of the Benjamin-Ono equation, Math. Ann. **340** (2008), 497–542.

[C] L. Carleson. *Some Analytical Problems Related to Statistical Mechanics*, Lecture Notes in Math. Springer-Verlag **779** (1979), 9–45.

[Ca] K. M. Case. Benjamin-Ono related equations and their solutions, Proc. Nat. Acad. Sci. USA **76** (1979), 1–3.

[Car] E. Carneiro. *A sharp inequality for the Strichartz norm*, Int. Math. Res. Notices **16** (2009), 3127–3145.

[Cb] V. Combet. Construction and characterization of solutions converging to solitons for supercritical gKdV equations, Diff. and Int. Eqs. **23** (2010), 513–568.

[CDKS] J.E. Colliander, J.-M. Delort, C.E. Kenig, and G. Staffilani. Bilinear estimates and applications to 2D NLS, Trans. A.M.S. **353** (2001), 3307–3325.

[CGD] K.W. Chow, R.H.J. Grimshaw, and E. Ding. Interaction of breathers and solitons in the extended Korteweg-de Vries equation, Wave Motion **43** (2005), 158–166.

[CH] R. Camassa and D.D. Holm. An integrable shallow water equation with peaked solutions, Phys. Rev. Lett. **71** (1993), 1661–1664.

[Ch1] H. Chihara. The initial value problem for the elliptic-hyperbolic Davey-Stewartson equation, J. Math. Kyoto Univ. **39** (1999), 41–66.

[Ch2] H. Chihara. Local existence for semilinear Schrödinger equations, Math. Japonica **42** (1995), 35–52.

[Ch3] H. Chihara. The initial value problem for Schrödinger equations on the torus, Int. Math. Res. Notices **15** (2002), 789–820.

[Ci] Y. Choi. *Well-posedness and scattering results of fifth order evolution equations*, Ph.D thesis, University of Chicago, 1994.

[CIKS] J. Colliander, A.D. Ionescu, C.E. Kenig, and G. Staffilani. Weighted low-regularity solutions of the KP-I initial-value problem, Discrete Contin. Dyn. Syst. **20** (2008), 219–258.

[CKS] W. Craig, T. Kappeler and W. Strauss. Microlocal dispersive smoothing for the Schrödinger equation, Comm. Pure Appl. Math. **48** (1995), 769–860.

[CKSTT1] J. Colliander, M. Keel, G. Staffilani, H. Takaoka, and T. Tao. A refined global well-posedness results for the Schrödinger equations with derivative, SIAM J. Math. Anal. **34** (2002), 68–86.

[CKSTT2] J. Colliander, M. Keel, G. Staffilani, H. Takaoka, and T. Tao. Resonant decompositions and the I-method for cubic nonlinear Schrödinger on $\mathbb{R}^2$, Disc. Cont. Dynam. Systems A 21 (2008), 665–686.

[CKSTT3] J. Colliander, M. Keel, G. Staffilani, H. Takaoka, and T. Tao. Global existence and scattering for rough solutions of a nonlinear Schrödinger equation on $\mathbb{R}^3$, Comm. Pure Appl. Math. **57** (2004), 987–1014.

[CKSTT4] J. Colliander, M. Keel, G. Staffilani, H. Takaoka, and T. Tao. Multi-linear estimates for periodic KdV equations, and applications, J. Funct. Anal. 211 (2004), 173–218.

[CKSTT5] J. Colliander, M. Keel, G. Staffilani, H. Takaoka, and T. Tao. Sharp global well-posedness results for periodic and non-periodic KdV and modified KdV on $\mathbb{R}$ and $\mathbb{T}$, Journal A.M.S. **16** (2003), 705–749.

[CKSTT6] J. Colliander, M. Keel, G. Staffilani, H. Takaoka, and T. Tao. Global well-posedness for KdV in Sobolev spaces of negative index, EJDE **26** (2001), 1–7.

[CKSTT7] J. Colliander, M. Keel, G. Staffilani, H. Takaoka, and T. Tao. Global well-posedness and scattering for the energy-critical nonlinear Schrödinger equation in $\mathbb{R}^3$, Annals of Math. **167** (2008), 767–865.

[Cl] M. Colin. On the local well-posedness of quasilinear Schrödinger equations in arbitrary space dimension, Comm. P.D.E. **27** (2002), 325–354.

[CM] T. Colin and G. Métivier. *Instabilities in Zakharov Equations for Laser Propagation in a Plasma*, Phase space analysis of partial differential equations, 63–81, Progr. Nonlinear Differential Equations Appl. **69**, Birkhäuser, Boston, MA, 2006.

[Cn] H. Cornille. Solutions of the generalized nonlinear Schrödinger equation in two spatial dimensions, J. Math. Phys. **20** (1979), 199–209.

[Co1] A. Cohen. Solutions of the Korteweg-de Vries equation from irregular data, Duke Math. J. **45** (1978), 149–181.

[CoK] A. Cohen and T. Kappeler. Solution to the Korteweg-de Vries equation with initial profile in $L_1^1(\mathbb{R}) \cap L_N^1(\mathbb{R}^+)$, SIAM J. Math. Anal. **18** (1987), 991–1025.

[Cr] M. Christ. Power series of a nonlinear Schrödinger equation, Mathematical aspects of nonlinear dispersive equations, Ann. of Math. Stud. **163**, Princeton Univ. Press, Princeton, NJ, (2007),131–155.

[CrCT1] M. Christ, J. Colliander and T. Tao. Asymptotics, frequency modulation, and low regularity ill-posedness for canonical defocusing equations, Amer. J. Math. **125** (2003), 1235–1293.

[CrCT2] M. Christ, J. Colliander, and T. Tao. A priori bounds and weak solutions for the nonlinear Schrödinger equation in Sobolev spaces of negative order, J. Funct. Anal **254** (2008), 368–395.

[CrCT3] M. Christ, J. Colliander, and T. Tao. Ill-posedness for nonlinear Schrödinger and wave equations, to appear Annales IHP.

[CrHoT] M. Christ, J. Holmer, and D. Tataru. Low regularity a priori bounds for the modified Korteweg-de Vries equation, Lib. Math. (N.S.) **32** (2012), 51–75.

[CrK] M. Christ and A. Kiselev. Maximal functions associated to filtrations, J. Funct. Anal. **179** (2001), 406–425.

[CrW] M. Christ and M. Weinstein. Dispersion of small amplitude solutions of the generalized Korteweg-de Vries equation, J. Funct. Anal. **100** (1991), 87–109.

[CS] P. Constantin and J. C. Saut. Local smoothing properties of dispersive equations, Journal A.M.S. **1** (1989), 413–446.

[CST] J. Colliander, G. Staffilani and H. Takaoka. Global well-posedness for KdV below L^2, Math. Res. Lett. **6** (1999), 755–778.

[CSU] N-H. Chang, J. Shatah and K. Uhlenbeck. Schrödinger maps, Comm. Pure Appl. Math. **53** (2000), 590–602.

[Ct] M. Cotlar. A general interpolation theorem for linear operators, Revista Matemática Cuyana, **1** (1955), 57–84.

[CVV] T. Cazenave, L. Vega and M.C. Vilela, A note on the nonlinear Schrödinger equation in weak L^p spaces, Commun. Contemp. Math. **3** (2001), 153–162.

[Cz1] T. Cazenave. *An Introduction to Nonlinear Schrödinger Equations*, Textos de Métodos Matemáticos **22**, Universidade Federal de Rio de Janeiro, 1989.

[Cz2] T. Cazenave. *Semilinear Schrödinger Equations*, Courant Lectures Notes 10, AMS 2003.

[CzL] T. Cazenave and P.-L. Lions. Orbital stability of standing waves for some nonlinear Schrödinger equations, Comm. Math. Phys. **85** (1982), 549–561.

[CzW1] T. Cazenave and F. Weissler. Rapidly decaying solutions of the nonlinear Schrödinger equation, Comm. Math. Phys. **147** (1992), 75–100.

[CzW2] T. Cazenave and F. Weissler. The Cauchy problem for the nonlinear Schrödinger equation in H^1, Manuscripta Math. **61** (1988), 477–494.

[CzW3] T. Cazenave and F. Weissler. *Some Remarks on the Nonlinear Schrödinger Equation in the Critical Case*, Lecture Notes in Math. **1394**, Springer, Berlin, (1989), 18–29.

[CzW4] T. Cazenave and F.B. Weissler. The Cauchy problem for the critical nonlinear Schrödinger equation in H^s, Nonlinear Anal. TMA **14** (1990), 807–836.

[D1] B. Dodson. Global well-posedness and scattering for the defocusing, L^2-critical, nonlinear Schrödinger equation when $d \geq 3$, Journal A.M.S. **25** (2012), 429–463.

[D2] B. Dodson. Global well-posedness and scattering for the defocusing, L^2-critical, nonlinear Schrödinger equation when $d = 2$, preprint.

[D3] B. Dodson. Global well-posedness and scattering for the defocusing, L^2-critical, nonlinear Schrödinger equation when $d = 1$, preprint.

[DaR] K.D. Danov and M.S. Ruderman. Nonlinear waves on shallow water in the presence of a horizontal magnetic field, Fluid Dynamics, **18** (1983), 751–756.

[DH] K. Datchev and J. Holmer. Fast soliton scattering by attractive delta impurities, Comm. PDE **34** (2009), no. 7–9, 1074–1113.

[DHR] T. Duyckaerts, J. Holmer and S. Roudenko. Scattering for the non-radial 3D cubic nonlinear Schrödinger equation, Math. Res. Lett. **15** (2008), 1233–250.

[DJ] P.G. Drazin and R.S. Johnson. *Solitons: An Introduction*, New York: Cambridge University Press, 1989.

[DK] B. Dahlberg and C.E. Kenig. A note on the almost everywhere behavior of solutions to the Schrödinger equation, Lecture Notes in Math. **908**, Springer, Berlin-New York, (1982), 205–209.

[Dl] J. Dollard. Asymptotic convergence and the Coulomb interaction, J. Math. Phys. **5** (1964), 729–739.

[DM] T. Duyckaerts and F. Merle. Dynamic of threshold solutions for energy-critical NLS, Geom. Funct. Anal. **18** (2009), 1787–1840.

[Do1] S. Doi. Remarks on the Cauchy problem for Schrödinger-type equations, Comm. PDE **21** (1996), 163–178.

[Do2] S. Doi. On the Cauchy problem for Schrödinger type equations and the regularity of the solutions, J. Math. Kyoto Univ. **34** (1994), 319–328.

[DR] V.D. Djordjevic and L.G. Redekopp. On two-dimensional packets of capillary-gravity waves, J. Fluid Mech. **79** (1977), 703–714.

[DRu] T. Duyckaerts and S. Roudenko. Threshold solutions for the focusing 3D cubic Schrödinger equation, Rev. Mat. Iberoam. **26** (2010), 1–56.

[DS] A. Davey and K. Stewartson. On three dimensional packets of surface waves, Proc. Roy. Soc. London, Ser. A **338** (1974), 101–110.

[Du] J. Duoandikoetxea. *Fourier Analysis*, Graduate Studies in Mathematics, **29**. Ame. Math. Soc. Providence, RI, 2001.

[DW] W. Ding and Y. Wang. Local Schrödinger flow into Kähler manifolds, Sci. China Ser. A **44** (2001), 1446–1464.

[Dx] D. Dix. Nonuniqueness and uniqueness in the initial-value problem for Burgers' equation, SIAM J. Math. Anal. **27** (1996), 708–724.

[E] M.J. Esteban. Existence d'une infinité d'ondes solitaires pour des équations de champs non linéaires dans le plan, Ann. Fac. Sci. Toulouse Math. **2** (1980), 181–191.

[EKPV1] L. Escauriaza, C.E. Kenig, G. Ponce, and L. Vega. Convexity properties of solutions to the free Schrödinger equation with gaussian decay, Math. Res. Lett. **15** (2008), 957–971.

[EKPV2] L. Escauriaza, C.E. Kenig, G. Ponce, and L. Vega. Hardy's uncertainty principle, convexity and Schrödinger evolutions, J. Eur. Math. Soc. (JEMS) **10** (2008), 883–907.

[EKPV3] L. Escauriaza, C.E. Kenig, G. Ponce, and L. Vega. On uniqueness properties of solutions of the k-generalized KdV equations, J. Funct. Anal. **244** (2007), 504–535.

[ES] W. Eckhaus and P. Schuur. The emergence of solitons of the Korteweg-de Vries equation from arbitrary initial conditions, Math. Methods Appl. Sci. **5** (1983), 97–116.

[F] G. Folland. *Introduction to Partial Differential Equations*, Princeton Univ. Press, Princeton, N.J. 1976.

[Fa] L.G. Farah. Global rough solutions to the critical generalized KdV equation, J. Diff. Eqs. **249** (2010), 1968–1985.

[FaLP] L.G. Farah, F. Linares and A. Pastor. The supercritical generalized KdV equation: Global well-posedness in the energy space and below, Math. Res. Lett. **18** (2011), 357–377.

[FaPa] L.G. Farah and A. Pastor. On well-posedness and wave operator for the gKdV equation, Bull. Sci. Math. **137** (2013), 229–241.

[FePaUl] E. Fermi, J. Pasta, and S. Ulam. Studies of Nonlinear Problems I, Los Alamos Report LA1940 (1955). In Nonlinear Wave Motion, edited by A. C. Newell, 143–156. Providence, RI. AMS, 1974.

[Ff] C. Fefferman, Inequalities for strongly singular convolution operators, Acta Math. **124** (1970), 9–36.

[FFFP] L. Fanelli, V. Felli, M. Fontelos, and A. Primo. Time decay of scaling invariant electromagnetic Schrödinger equation on the plane, Comm. Math. Phys. **324** (2013), 1033–1067.

[FGr] Y. F. Fang and M. G. Grillakis. On the global existence of rough solutions for the cubic defocussing Schrödinger equation in $\mathbb{R}^{2+1}$, J. Hip. Diff. Eqs. **4** (2007), 233–257.

[FLP1] G. Fonseca, F. Linares, and G. Ponce. Global existence for the critical generalized KdV equation, Proc. A.M.S. **131** (2003), 1847–1855.

[FLP2] G. Fonseca, F. Linares, and G. Ponce. Global well-posedness for the modified Korteweg-de Vries equation, Comm. P.D.E. **24** (1999), 683–705.

[FLP3] G. Fonseca, F. Linares and G. Ponce. The IVP for the Benjamin-Ono equation in weighted Sobolev spaces II, J. Funct. Anal., **262** (2012), 2031–2049.

[FLP4] G. Fonseca, F. Linares and G. Ponce. *On persistence properties in fractional weighted spaces*, to appear in Proc. A.M.S.

[Fm] A. Friedman. *Partial Differential Equations*, Holt, Rinehart and Winston, New York, 1976.

[Fo] G. Fonseca. Growth of the H^s-norm for the modified KdV equation, Diff. and Int. Eqs. **13** (2000), 1081–1093.

[FoPo] G. Fonseca and G. Ponce. The IVP for the Benjamin-Ono equation in weighted Sobolev spaces, J. Funct. Anal. **260** (2011), 436–459.

[Fr] F.G. Friedlander. *Introduction to the Theory of Distributions*, Cambridge University Press, New York 1982.

[Fs D. Foschi. Maximizers for the Strichartz inequality, J. Eur. Math. Soc. (JEMS) bf 9 (2007), 739–774.

[FS1] A.S. Fokas and L.Y. Sung. On the solvability of the N-wave, Davey-Stewartson and Kadomtsev-Petviashvili equations, Inverse Problems **8** (1992), 673–708.

[FS2] A.S. Fokas and L.Y. Sung. The Cauchy problem for the Kadomtsev-Petviashvili-I equation without the zero mass constraint, Math. Proc. Cambridge Philos. Soc. **125** (1999), 113–138.

[FSa] A.S. Fokas and P.M. Santini. Dromions and a boundary value problem of the Davey-Stewartson I equation, Physica D **44** (1990) 99–130.

[FXC] D. Fang, J. Xie, and T. Cazenave. Scattering for the focusing energy-subcritical NLS, Sci. China Math. **54** (2011), 2037–2062.

[GaO] L. Grafakos and S. Oh. Kato-Ponce inequality, Comm. PDE **39** (2014), 1128–1157.

[Gb] M. Goldberg. Dispersive estimate for the three-dimensional Schrödinger equation with rough potentials, Amer. J. Math. **128** (2006) 731–750

[G1] Y. Giga. Solutions of the semilinear parabolic equations in L^p and regularity of weak solutions of the Navier-Stokes system, J. Diff. Eqs. **62** (1986), 186–212.

[G2] R.T. Glassey. On the blowing up of solutions to the Cauchy problem for nonlinear Schrödinger equations, J. Math. Phys. **18** (1977), 1794–1797.

[GGKM] C.S. Gardner, J.M. Greene, M.D. Kruskal and R.M. Miura. A method of solving the Korteweg-de Vries equation, Phys. Rev. Lett. **19** (1967), 1095–1097.

[GHW] R.H. Goodman, P.J. Holmes, and M.I. Weinstein. Strong NLS soliton-defect interactions, Phys. D **192** (2004), 215–248.

[GKM] C.S. Gardner, M.D. Kruskal and R. Miura. Korteweg-de Vries equation and generalizations II. Existence of conservation laws and constants of motions, J. Math. Phys. **9** (1968), 1204–1209.

[Gl1] M. Grillakis. On nonlinear Schrödinger equations, Comm. P.D.E. **25** (2000), 1827–1844.

[Gl2] M. Grillakis. Regularity and asymptotic behavior of the wave equation with critical nonlinearity, Ann. of Math. **132** (1990), 485–509.

[Gl3] M. Grillakis. Regularity for the wave equation with a critical nonlinearity, Comm. Pure Appl. Math. **45** (1992), 749–774.

[GM] L. Glangetas and F. Merle. Existence of self-similar blow-up solutions for Zakharov equation in dimension two. I. Concentration properties of blow-up solutions and instability results for Zakharov equation in dimension two. II Comm. Math. Phys. **160** (1994), 173–215, 349–389.

[GO] O.B. Gorbacheva and L.A. Ostrovsky. Nonlinear vector waves in a mechanical model of a molecular chain, Physica D **8** (1983), 223–228.

[GPS] A. Grünrock, M. Panthee, and J. Drumond Silva. A remark on global well-posedness below L^2 for the gKdV-3 equation, Diff. and Int. Eqs. **20** (2007), 1229–1236.

[Gr1] A. Grünrock. Some local well-posedness results for nonlinear Schrödinger equations below L^2, preprint.

[Gr2] A. Grünrock. A bilinear Airy type estimate with application to the 3-gkdv equation, Diff. and Int. Eqs. **18** (2005), no. 12, 1333–1339.

[Gr3] A. Grünrock. Bi- and trilinear Schrödinger estimates in one space dimension with applications to cubic NLS and DNLS, Int. Math. Res. Notices **41** (2005), 2525–2558.

[GrV] A. Grünrock and L. Vega. Local well-posedness for the modified KdV equation in almost critical $\widehat{H_s^r}$-spaces, Trans. A.M.S. **361** (2009), 5681–5694.

[GS] J.-M. Ghidaglia and J.-C. Saut . On the initial value problem for the Davey-Stewartson systems, Nonlinearity **3** (1990), 475–506.

[GSch] M. Goldberg and W. Schlag. Dispersive estimates for Schrödinger operators in dimensions one and three, Comm. Math. Phys. **251** (2004), 157–158.

[GSS] M. Grillakis, J. Shatah and W. Strauss. Stability theory of solitary waves in the presence of symmetry I, II, J. Funct. Anal. **74** (1987), 160–197. **94**, (1990), 308–348.

[Gq] Q. Guo, Nonscattering solutions to the L^2-supercritical NLS equations, preprint.

[Gu] Z. Guo, Global well-posedness of the Korteweg-de Vries equation in $H^{-3/4}(\mathbb{R})$, J. Math. Pures Appl. (9) **91** (2009), 583–597.

[GT] J. Ginibre and Y. Tsutsumi. Uniqueness of solutions for the generalized Korteweg-de Vries equation, SIAM J. Math. Anal. **20** (1989), 1388–1425.

[GTV] J. Ginibre, Y. Tsutsumi and G. Velo. On the Cauchy problem for the Zakharov system, J. Funct. Anal. **151** (1997), 384–436.

[GV1] J. Ginibre and G. Velo. On the class of nonlinear Schrödinger equations, J. Funct. Anal. **32** (1979), 1–32, 33–72.

[GV2] J. Ginibre and G. Velo. Scattering theory in the energy space for the class of nonlinear Schrödinger equations, J. Math. Pures Appl. **64** (1985), 363–401.

[GV3] J. Ginibre and G. Velo. Time decay of finite energy solutions of the nonlinear Klein-Gordon and Schrödinger equations, Ann. Inst. H. Poincaré Phys. Théor. **43** (1985), no. 4, 399–442.

[GVi] M. Goldberg and M. Visan. A counter-example to dispersive estimates for the Schrödinger operators in higher dimensions, Comm. Math. Phys. **266** (2006) 211–238.

[H] G.H. Hardy. A theorem concerning Fourier transform, J. London Math. Soc. **8** (1933), 227–331.

[HaHK] M. Hadac, S. Herr and H. Koch. Well-posedness and scattering for the KP-II equation in a critical space, Ann. Inst. H. Poincaré Anal. Non Linéaire **26** (2009), 917–941.

[H1] N. Hayashi. Global existence of small analytic solutions to nonlinear Schrödinger equations, Duke Math. J. **61** (1991), 575–592.

[H2] N. Hayashi. Local existence in time of small solutions to the Davey-Stewartson systems, Ann. Inst. H. Poincaré Phys. Théor. **65** (1996), 313–366.

[H3] N. Hayashi. Local existence in time of solutions to the elliptic-hyperbolic Davey-Stewartson system without smallness condition on the data, J. Anal. Math. **73** (1997), 133–164.

[H4] N. Hayashi. Local existence in time of small solutions to the Ishimori system, preprint.

[H5] N. Hayashi. Global existence of small analytic solutions to nonlinear Schrödinger equations, Duke Math. J. **60** (1990), 717–727.

[He] L. I. Hedberg. On certain convolution inequalities, Proc. A.M.S. **36** (1972), 505–510.

[HH1] N. Hayashi and H. Hirata. Global existence and asymptotic behaviour in time of small solutions to the elliptic-hyperbolic Davey-Stewartson system, Nonlinearity **9** (1996), 1387–1409.

[HH2] N. Hayashi and H. Hirata. Local existence in time of small solutions to the elliptic-hyperbolic Davey-Stewartson system in the usual Sobolev space, Proc. Edinburgh Math. Soc. **40** (1997), 563–581.

[Hi1] R. Hirota. Exact solutions to the Korteweg-de Vries equation for multiple collisions of solitons, Phys. Rev. Lett. **27** (1971), 1192–1194.

[Hi2] R. Hirota. Exact envelope soliton solutions of a nonlinear wave equation, J. Math. Phys. **14** (1973), 805–809.

[HMZ] J. Holmer, J. Marzuola, and M. Zworski. Fast soliton scattering by delta impurities, Comm. Math. Phys. **274** (2007), 187–216.

[HN1] N. Hayashi and P. Naumkin. Large time asymptotics of solutions to the generalized Korteweg-de Vries equation, J. Funct. Anal. **159** (1998), 110–136.

[HN2] N. Hayashi and P. Naumkin. On the modified Korteweg-de Vries equation, Math. Phys. Analysis and Geometry **4** (2001), 197–227.

[HN3] N. Hayashi and P. Naumkin. On the Davey-Stewartson and Ishimori systems, Math. Phys. Anal. Geom. **2** (1999), 53–81.

[HNT1] N. Hayashi, K. Nakamitsu and M. Tsutsumi. Nonlinear Schrödinger equations in weighted Sobolev spaces, Funkcial. Ekvac. **31** (1988), 363–381.

[HNT2] N. Hayashi, K. Nakamitsu and M. Tsutsumi. On solutions of the initial value problem for the nonlinear Schrödinger equations, J. Funct. Anal. **71** (1987), 218–245.

[HNT3] N. Hayashi, K. Nakamitsu and M. Tsutsumi. On solutions of the initial value problem for the nonlinear Schrödinger equations in one space dimension, Math. Z. **192** (1986), 637–650.

[HO] H. Hayashi and T. Ozawa. Remarks on nonlinear Schrödinger equations in one space dimension, Diff. and Int. Eqs. **2** (1994), 453–461.

[Ho1] L. Hörmander. Estimates for translation invariant operators in Lp spaces, Acta Math. **104** (1960), 93–104.

[Ho2] L. Hörmander. *The Analysis of Linear Partial Differential Operators III*, Springer, New York 1984.

[HPR] J. Holmer, G. Perelman, and S. Roudenko. A solution to the focussing 3d NLS that blows up on a contracting sphere, to appear in Trans. A.M.S.

[HR1] J. Holmer and S. Roudenko. A sharp condition for scattering of the radial 3D cubic nonlinear Schrödinger equation, Comm. Math. Phys. **282** (2008), 435–467.

[HR2] J. Holmer and S. Roudenko. Blow up solutions on a sphere for the 3D quintic NLS in the energy space, Anal. PDE **5** (2012), 475–512.

[HR3] J. Holmer and S. Roudenko. A class of solutions of the 3D cubic nonlinear Schrödinger equation that blow upon a circle, Appl. Math. Res. Express. AMRX (2011), 23–94.

[HS] N. Hayashi and J-C. Saut. Global existence of small solutions to the Davey-Stewartson and the Ishimori systems, Diff. and Int. Eqs. **8** (1995), 1657–1675.

[HoZ] J. Holmer and M. Zworski. Slow soliton interaction with delta impurities, J. Mod. Dyn. **1** (2007), 689–718.

[HuZ] D. Hundertmark and V. Zharnitsky. *On sharp Strichartz inequalities in low regularity*, Int. Math. Res. Notices (2006), 18 pp.

[I] W. Ichinose. The Cauchy problem for Schrödinger type equations with variable coefficients, Osaka J. Math. **24** (1987), 853–886.

[IK1] A. Ionescu and C. E. Kenig. Global well-posedness of the Benjamin-Ono equation in low regularity spaces, Journal A.M.S. **20** (2007), 753–798.

[IK2] A. Ionescu and C. E. Kenig. Low regularity Schrödinger maps: global well-posedness in dimension $d \geq 3$, Comm. Math. Phys. **271** (2007), 523–559.

[IKT] A. D. Ionescu, C. E. Kenig and D. Tataru. Global well-posedness of the KP-I initial-value problem in the energy space, Invent. Math. **173** (2008), 265–304.

[ILP1] P. Isaza, F. Linares, and G. Ponce. On decay properties of solutions of the k-generalized KdV equation, Comm. Math. Phys. **324** (2013), 129–146.

[ILP2] P. Isaza, F. Linares, and G. Ponce. *Propagation of regularity and decay of solutions to the k-generalized Korteweg-de Vries equation*, to appear in Comm. P.D.E.

[IM1] P. Isaza and J. Mejía. Local and global Cauchy problems for the Kadomtsev-Petviashvili(KP-II) equation in Sobolev spaces of negative indices, Comm. P.D.E. **26** (2001), 1027–1054.

[IM2] P. Isaza and J. Mejía. Global solution for the Kadomtsev-Petviashvili (KP-II) equation in Sobolev spaces of negative indices, Comm. P.D.E. **26** (2001), 1027–1054.

[IMS] P. Isaza, J. Mejía and V. Stallbohm. A regularity theorem for the Kadomtsev-Petviashvili equation with periodic boundary conditions, Nonlinear Anal. **23** (1994), 683–687.

[IN] R. Iório and W. Nunes. On equations of KP-type, Proc. Roy. Soc. Edinburgh, Section A **128** (1998), 725–743.

[Io1] R. Iório. On the Cauchy problem for the Benjamin-Ono equation, Comm. P.D.E. **11** (1986), 1031–1081.

[Io2] R. Iório. KdV, BO and friends in weighted Sobolev spaces, Functional-analytic methods for partial differential equations (Tokyo, 1989), 104–121, Lecture Notes in Math. **1450**, Springer, Berlin, 1990.

[Io3] R. Iório. Unique continuation principles for the Benjamin-Ono equation, Diff. and Int. Eqs. **16** (2003), 1281–1291.

[Is1] Y. Ishimori. Multivortex solutions of a two dimensional nonlinear wave equation, Progr. Theor. Phys. **72** (1984), 33–37.

[Is2] Y. Ishimori. Solitons in a one-dimensional Lennard/Mhy Jones lattice, Progr. Theoret. Phys. **68** (1982), 402–410.

[JK] C. Jones and T. Küpper. On the infinitely many solutions of a semilinear elliptic equation, SIAM J. Math. Anal. **17** (1986), 803–836.

[JN] F. John and L. Nirenberg. On functions of bounded mean oscillation, Comm. Pure Appl. Math. **14** (1961), 415–426.

[JP] G. James and D. Pelinovsky. Gaussian solitary waves and compactons in Fermi-Pasta-Ulam lattice with Herztian potentials, Proc. R. Soc. Lond. Ser. A Math. Phys. Eng. Sci. 470 (2014) 20130462, 20pp.

[JSS] J. Journé, A. Soffer, and C. Sogge. $L^p - L^{p'}$ estimates for the time dependent Schrödinger operators, Bull. A.M.S. **23** (1990), 519–524.

[K1] T. Kato. Wave operator and similarity for some non-self-adjoint operators, Math. Ann. **162** (1966), 258–279.

[K2] T. Kato. On the Cauchy problem for the (generalized) Korteweg-de Vries equation, Adv. in Math. Supp. Stud., Stud. in Appl. Math. **8** (1983), 93–128.

[K3] T. Kato. Nonlinear Schrödinger equations, Ann. Inst. Henri Poincaré, Phys. Théor. **46** (1987), 113–129.

[K4] T. Kato. On nonlinear Schrödinger equations, Lecture Notes for Physics. Nordic Summer School, 1988.

[K5] T. Kato. On the Korteweg-de Vries equation, Manuscripta Math **29** (1979), 89–99.

[Ka] O. Kavian. A remark on the blowing-up of solutions to the Cauchy problem for nonlinear Schrödinger equations, Trans. A.M.S. **299** (1987), 193–203.

[KdV] D.J. Korteweg and G. de Vries. On the change of form of long waves advancing in a rectangular canal, and on a new type of long stationary waves, Philos. Mag. **39** (1895), 422–443.

[Ke] C.E. Kenig. On the local and global well-posedness theory for the KP-I equation, Ann. Inst. H. Poincaré Anal. Non Linéaire, **21** (2004), 827–838.

[KeKo] C.E. Kenig and K.D. Köenig. On the local well-posedness of the Benjamin-Ono and modified Benjamin-Ono equations, Math. Res. Lett. **10** (2003), 879–895.

[KePi] C.E. Kenig and D. Pilod. Well-posedness for the fifth-order KdV equation in the energy space, to appear Trans. A.M.S.

[KF] S. Kruzhkov and A. Faminskii. A generalized solution for the Cauchy problem for the Korteweg–de Vries equation, Math. USRR, Sbornik **48** (1984), 93–138.

[Kg] H. Kumano-go. *Pseudo-differential Operators*, MIT Press, Cambridge 1981.

[Ki1] N. Kishimoto. Well-posedness of the Cauchy problem for the Korteweg-de Vries equation at the critical regularity, Comm. Pure Appl. Anal. **7** (2008), 1123–1143.

[Ki2] N. Kishimoto. Low-regularity bilinear estimate for a quadratic nonlinear Schrödinger equation, J. Diff. Eqs. **247** (2009), 1397–1439.

[Ki3] N. Kishimoto. Local well-posedness for the Cauchy problem for the quadratic Scrödinger equation with nonlinearity $\overline{u}^2$, Comm. Pure Appl. Anal. **7** (2008), 1123–1143.

[KiT] N. Kishimoto. and K. Tsugawa. Local well-posedness for quadratic nonlinear Schrödinger equations and the "good" Boussinesq equation, Diff. and Int. Eqs. **23** (2010), 463–493.

[Kl1] S. Klainerman. Uniform decay estimates and the Lorentz invariance of the classical wave equation, Comm. Pure Appl. Math **38** (1985), 321–332.

[Kl2] S. Klainerman. Global existence for nonlinear wave equations, Comm. Pure Appl. Math. **33** (1980), 43–101.

[KlM] S. Klainerman and M. Machedon. Space-time estimates for the null forms and the local existence theorem, Comm. Pure Appl. Math. **46** (1993), 1221–1268.

[KM] H. Koch and J. Marzuola. Small data scattering and soliton stability in $\dot{H}^{-\frac{1}{6}}$ for the quartic KdV equation, Anal. PDE **5** (2012), 145–198.

[KM1] C.E. Kenig and F. Merle. Scattering for $\dot{H}^{1/2}$ bounded solutions to the cubic defocusing NLS in three dimensions, Trans. A.M.S. **362** (2010) 1937–1962.

[KM2] C.E. Kenig and F. Merle. Global well posedness, scattering and blow up for the energy-critical, focusing, nonlinear Schrödinger equation in the radial case, Invent. Math. **166** (2006), 645–675.

[KMa] B.G. Konopelchenko and B. Matkarimov. Inverse spectral transform for the Ishimori equation. I. Initial value problem, J. Math. Phys. **31** (1990), 2737–2746.

[KN] D.J. Kaup and A.C. Newell. An exact solution for the derivative nonlinear Schrödinger equation, J. Math. Phys. **1** (1978), 798–801.

[KP] B.B. Kadomtsev and V.I. Petviashvili. On the stability of solitary waves in weakly dispersive media, Sov. Phys. Dokl. **15** (1970), 539–541.

[KPo] T. Kato and G. Ponce. Commutator estimates and the Euler and Navier-Stokes equations, Comm. Pure Appl. Math. **41** (1988), 891–907.

[KPRV1] C.E. Kenig, G. Ponce, C. Rolvung, and L. Vega. Variable coefficient Schrödinger flows for ultrahyperbolic operators, Adv. in Math. **196** (2005), 373–486.

[KPRV2] C.E. Kenig, G. Ponce, C. Rolvung, and L. Vega. The general quasilinear ultrahyperbolic Schrödinger equation, Adv. in Math. **206** (2006), 402–433.

[KPV1] C.E. Kenig, G. Ponce and L. Vega. On the (generalized) Korteweg-de Vries equation, Duke Math. J., **90** (1989) 585–610.

[KPV2] C.E. Kenig, G. Ponce, and L. Vega. Oscillatory integrals and regularity of dispersive equations, Indiana U. Math. J. **40** (1991), 33–69.

[KPV3] C.E. Kenig, G. Ponce, and L. Vega. Small solutions to non-linear Schrödinger equation, Ann. Inst. H. Poincaré Anal. Non Linéaire **10** (1993), 255–280.

[KPV4] C.E. Kenig, G. Ponce, and L. Vega. Well-posedness and scattering results for the generalized Korteweg-de Vries equation via contraction principle, Comm. Pure Appl. Math. **46** (1993), 527–620.

[KPV5] C.E. Kenig, G. Ponce, and L. Vega. On the ill-posedness of some canonical dispersive equations, Duke Math. J. **106** (2001), 617–633.

[KPV6] C.E. Kenig, G. Ponce and L. Vega. A bilinear estimate with applications to the KdV equation, Journal A.M.S. **9** (1996), 573–603.

[KPV7] C.E. Kenig, G. Ponce and L. Vega. The Cauchy problem for the Korteweg-de Vries equation in Sobolev spaces of negative indices, Duke Math. J. **71** (1993), 1–21.

[KPV8] C.E. Kenig, G. Ponce, and L. Vega. On the Zakharov and Zakharov-Schulman systems, J. Funct. Anal. **127** (1995), 204–234.

[KPV9] C.E. Kenig, G. Ponce, and L. Vega. On the initial value problem for the Ishimori system, Ann. Henri Poincaré **1** (2000), 341–384.

[KPV10] C.E. Kenig, G. Ponce, and L. Vega. The Cauchy problem for quasi-linear Schrödinger equations, Invent. Math. **158** (2004), no. 2, 343–388.

[KPV11] C.E. Kenig, G. Ponce and L. Vega. On the generalized Benjamin-Ono equation, Trans. A.M.S. **342** (1994), 155–172.

[KPV12] C.E. Kenig, G. Ponce and L. Vega. Quadratic forms for the 1-D semilinear Schrodinger equation, Trans. A.M.S. **346** (1996), 3323–3353.

[KPV13] C.E. Kenig, G. Ponce and L. Vega. Higher-order nonlinear dispersive equations, Proc. A.M.S. **122** (1994), 157–166.

[KPV14] C.E. Kenig, G. Ponce and L. Vega. Smoothing effects and local existence theory for the generalized nonlinear Schrödinger equation, Invent. Math. **134** (1998), 489–545.

[KPV15] C.E. Kenig, G. Ponce and L. Vega. A theorem of Paley-Wiener type for the Schrödinger evolutions, Ann. Scient. Éc. Norm. Sup. **47** (2014), 539–557.

[KpTo] T. Kappeler and P. Topalov. Global wellposedness of KdV in $H^{-1}(\mathbb{T},\mathbb{R})$, Duke Math. J. **135** (2006), 327–360.

[KR] C.E. Kenig and A. Ruiz. A strong type (2, 2) estimate for a maximal operator associated to the Schrödinger equation, Trans. A.M.S. **280** (1983), 239–246.

[KrS] J. Krieger and W. Schlag. Stable manifolds for all monic supercritical focusing nonlinear Schrödinger equations in one dimension Journal A.M.S. **19** (2006), no. 4, 815–920.

[KSC] C.F.F. Karney, A. Sen and F.Y.F. Chu. Nonlinear evolution of lowe hybrid waves, Phys. Fluids **22** (1979), 940–952.

[KST] J. Krieger, W. Schlag, and D. Tataru. Slow blow-up solutions for the H1(R3) critical focusing semilinear wave equation, Duke Math. J. **147** (2009), 1–53.

[KT] C.E. Kenig and H. Takaoka. Global well-posedness of the modified Benjamin-Ono equation with initial data in $H^{1/2}$, Int. Math. Res. Notices (2006) Art. ID 95702.

[KT1] M. Keel and T. Tao. Endpoint Strichartz estimates, Amer. J. Math. **120** (1998), 955–980.

[KT2] M. Keel, T. Roy, and T. Tao. Global well-posedness of the Maxwell-Klein-Gordon equation below the energy norm, Discrete Contin. Dyn. Syst. **30** (2011), 573–621.

[KTa1] H. Koch and D. Tataru. Personal communication.

[KTa2] H. Koch and D. Tataru. A priori bounds for the 1D cubic NLS in negative Sobolev spaces, Int. Math. Res. Notices **16** (2007), Art. ID rnm053.

[KTa3] H. Koch and D. Tataru. Energy and local energy bounds for the 1-D cubic NLS equation in $H^{-1/4}$, Ann. Inst. H. Poincaré Anal. Non Linéaire **29** (2012), 955–988.

[KTV] R. Killip, T. Tao, and M. Visan. The cubic nonlinear Schrödinger equation in two dimensions with radial data, J. Eur. Math. Soc. (JEMS) **11** (2009), 1203–1258.

[KTz] H. Koch and N. Tzvetkov. On the local well-posedness of the Benjamin-Ono equation $H^s(\mathbb{R})$, Int. Math. Res. Notices **26** (2003) 1449–1464.

[KV1] R. Killip and M. Visan. The focusing energy-critical nonlinear Schrödinger equation in dimensions five and higher, Amer. J. Math. **132** (2010), 361–424.

[KV2] R. Killip and M. Visan. Energy supercritical NLS: critical $\dot{H}^s$-bounds imply scattering, Comm. PDE **35** (2010) 945–987.

[KVZ] R. Killip, M. Visan and X. Zhang. Energy-critical NLS with quadratic potentials, Comm. P.D.E. **34** (2009), 1531–1565.

[Kw1] K.M. Kwong. Uniqueness of positive solutions of $\Delta u - u + u^p = 0$ in $\mathbb{R}^n$, Arch. Rat. Mech. Anal. **105** (1989), 243–266.

[Kw2] S. Kwon. On the fifth order KdV equation: local well posedness and lack of uniform continuity of the solution map, Electron. J. Differential Equations (2008), 15 pp.

[KY] T. Kato and K. Yajima. Some examples of smooth operators and the associated smoothing effect, Reviews in Math Physics **1** (1989), 481–496.

[Kz] M. Kunze. On the existence of a maximizer for the Strichartz inequality, Comm. Math. Phys. **243** (2003), 137–164.

[Lb] G. Lamb. *Elements of Soliton Theory*, Pure and Applied Mathematics, A Wiley-Interscience Publication, John Wiley & Sons Inc., New York, 1980.

[Le] S. Lee. *Pointwise convergence of solutions to Schrödinger equations in* $\mathbb{R}^2$, Int. Math. Res. Notices (2006), 1–21.

[LiPo] F. Linares and G. Ponce. On the Davey–Stewartson systems, Ann. Inst. H. Poincaré Anal. Non Linéaire **10** (1993), 523–548.

[LiS] F. Linares and M. Scialom. On the smoothing properties of solutions to the modified Korteweg-de Vries equation, J. Diff. Eqs. **106** (1993), 141–154.

[LmPo] W.-K. Lim and G. Ponce. On the initial value problem for the one dimensional quasilinear Schrödinger equation, SIAM J. Math. Anal. **34** (2002), 435–459.

[LP] P.-L. Lions and B. Perthame. Lemmes de moments, de moyenne et de dispersion, C. R. Acad. Sci. Paris Sér. I Math. **314** (1992), 801–806.

[Lp] W. Littman. The wave operator and L^p-norms, J. Math. & Mech. **12** (1963), 55–68.

[LPS] F. Linares, G. Ponce and J.-C. Saut. On a degenerate Zakharov system, Bull. Braz. Math. Soc. (N.S.) **36** (2005), 1–23.

[LPSS] M.J. Landman, G. Papanicolaou, C. Sulem, and P.-L. Sulem. Rates of blow up for solutions of the nonlinear Schrödinger equation at the critical dimension, Phys. Rev. A (3) **38** (1988), 3837–3842.

[LS] J.E. Lin and W.A. Strauss. Decay and scattering of solutions of a nonlinear Schrödinger equation, J. Funct. Anal. **30** (1978), 245–263.

[Lx1] P. Lax. Almost periodic solutions of the KdV equation, SIAM Rev. **18** (1976), 351–375.

[Lx2] P. Lax. Integrals of nonlinear equations of evolution and solitary waves, Comm. Pure Appl. Math. **21** (1968), 467–490.

[M] B. Marshal. Mixed norm estimates for the Klein-Gordon equation, Proc. Conf. in honor of A. Zygmund, Wadsworth Int. Math ser. (1981), 638–639.

[Ma] Y. Martel. Asymptotic N-soliton-like solutions of the subcritical and critical Korteweg-de Vries equations, Amer. J. Math. **127** (2005), 1103–1140.

[McS] H. McKean and J. Shatah. The nonlinear Schrödinger equations and the nonlinear heat equation: reduction to linear form, Comm. Pure Appl. Math. **44** (1991), 1067–1080.

[Me1] F. Merle. Limit of the solution of a nonlinear Schrödigner equation at blow-up time, J. Funct. Anal. **84** (1989), 201–214.

[Me2] F. Merle. Construction of solutions with exactly k blow-up points for the Schrödinger equation with critical nonlinearity, Comm. Math. Phys. **129** (1990), 223–240.

[Me3] F. Merle. Determination of blow-up solutions with minimal mass for nonlinear Schrödinger with critical power, Duke Math. J. **69** (1993), 427–453.

[Me4] F. Merle. Existence of blow-up solutions in the energy space for the critical generalized KdV equation, Journal A.M.S. **14** (2001), 555–578.

[Me5] F. Merle. Lower bounds for the blowup rate of solutions of the Zakharov equation in dimension two, Comm. Pure Appl. Math. **49** (1996), 765–794.

[Me6] F. Merle. Blow-up results of virial type for Zakharov equations, Comm. Math. Phys. **175** (1996), 433–455.

[MeRa1] F. Merle and P. Raphael. Sharp upper bound on the blow up rate for the critical nonlinear Schrödinger equation, Geom. Funct. Anal. **13** (2003), 591–642.

[MeRa2] F. Merle and P. Raphael. The blow up dynamic and upper bound on the blow up rate for the critical nonlinear Schrödinger equation, Ann. of Math. (2) **161** (2005), 157–222.

[MG] H. Mc Gahagan. An approximation scheme for Schrödinger maps, Comm. PDE **32** (2007), 375–400.

[MGK] R. Miura, C. Gardner, and M. Kruskal. Korteweg-de Vries equation and generalizations. II. Existence of conservation laws and constants of motion, J. Math. Phys. **9** (1968), 1204–1209.

[Mi] T. Mizumachi. Large time asymptotics of solutions around solitary waves to the generalized Korteweg-de Vries equations, SIAM J. Math. Anal. **32** (2001), 1050–1080.

[MiT] T. Mizumachi and N. Tzvetkov. L^2-stability of solitary waves for the KdV equation via Pego and Weinstein's method, arXiv:1403.5321.

[MK] Y. Matsuno and D. J. Kaup. Initial value problem of the linearized Benjamin-Ono equation and its applications, J. Math. Phys. **38** (1997), 5198–5224.

[Mk] B. Muckenhoup. *Weighted norm inequalities for the Fourier transform*, Trans. A.M.S. **276** (1985), 729–742.

[MM1] Y. Martel and F. Merle. Instability of solitons for the critical generalized Korteweg-de Vries equation, Geom. Funct. Anal. **11** (2001), 74–123.

[MM2] Y. Martel and F. Merle. Asymptotic stability of solitons for subcritical generalized KdV equations, Arch. Rat. Mech. Anal. **157** (2001), 219–254.

[MM3] Y. Martel and F. Merle. A Liouville theorem for the critical generalized Korteweg-de Vries equation, J. Math. Pures Appl. **79** (2000), 339–425.

[MM4] Y. Martel and F. Merle. Blow up in finite time and dynamics of blow up solutions for the L^2-critical generalized KdV equation, Journal A.M.S. **15** (2002), 617–664.

[MM5] Y. Martel and F. Merle. Stability of blow-up profile and lower bounds for blow-up rate for the critical generalized KdV equation, Ann. of Math. **155** (2002), 235–280.

[MM6] Y. Martel and F. Merle. Nonexistence of blow-up solution with minimal L^2-mass for the critical gKdV equation, Duke Math. J. **115** (2002), 385–408.

[MM7] Y. Martel and F. Merle. Multisolitary waves for nonlinear Schrödinger equations, Ann. Inst. H. Poincaré Anal. Non Linéaire **23** (2006), 849–864.

[MM8] Y. Martel and F. Merle. Description of two soliton collision for the quartic gKdV equation, Ann. of Math. (2) **174** (2011), 757– 857.

[MMR1] Y. Martel, F. Merle, and P. Raphael. Blow up for the critical gKdV equation I: dynamics near the soliton, Acta Math. **212** (2014), 59–140.

[MMR2] Y. Martel, F. Merle, and P. Raphael. Blow up for the critical gKdV equation II: minimal mass dynamics, arXiv 1204. 4624.

[MMR3] Y. Martel, F. Merle, and P. Raphael. Blow up for the critical gKdV equation III: exotic regimes, to appear in Ann. Sc. Norm. Super. Pisa Cl. Sci.

[MMT] Y. Martel, F. Merle and T-P, Tsai. Stability and asymptotic stability in the energy space of the sum of N solitons for subcritical gKdV equations, Comm. Math. Phys. **231** (2002), 347–373.

[MMTa1] J. Marzuola, J. Metcalfe and D. Tataru. Strichartz estimates and local smoothing estimates for asypmtotically flat Schrödinger equations, J. Funct. Anal. **255** (2008), 1497–1553.

[MMTa2] J. Marzuola, J. Metcalfe and D. Tataru. Quasilinear Schrödinger equations I : small data and quadratic interactions, Adv. Math. **231** (2012), 1151–1172.

[MMTa3] J. Marzuola, J. Metcalfe and D. Tataru. Quasilinear Schrödinger equations II : small data and cubic interactions, Kyoto Journal of Mathematics, **54** (2014), 529–546

[Mn] C. Muñoz. On the inelastic two-soliton collision for gKdV equations with general nonlinearity, Int. Math. Res. Notices (2010) no. 9, 1624–1719.

[Mo1] L. Molinet. Global well-posedness in L^2 for the periodic Benjamin-Ono equation, Amer. J. Math. **130** (2008), 635–683.

[Mo2] L. Molinet. A note on ill-posedness for the KdV equation, Diff. and Int. Eqs. **24** (2011), 759–765.

[MoPi] L. Molinet and D. Pilod. The Cauchy problem for the Benjamin-Ono equation in L^2 revisited. Anal. PDE **5** (2012), 365–395.

[MPSS] D. McLaughlin, G. Papanicolaou, C. Sulem, and P.L. Sulem. The focusing singularity of the cubic Schrödinger equation, Phys. Rev. A **34** (1986), 1200–1210.

[MPTT] C. Muscalu, J.Pipher, T. Tao, and C. Thiele, Bi-parameter paraproducts. Acta Math. **193** (2004), 269–296.

[MR1] L. Molinet and F. Ribaud. Well-posedness results for the generalized Benjamin-Ono equation with small initial data, J. Math. Pures Appl. **83** (2004), 277–311.

[MR2] L. Molinet and F. Ribaud. Well-posedness results for the generalized Benjamin-Ono equation with arbitrary large initial data, Int. Math. Res. Notices, **70** (2004), 3757–3795.

[MRS] F. Merle, P. Raphael, and J. Szeftel. On collapsing ring blow up solutions to the mass supercritical NLS, to appear in Duke Math. J. **163** (2014), 369–431.

[MSa] J. Maddocks and R. Sachs. On the stability of KdV multi-solitons, Comm. Pure Appl. Math. **46** (1993), 867–901.

[MSm] S.J. Montgomery-Smith. Time decay for the bounded mean oscillation of solutions of the Schrödinger and wave equations, Duke Math. J. **91** (1998), 393–408.

[MST1] L. Molinet, J.-C. Saut and N. Tzvetkov. Well-posedness and ill-posedness results for the Kadomtsev-Petviashvili-I equation, Duke Math. J. **115** (2002), 353–384.

[MST2] L. Molinet, J.-C. Saut and N. Tzvetkov. Global well-posedness for the KP-I equation, Math. Ann. **324** (2002), 255–275.

[MST3] L. Molinet, J.-C. Saut and N. Tzvetkov. Ill-posedness issues for the Benjamin-Ono and related equations, SIAM J. Math. Anal. **33** (2001), 982–988.

[MSWX] C. Miao, S. Shao, Y. Wu, and G. Xu. The low regularity global solutions for the critical generalized KdV equation, Dyn. Partial Differ. Equ. **7** (2010), 265–288.

[MT] F. Merle and Y. Tsutsumi. L^2–concentration of blow-up solutions for the nonlinear Schrödinger equation with critical power nonlinearity, J. Diff. Eqs. **84** (1990), 205–214.

[Mu] R. Miura. The Korteweg-de Vries equation: a survey of results, SIAM Rev. **18** (1976), 412–459.

[Mu1] R. Miura. Korteweg-de Vries equation and generalizations. I. A remarkable explicit nonlinear transformation, J. Math. Phys. **9** (1968), 1202–1204.

[MV] F. Merle and L. Vega. L^2 stability of solitons for KdV equation, Int. Math. Res. Notices **13** (2003), 735–753.

[MVV1] A. Moyua, A. Vargas, and L. Vega. Restriction theorems and maximal operators related to oscillatory integrals in $\mathbb{R}^3$, Duke Math. J. **96** (1999), 547–574.

[MVV2] A. Moyua, A. Vargas, and L. Vega. Schrödinger maximal function and restriction properties of the Fourier transform, Internat. Math. Res. Notices, **16** (1996), 793–815.

[Mz] S. Mizohata. *On the Cauchy problem*, Notes and Reports in Mathematics in Science and Engineering **3**, Academic Press Inc., Orlando, FL, 1985.

[N] A.C. Newell. *Solitons in Mathematics and Physics*, Regional Conference series in Applied Math. **48**, SIAM, 1985.

[Na] K. Nakanishi. Energy scattering for nonlinear Klein-Gordon and Schrödinger equations in spatial dimensions 1 and 2, J. Funct. Anal. **169** (1999), 201–225.

[Nh] J. Nahas. A decay property of solutions to the k-generalized KdV equation, Adv. Diff. Eqs. **17** (2012), 833–858.

[NhPo1] J. Nahas and G. Ponce. On the persistence properties of semilinear Schrödinger equations, Comm. P.D.E. **34** (2009), 1–20.

[NhPo2] J. Nahas and G. Ponce. On the persistence properties of solutions of nonlinear dispersive equations in weighted Sobolev spaces, RIMS Kokyuroku Besstatsu (RIMS Proc.) (2011), 23–36

[NSc] K. Nakanishi and W. Schlag. Global dynamics above the ground state energy for the cubic NLS equation in 3D, Calc. Var. Partial Differential Equations **44** (2012), no. 1–2, 1–45.

[NSU] A. Nahmod, A. Stefanov and K. Uhlenbeck. On Schrödinger maps, Comm. Pure Appl. Math. **56** (2003), 114–151.

[NSVZ] A. Nahmod, J. Shatah, L. Vega, and C. Zeng. Schrödinger maps and their associated frame systems, Int. Math. Res. Notices **21** (2007) Art, ID rnm088, 29 pp.

[NT] H. Nawa and M. Tsutsumi. On blow up for the pseudo conformal invariant nonlinear Schrödinger equations, Funk. Ekv. **32** (1989), 417–428.

[NTT1] K. Nakanishi, H. Takaoka and Y. Tsutsumi. Counterexamples to bilinear estimates related with the KdV equation and the nonlinear Schrödinger equation, Methods Appl. Anal. **8** (2001), 569–578.

[NTT2] K. Nakanishi, H. Takaoka and Y. Tsutsumi. Local well-posedness in low regularity of the mKdV equation with periodic boundary condition, Discrete Contin. Dyn. Syst. **28** (2010), 1635–1654.

[OgT] T. Ogawa and Y. Tsutsumi. Blow-up of H^1 solutions for the one-dimensional nonlinear Schrödinger equation with critical power nonlinearity, Proc. A.M.S. **111** (1991), 487–496.

[Ol] P.J. Olver. *Hamiltonian and non-Hamiltonian Models for Water Waves*, Lecture Notes in Phys. **195**, 273–290, Springer, Berlin, 1984.

[On] H. Ono. Algebraic solitary waves in stratified fluids, J. Phys. Soc. Japan **39** (1975), 1082–1091.

[OT1] T. Ozawa and Y. Tsutsumi. Space-time estimates for null gauge forms and nonlinear Schrödinger equations, Diff. and Int. Eqs **11** (1998), 201–222.

[OT2] T. Ozawa and Y. Tsutsumi. Existence and smoothing effect of solutions for the Zakharov equations, Publ. Res. Inst. Math. Sci. **28** (1992), 329–361.

[OT3] T. Ozawa and Y. Tsutsumi. The nonlinear Schrödinger limit and the initial layer of the Zakharov equations, Diff. and Int. Eqs. **5** (1992), 721–745.

[Oz] T. Ozawa. Exact blow-up solutions to the Cauchy problem for the Davey-Stewartson systems, Proc. Roy. Soc. London, Ser. A **436** (1992), 345–349.

[P1] H. Pecher. Nonlinear small data scattering for the wave and Klein-Gordon equation, Math Z **185** (1985), 261–270.

[P2] H. Pecher. Global well-posedness below energy space for the 1-dimensional Zakharov system, Internat. Math. Res. Notices **19** (2001), 1027–1056.

[Pd] D. Pilod. On the Cauchy problem for higher-order nonlinear dispersive equations, J. Diff. Eqs. **245** (2008), 2055–2077.

[Pe1] G. Perelman. On the formation of singularities in solutions of the critical nonlinear Schrödinger equation, Ann. Henri Poincaré **2** (2001), 605–673.

[Pe2] G. Perelman. Asymptotic stability of multisoliton solutions for nonlinear Schrödinger equation. Comm PDE **29** (2004), 1051–1095.

[Pi] H.R. Pitt. Theorems on Fourier series and power series, Duke Math. J. **3** (1937), 747–755.

[Pl] F. Planchon. On the Cauchy problem in the Besov spaces for a nonlinear Schrödinger equation, Commun. Contemp. Math. **2** (2000), 243–254.

[Po] G. Ponce. On the global well-posedness of the Benjamin-Ono equation, Diff. and Int. Eqs. **4** (1991), 527–542.

[Pp] M. Poppenberg. On the local wellposedness for quasilinear Schrödinger equations in arbitrary space dimension, J. Diff. Eqs. **172** (2001), 83–115.

[PSS] A. Patera, C. Sulem, and P.L. Sulem. Numerical simulation of singular solutions to the two dimensional cubic Schrödinger equation, Comm. Pure Appl. Math. **37** (1984), 755–778.

[PW] R. Pego and M. Weistein. Asymptotic stability of solitary waves, Comm. Math. Phys. **164** (1994), 305–349.

[Ra1] P. Raphael. Stability of the log-log bound for blow up solutions to the critical non linear Schrödinger equation Math. Ann. **331** (2005), 577–609.

[Ra2] P. Raphael. Existence and stability of a solution blowing up on a sphere for the L^2 supercritical nonlinear Schrödinger equation, Duke Math. J. **134** (2006), 199–258.

[Rd] W. Rudin. *Real and Complex Analyis*, McGraw-Hill, New York (1986).

[RH] P. Rosenau, and J.M. Hyman. Compactons: Solitons with Finite Wavelength, Phys. Rev. Letters, **70** (1993) 564–567.

[RS] I. Rodnianski and W. Schlag. Time decay for solutions of the Schrödinger equations with rough and time-dependent potentials, Invent. Math. **155** (2004), 451–513.

[Ru] J. Rauch. *Partial Differential Equations*, Graduate Texts in Mathematics **128**, Springer-Verlag, New York, 1991.

[RuR] J. Rauch and M. Reed. Nonlinear microlocal analysis of semilinear hyperbolic systems in one space dimension. Duke Math. J. **49** (1982), 397–475.

[RV] A. Ruiz and L. Vega. Local regularity of solutions to wave equations with time-dependent potentials, Duke Math. J. **76**, (1994) 913–940.

[RVi] E. Ryckman and M. Visan. Global well-posedness and scattering for the defocusing energy-critical nonlinear Schrödinger equation in $\mathbb{R}^{1+4}$, Amer. J. Math. **129** (2007), 1–60.

[RZ] L. Robbiano and C. Zuily. Strichartz estimates for the Schrödinger equation with variable coefficients, Mém. Soc. Math. Fr. (N.S.) No. 101-102 (2005), vi+208 pp.

[S1] E.M. Stein. The characterization of functions arising as potentials, Bull. A.M.S. **67**, (1961), 102–104.

[S2] E.M. Stein. *Singular Integrals and Differentiability Properties of Functions*, Princeton Univ. Press, Princeton, N.J. 1970.

[S3] E.M. Stein. *Oscillatory Integrals in Fourier Analysis*, Beijing Lectures in Harmonic Analysis. Princeton University Press (1986), 307–355.

[Sa] C. Sadosky. *Interpolation of Operators and Singular Integrals, an introduction to Harmonic Analysis*, Lecture Notes Pure and Appl. Math **14**, Marcel Dekker 1979.

[SaYa] J. Satsuma and N. Yajima. Initial value problems for one dimensional self-modulation of nonlinear waves in dispersive media, Suppl. Prig. Theor. Phys. **55** (1974), 284–306.

[Sc] P. Schuur. Asymptotic analysis of soliton problems, Lecture Notes in Mathematics **1232** Springer-Verlag, Berlin, 1986.

[Sch] M. Schwarz Jr. The initial value problem for the sequence of generalized Korteweg-de Vries equations, Adv. in Math. **54** (1984), no. 1, 22–56.

[Scl1] W. Schlag. Dispersive estimates for Schrödinger operators in dimension 2, Comm. Math. Phys. **257** (2005), 87–117.

[Scl2] W. Schlag. Stable manifolds for an orbitally unstable nonlinear Schrödinger equation, Ann. of Math. (2) **169** (2009), no. 1, 139–227.

[SCMc] A. Scott, F. Chu, and D. McLaughlin. The soliton: a new concept in applied science, Proc. IEEE **97** (1973), 1143–1183.

[Se] I.E. Segal. Space time decay for solutions of wave equations, Advance in Math **22** (1976), 305–311.

[SG] C. Shen and B. Guo. Almost conservation law and global rough solutions to a nonlinear Davey-Stewartson, J. Math. Anal. Appl. **318** (206), 365–379.

[ShS] J. Shatah and M. Struwe. Regularity results for nonlinear wave equations, Ann. of Math. **138** (1993), 503–518.

[SiT] T. Simon and E. Taflin. Wave operators and analytic solutions for systems of nonlinear Klein-Gordon and nonlinear Schrödinger equations, Comm. Math. Phys. **99** (1985), 541–562.

[Sj] P. Sjölin. Regularity of solutions to the Schrödinger equation, Duke Math. J. **55** (1987), 699–715.

[Sl] D. Salort. Dispersion and Strichartz inequalities for the one-dimensional Schrödinger equation with variable coefficients, Int. Math. Res. Notice **11** (2005), 687–700.

[Sr1] W.A. Strauss. *Nonlinear Equations*, Regional conference Series in Math. **73**, A.M.S. (1989).

[Sr2] W. Strauss. Existence of solitary waves in higher dimensions, Comm. Math. Phys. **55** (1977), 149–162.

[SS] E.M. Stein and R. Shakarchi. *Complex Analysis*, Princeton Lectures in Analysis, Princeton University Press, (2003).

[SS1] C. Sulem and P.-L. Sulem. Quelques résultats de régularité pour les équations de la turbulence de Langmuir, C. R. Acad. Sci. Paris Sér. A-B **289** (1979), A173–A176.

[SS2] C. Sulem and P.-L.Sulem. *The Nonlinear Schrödinger Equation*, Applied Mathematical Sciences **139**, Springer-Verlag, New York, 1999.

[SSB] P. -L. Sulem, C. Sulem and C. Bardos, *On the continuous limit for a system of classical spins*, Comm. Math. Phys. **107** (1986), 431–454.

[SSS] A. Sidi, C. Sulem, and P.-L. Sulem. On the long time behaviour of a generalized KdV equation, Acta Appl. Math. **7** (1986), 35–47.

[ST] J.-C. Saut and R. Temam. Remarks on the Korteweg-de Vries equation, Israel J. Math. **24** (1976), 78–87.

[St1] J.-C. Saut. Remarks on the generalized Kadomtsev-Petviashvili equations, Indiana Univ. Math. J. **42** (1993), 1011–1026.

[St2] J.-C. Saut. Quelques généralisations de l'équation de Korteweg-de Vries. II, J. Diff. Eqs. **33** (1979), 320–335.

[Sta1] G. Staffilani. On the growth of high Sobolev norms of solutions for KdV and Schrödinger equations, Duke Math. J. **86** (1997), 109–142.

[Sta2] G. Staffilani. On solutions for periodic generalized KdV equations, Internat. Math. Res. Notices **18** (1997), 899–917.

[Sta3] G. Staffilani. KdV and Almost Conservation Laws, Harmonic analysis at Mount Holyoke (South Hadley, MA, 2001), Contemp. Math., 320, Amer. Math. Soc., Providence, RI, (2003), 367–381.

[Str1] R.S. Strichartz. Multipliers on fractional Sobolev spaces, J. Math. Mech. **16** (1967), 1031–1060.

[Str2] R.S. Strichartz. A priori estimates for the wave equations and some applications, J. Funct. Anal. **5** (1970), 218–235.

[Str3] R.S. Strichartz. Restriction of Fourier Transform to quadratic surfaces and decay of solutions of wave equations, Duke Math J. **44** (1977), 705–714.

[StTa] G. Staffilani and D. Tataru. Strichartz estimates for the Schrödinger operator with nonsmooth coefficients, Comm. PDE **27** (2001), 1337–1372.

[Stw] M. Struwe. Globally regular solutions to the u^5 Klein-Gordon equation, Ann. Scuola Norm. Sup. Pisa Cl. Sci. **15** (1989), 495–513.

[Su1] L.Y. Sung. An inverse scattering transform for the Davey-Stewartson II equations. III, J. Math. Anal. Appl. **183** (1994), 477–494.

[Su2] L.Y. Sung. The Cauchy problem for the Ishimori equation, J. Funct. Anal. **139** (1996), 29–67.

[Sy] A. Soyeur. The Cauchy problem for the Ishimori equations, J. Funct. Anal. **105** (1992), 233–255.

[SW] E.M. Stein and G. Weiss. *Introduction to Fourier Analysis on Euclidean Spaces*, Princeton Univ. Press, Princeton N.J. 1971.

[SWe] S. Schochet and M. Weinstein. The nonlinear Schrödinger limit of the Zakharov equations governing the Langmuir turbulence, Comm. Math. Phys. **106** (1986), 569–580.

[T] M. Taylor. *Pseudo Differential Operators*, Princeton Univ. Press, Princeton, N.J. 1981.

[T1] Y. Tsutsumi. L^2 solutions for nonlinear Schrödinger equations and nonlinear groups, Funk. Ekva. **30** (1987), 115–125.

[T2] Y. Tsutsumi. Global strong solutions for nonlinear Schrödinger equations, Nonlinear Anal. **11** (1987), 1143–1154.

[T3] Y. Tsutsumi. Rates of L^2-concentration of blow-up solutions for the nonlinear Schrödinger equation with critical power, Nonlinear Anal. TMA **15** (1990), 545–565.

[Ta1] J. Takeuchi. A necessary condition for the well-posedness of the Cauchy problem for a certain class of evolution equations, Proc. Japan Acad. **50** (1974), 133–137.

[Ta2] S. Tarama. Analyticity of solutions of the Korteweg-de Vries equation, J. Math. Kyoto Univ. **44** (2004), no. 1, 1–32.

[Td] D. Tataru. Parametrices and dispersive estimates for Schrödinger operators with variable coefficients, to appear in Amer. J. Math.

[TF1] M. Tsutsumi and I. Fukuda. On solutions of the derivative nonlinear Schrödinger equation. Existence and uniqueness theorem, Funk. Ekva. **23** (1980), 259–277.

[TF2] M. Tsutsumi and I. Fukuda. On solutions of the derivative nonlinear Schrödinger equation II, Funk. Ekva. **24** (1981), 85–94.

[Th] L. Thomann. Low regularity for a quadratic Schrödinger equation on $\mathbb{T}$, Diff. and Int. Eqs. **24** (2011), 1073–1092.

[Tk1] H. Takaoka. Well-posedness for the Zakharov system with the periodic boundary condition, Diff. and Int. Eqs. **12** (1999), 789–810.

[Tk2] H. Takaoka. Well-posedness for the Kadomtsev-Petviashvili II equation, Adv. Diff. Eqs. **5** (2000), 1421–1443.

[Tm] P. Tomas. A restriction theorem for the Fourier transform, Bull. A.M.S. **81** (1975), 477–478.

[To1] T. Tao. Spherically averaged endpoint Strichartz estimates for the two-dimensional Schrödinger equation, Comm. PDE **25** (2000), 1471–1485.

[To2] T. Tao. A sharp bilinear restrictions estimate for paraboloids, Geom. Funct. Anal. **13** (2003), 1359–1384.

[To3] T. Tao. Multilinear weighted convolution of L^2-functions, and applications to nonlinear dispersive equations, Amer. J. Math. **123** (2001), 839–908.

[To4] T. Tao. Global well-posedness of the Benjamin-Ono equation in $H^1(\mathbb{R})$, J. Hyperbolic Differ. Equ. **1** (2004), 27–49.

[To5] T. Tao. Global well-posedness and scattering for the higher-dimensional energy-critical non-linear Schrödinger equation for radial data, New York J. Math. **11** (2005), 57–80

[To6] T. Tao. Scattering for the quartic generalised Korteweg-de Vries equation, J. Diff. Eqs. **232** (2007), 623–651.

[To7] T. Tao. Nonlinear Dispersive Equations. Local and Global Analysis, CBMS Regional Conference Series in concentration Mathematics, **106**, AMS (2006).

[To8] T. Tao. A (concentration) compactness attractor for high dimension non-linear Schrödinger equations, preprint.

[Ts] M. Tsutsumi. Nonexistence and instability of solutions of nonlinear Schrödinger equations, unpublished.

[TT] H. Takaoka and N. Tzvetkov. On the local regularity of the Kadomtsev-Petviashvili-II equation, Int. Math. Res. Notices **2** (2001), 77–114.

[TTs] H. Takaoka and Y. Tsutsumi. Well-posedness of the Cauchy problem for the modified KdV equation with periodic boundary condition, Int. Math. Res. Notices **56** (2004), 3009–3040.

[TV] T. Tao and A. Vargas. *A bilinear approach to cone multipliers II, Applications*, Geom. Funct. Anal. **10** (2000), 216–258.

[TVZ] T. Tao, M. Visan and X. Zhang. Global well-posedness and scattering for the mass-critical nonlinear Schrödinger equation for radial data in higher dimensions, Duke Math. J. **140** (2007), 165–202.

[Tz1] N. Tzvetkov. On the Cauchy problem for Kadomtsev-Petviashvili equation, Comm. PDE **24** (1999), 1367–1397.

[Tz2] N. Tzvetkov. Global low-regularity solutions for Kadomtsev-Petviashvili equation, Diff. and Int. Eqs. **13** (2000), 1289–1320.

[V] L. Vega. Schrödinger equations: pointwise convergence to the initial data, Proc. A.M.S. **102** (1988), 874–878.

[Ve] S. Vento, Well-posedness for the generalized Benjamin-Ono equations with arbitrary large initial data in the critical space, Int. Math. Res. Notices (2010), 297–319.

[Vi1] M.C. Vilela. Regularity of solutions to the free Schrödinger equation with radial initial data, Illinois J. Math. **45** (2001), 361–370.

[Vi2] M.C. Vilela. Las estimaciones de Strichartz bilineales en el contexto de la ecuación de Schrödinger, Ph.D thesis, Universidad del Pais Vasco, Spain 2003.

[Vs] M. Visan. The defocusing energy-critical nonlinear Schrödinger equation in higher dimensions, Duke Math. J. **138** (2007), 281–374.

[VV] A. Vargas and L. Vega. Global well-posedness for 1D non-linear Schrödinger equation for data with an infinite L^2 norm. J. Math. Pures Appl. **80** (2001), 1029–1044.

[W] M. Weinstein. On the structure and formation of singularities of solutions to nonlinear dispersive evolution equations, Comm. PDE. **11** (1986), 545–565.

[W1] M. Weinstein. *On the Solitary traveling wave of the generalized Korteweg-de Vries equation*, Lectures in Appl. Math. **23** (1986), 23–30.

[W2] M. Weinstein. Modulational stability of ground states of nonlinear Schrödinger equations, SIAM J. Math. Anal., **16** (1985), 472–491.

[W3] M. Weinstein. Nonlinear Schrödinger equations and sharp interpolation estimates, Comm. Math. Phys. **87** (1983), 567–576.

[W4] M. Weinstein. Lyapunov stability of ground states of nonlinear dispersive evolution equations, Comm. Pure Appl. Math. **39** (1986), 51–67.

[Wa] M. Wadati. The modified Korteweg-de Vries equation, J. Phys. Soc. Japan, **47** (1972), 1681.

[Wd] R. Weder. $L^p - L^{p'}$ estimates for the Schrödinger equation on the line and inverse scattering for the Schrödinger equation with potential, J. Funct. Anal. **170** (2000), 37–68.

[X] S.L. Xiong. An analytic solution of Burgers-Korteweg-de Vries equation, Chin. Sci. Bull. **34** (1989), 1158–1162.

[Y] K. Yajima. Existence of solutions for the Schrödinger evolution equations, Comm. Math. Phys. **110** (1987), 415–426.

[Yo] K. Yosida. Functional Analysis, (6th edition), Springer-Verlag 1980.

[Z] A. Zygmund. On Fourier coefficients and transforms of functions of two variables, Studia Math. **50** (1974), 189–201.

[Za] N. J. Zabusky. A Synergetic Approach to Problems of Nonlinear Dispersive Wave Propagation and Interaction, Nonlinear Partial Diff. Eqs. Ed. W.F. Ames N.Y. Academic Press (1967), 223–258.

[ZaKr] N. J. Zabusky and M. D. Kruskal. Interaction of solitons in a collisionless plasma and recurrence of initial states, Physical Review Letters **15** (1965), 240–243.

[Zk] V.E. Zakharov. Collapse of Langmuir waves, Sov. Phys. JEPT **35** (1972), 908–914.

[ZK] V.E. Zakharov and E.A. Kuznetsov. Multi-scale expansions in the theory of systems integrable by the inverse scattering method, Physica D **18** (1986), 455–463.

[ZS] V.E. Zakharov and A.B. Shabat. Exact theory of two dimensional self modulation of waves in nonlinear media, Sov. Phys. J.E.T.P. **34** (1972), 62–69.

[ZSh] V. E. Zakharov and E. I. Shulman. Degenerated dispersive laws, motion invariant and kinetic equations, Physica D (1980), 185–250.

Index

F. Linares, G. Ponce, *Introduction to Nonlinear Dispersive Equations*,
Universitext, DOI 10.1007/978-1-4939-2181-2

CPSIA information can be obtained
at www.ICGtesting.com
Printed in the USA
LVHW081548221222
735789LV00004B/124

-family studies

THE KINGS OF MISSISSIPPI

The Kings of Mississippi examines how a twentieth century black middle class family navigated life in rural Mississippi. The book introduces seven generations of a farming family and provides an organic examination of how the family experienced life and economic challenges as one of few middle-class black families living and working alongside the many struggling black and white sharecroppers and farmers in Gallman, Mississippi. Family narratives and census data across time and a socio-ecological lens help assess how race, religion, education, and key employment options influenced economic and non-economic outcomes. Family voices explain how intangible beliefs fuelled socioeconomic outcomes despite racial, gender, and economic stratification. The book also examines the effects of stratification changes across time, including: postmigration; inter- and intra-racial conflicts and compromises; and, strategic decisions and outcomes. The book provides an unexpected glimpse at how a family's ethos can foster upward mobility into the middle-class.

Sandra L. Barnes is a Sociology Professor in the Department of Human and Organizational Development at Vanderbilt University and the first female African American Assistant Vice Chancellor. She is the author of *Empowering Black Youth of Promise* (2016), *Live Long and Prosper, and The Costs of Being Poor* (2005).

Benita Blanford-Jones develops and leads several urban youth empowerment and educational mentoring programs. She also holds a Bachelor's degree in Sociology and a Master's degree in Human Services Administration from Indiana University Northwest.

CAMBRIDGE STUDIES IN STRATIFICATION ECONOMICS: ECONOMICS AND SOCIAL IDENTITY

Series Editor

William A. Darity Jr., *Duke University*

The Cambridge Studies in Stratification Economics: Economics and Social Identity series encourages book proposals that emphasize structural sources of group-based inequality, rather than cultural or genetic factors. Studies in this series will utilize the underlying economic principles of self-interested behavior and substantive rationality in conjunction with sociology's emphasis on group behavior and identity formation. The series is interdisciplinary, drawing authors from various fields, including economics, sociology, social psychology, history, and anthropology, with all projects focused on topics dealing with group-based inequality, identity, and economic well-being.

The Kings of Mississippi

Race, Religious Education, and the Making of a Middle-Class Black Family in the Segregated South

SANDRA L. BARNES
Vanderbilt University, Tennessee

BENITA BLANFORD-JONES
Independent Scholar

CAMBRIDGE
UNIVERSITY PRESS

University Printing House, Cambridge CB2 8BS, United Kingdom

One Liberty Plaza, 20th Floor, New York, NY 10006, USA

477 Williamstown Road, Port Melbourne, VIC 3207, Australia

314–321, 3rd Floor, Plot 3, Splendor Forum, Jasola District Centre,
New Delhi – 110025, India

79 Anson Road, #06–04/06, Singapore 079906

Cambridge University Press is part of the University of Cambridge.

It furthers the University's mission by disseminating knowledge in the pursuit of education, learning, and research at the highest international levels of excellence.

www.cambridge.org
Information on this title: www.cambridge.org/9781108424066
DOI: 10.1017/9781108539654

First published 2019

Printed and bound in Great Britain by Clays, Elcograf S.p.A.

A catalogue record for this publication is available from the British Library.

Library of Congress Cataloging-in-Publication Data
Names: Barnes, Sandra L., author. | Blanford-Jones, Benita, 1973– author.
Title: The Kings of Mississippi : race, religious education, and the making of a
middle-class black family in the segregated South / Sandra L. Barnes,
Benita Blanford-Jones.
Description: Cambridge, United Kingdom ; New York, NY : Cambridge University Press,
2019. | Series: Cambridge studies in stratification economics : economics and social identity |
Includes bibliographical references and index.
Identifiers: LCCN 2018041244 | ISBN 9781108424066
Subjects: LCSH: African American families – Mississippi – Social conditions – 20th century.
| Middle class African Americans – Mississippi – Social conditions – 20th century. | Middle
class families – Mississippi – Social conditions – 20th century. | King family.
Classification: LCC E185.93.M6 B36 2019 | DDC 306.85/0899607307620904–dc23
LC record available at https://lccn.loc.gov/2018041244

ISBN 978-1-108-42406-6 Hardback
ISBN 978-1-108-43933-6 Paperback

To black mothers and other mothers everywhere. No longer mules of the world, but its anchors.

Contents

Figures

Tables

Tables

Introduction

A Black Family from Mississippi as a Socio-Ecological Phenomenon

The King family was a twentieth-century anomaly – a middle-class black family living in rural Mississippi.[1] Academic studies, mainstream writing, and anecdotes corroborate the same reality – that blacks living in the historic South experienced deleterious conditions due to racism, segregation, and de jure as well as de facto discrimination. Whether prior to or during Reconstruction or as a result of Jim Crow, they were subjected to profound and unrelenting economic, political, legal, and social oppression, often accompanied by the threat of violence, particularly lynching.[2] How did black families navigate these systemic, oppressive conditions daily?[3] What strategies did they use? And how could becoming middle class be possible?

This book presents the lives and experiences of seven generations of a black family that originated in Mississippi. Limited mixed-methodological, multi-disciplinary research has been performed on this topic. This book is one response to this omission. We rely on sociology and ecology (or a socio-ecological lens) as well as their own voices to examine how race, religion, education, and their intersection as a familial ethos influenced economic and non-economic outcomes of the King family. Empirical reports document the context. Narratives explain how intangible beliefs linked to religious and secular education fueled socioeconomic outcomes – under the ever-present specter of racial, gender, and economic stratification.[4] The Kings did not reflect the "well-scrubbed black middle class," but experienced economic challenges as they lived and worked alongside the many struggling black and white sharecroppers and farmers in Gallman, Mississippi.[5] Family members do not romanticize their lives and experiences, but rather candidly describe trials and triumphs, inter- as well as intra-racial problems, and views and values that help expand our understanding of what constituted middle-class living for rural blacks during that period.[6]

A DEFINITION OF THE MIDDLE CLASS

This analysis operationalizes "middle class" in a certain way while also acknowledging alternative definitions as well as the elusiveness of this broad socioeconomic classification.[7] Because their lives played out in largely agrarian locations, we focus on a definition of middle class that is linked to dynamics upon which people were stratified in farming communities: home and land ownership; high school completion (considered the principal educational milestone in rural spaces at the time that often eluded sharecroppers); perceived economic security; mindset about one's class position; formal and informal networks; belief in delayed gratification; fears about downward mobility; and efforts to stave off poverty. Moreover, based on an ecology of segregation, it is important to consider how a middle-class position was used to navigate its menacing, potentially debilitating effects.

Thus a three-pronged definition of middle class is employed here that includes a structural position in the broader system of class stratification based on social group position between labor and capital; economic position, narrowly interpreted, based on factors such as relative wealth, income, educational attainment, and/or occupational status; and certain beliefs associated with this class position such as delayed gratification and respectability norms associated with both religious and formal education. We do not dispute other possible operationalizations. This definition reflects the agricultural context in which the early generations of the King family lived, challenges notions about what constituted the middle class during that period, and creates an opportunity to broaden scholarly research on alternate ways to consider stratified spaces.

To our knowledge, few studies have presented a *generational narrative* that explains how some black rural families negotiated racially charged locales in search of security, stability, and safety. Moreover, this counter-narrative centers the potentially mediating effects of religion, education, and their nexus (referred to here as "religious education") in fostering beneficial decisions and outcomes.[8] But the King family did not achieve their goals in isolation. Just as chronic inequality and oppression were part of the historic landscape in Mississippi, so were predominately black churches and black schools designed to help blacks survive with pride, courage, and dignity.[9] *Kings of Mississippi* documents one black family's attempt to be adaptive and resilient during discriminatory times and in discriminatory spaces. And just as the economics of what it means to be middle class is important, the *how* and *why* of their experiences are equally salient and central here.

BOOK OBJECTIVES

To be clear, *Kings of Mississippi* is not a tale about grit. Nor is it a story of exceptionalism, but rather chronicles a family's attempt to navigate racial and economic disparities in a highly stratified society. And in sharing their story, the Kings provide a glimpse into dimensions of the black experience in the United States circa 1900–2015. National and local economies provide a backdrop on the experiences, perspectives, and responses to racism, poverty, discrimination, oppression, and other inequities in Gallman, Mississippi, and post-migration. Some of the accounts by King family members are colorful; some are grave. Individuals present vivid, nuanced portraits and recollections of twentieth-century life in the segregated South; strategic inter- and intra-racial interactions for family survival; and sacrifice as well as strain during migrations. Central in their history is the role of religion, as well as formal and informal education, and their intersection as buoys to traverse societal stratification and broader social change.[10]

The following questions are considered: (1) How were local blacks and whites stratified in Gallman, Mississippi, based on factors such as race, class, and gender? (2) What was the nature of inequality as experienced, in general, by blacks in the area? (3) How did the Kings understand, experience, and navigate life in the South and, for some members, migration North and to the Southwest? (4) What kinds of inter- and intra-racial conflicts and/or compromises did they experience? and (5) How did formal as well as informal education, religion, organizations, and social structures influence family decisions and outcomes? Answers to these queries will also help explain how black churches and schools could influence attitudes and actions in ways that were both ideological and practical. Personal stories, family reflections, and empirical data inform this multi-disciplinary endeavor. The legacy of the King family has academic and applied import by illuming, analyzing, and documenting successes, pit-falls, and impasses that individuals believe shaped and continue to shape their family's lives.

RELIGIOUS EDUCATION AS A SOCIAL, CULTURAL, AND ECONOMIC BUOY

King family members posit that religious education was just as salient as formal education for their upward mobility. Its espousal and application varied, yet Christianity specifically emerged as a central motivator for action or, in some instances, inaction. Although this project presents economic, political, and social contexts in which the Kings found themselves, it also documents biblically based proactive and sometimes reactive responses to challenges. Narratives, many as reflections by older family members and peers, illustrate attitudes and actions among the Kings who endeavored to achieve their goals *despite* harrowing constraints. Equally important, religious education helps explain why certain family members believed that they would achieve economic stability – even though most blacks around them had not done so and most whites directly and tacitly endeavored to squelch these aspirations. The explicit use of scripture and religious practices as cultural tools is central to their narratives. Explanations linked to biblical practices of prayer, obedience, peacekeeping, and the Protestant Ethic provide further insight about family decision-making processes. In addition to assessing the effects of practical application of biblical dictates, insights emerge about inter-faith differences and outcomes based on variations in factors such as education, age, and gender. This book also assesses the role of religion in positioning the King family as "good Christian Negroes" and, later, "good Christian blacks" to their black and/or white counterparts. In addition, we consider how perceived deservedness, largely based on the matriarch's "religious reputation" among influential whites, provided crucial alliances as well as a protective mechanism against animosity and physical violence from local whites for whom racism was a way of life. But how is *religious education* defined here and how did it influence this family?

Religious Education Defined

For the Kings and their progenitors, religious education was steeped in Christianity and instructionally informed by the bible. However, this concept extends beyond "The Book" to include context-specific values and practices. As a guiding process, it was decidedly conservative in its stance on family expectations and gender roles, particularly among the four earliest generations of Kings. Prescriptions around work and the appropriate work ethic were clear. However, in contrast, religious education often manifested in views that were unexpected for the time, particularly among certain female family members. Formative biblical tenets included the Virtuous Woman of Proverbs 31, Greatest Commandment in Matthew 22:34–40, and parental instructions in Proverbs 22:6. Yet race-based differences in religious education are best summarized in *All God's Dangers* (Rosengarten 1974): "It's just a different performance from one church to another. Same God, but they [black Christians] serves Him different; different enough for you to set and look it through, and hear the difference in the sermons that's preached and the songs that's sung" (p. 455).

The family did not seem to be beholden to a particular theology. Yet key tenets of their religious education were decidedly practical and stressed common sense and core Christian principals. We contend that these dual practical and religious dimensions enabled scriptural application in ways the Kings believed to be authentic and empowering. Moreover, how the Kings typically applied scripture had a decidedly socio-ecological thrust because it reflected the specific contexts and circumstances in which they found themselves. So although core tenants existed, some aspects of them could be adapted. Lastly, religious education as illustrated by the Kings reflected a certain comfortability combining the bible with formal and/or informal educational practices. In this way, "if any would not work, neither should he eat" in 2 Thessalonians 3:10 took on a specific charge and fueled the drive among some family members to earn as much formal education as possible to best compete for gainful employment.[11] And just as such beliefs and behavior could have a positive impact, like other ideologies, they could also be used as a chastening rod. Readers should note that this definition does not suggest that every King family member understood, espoused, or applied religious education identically. To the contrary, this definition reflects a dynamic, broad-based rubric that was differentially employed by family members and thus differentially shaped their lives, in general, and socioeconomic trajectories, in particular. As becomes evident, religious education reflected a broad constellation of formal and informal instruction and guidance that was the bedrock for the daily lives of the King family in religious, economic, social, political, cultural, and practical ways.

ABOUT THE KING FAMILY: PERFORMING A BLACK FAMILY GENERATIONAL STUDY

The King family lived in the rural farming community of Gallman, Mississippi, in the early and mid-1900s (details on both the town and state are provided in Chapter 1). Like most blacks, they were initially sharecroppers on white-owned farms.[12] Yet a series of events gradually altered the family's socioeconomic trajectory. These include: the family matriarch, Irma's, dream of owning her own land and livestock; military stints of her husband and family patriarch, George Franklin, and other family members; inter-racial networks forged by Irma as a domestic;[13] high school graduations of their children;[14] and "migration spells" by relatives who relocated North and Southwest.

As a result of such experiences, by the mid-1900s, the King family owned 20 acres of land including a fishing pond (would later become 40 acres), a house, numerous livestock, and several vehicles. We are not suggesting that the King family was the *only* middle-class black family in Gallman, that every family member fared similarly, or that their lives paralleled those commonly associated with the black middle class in other studies, but rather that the family's experiences provide an important portrait for understanding some of the religious, economic, and social dynamics that led to a certain degree of economic security in the South.[15] The *how* of their story is also important. Some economists postulate that "individual's social capital, or family and community background, yields nontradable social relations and cultural attributes that include economic well-being . . . social capital is a vital element of inter-generational transmission of socio-economic status" (Mason 2003: 60). For the Kings, in addition to their homestead, the most salient social capital took the form of religious as well as secular education and practical trades. We attempt to uncover how members of this family carved out a niche in this farming community that enabled them to weather an extremely harsh, often unpredictable racial, social, and economic climate.[16]

Black families have historically been mainstays in black communities; so have black churches and black schools. Many studies suggest that these three entities are *stewards* of the black community.[17] Yet their effects vary and had differential effects on the Kings. These sites, black churches and black schools, were selected here because of their historic interconnectedness in the broader black community and often under-estimated influence on efforts to forge socioeconomic stability.[18] We seek to inform existing literature with nuanced portraits of black family life as part of these institutions. Just as inequality is understood and experienced differently – even within the same family – beliefs, priorities, and responses associated with social problems like segregation, racism, discrimination, and poverty had differential effects on members of this family. Moreover, conditions may appear different when one is experiencing them rather than reflecting upon them years later. Our research considers the dynamics of oppression and resistance based on narratives from people who lived through egregious conditions and from relatives who would later benefit. We also consider how beliefs and behavior cultivated in the South over time became central to survival and successes there and in other locations.

Applying a Socio-Ecological Lens to the Study of Black Family Life in the Segregated South

A socio-ecological theoretical framework combines aspects of sociology and ecology to examine phenomena. First, this lens focuses on the influence of environmental factors that are historic and structural across time. In this book, ecology helps frame the broader social context and the King family's place in it.[19] Additionally, a sociological lens enables the study of, among other dynamics, the influence of religion and education on the King's behavior and beliefs. Additionally, literature from the sociology of religion enables us to consider effects, motivators, and explanations for family members in ways uncommon in ecology. This two-fold lens also focuses on the intrinsic and extrinsic impact of religiosity, particularly Christian edicts, that guided and continue to guide most of the Kings. Insights from the sociology of education help assess how organized and informal educational tenets and practices influenced the family ethos (refer to Figure I.1). This multi-faceted model is employed to better understand individual- to structural-level experiences and outcomes. For example, in addition to assessing how family members were influenced *daily* by factors such as racism, sexism, segregation, and the specter of poverty that were systemic/macro in origin, it is also crucial to focus on the effects of the local black church and schools, black migration, labor market patterns, the military, and racial tensions about group position and property attainment across time.[20]

However, unlike the traditional ecological emphasis on human or childhood development, this analysis concentrates on *individual* as well as *collective familial development*. Although ecological theory has been used to study environments such as schools, to our knowledge, it has not included the additional sociological lenses on the black family, black community, black religiosity, and the specific impact of religious education. Furthermore, ecological theory is often utilized to study micro-level developmental change. We too examine individual experiences, in this instance, those of King family members. However, we also document broader dynamics linked to black churches and black schools as *familial assets* – their ideologies, cultures, and programs that the Kings suggest were purposefully harnessed to help meet their goals and foster dignity and respect in a society designed to deny these basic rights (refer to the appendix about the research methods used here).

Although the family matriarch, Irma, never left Gallman, Mississippi, many of her siblings migrated circa 1920–1950 to the following locales:

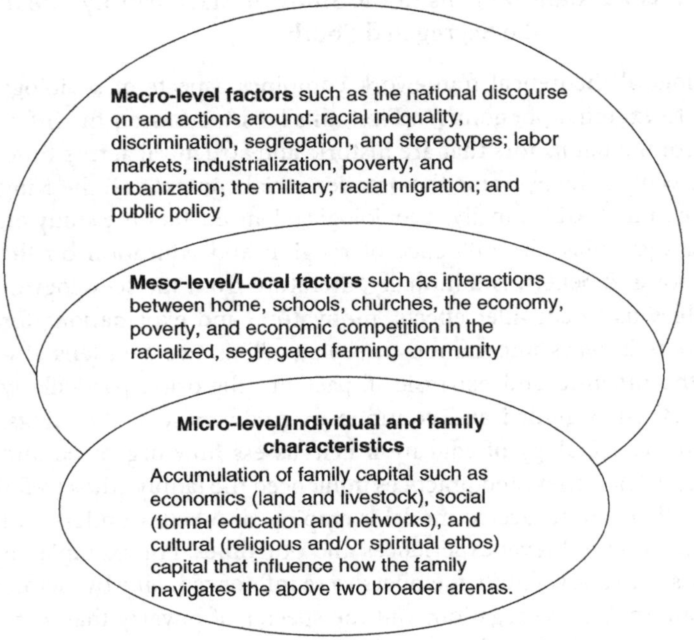

Macro-level factors such as the national discourse on and actions around: racial inequality, discrimination, segregation, and stereotypes; labor markets, industrialization, poverty, and urbanization; the military; racial migration; and public policy

Meso-level/Local factors such as interactions between home, schools, churches, the economy, poverty, and economic competition in the racialized, segregated farming community

Micro-level/Individual and family characteristics
Accumulation of family capital such as economics (land and livestock), social (formal education and networks), and cultural (religious and/or spiritual ethos) capital that influence how the family navigates the above two broader arenas.

Figure I.1: Socio-Ecological Model for *Kings of Mississippi*
Key: Application and Extension of Bronfenbrenner (1977, 1979, 1986)

Jackson, Mississippi; Chicago, Illinois; Detroit, Michigan; Toledo, Ohio; St. Louis, Missouri; and, Denver, Colorado.[21] Several decades later, her children would all end up in Gary, Indiana. These seven socio-ecological sites are of particular interest because their demographic and social distinctions played important roles in the mobility experiences of the Kings and their offspring. In addition, inter- and intra-racial comparisons of factors commonly associated with socioeconomic status such as income, education, and poverty become particularly salient for family members who migrated to urban areas. And just as resettlement patterns directly and indirectly affected the Kings, local decisions also shaped their lives, particularly in Copiah County, where many of the first four generations of family members initially lived.[22] Narratives and empirical information help gauge dynamics such as farm ownership by race and racial education gaps. A family tree and timeline help document lineage, offspring, and related economic outcomes. The way family members *understood their class position* and explained their economic trajectories is equally important.

ABOUT THE BOOK AND ITS CONTRIBUTIONS

How do race, place, class, gender, and religion play out in the daily lives of one black family across time? Do their experiences help to better understand black families more broadly? This study extends existing research by focusing on these factors and their intersection to examine anew black family and Black Church dynamics. This research cannot do justice to every aspect of black life for the Kings and their offspring, but rather documents some of the complexities evident in the historic black experience in light of religion-informed motivations and behavior.

Chapter 1 provides background information about the King family matriarch's abiding wish – to own land and a milk cow. Despite the ever-present shadow of segregation and resulting inequities and discrimination, her simple desire shaped the family ethos, behavior, and socioeconomic outcomes. Moreover, the chapter positions religion and secular education as important features in the lives of many blacks in the rural South and the primary influencers for the Kings. Generational narratives, demographic and military statistics, and the family typology help explain seminal experiences, decisions, and dangers as family members pursued their understanding of economic and non-economic facets of middle-class status. The several Great Migrations provide the context for Chapter 2 as King family members trekked to new spaces in northern and southwestern cities in search of employment opportunities and to escape Jim Crowism. Decisions to leave Mississippi were just as strategic as those to remain. Using life on the family's homestead as the point of departure, the chapter recognizes a linked-fate mentality that helped sustain relatives economically, emotionally, and practically. Employment and labor market statistics also help explain how structural dynamics linked to segregation influenced efforts to be agentic. These quantitative and qualitative results document the relationship between behavior, beliefs, and economic outcomes to develop and reinforce customs such as self- and group-efficacy.

Various forms of education that were central to this family's socioeconomic and personal outcomes are juxtaposed in Chapter 3. Major life choices associated with acquiring social capital, such as formal education, espousing religiosity, and navigating rural spaces, influenced personal and familial outcomes. Family experiences illustrate the seminal effects of formal education and training. The broader educational landscape of such outcomes for whites and blacks provides the ecological framework to assess educational and occupational outcomes for the Kings. In Chapter 4, *blurred* and *bordered socio-ecological boundaries* are introduced as pivotal concepts to assess some of the non-traditional religious, educational, and practical approaches used by family members to make decisions and address challenges. Moreover, the concept of *boundary crossing* is presented as a strategic decision-making process to help navigate difficult situations. The impact of familial dictates on these same decisions is also considered, including examples of how the three above-noted concepts provide insight into the complex interplay between life in Mississippi that often resulted in tensions – and incongruent beliefs and behavior. Based on adherence to *religious education*, specific guiding scriptures reinforce the family ethos and encourage generational outcomes such as educational attainment, delayed gratification, and belief in the Protestant Ethic, and, in some instances, to stymie inappropriate lifestyles. Using the family's local church in Gallman, Clear Creek No. 1 Missionary Baptist Church, as the point of reference, the chapter examines the influence of religious education on early generations of Kings.

Chapter 5 details varied post-migration experiences for King offspring as urban living influenced both the life chances and quality of life of later generations of family members who were no longer under the direct influence of the family patriarch and matriarch. The concept *migration spells* and the continued importance of the family homestead illustrate another less often considered outcome of middle-class status for blacks. Migration spells are broadly defined here as strategic time periods during which family members lived on the King's farm in Mississippi or with other relatives for economic, non-economic, and practical reasons. According to family members, these living arrangements, some short-term and others much longer in length, enabled many of them to navigate challenges and/or help establish or re-establish themselves financially in ways that would have otherwise been extremely difficult. The chapter also considers the varying mediating effects of religious education and the manufacturing sector on how individuals responded to social forces associated with employment discrimination and urban conditions as well as how more recent generations understand religious education.

How does the King family experience inform us about the contemporary black family? What are some of the implications? The concluding chapter responds by revisiting the life and legacy of the family matriarch, Irma. It considers how her ideology about religion, education, and family ultimately informed the family's ethos and outcomes associated with middle-class status. Additionally, by broadly comparing and contrasting demographics between contemporary black and white families, the dual realities of routinization and resistance among blacks are examined to suggest broader predictions about black family approaches to survive and thrive in the face of continued socioeconomic challenges and inequities in the USA. Lastly, a *new millennium DuBoisian Mode of Inquiry* grounds a discussion of the impact of black family culture apparent among the Kings to inform a more culturally sensitive socio-ecological theory as a nuanced lens to better understand, equip, and empower contemporary black families.

1

"My Own Land and a Milk Cow": Race, Space, Class, and Gender as Embedded Elements of a Black Southern Terrain

Irma only wished for two things – to own her own land and a milk cow. She remembered the harsh conditions during her childhood – a dirt-poor life in a shack with dirt floors. Food was scarce, but work was constant – all to make the white landowner wealthy. Irma wanted better for her children. She prayed to God for grace and mercy; she had faith that God would respond. So despite the continual menace of segregation and immediacy of poverty, Irma worked to make her dream a reality. To her, land and livestock represented family security, stability, and safety. Unbeknownst to Irma at the time, her simple dream was the impetus behind what would eventually become a middle-class life for her family and children.

Scholars suggest that being middle class is as much about mindset as material possessions.[1] This chapter introduces key King family members, their personal convictions, and several pivotal decisions that fostered their evolution into the middle class. James and Loren King, their daughter, Janice King, as well as Janice's oldest living daughter, Irma, and her husband, George (Big Papa and Big Mama, as they were affectionately called), are central to the account. The latter two individuals would become the family's twentieth-century patriarch and matriarch. Family members and peers who were youth or young adults in the early to mid-1900s describe navigating segregated spaces when being considered a "model Negro family" or "uppity Negroes" could have dramatically different consequences. Their stories illustrate how the Kings understood race, class, gender, and place as well as the economic and non-economic dynamics they associate with the middle class. Moreover, religious education as largely informed by Christianity, emerges as the ideological thread running through the King family ethos. Ecology is also central given the daily exposure to inequality in predominately white spaces and family members' attempts to intentionally develop socio-psychological counter-narratives and buffers in homes, churches, and schools as well as economic buoys via farming and the military. Mississippi and Gallman, in particular, provide the historical backdrop for this family's experiences. And thick descriptions of life in a bifurcated environment, including labor markets, illumine family decision-making.[2]

MISSISSIPPI, COPIAH COUNTY, GALLMAN, AND HAZLEHURST – THE IMPORTANCE OF LAND

Named after the river that runs along its western border, in 1817, Mississippi became the twentieth state in the Union. Rather than duplicate existing literature on its economic, political, social, and cultural history, the following summary provides the socio-ecological context that informs early King family experiences.[3] We focus on how land – in the state, county, and locally in Gallman-influenced power dynamics as well as individual and group decisions, experiences, and outcomes for the Kings (refer to Figure 1.1). Cotton and the slave labor of blacks fueled the economy in Mississippi during the first half of the nineteenth century.[4] Succession in 1861, rather than acquiesce, meant Mississippi experienced severe economic hardship during and after the Civil War and reflected the continued political conservatism for which the state is still known. According to census figures, in 1900, about 89.7 percent of blacks living in the USA lived in the South. More than 75 percent of them lived in rural areas as compared to whites. Additionally, of the 1,551,270 residents in 1900 in Mississippi, 907,630 or 58.5 percent were black.[5]

Despite its location in the Bible Belt,[6] whites in Mississippi espoused and implemented some the harshest discrimination[7] and violence against blacks in US history, as they attempted to maintain a legacy of control via politico-economic sanctions, disenfranchisement, social norms, and terror. The economic benefits of racial oppression were also apparent. The following grave recollection in *All God's Dangers* (1974) corroborates this bifurcated southern ecology and references dynamics germane in this book – black families, church, class position, and segregation:

> Niggers was scared . . . white folks in this country didn't allow niggers to have no organization . . . and watched you . . . didn't allow you to associate in a crowd, unless it was your family or your church . . . they was white people; they classed theirselves over the colored. (Pp. 297–8)

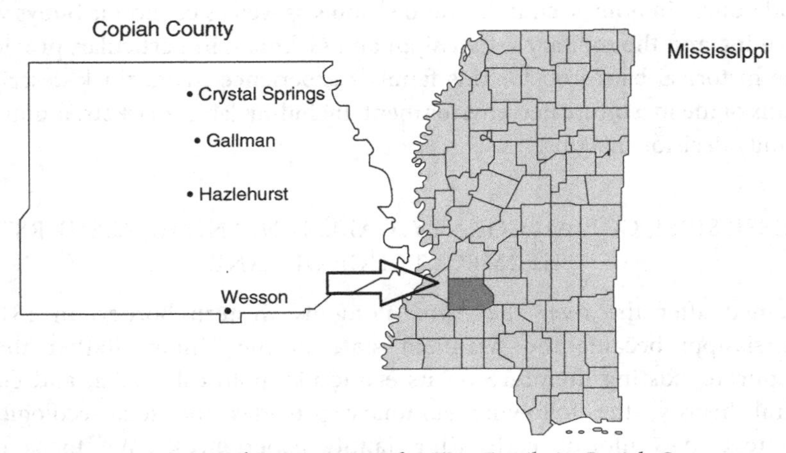

Figure 1.1: Map of Key Locations for King Family in Copiah County (Gallman, Hazlehurst, Crystal Springs, and Wesson, Mississippi)

Increasing numbers of economists, particularly neo-Marxists, continue to be critical of theorists who "discount the profitability of discrimination to capitalists ... when we view white worker's income and employment prospects relative to those of their African American counterparts, being white still pays" (Williams 1993: 221, 227). Continued ecological challenges from 1900 to the 1930s, including cotton crop losses, floods, droughts, and a boll weevil infestation, meant Mississippi became less attractive to black residents and white northern investors.[8] Several decades later, the labor market for blacks followed a common pattern that would continue post-migration:

> For at least 75 years after emancipation, the vast majority of black families worked in Southern agriculture ... The majority of black women worked in the fields, with the male head of the extended family unit receiving the wages earned by the family unit ... the other primary occupation for black women's wage labor was domestic work. Young black girls were prepared by their families for domestic work ... historically the classic pattern of employment for African American men and women has been higher paying yet less secure work for black men as contrasted with lower-paying, more plentiful work for black women. (Collins 1990: 53–4, 59)

Mississippi's gendered, segregated labor market meant either low wages or no wages for black males and employment akin to servitude for their female counterparts. Moreover, segregation at every level of society resulted in deleterious conditions, poor quality of life, and constrained life chances for blacks that manifested negatively, but differently, for males and females.[9] In this way, according to Cotton (1993):

> Labor market outcomes are generated in a racially collusive, oligopolistic environment where white employers and white workers are mutually benefitted by the underpayment and underutilization of black workers. (P. 200)

Several migrations by blacks during the early and middle of the 1900s as well as subsequent Civil Rights Movement activism resulted in a shift in Mississippi's racial demographics. This out-migration meant that, for states ranked by population in 1900, Mississippi was ranked 20th. By 1950, it was 26th and by 2000, Mississippi was ranked 31st. The exodus of over 400,000 blacks has meant that whites are currently the majority. Thus the state's early dependence on cheap black labor and continued focus on agrarian rather than industrial and infrastructure development resulted in economic challenges before and long after the Civil War.

The towns of Gallman, Crystal Springs, and Hazlehurst where the Kings lived and/or frequented were part of Copiah County. The county was established in 1823. Its initial county seat was established in 1824 in Gallatin, but was moved to Hazlehurst in 1872. As presented in Table 1.1, from the early to mid-1900s, the county's population ranged from about 29,000 to 36,000 residents. Relatively similar percentages of females and males lived in the county. The total number of farms each decade ranged from about 3,500 to 5,300. White farm owners declined during this fifty-year period – peaking at 72.2 percent in 1920 and lowest at 23.3 percent the following decade. Non-white farm ownership rates (10.6 to 31.1 percent) were relatively lower than their white peers, but vacillated less dramatically. Although politically and economically disenfranchised, non-whites were the numeric majority in Copiah County. The relatively greater presence, post-slavery farming successes, and initial political inroads of blacks, resulted in white animosity and fear that fueled Jim Crowism and segregation.

Hazlehurst and Gallman, Mississippi, as Socio-Ecological Sites

Hazlehurst is located approximately 30 miles south of the state capital of Jackson, Mississippi.[10] Founded on November 3, 1986, by European settlers, the city was named after Colonel George H. Hazlehurst, an engineer responsible for constructing the New Orleans, Jackson and Great Northern Railroad in November 1865. The railway fueled growth and ultimately led

Table 1.1 *Demographics for Copiah County, Mississippi (1900–1950)*

	1900	1910	1920	1930	1940	1950
Total Population	34,395	35,914	28,672	31,614	33,974	30,493
White (#, %)	16,355 (47.6)	15,927 (44.4)	13,567 (47.3)	16,000 (50.6)	16,107 (47.4)	14,210 (46.6)
Non-White (#, %)	18,040 (52.4)	19,981 (55.6)	15,102 (52.7)	15,612 (49.4)	17,863 (52.6)	16,283 (53.4)
Male (#, %)	16,981 (49.4)	17,666 (49.2)	14,188 (49.5)	15,837 (50.1)	16,834 (49.6)	14,876 (48.8)
White Male (#, %)	8,001 (48.9)	7,919 (49.7)	6,826 (50.3)	8,173 (51.1)	8,051 (50.0)	6,945 (48.9)
Non-White Male (#, %)	8,976 (49.8)	9,743 (48.8)	7,359 (48.7)	7,663 (49.1)	8,783 (52.2)	7,931 (53.3)
Total # Farms	4,500	5,324	4,349	4,962	4,526	3,498
White Owned Farms (#, %)	1,261 (60.9)	1,370 (69.0)	1,303 (72.2)	1,158 (23.3)	1,976 (49.0)	1,205 (34.3)
Non-White Owned Farms (#, %)	446 (18.4)	617 (31.1)	501 (27.8)	528 (10.6)	872 (20.7)	573 (16.4)
Average Farm Value ($, building & land)	446	1,066	3,334	2,119	1,124	3,505
Average Farm Size (acres)	87	72	87	67	78	107

Key: Taken from Social Explorer Dataset (SE), Census 1920–1950, Digitally transcribed by Inter-university Consortium for Political and Social Research. Edited, verified by Michael Haines. Compiled, edited and verified by Social Explorer and United States Census of Agriculture (http://usda.mannlib.cornell.edu/usda/AgCensusImages/1950/01/22/1801/28534800v1p22_Ch2.pdf and http://usda.mannlib.cornell.edu/usda/AgCensusImages/1940/01/33/1266/Table-01.pdf). For the sake of parsimony, male statistics are provided (female percentages can be calculated by subtracting male figures from totals). The majority of "non-white" is black.

to the city becoming the county seat from the original seat in nearby Gallatin. On November 3, 1865, the State of Mississippi issued a municipal charter to Hazlehurst. According to census figures, the approximately 662 residents in 1870 has increased over six-fold to 4,009 in 2016–2017. In 1940, 3,124 persons lived in Hazlehurst; by 1950 the number had risen to 3,397 persons, an 8.7 percent increase from 1940.[11] Hazlehurst's economy was primarily agricultural, including cotton and vegetables. From the 1920s to the 1940s, it had one of the most lucrative tomato shipping industries in the country. By 1950, 795 of its residents were white males and 758 were non-white males; 892 were white females and 932 were non-white females.

Gallman is a small town founded in the 1950s (i.e., classified as a minor civil division) and located approximately five miles north of Hazlehurst.[12] The two locales are inextricably linked geographically, economically, politically, and socially as well as to the lives of the King family. Black residents in Gallman, often referred to traveling to Hazlehurst as "going to town." Gallman has achieved some notoriety due to its historical perseveration sites called the "Gallman Historic District."[13] Incorporated in 1896, the town's earliest European settlers were farmers, most who emigrated from North and South Carolina. The settlement was originally called Mullgrove, but was renamed Gallman in honor of Reverend Willard B. Gallman, a local Baptist minister and community leader. Gallman established a reputation because of the development of its transportation infrastructure. Stage and railroad services were situated there, increasing the area's significance.[14]

Railway accessibility, truck farming, and the fertility of the terrain positioned Gallman for agricultural prominence. By the early 1870s, local farmers produced crops such as peas, cabbage, beans, tomatoes, and asparagus. The train depot meant that farmers could easily ship large boxcars of vegetables to northern locales. Poor black and white sharecroppers would ensure that these foodstuffs were abundant and available for transport. In the 1880s, northern investors realized the lucrative potential of Copiah County and spurred a focus on tomatoes as its primary crop. By the early 1900s, Gallman and the larger Copiah County had become one of the top vegetable producers (especially of tomatoes) in the USA.[15] Yet this success was short-lived. For example, the town's tomato shipments in 1927, 1937, and 1938 were 175, 40, and 51 cars, respectively.[16] The town's economic prominence declined significantly as a result of two periods of population loss during the Great Migrations from about 1910 to 1920 and 1940 to 1970. Fewer blacks were available for cheap labor. Although not the political majority, blacks have represented at least 45.0 percent of the population for much of the twentieth century.

National and state-wide politically-driven racial inequality trickled down and shaped the lives of blacks and whites in the two locales. Poor white and black sharecroppers bore the brunt of economic and political inequities, yet race privilege for members of the former group meant escaping brutality and bloodshed. As in the labor market in general, historically, both Hazlehurst and Gallman were stratified first and foremost based on race; within this classification, residents were further sorted by class, gender, and age. Remnants of this history remain. Findings here support the importance of property and other assets in shaping one's class position. Segregation shaped inter-racial interactions, farming practices, schools, churches, and daily life; this analysis considers specific experiences among some of the earliest members of the King family. Early population demographics for Gallman, Hazlehurst, Crystal Springs, and Wesson are presented in Table 1.2.[17] We include the latter town because it was the birthplace of several early members of this family.

Table 1.2 *Population in Gallman, Hazlehurst, Crystal Springs, and Wesson, Mississippi (1870–1950)*

	Gallman	Hazlehurst	Crystal Springs	Wesson
1870	na	662	864	464
1880	83	463	915	1,707
1890	na	na	997	3,168
1900	189	1,579	1,093	3,279
1910	188	2,056	1,343	2,024
1920	176	1,762	1,395	885
1930	207	2,447	2,257	799
1940	180	3,124	2,855	837
1950	170	3,397	3,676	1,235

Key: Taken from Crook, Brenda. 1996. "Historic and Architectural Resources of Copiah County." *Mississippi: National Register of Historic Places. United States Department of the Interior National Park Service* and the Mississippi Statistical Summary of Populations. na = not available.

A Profile on Farming in the United States and Mississippi in the Early Twentieth Century

At the turn of the twentieth century, Mississippi was a predominantly rural farming space.[18] There were a total of 220,803 farms in the state. The average value of farmland in the USA in 1900 was $2,285 per farm or $15.59 per acre (including buildings). However, the average value of land in Mississippi was $520 per farm or $6.30 per acre. Only Alabama had a lower average farm value. In addition, the value of black farms in Mississippi in 1900 totaled $86,890,974; of that value, over $55 million was land and almost $16.5 million was livestock. According to Lorenzo Green and Carter G. Woodson (1930) in *The Negro Wage Earner*, in 1890, there were approximately 590,666 black farmers, planters, or overseers in the USA.[19]

In 1900, blacks controlled about 13.0 percent of all US farms for a total of over $200 million or a little less than $300 for each black farmer (U.S. Census 1975: cx).[20] When race is considered for that same year, 4,970,129 farms were operated by whites and 769,528 farms were operated by blacks. In Mississippi, in 1900, a disproportionate percentage of farms were actually operated by *black farmers* (128,679 versus 92,124 farms operated by whites). According to census reports, "The largest relative number of farms operated by colored farmers in the South central district was in Mississippi – 58.3%, a figure exceeded only in Hawaii."[21] A disproportionate percentage of blacks operated cotton, tobacco, and rice farms at that time. Black farmers were impressive both in terms of their industriousness and success as farmers:

> As the Negro started with nothing forty years ago, the small relative differences shown by the percentages ... [in terms of farm output, value, and size] give evidence of substantial progress in the past and are a hopeful augury of the future. (U.S. Census 1975: cvi)

Black farmers were also prudent in their business enterprises:

> [F]or all geographic divisions, the figures show that the Negro is becoming a farm owner along conservative lines. ... Negroes buy small farm homes, for which they can pay, and then rent additional land. This method gives them greater assurance of keeping what they first acquire than any other that could be adopted, and argues well for the future acquisition of farm lands by that race. (U.S. Census 1975: cv)

Like the Kings, black farmers realized the importance of securing property; implemented prudent processes for land acquisition; gradually expanded their holdings; and, strove to avoid debt. The following summary describes the promise of agrarian blacks at the turn of the century:

> The Negroes at the close of the Civil War were just starting out upon the career as wage-earners. They had no land and no experience as farm owners or tenants and none of them became farm owners by inheritance nor inherited money with which to purchase land. Of the 371,414 white farmers added since 1860 very many were the children of landowners and came into the possession of farmland, or the wherewithal to purchase the same, by inheritance. When this difference in the industrial condition of the two races in 1860 is taken into account, the fact that the relative number of owners among the Negro farmers in the South Atlantic states in 1900 was practically three-fourths as great as the relative number of owners among the white farmers of those states added in the same period, marks a noteworthy achievement. (U.S. Census 1975: cvii)

Although census reports reflected the racial titles of the day, they could not avoid acknowledging the economic advances of "the Negro." Black accomplishments post-slavery were particularly significant given:

> The end of slavery in the United States came with war and when at last the war had ended 4 million largely illiterate ex-slaves were thrust into a ravished and economically battered region. No effort was made at restitution, nor even any significant effort at investment in the ex-slaves aside from the modest and short-lived activities of the Freedman's bureau. (Darity and Myers 1998: 157)

This means that without substantial national or local assistance, blacks began to use their knowledge of agrarian spaces to build farms. But how did black farmers in general fare as compared to their white peers over time? Census documentation also answers this query:

> The tenure position of non-white farmers as a group improved markedly, even more than that of white farmers as a group in the period 1945 to 1950 in the South. Non-white owners increased 2.2%, comparing favorably with the 2.0% increase for white owners during this period. The decrease in tendency among non-white tenants was 23.0% as compared to 21.7% for white tenants. The number of non-white croppers decreased 26.7%, while the number of white croppers declined only 15.6%.[22] (Census of Population and Housing 1950: 21)

Multiple sources show that, despite systemic forces designed to derail their efforts, by the mid-1900s, southern blacks were beginning to out-pace whites in farm ownership. Table 1.3 further corroborates this pattern when one considers the representation of farm operators by race during the early to mid-1900s. During this time period, increasing percentages of blacks (21.40 percent in 1900 to 61.68 percent by 1969) became full owners of farms. White full farm ownership increased during this same period (57.40 percent to 67.60 percent), but not as dramatically as for their

Table 1.3 *Profile of Farm Operators (1900–1969)*

Year	Black (%)					White (%)				
Operator Type	Full Owner	Part Owner	Manager	Tenant	Cropper	Full Owner	Part Owner	Manager	Tenant	Cropper
1900	21.40	3.81	0.22	74.6	na	57.40	5.60	0.91	36.11	na
1910	19.69	4.85	0.13	75.32	na	52.28	7.79	0.68	39.24	na
1920	19.35	4.23	0.19	76.23	47.43	53.74	6.67	0.72	38.86	25.62
1930	15.93	4.71	0.09	79.26	56.22	44.84	7.83	0.71	46.62	35.11
1940	20.86	4.61	0.05	74.48	59.04	50.95	7.96	0.57	40.51	25.69
1945	24.19	4.25	0.07	71.50	56.82	60.84	7.46	0.58	31.12	25.57
1950	25.31	9.28	0.04	65.37	54.19	60.65	13.10	0.47	25.79	27.55
1954	28.02	10.95	0.08	60.95	56.72	61.78	16.20	0.50	21.52	26.90
1959	33.79	14.13	0.11	51.97	53.16	62.12	20.69	0.65	16.54	20.88
1964	38.36	16.93	0.08	44.63	na	62.09	22.92	0.59	14.40	na
1969	61.68	18.09	na	20.23	na	67.60	21.37	na	11.03	na

Key: Figures reflect percent of racial group. na = not available. Figures provided for years available. Taken from U.S. Department of Commerce: Bureau of the Census. 1975. "Historical Statistics of the United States Bicentennial Edition – Colonial Times to 1970."

black peers. When one considers part-time operators, the range for blacks (3.81 to 18.09 percent) was similar to whites (5.60 to 21.37 percent). Managers were negligible regardless of race. Yet tenancy fluctuated more greatly for white farmers than for blacks.[23] These trends illustrate the impressive advances made by black farmers during the 1900s. Moreover, their significant progress and improvements suggest that a central agenda of Reconstruction, segregation, and Jim Crow was to squelch the industriousness and successes to systematically and economically disenfranchise blacks. This insidious agenda had day-to-day implications for black families like the Kings.

THE KINGS: FOUNDATIONS FOR A FAMILY ECOLOGY AND ETHOS

This section introduces key members of the earliest known generation of the King family. Considerable time is spent exploring the early experiences of two individuals, George and Irma, because their lives and encounters provide the context for understanding this family's religious, educational, and socioeconomic decisions and resulting outcomes. Moreover, we focus on Irma's lineage because family members contend that she was central in cultivating the ideology and practices that ultimately resulted in the family's transition into the middle class. The King family tree in Figure 1.2 summarizes another contextual dimension that will inform the narratives and analyses in subsequent chapters.

Felix King was born in Virginia in 1824; his wife, Samantha, was born in Indiana in 1826. They had two children, James, born in 1859, and Emma, born in Mississippi in 1865. Little is recorded about these earliest Kings save that Felix, Samantha, and James were born into slavery and Emma was born the year slavery officially ended. The family began to more formally document subsequent generations of James's offspring. James and Loren King are the late nineteenth century family patriarch and matriarch (Figure 1.3). They represent the second of seven generations referenced in this study. Both were born in slavery and raised in Wesson, Mississippi, in November 1859 and September 1862, respectively.[24] Post-slavery, both relocated and eventually met in Gallman, Mississippi, where they spent the remainder of their lives. Although slavery had formally ended, their lives were still irrevocably influenced by the punitive conditions exacted by southern whites who hearkened for the "good old days" and continued to practice forced servitude. Rather than serendipity, the ecology of space shaped their initial encounter; James and Loren met in church. The time-consuming lives of sharecroppers meant that blacks were most likely to see and interact with friends and other family members at church.[25] James and Loren were no exception. James became a farmer and teacher; Loren was a homemaker. They married and had five children: Danny, Colvin, Bella, Mindy, and Janice.[26] Based on the book's focus, we follow the lineage of their last child, Janice. As might be expected, much of the King's early history is maintained in the family bible published in 1910.[27]

Janice King was born September 29, 1890, in Wesson, Mississippi, and died September 29, 1971 (Figure 1.4). Her documented history focuses on her religious affiliations, particularly involvement with Missionary Baptist churches that would influence subsequent family members. [28] Janice gave birth to twelve children. Her first marriage to Howard resulted in the following six children; Willard Leon, Cora H., Irma, Lily Mae, Margie Lou, and Jim. After Howard died, Janice moved from Wesson to Gallman, Mississippi, married Sam (called "Poppa" by his grandchildren), and had six more children; Gina Mae, Katherine Mae, Robbie Lee, Leddie Lee, Sinclair Jr., and Daniel Lee.[29] Their sizable family honored the Old Testament edict to be fruitful and multiply.[30] Equally important, it provided the needed human resources to sharecrop.[31] Although she spent most of her life as a poor sharecropper, Janice King would eventually have her own large garden as well as a plethora of chickens, geese, small chickens (affectionately called "bannies"), and hogs. Referred to as "Mama" by her grandchildren and "Mama Janice" by her great-grandchildren, Janice King's decisions, both sacred and secular, would shape the beliefs and behavior of her children, especially those of her oldest living daughter, Irma.

Irma King

Conceptualizing a Family Ethos

Born on November 24, 1914, in Wesson, Mississippi, Irma knew the harsh realities of segregation, racism, and abject poverty. As the oldest female, Irma was responsible for childcare for her younger siblings, especially the youngest five, Margie Lou, Gina Mae (simply called "Gert"), Katherine Mae, Robbie Lee (called "Robbie"), and Sinclair (called "Sam" or "June"). Irma's life in Mississippi, her maternal ties to her brothers and sisters, as well as the eventual migration of these five siblings, would be intricately linked to the middle-class lifestyle she would eventually help forge. Irma's early experiences of oppression and poverty fueled her personal life goals. The following observation parallels a large literature and illustrates common aspirations that are particularly germane to this study that can emerge irrespective of socioeconomic status: "poor people – males and females, blacks and whites, youths and adults . . . have . . . as high life aspirations as do the nonpoor and want the same things, among them a good education and a nice place to live" (Cotton 1993: 197). Despite Irma's humble beginnings, these same beliefs also informed her aspirations.

The following two parallel recollections describe the impetus for what became the King family ethos. This ideology was forged amidst the unrelenting turmoil of servitude. Yet Irma still wanted more. Her children contend that even her difficult upbringing did not alter Irma's faith in the Old Testament God who had a penchant for the marginalized.

Carla, Irma's 74-year-old middle daughter, corroborates the scenario that introduces this chapter and explains the socio-ecological backdrop from which the family ethos emerged:

> My Mul said that when she was growing up and after her father passed, then Momma Janice [Irma's mother, Janice King] had it so hard because she did live on the white man's place. My Mul said that they just had to work so hard and couldn't go to school so she had said that, when she got grown, there were two things she wanted – her own place and her own milk cow.

The endearment, *My Mul* (short for "My Mother"), used by Irma's children, originated when they were too young to pronounce the common maternal title, and continued even after they could. The above memory inculcates how daily travail linked to race, class, gender, and space resulted in the conceptualization of the King family ethos. For Irma, land meant stability and the ability to avoid being at the beck and call of whites. A cow represented another dimension of self-reliance in the form of milk and produce.[32] Irma realized that property could enable blacks to do more than eek out a mere existence as over-worked, underpaid servants. Additionally, her inability to pursue formal education as a child due to frequent seasonal interruptions caused her to believe that acquiring land would translate into education for her children. Her 77-year-old daughter, Martha, provides additional details:

> My Mul had always said that . . . when they grew up, they lived on the white people's place and . . . that's why they didn't really get a chance to go to school to get an education because they were sharecroppers. When the white man would plant all his crops, then the black people – most of them had a lot of kids in those days – then their kids would have to stay out of school to share the crops when they

were ready. So My Mul said that she didn't want her kids to have to grow up like that. She always wanted her own home so we wouldn't have to do that.

Irma's own early experiences sharecropping for white landowners instilled the desire to avoid the same fate for her children. Yet, as a young black woman with limited formal education, how would she realize this dream?

Irma married George Franklin in 1941 (Figure 1.5) and they had six children.[33] Two daughters, Bonnie and Helena, were born prematurely and died in infancy. Like their parents, the three remaining daughters, Martha Ellen, Janice Mae,[34] and Carla Reace, doted on their youngest and only brother, James Lewis. Post-marriage, Irma continued to cultivate and model the Christian tenets for which she would become well known:

She [Irma] was a quiet, humble person. She didn't believe in fussing and arguing. And she believed in helping people and sharing what she had with people ... She worked ... and she cooked our meals ... She would make sure we studied our Sunday school lesson. We had our little Sunday school books and we had to remember our Sunday school bible verses. So she would rehearse us over that and took us to church and Sunday school ... She read her bible and she believed in living a Christian life in terms of how she treated other people and in her talk and the way she lived. She didn't say much ... but her actions proved that – the way she taught us to make sure we got that foundation. (Martha, 77 years old)

Irma was consistently described as a quiet, reserved woman. At about 5′9″ tall, her sturdy, dignified frame was the result of decades of caretaking inside and outside the home. According to Martha, Irma's unassuming demeanor belied a fierce commitment to God, kin, and community. Seventy-five-year-old Janice concurs:

Although she had a fourth-grade education, My Mul used her skills to do other things like sew. She was also very knowledgeable, wise, and encouraging. She had to leave school to work in the fields and she didn't want that for us.

The above quote by Irma's second daughter describes her mother's self-sufficiency as well as her instrumental and expressive abilities. The comment also reminds readers of the bleak existences most southern blacks endured as prisoners to land they farmed, but did not control. Decades later, her great-grandson, 37-year-old carpenter, Kevin, attributes Irma's lifestyle and advice in helping him avoid unwise choices:

I've had dreams about Big Mama at critical times in my life ... I saw her spirit ... It came at times when I wanted to do something wrong or knowing that I should stand up for what was right. I made the decision not to waiver – doing the hard thing, which is the right thing, and knowing and trusting that the situation will work out. There have been several times when I could have been someone else, but

I didn't . . . I can either be blind to the systematic genocide of our people or I can be an anchor, knowing that the people behind me may not have the connections with the past or with our ancestors who aren't here anymore or even the elderly today that they don't respect. The three times I've had dreams about Big Mama all came at critical points. It was good for me. I've never had any other dreams about family members the way I've had dreams about her.

Kevin spent his youth living with Irma during a migration spell during which she served as his surrogate mother. Dreams of her have enabled him to make good decisions and avoid poor ones; correlate personal decisions with those of other people of African descent; and, respect his elders. Moreover, as an adult, Kevin recognizes his family's strengths and his responsibility to continue the tradition: "I know the blood line is strong. It was about sticking to it. There was no extra, they saw what was going on . . . to lead to success, independence, or the building of our nation or other people or even of our family." As an organic intellectual, Irma's family members suggest that negative circumstances did not undermine her godly lifestyle, aspirations, and expectations. However, aspirations were not enough to accomplish Irma's goals.

Seminal Decisions by an Unassuming Domestic

It may be difficult for many readers to consider black women who work as domestics as empowered. The intersection of race, class, and gender oppression would seem to preclude such possibilities – without romanticizing their experiences. *The Maid Narratives: Black Domestics and White Families in the Jim Crow South* by Van Wormer, Jackson III, and Sudduth (2012) presents this conundrum – whether black female domestics in the Jim Crow South were, "heroes who prevailed over hardship or victims of their time and circumstances . . . [as part of a southern] . . . caste system" (pp. 3–4). Yet black women-centered frameworks offer counterpoints regarding non-traditional ways in which seemingly disempowered individuals like domestics defied oppression in thoughtful, proactive ways.[35] Views by Irma's children and colleagues parallel the following quote that; "black women's contributions to black family well-being, such as keeping families together and teaching children survival skills . . . allow black women to develop cultures of resistance" (Collins 1990: 44).

Their altruistic and adaptive behavior did not negate the undue difficulties with which black female domestics contended. This form of women's work has a long history in the USA. For example, in 1890, there were approximately 9,248 black housekeepers and stewards on record in the USA, and 10,596 in 1900 (14.5 percent increase). By 1910, 11,620 blacks filled these positions. Moreover, by 1920, 15,161 blacks were housekeepers or stewards (Green and Woodson 1930). Additionally, the 1950 Mississippi census shows that of the 87,946 non-white employed females, of which the majority were black, about 30,833 of them were "living out" private household workers (i.e., domestics who did not live with their white employers).[36] When one considers her double duties as a paid domestic for a white employer and unpaid domestic role for her own family, Irma, and other black women like her, represented an early rural, under-acknowledged version of the second-shift long before the concept was chronicled by white writers.[37]

Females in these posts were overworked and underpaid. Intra-racial wage differentials based on gender and region were the norm:

Negroes in domestic service received wages ranging from 50 cents to $3.50 a week in such cities as Atlanta and Nashville ... [in] Virginia, in 1898, Negro men received from $8.00 to $10.00 a month, the women $1.00 to $5.00, according to age and work, a general servant in an ordinary family received $4.00. In the Black Belt in 1902 Negroes received $4.00 to $6.00 a month for cooking and $1.50 to $3.00 a month for nursing ... In North Carolina wages ranged from $4.00 to $6.00, and from $3.00 to $8.00 in Richmond. This inadequate remuneration served to turn the whites generally from such employment for the higher remuneration of other spheres ... this meager wage scale also caused the Negro house servant and waiter to take the first opportunity to go North where wages for domestic service were four or five times greater. (Green and Woodson 1930: 81)

The above range of monthly wages (about $2.00–$10.00) translates to about $24–$120 annually as compared to average annual incomes of $479–$593 in 1900 for posts in manufacturing, agriculture and forestry, construction, railroad work, and communications.[38] These same scholars note challenges for this gendered workgroup: "the economic proscription of the Negro women workers is demonstrated by the fact that in 1900 out of every 100 Negro women and girls working for wages, 96 were either field workers, house servants, waitresses, or laundresses" (Green and Woodson 1930: 77).

Yet for Irma, a culture of resistance meant developing a plan to realize her childhood dream of land and livestock. The following series of quotes describe the decisive decision that altered the King family's socioeconomic trajectory:

When my daddy went to the service, he would send an allotment back to My Mul. So Big Poppa [George's father], who was my granddaddy, had lived back in the house down the lane. And for some reason, a plot of land became available and so My Mul had saved up the money and she wanted to wait until My Daddy came home from the service before buying the land. And Big Poppa told her – he called her *Girl* – he said, "Girl you take that money and go on and buy this land because when that boy comes home, he ain't gone do nothing, but mess that money up. And you ain't gonna have no house or nothing." So My Mul took that money and went on and bought the land. So that's how we ended up having our own property. And then the land behind My Mul [where Irma's father-in-law lived] was Aunt Hana's [George's sister]. She was in Detroit, but she wanted a place for her parents to stay. It had a house on it and twenty acres of land. And the place My Mul bought had a house on it and twenty acres of land. (Martha, 77 years old)

According to Martha, a combination of several structural forces and Irma's agency resulted in the acquisition of property for the Kings. George's military stint in World War II provided the necessary funds. He sent most of his monthly allotment or salary home; Irma saved it all. George's father, who seemed to understand the somewhat reckless nature often found in youth, encouraged Irma to buy the property in her husband's absence. As illustrated in a subsequent section, the economic benefits of military service were also central to mobility for several other King family members. Her daughter, Carla, extends this narrative:

When my father went to the service, [Irma] she already had two children and I think she was expecting me. So they sent her this little check. And she took that money and she saved it up, and her father-in-law told her to save it up and to buy her a place for the family. And so that's what she did . . . she didn't spend any of it and then she bought that house and twenty acres of land from Mr. Pryor, a white man.

A plot of land in the "bottoms" or "black" part of town became unexpectedly available. The purchase was made with a $500 down payment and two subsequent installments of $250 each (refer to Figure 1.7). Its locality meant that Irma's purchase of such a large parcel, which included a house, would not likely raise suspicions or concerns among local whites who coveted black success or who might believe that the Kings were becoming uppity Negroes by buying property. Yet her reputation as a "good Christian" facilitated the purchase and reflected another way in which blacks were stratified reputationally in Gallman that had economic implications.

All God's Dangers by Rosengarten (1974), another study about that historic period, provides a menacing reminder of what Irma potentially faced negotiating this purchase. According to its main character, Nate:

Whenever the colored man prospered too fast in this country under the old rulins, they [whites] worked every figure to cut you down, cut your britches off you . . . to take everything you worked for when you got too high . . . keep the dollar out of the niggers' hands . . . the white people was afraid . . . the money would make the nigger act too much like his own man. Nigger has a mind to do what's best for hisself, same as a white man. If he had some money, he just might do it. (Pp. 27, 264)

This social context meant Irma had to be discerning before, during, and after this transaction. According to family members, several gender-related dynamics facilitated the purchase. First, Irma worked as the domestic for an influential upper-class white family in Gallman. This weak social tie cultivated an inter-racial connection that helped make Irma's purchase possible.[39] Additionally, unlike Nate, a black rural sharecropper with aspirations of land ownership, Irma's gender and physical condition as a visibly pregnant mother of two children may have better positioned her for a favorable outcome than Nate describes in the next quote: "I had a hard way to go. I had men to turn me down, wouldn't let me have the land . . . wouldn't sell me guano, didn't want to see me have anything" (*All God's Dangers* 1974: 544). Inferring from Nate's comment, although Irma was making the purchase for her entire family, she may have been considered less a threat than a black male to local whites. Despite the potential consequences of white fragility, Irma realized that the benefits of homeownership outweighed possible dangers. Decades later, Shapiro (2004) correlates property, class position, educational outcomes, and upward mobility in ways that were intuitively apparent to Irma:

Most Americans accumulate assets through homeownership. Home equity accounts for roughly 44 percent of total measured net worth. Wealth built up in one's home is by far the most important financial reserve for middle-class families. In fact, home wealth accounts for 60 percent of the total wealth among America's middle-class . . . homeownership is a signature of the American Dream and . . . frames class status, family identity, and schooling opportunities . . . [it] provides the nexus for transformative assets of family wealth. (Pp. 107, 188)

This same scholar argues that blacks pay an 18 percent "segregation tax" meaning that their property tends to be worth less than their white counterparts.

Irma's purchase of a family homestead October 23, 1944, represented the first *tangible* step in what would become the King's transition into the middle class.[40] Moreover, based on their close relationship, George's sister, Hana, would eventually deed her twenty acres and house to Irma. The 2015 market value of their forty acres in Mississippi (excluding a house) is about $224,000.[41]

Domesticity and a Non-Traditional Trek to the Middle Class

Irma's domestic skills enabled her to maintain her own household; the need for additional income required her to also help maintain someone else's. Just as it is important to consider *what* constitutes middle class, *how* this status is maintained is also informative. As suggested by Golden (1983: 21), black domestics like Irma often parlayed challenging experiences into "a soft armor of survival." Moreover, Collins (1990) describes an outsider-within identity that required Irma to navigate an inter-racial position that included maternal dimensions. Such black women "not only performed domestic duties but frequently formed strong ties with the children they nurtured, and with the employers themselves" (p. 11). Donna, one of her granddaughters, describes Irma's long-term occupation: "My Big Mama cleaned house for Ms. Dina. It was a job that she did with pride" (53 years old, Certified Public Accountant [CPA]). In addition, Donna's mother explains the long relationship between these black and white families: "Mr. Roy and Dina Williams . . . he was in charge of the post office and the little store. Ms. Dina was a teacher. And my mother kind of raised their kids" (Martha, 77 years old).

According to family members, Irma did not complain about her job as a domestic, but believed in doing her best, no matter the type of work. She tried to instill that same work ethic in her children. As owner of the only store in Gallman, Roy, her employer, wielded considerable influence among whites and blacks alike. The use of the title "Mr." when referencing him illustrated his social position as a white man under Jim Crow and elevated status as a wealthy white man.[42] The same granddaughter describes going to work with her grandmother and witnessing economic exploitation first-hand:

> I begged and pleaded to go and finally . . . We headed to Ms. Dina's house early with Big Mama coaching me all the way. "Don't take nothing! Make sure you say, Yas'm and No ma'am and don't take nothing!" . . . [After the experience] I never asked to go and help out again. It wasn't worth the $1 or $2 dollars that she paid me that day to clean up for Ms. Dina. There was a huge pot of quarters by the door in the house where they just threw their spare change. Those quarters represented the overflow that my grandmother wasn't experiencing. Ms. Dina was very kind and I really liked her, but I hated that my grandmother cleaned toilets for chump change. I realized that the stories that I heard about slavery and the black/white divide were actually real. My grandmother was the maid. It made me want more for myself and my children.

Her brother recalls a similar incident: "in the summer, she [Irma] would take me to work with her. She would have me washing cars and cutting grass. She said, 'don't speak' or say 'yes ma'am' and 'yes sir.' Big Mama was working for the white man, she didn't want me to say anything out of place . . . she was dealing with bigotry" (Donald Jr., 56 years old, minister and administrator).

Implications of the nexus of racism, classism, and sexism inform the above difficult, disillusioning memories.[43] In reality, Irma and George were actually more economically stable than her granddaughter's family, but Donna and Donald Jr. were troubled by Irma's *occupation*, in particular, and the servile status of most blacks in Mississippi, in general. They also associated oppression with the precariousness of black life in the South, the low wages most blacks earned, and the complicity of seemingly *well-meaning* whites like Dina who participated in their oppression. Most importantly, as a young girl, Donna's realization of Irma's sacrifices for their family fostered her own lofty personal and familial goals.

As reported in *The Maid Narratives: Black Domestics and White Families in the Jim Crow South* (2012), the relationship between domestics and their employers could be complicated. Domestics tended to be extremely exploited, received few benefits, and were exposed to possible sexual harassment.[44] In contrast, Irma's children suggest that, for that historical period, the Williams family seemed to provide Irma with a relatively decent wage and also established and contributed to her retirement fund. Yet their wage did not alter the unequal system of oppression in which they colluded. And unlike most of the women documented in *The Maid Narratives* who eagerly left Mississippi, Irma stayed, and considered this locale her home. Some readers may imagine a docile Irma – a domestic resigned to her fate. However, despite their marginalized work roles, black women like Irma were known to be agentic; they cultivated their own positive self-images, special wisdom (or subjugated knowledges), and everyday forms of resistance.[45] One of Irma's granddaughters supports this claim:

> Years ago, Big Mama told me, "I've met a few nice white people, but sadly white people aren't even nice to each other, so don't expect them to be nice to you." I was in my early 20s when she shared this with me. It was the only time I heard Big Mama ever say anything negative about anybody. I now realize that she was just being honest and trying to prepare me for life. Sadly, these words of wisdom have been true time and time again. (Tanya, 53 years old, administrator)

Irma's everyday resistance also meant traversing predominately white spaces as a domestic without embracing this employment position as her identity. She realized that her family depended upon it. Despite Irma's unassuming nature, her advice given above meant that she recognized and experienced racism and had realistic expectations about southern whites. Her 52-year-old grandson, Calvin, also acknowledges her prudence: "Big Mama, you could always talk to her and she wasn't going to sugar coat things. You knew she was going to give you wisdom and that it was coming from motherly love." Moreover, her past role as surrogate mother for her own siblings seemed to prepare her for juggling multiple caretaking roles. Yet Irma's progenitors suggest that her life was fueled by spiritually and savvy rather than grit:

> She was just a strong woman and could take almost anything and survive off of it. I can remember when we didn't have nothing to eat, especially at night, you might go in and have a bowl of milk and bread. And that was all you had. But she knew how to take basically little or nothing and survive off it . . . She was a very religious person. Now she didn't talk religion all the time, but she was one of those persons who lived it instead of talking it . . . she was just a strong individual in all of our lives. (Carla, 74 years old)

According to the same daughter, Irma's decision-making was driven by concerns for her children's futures:

> My Mul never kept us at home not one day to work in the field. And when the white folks wanted us to pick beans she would ask us did we want to go . . . maybe Mr. Maynard was picking beans that Saturday . . . Most of the time, Janice [the middle daughter] would say "No." And she didn't make her go out there if she didn't want to do it. But I went so that I could make me some money.

Carla and her siblings could decide whether *they wanted to* gather crops to earn extra money. And just as her sister usually declined, Carla felt agentic and industrious because she could decide whether or not to do such work.

Women like Irma led complex lives.[46] Like many southern black women at the time, Irma worked as a domestic. Yet this was not her identity. Her family suggests that she made the distinction between how she earned a living and who she was as a person. Collins (1990) describes this wisdom that informs our socio-ecological focus here:

> Self is not defined as the increased autonomy gained by separating oneself from others. Instead, self is found in the context of family and community . . . by being accountable to others, African American women develop more fully human, less objectified selves . . . a larger self. There's a "self" of black people. (P. 105)

Using Collins' (1990) terminology, without fanfare and while striving to maintain peace with those around her, Irma rejected stereotypes and controlling images that strove to define her as a "maid." She made pragmatic decisions for her family, particularly her children. Yet we suggest that Irma's efforts should not be misconstrued as the result of grit. Despite what is described by her family and friends as unwavering spiritual strength, it is unimaginable that Irma did not expend inordinate physical, emotional, and psychological energies to navigate segregation, gendered roles in maintaining her family and farm-related chores, and her domestic job. Attributing Irma and the King family's middle-class successes to grit ignores and/or diminishes the debilitating tangible and intangible effects of racial oppression deeply embedded in the South; the myriad of blacks who likely made similar decisions without comparable outcomes; the intentionality needed by southern blacks to navigate economic, political, and social landmines daily; and, the quagmire that exceptionalism renders the humanity of both transgressors and the transgressed.

This King family matriarch, motivated by religious, economic, and other non-economic reasons, put in motion a family ethos and lifestyle that would become more stable and prosperous than most blacks and many whites in Gallman. Yet this reality cannot ignore what the Kings navigated (example, employment constraints), endured (example, micro-aggressions and their negative effects), and tacitly tolerated (example, systemic segregation) in ways that demarcate their socioeconomic status in the South from common ways of defining and understanding the middle class.

George Franklin

Protector/Provider and Gendered Presence

Although Irma was consistently described by her children, grandchildren, and contemporaries as the quiet force behind the family's slow trek to socioeconomic stability, her husband's involvement was essential to its success. George Franklin was born January 4, 1917, in Gallman, Mississippi (Figure 1.5). Standing well over six feet, two inches tall and physically fit from years of labor, George had a reputation as a strong presence in any space he entered. He embraced traditional gender roles and prided himself on being a protector and provider. Black men like George embraced the familial, economic, and religious expectations of these roles even when white society attempted to stymie their attempts:

> Persistent efforts to acquire land, upward social mobility . . . and pride of race were characteristic of the stable, male-headed Afro-American families . . . one measure of a male's status in the black community historically was his ability to provide for his family, to educate his children, and to support his church. (Berry and Blassingame 1982: 80)

The above assessment confirms rather than necessarily affirms this family form. However, like most Kings, these two scholars correlate socioeconomic stability with property, education, religious involvement, and nuclear families.

A man of few words, George was described as an unusually reserved, taciturn man whose stoicism subsided later in life, particularly in church settings and with his grandchildren. He also had a reputation as a leader and austere disciplinarian. One of his responsibilities was to enforce Irma's household directives:

When he spoke, you heard him. My Mul better not had to tell us to do something a second time because, I mean, if he raised his voice, you didn't *think* you were going to move, you *knew* you were going to move *then* [emphasis was hers]. Irma better not have to tell you that no more ... he had a rough voice. (Martha, 77 years old)

And although Irma usually deferred to him, George recognized her important, complementary role. Yet it was clear that their household was a male-dominated one:

Big Papa did not come home often while we were there to visit, but when he did, everything came to a halt. Big Mama would announce, "Here comes George!" and we would get to work ... taking his glass out of the freezer and filling it up with Kool-Aid, changing the channel to the local news, opening the curtains and making sure that everything was in place ... He was never mean, just distant ... a figure to be revered and respected. (Donna, 53 years old)

According to family members and peers, George had also established a reputation in the community for his mathematical astuteness and mechanical skills. His eldest daughter, Martha comments, "My daddy had a sixth grade education ... he could read and really do math ... so that helped him a lot in his schooling." A military stint provided additional social capital that George used upon returning to Gallman:

Once he got out of the service, My Daddy went to school to become a mechanic. He was a very good mechanic. I know he worked for Mr. Sojourner who had a mechanic shop ... Folks from all around the community would come and get him to work on their cars. And even the county road people, a lot of times when their machines would break down, they would come ... he could fix their equipment for them. He was very good at what he did.

Despite his initial limited formal education, the combination of George's seemingly natural aptitude, military technical training, and subsequent community college certification as a mechanic helped him establish an economic niche in the small town.

When considered contextually, census figures show that in Mississippi in 1950, about 228,156 non-white males over the age of 14 years old were employed; only 1,668 of them were mechanics.[47] Moreover, George's expertise crossed racial boundaries such that whites and blacks who needed vehicles and other machinery repaired sought him out. Family members also suggest that George strove to improve his education informally and was particularly interested in news and US affairs:

> Big Papa was teaching Sunday school one Sunday and I was so impressed to hear him talk about what was going on in the world. When I heard him say *Czechoslovakia* and actually pronounce it correctly, I was more than impressed. I was familiar with Czechoslovakia, but didn't know exactly where it was. I sat thinking, look at my grandfather stand there and show just how smart he was. This was a motivating factor in me to always want to learn as much as I could and value higher education. (Manuel, 49 years old)

The above experience reminded Manuel not to discount informal education or assume that blacks like George were not highly intelligent, astute, and knowledgeable about religious and secular affairs. As noted by another grandchild, later in life, George modeled religious education, particularly to his grandchildren:

> Religious education means having a firm religious foundation. Not just knowing who God is, but knowing how to have a relationship with God and how to apply that to your life . . . it became important as I watched my family. I actually only have a few memories of Big Papa and one is in church. We went to church every night that we were down there [visiting Gallman]. I remember him sitting off to the left with the Deacons . . . and he prayed . . . And seeing how important it [Christianity] was to them [her family], made it important to me. As I got older, praying and being active in my own church helped me have a good religious foundation. (Barbara, 43 years old, Program Director)

However, according to several of his daughters, Irma had to continually motivate George about farm tasks. Martha laughs as she recounts her parents' gendered morning rituals:

> My Daddy hated to get up in the morning. He always waited until the last minute to get up. He liked to drink coffee and smoke cigarettes in bed . . . He would get up and make a fire in the fireplace, put wood in the stove, and then get back in the bed. And My Mul would get up and make breakfast. She would bring him coffee in bed and he was in no hurry to get up . . . I don't know if he ever got to work on time. He would go to work as a mechanic and might come home late.

Rather than an indictment of her father's work ethic, Martha's comment is her attempt to provide a relative comparison between the workload of her parents inside and outside the household. For her, in addition to employment as a domestic, Irma bore the brunt of the household and child-rearing responsibilities in their home. Yet George's late nights at work evidenced commitment to the family's well-being. Although imbalanced gender roles were common during that period, Martha and her siblings were cognizant of their mother's centrality to their family's economic and non-economic successes.

Ownership of land and a milk cow positioned George and Irma for a certain degree of economic stability unknown to most blacks and many working-class whites locally. Although white privilege still enabled members of the latter group to capriciously dole out micro-aggressions on blacks, George and Irma's increasing material stability represented evidence that God was looking after their family. Additionally, property ownership provided an equally important benefit for which Irma had so persistently labored – agency. No longer would her children have to sharecrop for subsistence on a white man's property. More importantly, their educational pursuits would not be interrupted by the seasonal dictates of another man's farm. It can be said that the King's farm, livestock, and other material possessions resulted in a certain degree of self-sufficiency, especially desired by persons with a history of marginalization and oppression.[48] Carla compares and contrasts the relative benefits property afforded her family:

> It was uncommon for a whole family of children to graduate from high school like we did because, especially the boys, after you have stayed out of school and worked in the field and you have fallen so far behind, you just started doing something else instead of going to school.

The children of sharecroppers, especially males, were usually at the mercy of the whites on whose land they lived. These youth often eventually dropped out of school when they fell too far behind to catch up. In this way, land ownership directly influenced educational attainment for many rural blacks. And class, gender, and age could influence educational outcomes for such black youth as well. Thus there were multiple economic and non-economic benefits of the King family farm.

Navigating Racial and Economic Landscapes and Landmines

A mechanic by trade, George had a reputation as a savvy businessman.[49] This aspect of the provider role included broad responsibility for financial decisions on the farm. Yet it was not without its challenges, as black men were particularly vulnerable to white ire.

George's mathematical and social skills required him to avoid both being taken advantage of by unscrupulous whites and possible repercussions of being considered an uppity Negro:

He learned to be a mechanic in the service . . . but he also worked at the saw mill. He had a regular job after he got back from the service . . . He might not have had a lot of formal education, but he had good sense. You couldn't trick him out of nothing. These white folks, after they have their crops, like sweet potatoes, they would drive through the road and stop and ask if you wanted to buy some sweet potatoes. I remember one time . . . its four pecks in a bushel and this white man wanted to charge you a dollar a bushel or fifty cents a peck. So My Daddy said, "you might as well get the bushel because you save money." So he was smart enough to know when somebody was trying to outdo him. (Carla, 74 years old)

George juggled two jobs to maintain the family homestead. He was also required to walk a tight rope as a "good Christian Negro" to navigate the above types of interactions with whites. Military expectations for a black private first class usually meant manual labor. However, George's aptitude for mathematics and technical work enabled him to distinguish himself from many of his white and black peers. One grandchild recalls a similar example of George's economic savvy:

One day Big Papa was selling a tractor to a white man who came to look at it. I guess he thought Big Papa was stupid and tried to talk him down on the selling price. I can still clearly remember him telling the man, "the tractor is worth $500, and that is the only amount that I will take." Later, after the man went on a test drive, I heard him say, "let me go see where this man has gone with my tractor." I remember this valuable lesson when I am doing business with my company today. (Manuel, 49 years old)

George's offspring were aware of their family's class status – and the effort required to maintain it. They also remember important strategies they still rely on today. But values and material goods did not alter the reality of segregation or that continual work and prudent decisions were required to maintain their farm. Early years in poverty resulted in frugality for which George was reputed:

My favorite memory of Big Papa was when I was about 12 years old. We spent every summer in Mississippi the week of my birthday and I had never received a present . . . I decided to ask him if he would give me something. He responded that he would, but that he needed to get some change . . . maybe he wanted to give me twenty dollars . . . I had heard stories about his cheapness and how it was a struggle to get him to spend money, but . . . he had promised and I was going to hold him to it! . . . Every day I waited for Big Papa . . . At last, it was time for us to head back to the city, so I asked him if he remembered and he said, "yes." He slowly reached into his pocket with lots of pomp and circumstance and my eyes widened with anticipation. He opened his hand, picked out a quarter and gave it to me. I think he was proud! I didn't want to show any disappointment so I thanked him and got into the car holding back my tears. We laughed the entire trip home! A quarter!! But . . . I was the *only* one [of his grandchildren] to ever get a gift [but] . . . not even a $1 dollar bill [laughs again] . . . this memory has lasted me a lifetime. (Donna, 53 years old, executive and CPA)

The above whimsical narrative provides a vivid portrait of both George's influence on his grandchild and his reputation for being frugal. The CPI Inflation Calculator suggests that $0.25 in 1976 would have the buying power of $1.09 in November 2017. Yet George's cautiousness seems justified when one considers that, during her early years as a domestic, Irma was only paid $1.00 a week.

According to the above narratives, as the patriarch, George's role was central to the family's survival and successes. When considered within the broader ecological context, his position as a traditional protector/provider placed him at considerable risk from whites concerned that he, and his family, might be over-stepping their bounds:

In 1900 a black male could expect to live to be 32 years of age. A white male could expect to live to be 46. In 1950 a black male could expect to live to be 59 years of age while a white male had a life expectancy of 66 years. By 1970 there were still further increases in life expectancy rates for both whites and blacks. Black males could expect to live to be 61 while white males could expect to be 68. The ratios of life expectancy has remained virtually constant over this entire 20 years. Fixed and 90 percent. (Darity and Myers 1994: 135)

The lag in life expectancy for black males as a result of systemic factors such as differential healthcare, poor work and living conditions, and violence further inform the challenges George and black men like him were required to navigate historically.

At the time of their deaths, George and Irma had seventeen grandchildren and twenty-six great-grandchildren. George died of lung cancer June 28, 1989, in Gallman, Mississippi, at the age of 72 years old; Irma died October 4, 1994, of breast cancer at the age of 79 years old in Gary, Indiana. Both their voices ring loud and clear through the narratives and recollections of their offspring, friends, and peers. Figures 1.2–1.10 are portraits and other documents from the earliest generations of Kings.

Name (Number of Children, Grandchildren, Great-Grandchildren)			
		Felix and Samantha (2)	
		James and Loren (2)	Emma
A. Siblings	*A. Davey (4)*	*A. Janice Mae (12)*	
		B. Irma (6, 17, 26) C. Martha Ella (5, 9) C. Janice Mae (3, 6) C. Carla Reace (5, 7) C. Bonnie (deceased) C. Helena (deceased) C. James Lewis (4, 4) *B. Irma's Siblings* Willard (6, 7) Cora (0) Lily (1) Margie (0) Jim (0) Gina Mae (7, 32, 21) Katherine Mae (1, 2) Robbie Lee (1) Leddie Lee (0) Sinclair Jr. (5, 14, 16) Daniel Lee (6, 11, 5)	

Figure 1.2: King Family Tree
Key: James and Emma were siblings. A=Siblings, B=Siblings, C=Siblings.

Figure 1.3: James and Loren King, Unknown Dates. The image quality reflects the quality of the historical source image.

Figure 1.4: Janice King, Circa 1930. The image quality reflects the quality of the historical source image.

Figure 1.5: George and Irma, Circa 1960. The image quality reflects the quality of the historical source image.

Figure 1.6: Gina, Robbie, Irma, and Margie. The image quality reflects the quality of the historical source image.

MILITARY INFLUENCE AND THE KING FAMILY EXPERIENCE

A stint in the military during World War II provided the requisite funds for George and Irma to purchase twenty acres of farmland and a house; livestock purchases soon followed. Just as the military helped pave the way for their improved lifestyle, enlisting improved the economic outcomes of other family members as compared to many civilians in the South and North. The diverse influences of the armed forces on the Kings were a direct result of the systemic consequences of military conflicts that changed the political and social climate in this country. Particularly after World War II, the military was believed to provide opportunities for conscientious blacks.[50] However, scholarship is mixed about the beneficence of such service for blacks. For example, Berry and Blassingame

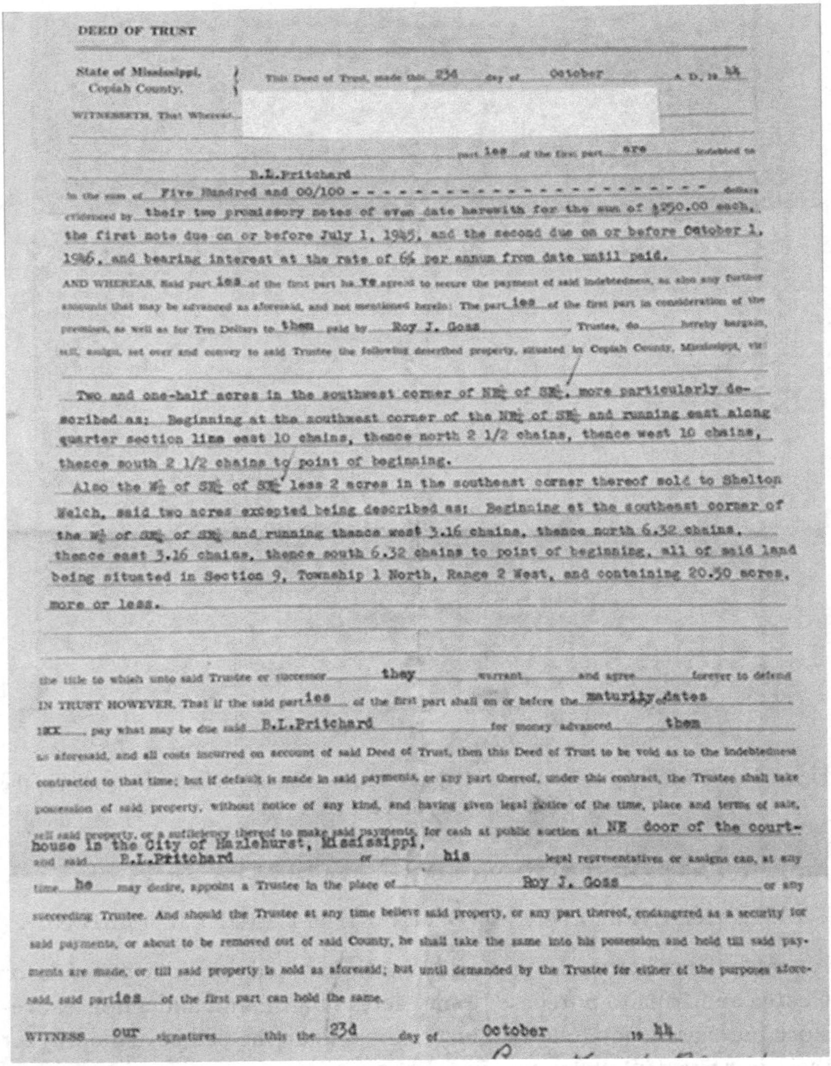

DEED OF TRUST

State of Mississippi, Copiah County. This Deed of Trust, made this 23d day of October A. D., 19 44

WITNESSETH, That Whereas

part ies of the first part are indebted to B.L.Pritchard

in the sum of Five Hundred and 00/100 - dollars evidenced by their two promissory notes of even date herewith for the sum of $250.00 each, the first note due on or before July 1, 1945, and the second due on or before October 1, 1946, and bearing interest at the rate of 6% per annum from date until paid.

AND WHEREAS, Said part ies of the first part ha ve agreed to secure the payment of said indebtedness, as also any further amounts that may be advanced as aforesaid, and not mentioned herein: The part ies of the first part in consideration of the premises, as well as for Ten Dollars to them paid by Roy J. Goss, Trustee, do hereby bargain, sell, assign, set over and convey to said Trustee the following described property, situated in Copiah County, Mississippi, viz:

Two and one-half acres in the southwest corner of NE¼ of SE¼, more particularly described as: Beginning at the southwest corner of the NE¼ of SE¼ and running east along quarter section line east 10 chains, thence north 2 1/2 chains, thence west 10 chains, thence south 2 1/2 chains to point of beginning.

Also the W½ of SE¼ of SE¼ less 2 acres in the southeast corner thereof sold to Shelton Welch, said two acres excepted being described as: Beginning at the southeast corner of the W½ of SE¼ of SE¼ and running thence west 3.16 chains, thence north 6.32 chains, thence east 3.16 chains, thence south 6.32 chains to point of beginning, all of said land being situated in Section 9, Township 1 North, Range 2 West, and containing 20.50 acres, more or less.

the title to which unto said Trustee or successor they warrant and agree forever to defend

IN TRUST HOWEVER, That if the said part ies of the first part shall on or before the maturity dates 19XX pay what may be due said B.L.Pritchard for money advanced them as aforesaid, and all costs incurred on account of said Deed of Trust, then this Deed of Trust to be void as to the indebtedness contracted to that time; but if default is made in said payments, or any part thereof, under this contract, the Trustee shall take possession of said property, without notice of any kind, and having given legal notice of the time, place and terms of sale, sell said property, or a sufficiency thereof to make said payments, for cash at public auction at NE door of the court-house in the City of Hazlehurst, Mississippi, and said B.L.Pritchard or his legal representatives or assigns can, at any time he may desire, appoint a Trustee in the place of Roy J. Goss or any succeeding Trustee. And should the Trustee at any time believe said property, or any part thereof, endangered as a security for said payments, or about to be removed out of said County, he shall take the same into his possession and hold till said payments are made, or till said property is sold as aforesaid; but until demanded by the Trustee for either of the purposes aforesaid, said part ies of the first part can hold the same.

WITNESS our signatures this the 23d day of October 19 44

Figure 1.7: Land Purchase Deed for King Farm. The image quality reflects the quality of the historical source image.

(1982) provide a troubling view of this relationship: "blacks were the only Americans who continually fought to serve in the military to prove their humanity. Ironically, they had to die in order to live . . . blacks became cannon fodder again and again to prove their loyalty to a nation which

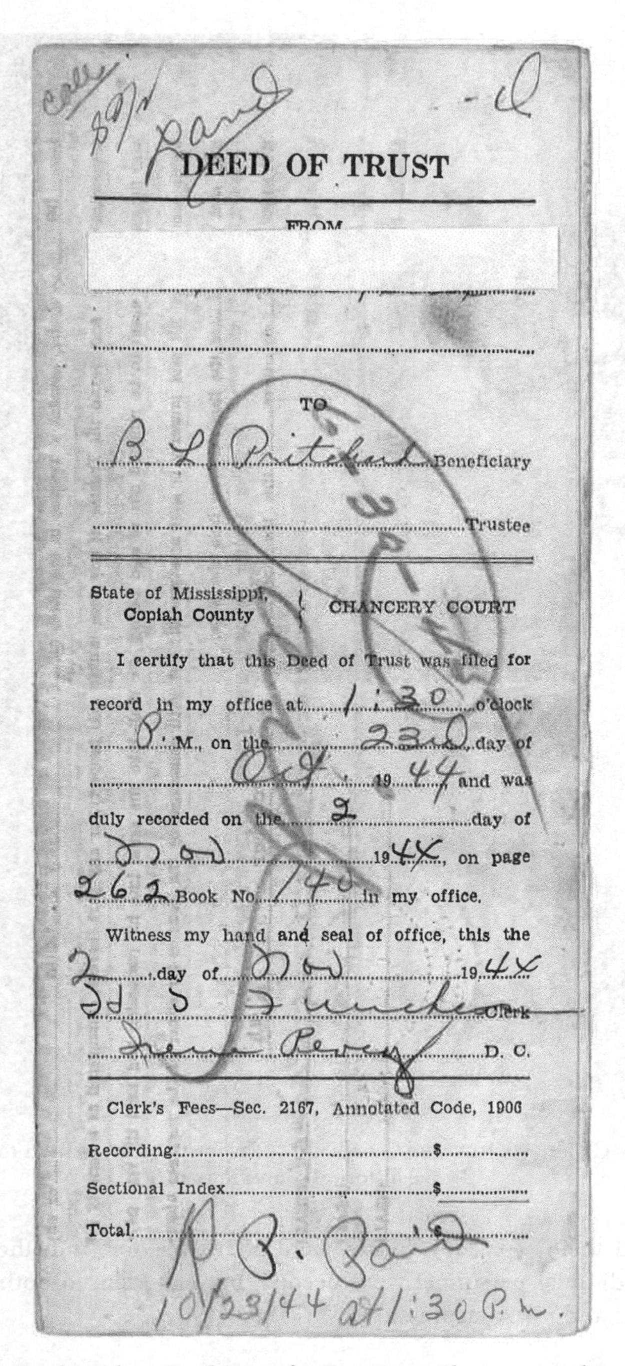

DEED OF TRUST

FROM

TO

B L Pritchard .... Beneficiary

.... Trustee

State of Mississippi, Copiah County } CHANCERY COURT

I certify that this Deed of Trust was filed for record in my office at 1:30 o'clock P. M., on the 23rd day of Oct, 19 44 and was duly recorded on the 2 day of Nov, 19 44, on page 262 Book No. 140 in my office.

Witness my hand and seal of office, this the 2 day of Nov 19 44

.... Clerk

.... D. C.

Clerk's Fees—Sec. 2167, Annotated Code, 1906

Recording........ $........

Sectional Index........ $........

Total........ $........

10/23/44 at 1:30 P.m.

Figure 1.8: Land Purchase Deed Cover for King Farm. The image quality reflects the quality of the historical source image.

Figure 1.9: Children of Irma and George, 1994. The image quality reflects the quality of the historical source image.

disowned them" (p. 295). Blacks were often welcome into these spaces when additional personnel were needed, but experienced both systemic inequities and micro-aggressions.[51]

Figure 1.10: Janice King, Circa 1960. The image quality reflects the quality of the historical source image.

Black Representation in the Military: Empirical and Experiential Outcomes

Limited representation, segregated living and work spaces, and differential treatment meant that, for many blacks, military service did not differ from civilian life. Although blacks have always had a military presence historically, their involvement increased during the time frame 1880–1920. For example, in 1890, there were approximately 2,782 black sailors, soldiers, and US Marines. The beginning of the twentieth century witnessed a 25.7 percent increase to 3,498 blacks. By 1910, 3,734 blacks held such positions. And by 1920, a 235 percent increase in black military involvement was evident; 12,511 blacks held these service posts (Green and Woodson 1930). When their presence is specifically considered across military units, Quarles (1987) posits that in the mid-1940s, the number of blacks in the armed services was small, "13,200 in the Army and 4,000 in the Navy" (p. 254). Yet black representation among officers was abysmal:

In the Navy by September 1945, the number of black male commissioned officers had increased from 12 to 52, 15 line and 37 staff officers, as compared to more than 70,000 white officers . . . conditions in the Marines and the Coast Guard differed little from those in the Navy. (Berry and Blassingame 1982: 327)

However, pressure and practicality meant that the Army disaggregated training facilities such that by the end of 1942, more than 1,000 blacks had graduated from officer training school to become second lieutenants. Integrated classes, but segregated eating tables and lounges were the norm in the Women's Auxiliary Army Corps (WAACs); in 1942, 36 of the 440 women graduates were black.[52] In 1943, after complaints by black press and clergy, black sailors were being trained in more than 50 different job categories in the Navy.[53] Yet training for blacks in the Air Force was lackluster at best.

Abolition of the draft in 1972 meant increasing numbers of blacks entered the military as a career that provided job opportunities absent in most civilian sectors. Military service could also become a *bridging environment* (for enlisted men more than officers), as skills and abilities honed during stints could be transferable to civilian careers.[54] Research generally suggests that the military provided families like the Kings with certain economic and non-economic benefits:

For many, if not all, ethnic minority men, military service can positively affect their subsequent chances in the opportunity structure of civilian society. Geographic mobility and personal independence, occupational training of various kinds, and experience in bureaucratic structure, all make it easier for the veteran to obtain those civilian jobs that provide better pay . . . for . . . blacks . . . there is an economic advantage to being a veteran . . . we wish to emphasize the importance of *bridging environments* [emphasis is ours] in fostering social mobility. (Browning et al 1973: 77, 80, 84)[55]

Moskos (1986) is convinced of similar benefits:

Blacks occupy more management positions in the military than they do in business, education, journalism, government, or any other significant sector of American society. The armed services still have race problems, but these are minimal compared with the problems that exist in other institutions, public and private. (n.p.)

So blacks who experienced these military-based benefits were often better prepared to navigate the constrained, segregated labor force. Moreover, such capital could have long-term benefits for veterans. The following comment summarizes the educational and occupational benefits military service had on George's life:

The military had a higher representation in precisely those occupational groups which between 1940 and 1960 registered the greatest gains in the labor force . . . among such groups [were] professional, technical, and kindred workers; managers and officials; clerical workers; and *mechanics* [emphasis added here] and repairmen. (Browning, Lopreato, and Poston Jr. 1973: 77)

Additionally, some black enlisted men were able to earn high school diplomas (especially during peace times) and secure social capital in the form of on-the-job training.

Seventy-seven-year-old Martha describes the on-the-job training typical for black enlisted man in World War II: "back in the day, when they got ready to send those young men to the service . . . it wasn't how much education you had or how much you knew because they were going to teach you what to do." Janice suggests that this specialized training helped George post-military: "after he got out of the Army, Big Papa went to Community College and got a certificate to fix cars." His technical skills helped cultivate a viable post-military livelihood for George and his family.[56] A World War II black enlisted man from Alabama had a similar story:

By going off to war, my brother and I brought honor to the family . . . we were often told that we were a credit to our race – never mind that both of us were assigned to menial service units in a segregated army consisting of black troops and white officers. We were doing our bit in the nation's struggle against tyranny abroad . . . and everyone knew that our service would make things better for our people after the war. An even more tangible reward for our military service was that my brother and I were able to send home a regular allotment from our steady if low pay which provided our mother the only savings she would ever know. In addition, our war savings bond and life insurance provided our first investments in the larger society. (Billingsley 1992: 188)

Additionally, census reports confirm the employment push out of low-paying occupations the military could provide some blacks: "according to the U.S. Census, in 1940, 70 percent of black workers were in farming or employed as domestic and unskilled labor. By 1954, this had declined to 50 percent . . . because of the war [World War II] and the technological revolution it hastened, there were major shifts in the occupational structure of the African-American community" (Billingsley 1992: 189). But how did these shifts translate into economic outcomes?

Most reports on the subject suggest that American soldiers were well paid in comparison to the average civilian. Table 1.4 provides a summary of allotments for the three military engagements in which King family members participated; World War II, the Korean Conflict, and the Vietnam

Table 1.4 *Median Wages for Enlisted Military and Rural Farm Wages by Race, Gender, and Year*

War/Conflict	Enlisted Annual Wage	Median Annual Income (Start and End Year)			
		1939+		***1945****	
World War II 1939–1945	$600	WM: $1,112 NWM: $460 Farmer: $373	WW: $676 NWW: $246 Domestic: $296	W: $1,483 NW: $549	
		1950		***1953***	
Korean Conflict 1950–1953	$936	W: $1,115 NW: $817 Farmer: $711	Domestic: $448	WM: $1,601 NWM: $567 Farmer: $482	WW: $672 NWW: $293 Domestic: $554
		1955		***1975***	
Vietnam War 1955–1975	$936 (1955)–$4,334 (1975)	WM: $1,484 NWM: $600 Farmer: $461	WW: $606 NWW: $295 Domestic: $502	W (families): $10,750 B (families): $5,467 WM-HOH: $14,384 BM-HOH: $10,372	WW-HOH: $5,623 BW-HOH: $4,302
		1980+		***2010***	
Non-War Period 1980–2010	$6,015–$15,919	W: $15,176 B: $11,939 Farmer: $10,193		WM: $44,200 BM: $32,916 Farmer (mean): $15,000–$17,499	WW: $35,568 BW: $30,784

Key: NA = Not available. W=White. NW=Non-White (blacks constitute the majority in this group). M=Men. W=Women. B=Black. HOH=Head of Household. *Mean annual amounts. Unless otherwise noted, individual wages are listed. 1939+ figures include rural and non-rural. Figures for 1939 from Table 24. Median Wage or Salary Income Persons 14 Years of Age and Over with Wage or Salary Income, by Color and by Major Industry Group, By Sex, for the United States: 1939 in 1949. Figures for 1945 from Current Population Reports: Consumer Income Series P-60, No. 2, Table 1. Percent Distribution of Families and Individuals by Total Money Income Level, by Color of Head for the United States, Urban and Rural: 1945. Figures for 1950 from Series G3 172–415 Median Money Wage or Salary Income of All Workers With Wage or Salary Income, and of Year-Round Full-Time Workers, by Sex, Race, and Major Occupation Group: 1939 to 1970. Figures for 1953 from U.S. Census Table 2: Color and Residence: Median Income in 1953 of Persons 14 Years of Age and over, By Sex, for the United States, Urban and Rural. Figures for 1955 from U.S. Census Table 2: Color and Residence: Median Income in 1955 of Persons 14 Years of Age and over, By Sex, for the United States, Urban and Rural. Figures for 1975 from Table 10. Sex, Race, and Spanish Origin of Head-Household by Total Money Income in 1975, by Nonfarm-Farm Residence and Regions and 1976 Current Population Reports: Farm Population U.S. Department of Commerce Bureau of the Census Series 27, No. 47, Table F: Income Characteristics of Farm and Nonfarm Families, by Race: 1974. Figures for 1980 from Race and Farm-Nonfarm Residence- Families and Unrelated Individuals, by Total Money Income in 1980. Figures for 2010 taken from U.S. Bureau of Labor Statistics. 2011. "Highlights of Women's Earnings in 2010: Report 1031" and Farmworker Justice (2014) and reflect average total individual income (includes income that some farmworkers earn from non-agricultural jobs).

War. Although annual wages would be expected to vary based on the source and how indicators are defined, this table provides a broad-based way of considering possible financial benefits of military versus civilian rural wages.[57] Median annual rural wages for civilian versus enlisted persons are compared during each conflict period. However, amounts for the military do not include other possible perks. When possible, we include rural farm wages to reflect agricultural locales common among the Kings. When available, wages for domestics and farmers are also included to reflect Irma and George's occupations and those of several of her sisters. For the sake of parsimony, civilian wages are provided for the beginning and ending years of each conflict as well as based on race and gender when possible. Results illustrate some of the monetary benefits enlisted blacks could realize as compared to civilians and support studies that it was relatively more *economically* advantageous for many blacks to enlist, especially before the 1980s.

First, black enlisted persons were generally paid the same as whites at the same rank. However, unequal presence by whites as officers translated into relatively higher salaries than for blacks, who were disproportionately represented among lower ranks. When median wages for newly enlisted persons during World War II are considered, they earned about $600 annually. Yet median incomes in 1939 in rural locations were $1,112, $676, $460, and $246 for white men, white women, non-white men, and non-white women, respectively. Based on these figures, in 1939, non-whites, farmers, and domestics would generally benefit economically from military enlistment. However, the annual wage for white women was only slightly above the mean military wage. Next, the 1945 median income for white civilians was about $1,483 and $549 for non-whites. This means that the civilian wage for rural whites in 1945 was almost three times that of their non-white counterparts. Additionally, these wages meant that it was not economically advantageous for most whites to enter the military. Yet a military stint could have been economically beneficial for rural blacks. This appears to have been the case for George and several other relatives.

When the Korean Conflict is considered, the 1950 median annual income of $936 for enlisted persons meant that only non-whites ($817), farmers ($711), and domestics ($448) could benefit financially from a military stint. A similar pattern was apparent in 1953 for white women, non-whites in general, farmers, and domestics. Moreover, military wages for non-white males and non-white females are between 1.65 to 3.2 times higher than median annual incomes for rural farm workers from these two groups. King family relatives Sinclair Jr., Horris Lee, and Donald Sr. participated in the Korean Conflict. Similar patterns are apparent when one considers the Vietnam War (1955–1975)[58] in which James Lewis and Fred Smith served. Yet by 1975, civilian incomes were more competitive. The range during the non-war period (i.e., all-volunteer period) of about 1980 to 2010 suggests fewer direct wage benefits for blacks.

In addition to possible financial benefits, the training that enlisted persons could experience (for example, George's training as a mechanic) was potentially transferable state-side. Later, some family members were also able to use the GI Bill to purchase homes, which represented a source of family wealth outside of the Mississippi homestead.[59] Readers should note that these results focus solely on economic considerations and don't account for non-economic reasons King family members and other blacks enlisted such as patriotism that are also reported in numerous studies.[60] Table 1.5 includes a list of King family veterans and thumbnail descriptions of their post-military careers. Certain family members segued military life into economic benefits and social capital in the form of education, experience, and on-the-job training that positioned them for the middle class and to pass on similar benefits to their children. Yet their experiences varied.

Beyond Economics: King Family Military Experiences

As Browning et al (1973) note: "most veterans, whatever their ethnic status, do not have fond memories of their service experience" (p. 83). Beyond economics, military benefits and outcomes for blacks like the Kings varied. For example, 74-year-old Carla describes her father's experience: "He was in World War II . . . He wasn't there too long because he got his index finger on his right hand shot off and he couldn't use a gun or rifle, so they sent him home." In such instances, just as enlisting could provide employment with educational and career opportunities, it could be at the expense of disrupted black families and poor health outcomes.[61] Some scholars caution against over-romanticizing military life and its advantages for minorities. George's daughter provides a parallel commentary about her father; "Momma said that the Army changed My Daddy. Before he went in, he never left her side – all sweet and stuff. When he got back, he was different . . . gone all the time" (Carla, 74 years old). Although George never discussed his military stint with his family, they noted a marked difference and emotional distance that they did not consider beneficial.[62] It was not until George became much older that he would again exhibit his pre-military behavior. Like his father, 69-year-old James Lewis, rarely discusses

Table 1.5 *Thumbnail Descriptions of King Family Military Experiences*

Name	Military Setting	Branch	Period	Post-Military Career
Daniel	World War II	Army	na	Airplane mechanic
George Franklin	World War II	Army	1953–1955	Mechanic
Sinclair Jr.	Korean Conflict	Army	1951–1953	Truck driver
Horris Lee	Korean Conflict	Army	1955–1959	Steel mill
Donald Sr.	Korean Conflict	Airforce	1955–1958	Steel mill^{+}
Fred	Vietnam	Airforce	1971–1974	Teacher/Coach
James Lewis	Vietnam	Marines	1968–1969	Steel mill
Donald Jr.	Non-war	Marines	1982–1984	Minister/Admin.
Donna	Non-war	Army	1983–1986	CPA/Admin.
Calvin Jr.	Non-war	Navy	1985–1989	Minister/Truck driver
Kevin Jr.	Non-war	Army	1999–2003	Carpenter

Key: The United States Armed Forces are made up of the five armed service branches: Air Force, Army, Coast Guard, Marine Corps, and Navy. Admin. = Administrator. na=not available.

his military experience. Yet he recounts several episodes when his family and a ritual from the Black Church tradition helped him through challenges:

When I was in the military, I got in some trouble so I called my mother to help me. She said that she was sorry to hear about it, but I must have gotten away from my home training [he laughs]. So she said she would help me – by praying for me. I didn't understand it then, but I understand it now.

According to the above account, instead of intervening on behalf of her son directly, Irma elected to support him spiritually via prayer. This same religious ritual and direct family support sustained him during another difficult period:

Religion helped me when I got wounded in Vietnam. I was paralyzed for about six months. When I got home, my parents told me everything was going to be alright. And they prayed for me.

James's experience and others like it, suggests that although the military could represent an important socioeconomic bridge, it usually did not come without costs. For James Lewis, a Purple Heart recipient, it meant losing half a year of his life recuperating from life-threatening injuries. What other ways did military stints affect families like the Kings given that blacks were disproportionately represented on the bottom rungs of the occupational ladder?[63]

George and Irma's nephew, Fred, recalls his military experiences and decision to only complete one tour:

> I was in the Airforce for three years. When Vietnam ended, those who were going back to school were let out three months early. I left December '74 and January '75, I went back to [name of HBCU]. The Airforce gave me more opportunities [reason for that choice over the Army]. I was a jet mechanic and I worked on the landing gear and flight controls. I loved mechanic work and the aircraft was the most challenging thing I thought I could do in the military. There weren't a whole lot of us [blacks doing that work]. My only regret was that I didn't stay in twenty years and get a larger retirement check. I was stationed in Charleston, SC. But it was the most prejudiced place I had ever been – Charleston and the military. That's why I didn't stay in the military. (Fred, 69 years old, retired teacher and coach)

Fred was Gina Mae's son. Although he enjoyed the challenge and responsibilities afforded him in the Airforce, chronic discrimination both in the military and the city where he was stationed hastened his departure. Although he regretted not making the military a career, Fred was able to return to college and complete both a Bachelor's degree in Health and Physical Education and Master's degree in Physical Education and Athletics Administration. Several decades later, another family member candidly describes some of the benefits and drawbacks of Army life:

> I joined the military in August 1983 and served three years active and one year in the Reserves. The benefits were that I learned discipline, got to travel, received money for college when I was honorably discharged and met my husband. The hardest thing ... there were so many healthy men looking for sex and sexual harassment was accepted back then. I went in as an E-3 and didn't get promoted to E-4 for a very long time. I wouldn't sleep with my sergeants and their sergeants had the same expectations. I'm glad I went, but I would never recommend it to another female. We're just not designed for that ... I joined the Army because I didn't fit at college [but] ... I have NO regrets ... I got out after I had [name of her first daughter] and never looked back. (Donna, 53 years old, CPA and executive)

As the only female member of the King family to join the military, Donna's quote above provides insight into gender differences in service life where, for her, the benefits still outweighed the challenges. She experienced marginalization based on gender rather than race. In addition to sexual harassment, rebuking continual unwanted sexual advances resulted in delayed promotions and an attenuated career. In contrast, Calvin's four years in the Navy were professionally and personally beneficial:

> I was at [college name] for a year majoring in Psychology . . . I realized I wasn't ready for college right then and I realized that the Navy would allow me to see the world. It was definitely different. Boot camp was like you see on television. Once you go to your ship and get stationed, it's like a regular job outside of the military structure besides the saluting and "Yes Sirs." I was stationed on the USS Iowa. I was in charge of the aviation store room – dealing with everything with the flight deck. We didn't land planes, but we landed helicopters – from gear, chains to secure helicopters, everything, I was in charge of it . . . because my first class liked me . . . Every time he needed special assignments, he would call me . . . I actually thought he didn't like me. So one day, I asked him, "Why you don't like me? What did I do?" He really looked puzzled and explained, "I get you to do things not because I don't like you, but because you do things right and I don't have to check behind you."

For 52-year-old minister and truck driver, Calvin, a Navy stint enabled him to grow developmentally. After dispelling concerns about racism from his white supervisor, he was given responsibilities that higher-ranking peers were not. However, newly married, long stints away from his wife resulted in Calvin' decision to leave the Navy.

Thirty-seven-year-old Kevin believed that the military would provide him with needed training, focus, and transition from family challenges:

> I was a little frustrated in my personal life . . . and as I headed toward college, I wanted to keep focused . . . I didn't want to mess up, plus I could get money for school. I looked at the pros and cons and made the decision . . . I had some conflict with mom for a few years . . . I was always on punishment . . . it didn't matter what the situation was. It got to the point where I didn't want to go home . . . I thought it would be a positive direction.

Yet Kevin describes how racism and rigid rules and regulations derailed his Army career:[64]

> I remember how people [whites] treated me on base, they were all friendly, but one time I was at a bar and I saw some of them. I went up and started talking and one guy began to disrespect me. The others just looked and laughed. I was like, "man, that's not cool." He just played me off. I was so mad, but I made myself stop so I wouldn't do anything to get in trouble. And this was a guy who was supposed to be kinda my friend. I was like "wow!" They can sure be different.

Kevin's experience of racist treatment by whites outside the barracks parallels the following comment by Moskos (1986): "interracial comity is stronger in the field than in the garrison, stronger on duty than off, and stronger on post than in the world beyond the base" (n.p.). Like Donna, Kevin became disillusioned when the ideals of equality lauded by the military did not bear out on a daily basis. Despite Calvin's positive experience, he chose to leave the military to spend more time with his family.

Prevailing anecdotes have been debunked that attributed differential outcomes for blacks in the military to lower qualifications and abilities.[65] The next comment about the pernicious effects of racism may be broadly applicable to other military settings:

It appears that being white in the Army is more important than doing well on universalistic criteria [i.e., based on objective standards] . . . the black enlisted man in the U.S. Army is subject to inequality which is not the result of failure to meet universalistic criteria, i.e., indirect impersonal institutions, but rather a result of the direct racist actions of real-life people. (Butler 1976: 817)

Calvin's experiences support the above academic finding and suggest the importance of practical skills: "one thing I learned in the military is just because you see people with a lot of rank on their shoulders, doesn't mean they are good at what they do, but rather that they could pass the test. You could be responsible and not pass the tests." Most troubling, military service has tended to require many blacks, particularly young enlisted males, to risk their lives in ways distinct from many of their white peers. Studies show disproportionate casualty rates among blacks and enlisted men from lower socioeconomic groups because they were more likely to be assigned to infantry and artillery units.[66] Yet economic incentives continue to have drawing power for certain blacks:

Even as late as the mid-1960s, blacks disproportionately volunteered for frontline combat and hazardous duty to receive more money. A black Sergeant explained that he volunteered for hazardous paratrooper jump duty, "because of the extra $55, because it was the elite part of the service . . . the job opportunities outside just weren't that good. The Army is taking care of me and my family." (Berry and Blassingame 1982: 332)

Lutz (2008) acknowledges the historic and contemporary commitment by blacks to military service: "African Americans have fought in every American war" (p. 170). They continue to do so. According to 2015 demographic military reports, 216,926 blacks are active duty members of the military. Their involvement as enlisted persons and officers in the Army, Navy, Marine Corps, and Air Force are: 94,423 and 12,085; 51,474 and 4,267; 18,548 and 1,101; and, 39,997 and 3,634, respectively.[67] Yet it has been argued that until blacks have the same opportunities for gainful employment and career trajectories afforded their white peers, military enlistment represents a questionable alternative.[68] According to the Kings who participated in the military, despite Truman's Executive Order 9981 in

1948, decades later, the military is still fraught with inequities against people of color and women. In support of their claims, a 2017 report based on Pentagon data from 2006 to 2015 shows that:

> Black service members were as much as two times more likely than white troops to face discipline in an average year . . . over the past decade, racial disparities have persisted in the military justice system without indications of improvement . . . in every branch . . . in an average year, black Marines were 2.6 times more likely than whites to receive a guilty finding at a general court martial . . . black airmen were 71% more likely than whites in the Air Force to face court martial or non-judicial punishment . . . black soldiers were 61% more likely to face court martial than whites in the Army; and black sailors were 40% more likely than whites in the Navy to be court martialed; that percentage is 32% for black Marines. (Brook 2017: n.p.)

These inequities were attributed to a lack of leadership diversity, stereotypes, and implicit biases that parallel racial challenges in the broader society. According to the same source, in 2016, approximately 78 percent of military officers were white; only 8 percent were black, yet, "the military has known about these numbers for decades and has done nothing about it" (Brook 2017: n.p.).

Conclusion: The Military and the Black Middle Class

Military involvement could provide benefits for blacks that were absent in the civilian world, including consistent, often higher, income; increased status as they showed their allegiance to an often unwelcoming country; more egalitarian interactions with whites; travel; increased career opportunities; heightened aspirations; activities that reinforced masculinity for males; greater likelihood of holding leadership positions; retirement benefits; and, later, the ability to purchase homes via the GI Bill. A tangential benefit for the black community meant jobs for certain blacks who remained behind in industrial locales. Yet these same individuals faced challenges such as underrepresentation among officers and in upper ranks, institutional inequality, and occupational discrimination.[69]

It appears that Berry and Blassingame (1982) best summarize this military service paradox: "the general pattern of black service from the colonial period to the 1970s was a refusal by whites to use blacks unless it was perceived as absolutely necessary to victory, minimum concessions towards favorable treatment during a crisis, and a retreat to outright oppression thereafter" (p. 341). The benefits of military stints for blacks remain debatable. Most Kings acknowledge the economic benefits of military involvement; non-economic advantages are less clear. Thus enlistment reflects another conundrum of how blacks can best navigate systemic constraints in ways that help meet their varied socioeconomic and ideological objectives.

CONCLUSION: LAND, LIVESTOCK, AND CLASS TRAJECTORIES

Post-slavery, most southern blacks began as sharecroppers eking out a meager existence; so were the Kings initially. Economic progress in the twentieth century, particularly for many southern blacks, was tied to migration and military conflict. During the early 1900s, black migrants often filled positions for white unskilled industrial workers who had enlisted in the Army. However, between 1930 and 1936, the Depression resulted in fifty percent of skilled blacks losing their jobs.[70] Studies suggest that some of the greatest economic gains for blacks occurred between about 1940 and 1950 as a result of World War II.[71] Although blacks were generally excluded from skilled jobs, a pattern was common of hiring and firing blacks based on white labor force presence.[72] Yet Berry and Blassingame (1982) argue, "in the South the color caste occupational structure changed little during the war" (p. 200). It is during these tumultuous times that early generations of Kings lived. This chapter broadly reminds readers of the southern black experience, specifically focusing on key members of the King family who lived in Gallman, Mississippi. Statistics and historical accounts here illustrate the legacy of segregation and discriminatory practices that were the backdrop for this family's experiences. The socio-ecological context for their lives is further discussed in subsequent chapters.

Gallman, Mississippi, was a small segregated farming area consisting of one white-owned grocery store and gas station. The lives of most residents, both black and white, were linked to land – either for cash crops or subsistence. Yet a seemingly innocuous wish by a domestic, Irma, jettisoned one family's economic trajectory. Inter- and intra-racial as well as class distinctions were evident based on whether black families owned or rented on "the white man's land." Family members recall several pivotal choices, informed by a religious ethos that guided their early steps toward the middle class. Some economists point to family class status rather than family values as the grounding mechanism for economic stability; "regardless of race or sex, family behavior is substantially less important for determining socio-economic status than family class background. As a rule, childhood family socio-economic status explains twice as much of the variation in socio-economic outcomes among individuals of a particular race-sex group as childhood family values" (Mason 2003: 63). As shown in subsequent chapters, the King's experiences help nuance this statement and illumine how both class position *and* Christian-based family values were pivotal for their upward mobility.

We also begin to show evidence of religious education, fueled by Irma's early years in poverty, as a motivator for economic decision-making. Yet religiosity alone does not explain these initial economic outcomes. As Darity and Myers (1994) purport, "the development of family life on an institutional basis was closely tied up with the accumulation of property in these [middle-class] families" (p. 141). As the Kings improve their homestead, it will be important to further understand how intangible dynamics linked to religious education correlate with their middle-class attitudes and actions.

2

"Bikes or Lights": Familial Decisions in the Context of Inequality

Growing up, Irma was more like a surrogate mother to Sinclair Jr. than an older sister. Their family lived in abject poverty. His widowed mother had spent most of her time sharecropping. When Irma wasn't working in the field herself, she was responsible for taking care of her younger siblings, cooking, and cleaning for the entire family. After completing high school and a stint in the Army, Sinclair (called "June") migrated from Mississippi to Denver, Colorado, in 1953 in search of gainful employment. He married and made a career as a truck driver for the US Postal Service. During an annual trip to Mississippi to visit Irma, her children asked Uncle June for a gift. They wanted bicycles! They had seen local white children riding them, but to have their own was a pipedream. They knew their frugal father would never use money so lavishly. To him, money should either be saved or invested in the farm – and it still lacked basic amenities like running water and electricity. June promised Irma that he would send money to purchase bikes. His nieces and nephew were elated! Several weeks later, the money arrived by mail. Irma remembered its original intent. She also knew the family's pressing needs. So she made a decision. Instead of buying bikes ... she bought lights. Irma used the money from June to purchase electricity for the farm.

The concept *linked fate* is central to understanding seminal decisions made by the King family. It is not happenstance that the notion of linked fate has historically been tied to the black experience, in general, and the black religious experience, in particular. A linked-fate mentality suggests that members of a group believe that their successes and/or failures are indelibly connected. Although individuals may evaluate their own accomplishments and/or setbacks personally, inevitably their views about such outcomes are correlated with those of other group members.[1] Informed by the introductory narrative, Chapter 2 chronicles how the Kings prioritized objectives and responded to negative social conditions around rural living, migration, and economics, as well as some of the effects of structural forces and personal choices on quality of life and life chances. Also integral are the contrasts and tensions associated with what constitutes being middle class in the rural South based on economic and non-economic factors.

The above "bikes versus lights" scenario represents an inter-generational example of a linked-fate mentality, as Irma made a deliberate, practical decision to improve the family's quality of life. Other instances of similar choices also emerged. For example, Janice King sharecropped "like a man," which prevented her family from experiencing homelessness. Irma's role as surrogate mother meant Janice King could focus on being the family provider. And Irma's relationship with and sacrifices for her younger siblings meant her brother June could complete high school, relocate, and secure gainful employment as well as eventually assist his older sister financially. This chain of events resulted in the family's ability to secure electricity in the mid-1950s when the vast majority of blacks and working-class whites in Gallman did not have this convenience (note: the average price of monthly usage from 1960–1970 for farms was $4.04–$4.09 for 100 kilowatts per hour).[2] Personal decisions are not considered in isolation; a linked fate fosters familial responsibility and accountability. In this way, a linked fate becomes a mechanism to better understand pivotal family choices that resulted in middle-class outcomes for the Kings.

Other King family members also made sacrifices in the quest for economic stability. Their mobility was linked to the experiences and work ethics of relatives who migrated to southwestern and northern cities. Family members describe strategic and periodic migration (we refer to them as "migration spells" and discuss them in detail in Chapter 5), that enabled them to support each other and establish extended families in cities like Denver, St. Louis, Chicago, Toledo, and Detroit. Like so many blacks during the early to mid-1900s, several of Irma's brothers and sisters migrated to escape segregation and in search of opportunities. Narratives show that the King farm represented a harbor as family members migrated back and forth across various socio-ecological spaces.

FARM LIFE AMONG THE RURAL BLACK MIDDLE CLASS IN GALLMAN, MISSISSIPPI

A certain mindset, combination of material possessions, and social capital helped position the offspring of James and Loren King, George and Irma, to become middle-class blacks in the segregated, rural town of Gallman, Mississippi. A homestead also enabled them to stave off common dimensions of poverty – homelessness, hunger, uncertainty, and insecurity. In addition, by producing their own crops and livestock, they became self-sufficient. Yet, even their own farm and economic assets did not mean escape from the long, brutal reach of Jim Crowism and its associated disparities.

The Slow Trek Toward Middle Class

According to Willis in *Forgotten Time: The Yazoo-Mississippi Delta after the Civil War* (2000), the stage for economic inequities was set during and post-Civil War and enveloped former slaves as well as poor whites. Labor force participation statistics in Table 2.1 inform our understanding of the economic terrain that the early Kings navigated. The agricultural focus in Mississippi resulted in unique proximal dynamics among blacks and whites not as evident in certain other states. During the early to mid-twentieth century, similar numbers and percentages of whites and non-whites lived and were employed in Mississippi. From 1920 to 1950, at least 78 percent of non-white and white males 14 years old and older were in the labor force. However, relatively greater percentages of non-white males were employed. Females were considerably underrepresented, at least by a factor of two, as compared to males. However, an intra-gender comparison shows greater relative representation by non-white females aged 14 years old or older as compared to white females (for example, in 1920, 45.3 percent versus 15.2 percent). However, by the mid-1900s, similar percentages of white (21.8 percent) and non-white (28.6 percent) females were in the labor market. These trends illustrate some of the employment dynamics that sustained both black and white rural families.

During the early 1900s, a considerable number of blacks in Mississippi sharecropped and later purchased farmland. However, chronic Jim Crowism, a poor economy, escalating debt, low cotton prices, and the inability to secure credit meant that many of these same black entrepreneurs lost their farms and, by the mid-1900s, were sharecropping again. It is in this precarious environment that Irma purchased property. And although George found an economic niche in Gallman, Mississippi, as a mechanic and plant worker, his foray into farming had mixed results initially. According to his eldest daughter, the farm's success was largely a result of the efforts of Irma and her daughters:

He tried to farm, but he would not come to help us (his wife and children) gather the stuff. And on the days he was supposed to go get help, my Mul would say, "Now George you know we were going to pick this stuff. Why didn't you get your help

Table 2.1 *Labor Force Participation Statistics by Race, Gender, and Year (1920–1950)*

Year	Male				Female			
Race	Population	Number	Percent of Total Pop	Percent of Pop. 14+	Population	Number	Percent of Total Pop.	Percent of Pop. 14+
				White				
1920	433,396	245,936	54.4	85.3	420,566	41,027	9.8	15.2
1930	505,615	289,156	57.2	85.4	492,462	56,149	11.4	16.9
1940	556,157	313,726	56.4	79.1	550,170	73,686	13.4	18.6
1950	595,680	331,253	55.6	78.1	592,952	93,294	15.7	21.8
				Non-White				
1920	463,728	265,363	57.2	89.8	472,928	139,542	29.5	45.3
1930	499,526	302,238	60.5	91.7	512,218	161,938	31.6	47.1
1940	528,325	298,273	56.5	84.4	549,144	122,777	22.4	32.8
1950	481,111	238,141	49.5	79.5	509,171	94,208	18.5	28.6

Key: Pop. = "Non-White" includes other groups such as Hispanics and Asians, most persons in this category are black: 14+ means ages fourteen and older: Figures on the actual table for persons and gainful workers 14 years old and over for 1920 and 1930 include unknown age. Taken from "U.S. Census (1950) Labor Force, 1950 and 1940, and Gainful Workers, 1930 and 1920, by Color and Sex, for the State of Mississippi."

yesterday?" So she finally told him, "If you don't get us some help, me and my children are not going to gather this stuff anymore." (Martha, 77 years old)

Bottomland was notoriously difficult to farm. It was speculated as the reason whites did not covet it, but rather made it available to blacks. And even after accomplishing the goal of purchasing property, the Kings had to put a great deal of work into their new homestead. According to Martha, "It was hard because we didn't have running water, we didn't have no lights, no phone and we used an outhouse. Our house was just a shell of a house with a tin top on it." Yet they could still afford to pay other sharecroppers. And as a result of Irma's prodding, George worked the farm more concertedly:

We had cotton one year, we had peanuts, a couple of years. And he did what he had to do to turn the peanuts up on the road and then bring them out there for us while we sat under a shade tree. We would sit there and pick the peanuts out ... I never remember him sitting down pulling a peanut off the vine or pulling cotton from the boll ... we would raise cabbage and beans too. (Martha, 77 years old)

From the early 1800s to about 1860, Mississippi was the largest cotton-producing state in the USA. Moreover, "Copiah County ranked 14th of the state's 79 counties in the production of cotton" (Crook 1996: 21). In addition to providing foodstuff for their family, the sale of this staple as well as peanuts, cabbage, and beans would mean that in the 1940s, the King farm had the potential to become quite financially productive. For example, census figures show that 1940 to 1950, value ranges of farm items that the King family produced were: peanuts ($3.30–$10.90 per pound);[3] cotton ($9.89–$40.07 per pound);[4] cattle ($7.56–$23.30 per hundred pounds);[5] hogs ($5.39–$18.00 per hundred pounds);[6] and, chickens ($13.00–$22.20 per pound).[7] Peanuts and cotton could be especially lucrative. Yet continually planting cotton could severely damage land. A prudent farmer alternated between other crops. In Copiah County, "many farmers planted two cash crops: early vegetables in the spring followed by cotton, corn, sweet potatoes, and sugar cane in the late summer and fall" (Crook 1996: 21). Although George considered himself the breadwinner, his daughters suggest that a linked-fate mindset enabled their family to survive during those initially challenging years cultivating the farm. Moreover, Martha's comments should not be considered indictments, but rather candid assessments about the collective muscle necessary to maintain a farm. Another family member explains the multifaceted importance of land:

It was good to go to the field and raise your own dinner. We didn't have to run to the store like we do today. We raised peas, butter beans, pole beans, sweet potatoes and everything. We raised our vegetables in the field. We never did go to the store to buy nothing. We never did go buy a chicken; we raised our chickens. We raised eggs. We raised hogs. We would have lard under the table. We would pack that meat down with salt. We had sausage in the smoke house. For lunch, we would take salt meat and sugar butter biscuits. (Janice, 75 years old)

To Janice, "going to the field" meant blacks owned land that provided self-efficacy and afforded a certain degree of independence as compared to sharecroppers. The ability to raise crops and livestock also meant blacks could avoid possibly being over-charged for staples in town and negative interactions with whites. She proudly describes the ingenuity by which her family and other black landowners were able to store food and sustain themselves. Her descriptions are direct and concise. Moreover, the last lines allude to an indirect benefit of land ownership. Descriptions of lunch foodstuffs illustrate that children were able to attend school consistently – a luxury not typically afforded the children of black and white sharecroppers whose educations were often interrupted to tend crops. Martha also connects the family farm to George's commitment to education:

He never kept us out of school to do any farm work. Everything we couldn't do on a Saturday or in the evening after school just went undone. And the only time we stayed out of school was when we were sick. We had a lot of problems with our tonsils. But even some of those days we went to school with a sore throat. He meant for us to get an education.

In addition to setting clear parameters about when and how their children could complete farm-related chores, George set equally clear educational goals. Older family members point to this educational edict – high school completion – as another feature associated with the middle class in the rural South. And by doing so, these parents prioritized activities such as secular and religious education over strictly economic outcomes. George also worked to improve their property:

At that time there wasn't a pond on the farm, so My Daddy had one dug. And it didn't have running water at that time. When we first started out, we didn't have pigs or cows. We always had chickens. But later on, my father accumulated quite a number of horses, cows, and pigs and we just had plenty of food and plenty of meat. Once a year, they would kill a pig and a young calf. (Carla, 74 years old)

The above quote describes the progress George and Irma made as farmers. Over time, they were able to expand the number and type of livestock considerably as well as pay to have a fishing pond built on the front acreage of the property. Moreover, the annual slaughter of several livestock in preparation for winter reflected a certain degree of financial stability as compared to most local poor and working-class blacks and whites. The collective efforts of the Kings, adults and children alike, resulted in progress toward their goal of economic stability and educational attainment. Yet just as they acquired certain facets of a middle-class lifestyle, others escaped them.

Life of Luxury and Lack: Complexities Around Class and Racial Stratification

According to scholars, the middle class can be an elusive classification just as likely characterized by economic and non-economic expectations for membership as by fears and concerns about downward mobility.[8] Narratives in this section illustrate the juxtaposition of these two disparate dimensions of black southern life.

Land-Based Luxuries and the Southern Black Middle Class

For some individuals, the middle class appears to be as much a state of mind as conspicuous consumption to illustrate one's relative position to other people. Lavish spending did not seem important to George and Irma. Yet their vacillating positionality between relative luxury, as compared to sharecroppers and many other locals, as well as lack due to inequality made for complex lives. For example, when assessing their relative economic position, Irma had a reputation for sharing with struggling neighbors:

> I can remember that we would pick black berries and dew berries and My Mul would take them over to this ladies' house who had a lot of children, but who did not have a husband. My mother would can stuff in jars ... And when the little children would see the car coming, they would say, "Here comes Aunt Irma with our food." I was about 11 or 12 years old. I didn't quite understand why we were always giving away the food and stuff ... she probably had about six or seven children. Momma did it because [woman's name] did not have no way of taking care of herself and her children. I doubt if there was anything such as welfare or food stamps back then for people of color and so the woman had all these small kids, so she couldn't get out and work or nothing. So my mother was just the type of person who always liked to help somebody in need. (Carla, 74 years old)

According to the above remark, for Irma, owning a farm also represented an opportunity to share its output. The Social Security Act was passed on June 17, 1935, and Aid to Families with Dependent Children (AFDC) in 1940 to provide financial relief and food stamps to meet basic needs. Yet segregation also affected these allotments such that blacks were either unaware of and/or excluded from such government aid. The absence of government subsidies for rural blacks or family support, meant single parent families, the elderly, and children, were likely to experience extreme poverty.[9] Williams (1993) corroborates these practices: "southern states routinely excluded black women from the AFDC roles" (p. 221). This appears to have been the case in Gallman. Although it was frustrating for Carla at the time, she came to understand her mother's monthly apportions to extended family and other neighbors as another example of a linked-fate mentality in passages such as Proverbs 28:27, Psalm 41:1, James 1:27, Ezekiel 16:49, Matthew 19:21, Mark 10:21, and Luke 14:13 to help the poor, particularly widows and orphans. Furthermore, segregation made such donations imperative.[10] Yet without proper planning, such philanthropic efforts could financially drain their own family:

> We took everything [donating to neighbors in need] because back then, you canned everything in the jar, including meat. If you killed a hog or anything like that, you canned that in the jar ... so beans, tomatoes, and meat, and berries, we just took whatever she had canned up . . . we did this probably once a month ... But my father raised some stuff too so everything we took over there was not in a jar. So when we chopped cabbage from our garden, we would take fresh food ... either from the farm or we would pick it out of the woods. (Carla, 74 years old)

Another family member, nephew Fred, also recalls their varied skills that helped establish them economically, "Uncle George taught me how to do mechanic work. That's what he did, besides farming. Aunt Irma had a whole storage unit of things that she canned . . . They were very religious. I loved them. Irma and Aunt Robbie were my favorite aunts on my mother's side" (Fred, 69 years old). Yet even the most prudent farmer was still at the mercy of socio-ecological uncertainty linked to weather, insects, unresponsive crops as well as unexpected intra-family stresses and strains. Although George and Irma believed God would take care of them, it was also necessary to expect the best, but prepare for the worst. Crops, livestock, and well as foraged nuts and berries on their homestead all became ways to maximize resources.

Black landowners who wished to earn additional spending money could count on white-owned crops to pick; black sharecroppers were forced to comply. Everlyn was born in October 1935. Her mother, Ceola, was married to George's brother, Robert. As a child, her family were sharecroppers on white-owned land:

I know when I about 7 years old, I remember how we had to go to the field and work . . . The white folks would tell us to go on a big old long truck. And we had to go to the field to work when he said go to work . . . I remember when me and my brothers and sisters had to get up and go to the field, and if we had our clothes on to go to town, we could not go. Uncle Daniel was driving the truck (to take them to town) and if the white man came on a horse or mule and told us to go to the field, we had to go sharecrop . . . That made me feel bad . . . I might be wrong, but I say, "I'd rather be dead than to be your slave." We were slaves back then.

Sharecropping, as the chief economic enterprise, was dependent on low-wage labor by blacks of all ages or by blacks who lived on and tended white-owned farms. The precarious nature of agrarian life meant the poorest blacks were at the mercy of white landowners who could demand their labor and disrupt their days. Everlyn Mae vividly described the emotionally and psychologically debilitating effects of such control where little distinguished between chattel slavery and servitude.[11]

For the Kings, ownership of property and livestock were common measures of being middle class in Gallman. Another story by Nate, a black tenant farmer in Alabama, illustrates the significance of livestock:

My cows brought me two hundred and seventy dollars. Both of em was dandy good cows, but bags like good dairy cows . . . So eighty dollars for that bull yearling and eighty-five dollars for that heifer – that gave me a piece of change. (Rosengarten 1974: 522)

Nate further describes his ability to negotiate the sale of an under-performing milk cow from a local white man for $100 to $175. Like Nate, Irma realized early on that blacks who were able to secure even one cow could, over-time, produce livestock that would be economically beneficial. White farmers knew this as well. For example, in the early 1900s, although eight months pregnant, Beth Anne, a King family in-law, was jailed in Jackson, Mississippi, for beating a white man with a broom because he was trying to steal her cow. However, Beth Anne was released without prosecution because of her physical condition and incessant screaming and complaining. Property ownership positioned the Kings more favorably than many of their peers, both black and white, but could not guarantee security in other equally important ways.

Navigating Extra-Class Constraints

King family members suggest that, despite certain material possessions, structural dynamics required vigilance to navigate uncertainty associated with being black in the South. Although accounts in the prior section illustrate some of the ways the Kings constituted a middle-class family, their lifestyle was fragile, and they still faced many of the same resource constraints as other southern blacks:

> Big Papa [George] would play the piano . . . honey dripper stuff . . . guitar too. They had a piano. I used to play it all the time. I would sit up there and look at the pictures (on top of the piano), [but] we had to wash down in the creek and start a fire. Anywhere we found a creek, we washed. (Everlyn Mae, 81 years old)

The above memory describes benefits as well as challenges of rural life in Gallman, Mississippi. Everlyn mentions some of the material possessions that Irma and George had (example, a piano, guitar, and montage of professionally made family portraits) that were uncommon for most rural blacks at the time. Yet the lack of certain basic resources (example, running water, in general, and hot water, in particular) meant additional work to complete important domestic chores. The juxtaposition of relative luxuries and lack reflected the precarious socioeconomic tensions of blacks in the South.

Even familial funds could not combat systemic-level constraints that prevented rural blacks from accessing certain public resources. Below, Carla, describes a life-saving migration spell:

> I was probably about four and I got sick. I couldn't swallow. My Mul said I wasn't eating. I had a high fever. And they didn't have no money to take you to the doctor and so Aunt Hana [George's sister] came down from Detroit and got me on the train and took me back to Detroit where she could put me up under a doctor. I had strep throat, but I got well. But that was My Mul's prayer . . . I stayed in Detroit about a year because after I got well, Aunt Hana was not gonna bring me home . . . until My Mul told her that she was gonna put the law on her [she laughs]. Then she brought me back . . . I guess I didn't want to go back [to Mississippi] . . . you had an inside toilet and all that kind of stuff [in Detroit].

Regardless of their class position, in the mid-1900s, segregation and racism meant that most rural blacks did not have access to consistent medical care. The King family knew this reality well. Two of Irma and George's daughters had already died as infants for lack of hospital care. Without an intervention, Carla would have likely met the same fate. According to the above story, recalled to Carla by Irma, based on prayer and wisdom, Irma reached out to extended family in the North who could provide a better chance that Carla would get proper medical attention. Financially stable as an owner of rental property, George's sister, Hana, espoused a linked-fate mentality and eagerly helped her relatives. In addition to receiving medical care, Carla was exposed to amenities that were initially absent on their farm. So despite certain material possessions, George and Irma were not wealthy nor exempt from racial constraints prevalent in the South.

George and Irma had to be extremely frugal to maintain their farm. They had complementary reputations – her for being resourceful and him for being thrifty. According to Janice, "Big Mama could sew ... She made quilts on her hand (i.e., without a sewing machine) ... she made our clothes, and although cheap, Big Papa would purchase our shoes." Her older sister, Martha, elaborates:

> We didn't pack our lunch. My Daddy would give us $0.50 a week for lunch money in high school. In elementary-middle school we took our lunch ... My Daddy knew he was going to give us the money and had it in his pocket, but he would never give it to us unless we asked. And then he took his time. We could go to the store and get ten cookies for a nickel and a pop for a nickel. That was my lunch for four years. Sometimes our class would sell sandwiches for $0.20 and My Mul would give me money and I would have a really good lunch.

In addition to traditional gender roles, the above recollection alludes to middle-class status. Both the ability to purchase shoes for his children and provide weekly lunch money during their high school years per child was uncommon. According to the CPI Inflation Calculator, $0.50 in 1950 would be worth about $5.25 in 2017. That George and Irma made such provisions for their children, illustrates the family's relative economic status. However, Martha notes that they only had this luxury – purchasing lunches – during high school once the farm was on solid footing. Yet, studies suggest that a common fear among the middle class is downward mobility.[12] So even after their farm was financially sound, the specter of poverty meant that George and Irma continued to live frugally even into old age.

A linked-fate mentality also meant that friendship ties could be used to extend limited family resources. The next quote juxtaposes various ways in which this family was fortunate, but still lacked basic amenities:

> Once My Daddy made it back from the service, he had a truck that he drove and then my mother had a car ... although blacks did have some transportation, but not usually two vehicles in one family ... But it was some time later ... years later, when we got running water. We had to haul our water and we had great big old containers. Mr. Roy [Irma's white employer] never did say nothing when we would fill the containers up at the store and then haul them back home. And then when it rained, we caught water in big old pails. When we got running water ... They brought it down through the road we lived on and that's all the far they went. Now *you* had to dig that trench, lay the pipes, and run them from the road to your house yourself ... My Daddy did this by himself and with his nephew.

Carla describes the process by which they collected water for years. Regardless of race, in-door plumbing was uncommon. Census reports show that, in 1940, in a population of 33,974 in Copiah County, only 1,376 residents or 16.7 percent, had running water in-house. Carla's scenario further illustrates nuanced class dynamics, as the Kings could afford property and multiple vehicles, but did not have running water in-house until about 1960. However, this dilemma was driven more by broader race-based infrastructure constraints than household finances. Equally important, this scenario presents another dimension of the complex way blacks and whites were stratified in Gallman. Friendship ties and sweat equity with white and black neighbors enabled George to initially pump and haul water for his family, and later, build the necessary infrastructure to secure running water. The latter accomplishment required George's skills as a mechanic. Next, Carla's reference to the introductory scenario contrasts it with memories of how they maintained sanitary conditions in the absence of running water:

> We had electricity before we had water. We got that [electricity] when I was a teenager . . . I remember that we used to listen at the radio. And you didn't do that all day because she [Irma] still had to pay that bill. They had certain things they [their parents] would let you listen to and we would always listen on Sunday morning before we went to church because that's when you got all your news . . . but we still had an out-house. They had lime that you could buy, it was white, and you poured that over into the stall and it would kill germs . . . We had chamber pots in the house and if you had to use the bathroom during the night, that's what you would use and in the morning, you would empty them in the outhouse.

Electricity meant the Kings no longer had to use kerosene lamps; in-house water would come several years later. Irma and George had to prioritize these two conveniences. Irma's desire for electricity as well as the somewhat easier, inexpensive, but labor intensive process by which water was secured, justified delaying plumbing. Electricity also meant increased communication with the outside world, including listening to "church" music, preaching, and news on the radio. Electricity and its related uses were uncommon. For example, 1940 census figures show that only 3,058 residents of Copiah County (37.1 percent) had a radio in their houses. Although she has multiple televisions now, Carla continues the Sunday tradition of listening to church services on the radio.

Yet no matter the challenges they faced in maintaining their farm, the King children were very aware of the economic and non-economic benefits property afforded:

We had 20 acres ... [name of one of Irma's peers] did not own her own land. She stayed on the white man's place ... it meant that you did not have a place of your own and you had to work in their fields, picking whatever they planted ... her older children were old enough to work in the white man's field, otherwise she couldn't have stayed in his house ... Now, if they [white landowners] came around and asked you to work in the field and you didn't live on their place, you got paid. 'Cause we have worked in the field, but they paid you so much a bushel. I think the top price was like $0.35 cents a bushel. (Carla, 74 years old)

By broadly delineating the lives of blacks who did or did not own land, the above comment does not present the former as a panacea, but rather the relatively more dire conditions of the latter. Tenants usually sharecropped in exchange for room and board – with the promise of wages if profits were gained. Yet whites rarely lived up to these arrangements, meaning black sharecroppers were often tied indefinitely to land they would never own, waiting for revenue they would never receive.[13] In contrast, black and white day laborers were paid, albeit meagerly. In this way, black landowners had agency that sharecroppers did not. And for Carla, such work was a choice that enabled her to contribute to her family's income and feel a sense of self-sufficiency. She continues:

If the beans were plentiful you could make good money. I remember one day, I picked like 20 bushels, but the beans had to be very good for you to pick a lot of bushels ... I would divide some of the money up with my mother and then I spent the rest of it ... on clothes. I definitely remember buying me a pair of shoes ... and stuff to wear to school ... I was about 14 or 15 years old.

In addition to earning a daily wage that she considered substantial at the time (about $7.00 in 1957 or $61.23 for 20 bushels in 2017), Carla's decision-making reflected a linked-fate mentality, as she shared her income with her mother and purchased personal items such that her parents would not have to do so. This scenario also contrasts the sobering reality that sharecroppers were able to stave off homelessness, but were rarely able to accumulate enough money to escape poverty. Thus many rural blacks were trapped in a chronic cycle of penury and uncertainty. And like her younger sister, 77-year-old Martha understood both the relative nature of poverty, and the stark differences in quality of life and life experiences of blacks and whites in Gallman:

I never thought we were poor because we always had somewhere to stay. We always had clothes to wear even though sometimes they were hand-me-downs. They were *good* hand-me-downs and we always looked nice and our other friends did not have their own homes and some of them did not have cars and things. So I didn't feel poor ... We weren't rich either ... but we were better off than some white people too. I knew there was a difference [between races]. But growing up in that era, you knew what you could do and what you couldn't do. So you just did that ... but I knew I wasn't going to stay down there and just babysit white kids.

According to the above comment, George and Irma's family walked an economic tightrope between classes in Mississippi. Their assets such as land, vehicles, and electricity represented a certain degree of relative middle-class living most blacks and working-class whites in Gallman did not have. Yet frugality was required to maintain their precarious socioeconomic position. And like all blacks in the South, no matter their economic status, segregation was a way of life that, like poverty, they strategically strove to outrun.

Strides Away from Poverty and Toward Middle Class: Siblings and Social Capital

Although this analysis uses the experiences of Irma and George as direct points of departure for understanding this family's broad class trajectory, other family members made decisions that resulted in middle-class outcomes. This section considers some of their lives, especially siblings who were essentially raised by Irma. According to family members, Irma's beliefs and behavior (i.e., what she did) as well as what she did not do (i.e., migrate from Mississippi), are correlated with the subsequent middle-class lifestyles most of her younger siblings experienced post-migration. The map in Figure 2.1 summarizes the migratory patterns of Janice King's children; Figure 2.2 provides thumbnail descriptions of their experiences. Although they left Mississippi, Janice King's younger children remained emotionally close to Irma and the family homestead; they remembered her sacrifices during their early years in poverty. Moreover, the family farm represented a compound where adults and/or their children could have migration spells. Martha summarizes this linked-fate dynamic: "By her being the oldest girl, she had to help Momma Janice out once her first husband died and had to help with the other children. And that's why they looked up to her. They looked up to her like a second mother." So although their treks toward middle class led most of them away from Mississippi, their ties to Irma and the family farm were abiding.

As was typical at the time, several Kings migrated to the large northern cities of Chicago, Illinois (Katherine Mae), Toledo, Ohio (Gina Mae), and St. Louis, Missouri (Martha Lou) that promised better jobs and less overt racism.[14] Others who wanted to *escape* Gallman, but not venture too far,

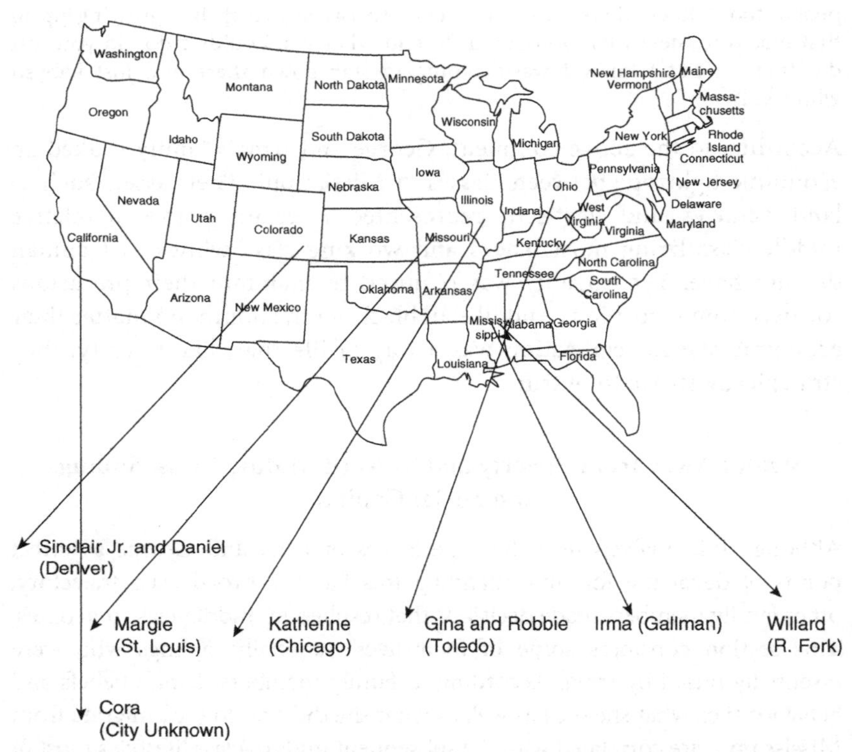

Figure 2.1: US Map and Migratory Experiences of King Family Members, Circa 1940–1960

Key: Destinations reflect final locations where Kings migrated and do not reflect other locales family members may have also lived short-term. R. Fork =Rolling Fork.

migrated to Jackson, Mississippi (Robbie Lee).[15] And still other family members ventured to areas less frequented by blacks (Sinclair Jr.). Despite tapping into the ties of other family members for housing and their weak social ties to locate employment, many of them ended up competing with a myriad of other minimally educated blacks for entry-level positions and navigating life in segregated urban areas.[16] However, low-wage jobs were still better than what was available in Gallman, Mississippi. Did certain demographic dynamics influence their success?

Figure 2.3 illustrates broad demographic differences across six post-migration locales where many family members relocated. The curves represent the ratios of populations based on gender (male and female),

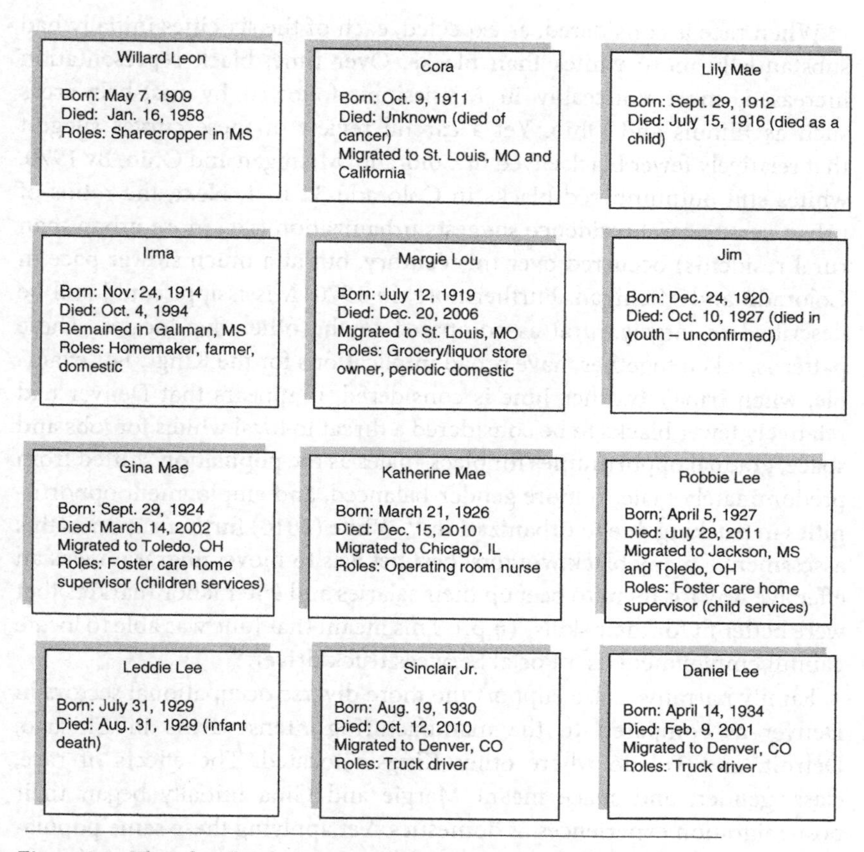

Figure 2.2: Thumbnail Descriptions of Janice King's Children (Sept. 29, 1890–Sept. 29, 1971)
Key: Information provided by family narratives, county records, obituaries, and other family records.

race (white and black), and residence (urban and rural) from 1860 to 1970. Considering relative relationships between these three characteristics will help understand some of the intersecting influences on this family's socioeconomic outcomes. A review of the six gender curves show relatively similar ratios of males to females in Colorado, Illinois, Michigan, Mississippi, Ohio, and Missouri over the designated period. However, further review shows both a greater initial male-to-female proportion (ratio of 16.5) and relatively greater male presence in Colorado from 1860 to 1900 than in the other five states.

When race is considered, as expected, each of the six cities initially had substantially more whites than blacks. Over time, black representation increased, most noticeably in Mississippi, followed by rust-belt areas such as Illinois and Ohio. Yet a careful review of these curves suggest that relatively fewer blacks lived in Colorado, Michigan and Ohio. By 1970, whites still outnumbered blacks in Colorado 32 to 1. Next, the ratios of urban versus rural residence suggests urbanization (i.e., more urban than rural residents) occurred over this century, but at a much slower pace in Colorado and Michigan. Furthermore, in 1970, Mississippi could still be described as largely rural as compared to the other five locales. These patterns, taken together, have broad implications for the Kings. For example, when Irma's brother June is considered, it appears that Denver had relatively fewer blacks to be considered a threat to local whites for jobs and space, gradual opportunities for black males as the population shifted from predominately male to more gender-balanced, and employment opportunities in trucking due to urbanization.[17] White (2016) further supports this assessment: "many black workers' best bet was to move: migration was an effective way for them to beef up their salaries and enter labor markets that were better fit for their skills" (n.p.). This meant that June was able to locate gainful employment as a Postal Services truck driver.[18]

Family narratives also support the more diverse occupational sectors in Denver as compared to the manufacturing-intense areas in Chicago, Detroit, and Toledo where other Kings relocated. The effects of race, class, gender, and space meant Margie and Gina initially began their post-migration experiences as domestics. Yet applying these same population trends meant Katherine's decision to relocate to the quickly urbanizing city of Chicago and to enter the medical field helped position her for upward mobility. In addition, Katherine's aspirations meant that she left Mississippi hardly able to read and write, but went on to complete nursing school and become an operating room nurse, "in a women's sector with little return to potential experience, but with substantial returns to education" (Dickens and Lang 1993: 168).

These same demographics inform Martha Lou's ability to proactively respond to the daily round needs of urban blacks in St. Louis, Missouri, by opening a grocery/liquor store. It should also be noted that Margie, Gina, and Katherine's ability to locate marriageable males further fortified them economically in a way that did not occur for Robbie.[19] In addition, George's sister, Hana, took advantage of urbanization and limited housing for blacks in Detroit by providing rental property. As detailed in a subsequent chapter, other family members found gainful manufacturing jobs in rust-belt cities in and around Illinois. Human capital aside, the above demographic trends illustrate how life chances and socioeconomic outcomes for the Kings were affected by where they migrated and opportunities upon arrival. Moving "off the beaten" migration path and/or venturing into employment sectors that were less chartered by blacks resulted in employment options that affected their ability to navigate life post-migration for many of Irma's siblings.[20]

AN ECOLOGY OF SEGREGATION AND SPACE: HISTORIC DISCRIMINATION AND RELATIVE ADVANTAGE

Scholarship is clear – segregation and Jim Crowism resulted in debilitating and inhumane conditions for southern blacks. Spear (1967) notes: "Jim Crow legislation, inferior schools, legal injustice, and lynchings were ever present factors in southern Negro life. Both economic and social conditions ... contributed to ... 'accumulated migration potentialities'" (p. 131).[21] These dynamics cultivated an *ecology of segregation* where spatial usage, infrastructure, and generational practices such as inheritances helped cultivate and maintain inter-racial disparities. As well as considering ecology, it is equally important to assess how blacks navigated this adverse environment. Whites cast the differential experiences and outcomes associated with inequality as something that both they and blacks deserved – simply by virtue of race:

> Back in the day, they didn't call it racism like they call it today. They called it segregation. In town, they had a water fountain or a bath room – the black people had to go to the courthouse and go downstairs to the bathroom. They had the "White" and the "Colored" signs. And at the train station, it was a two-sided building. It was a nice building, but on one side they had "Colored." So you knew if you had to catch the train, you had to go on that side of the building and white people sat on the other side. Now, I don't know what they had over there (the white side of the building). Then when the train came, the conductor would be out there and then the white people were in the coaches upfront. They had one up there for white people and one back there for black people. (Martha, 77 years old)

Although she describes this history matter-of-factly, Martha's comment above conjured up somber feelings edged with anger. The culture or "way of life" in Gallman meant both races knew "their place," but signage and other iconography served to continuously remind blacks who might imagine life differently and for any whites who might feel guilty about their unmerited favor.[22] Most whites took great pains to secure and maintain the seemingly commonplace bifurcated life such that infrastructure,

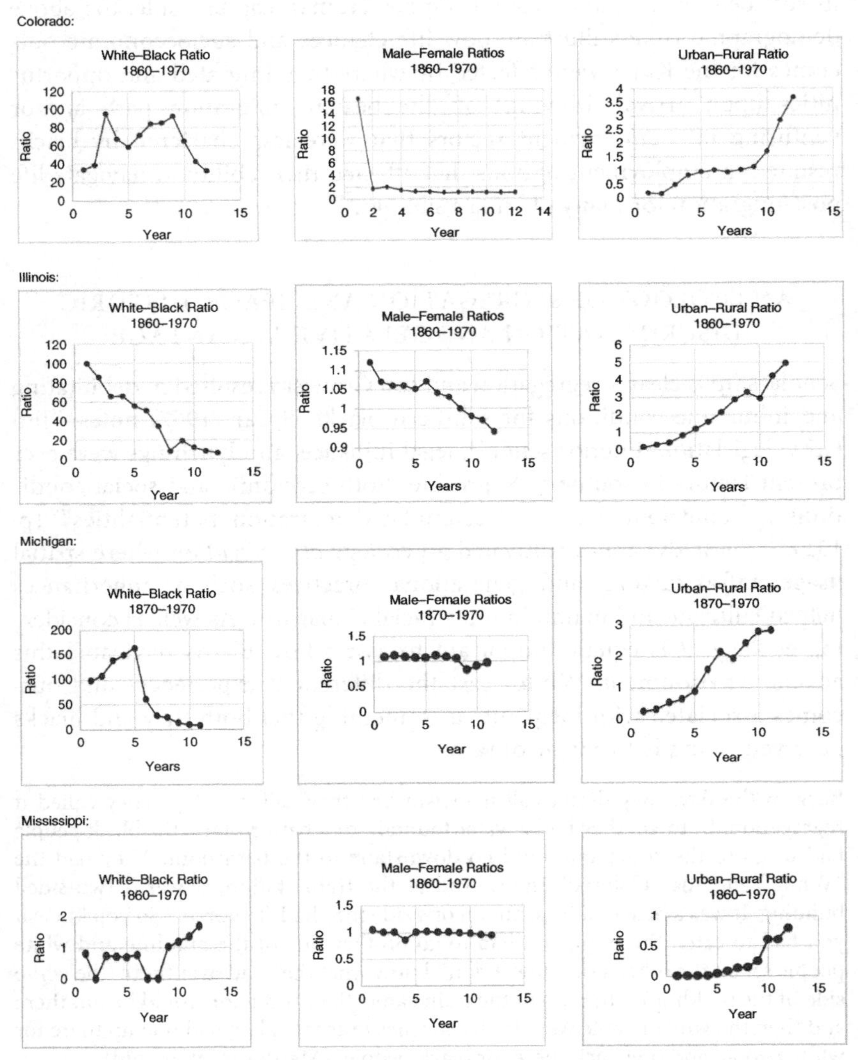

Figure 2.3: Curves of Demographic Patterns by Gender, Race, and Residence (US Census 1860–1970)

processes, and daily round activities reinforced segregation. Everlyn, extends this description:

We had to go through a whole lot back then. Black folk could not go to town and buy a Coke. You had to drink a belly washer [a cheap soda] . . . Back then you had

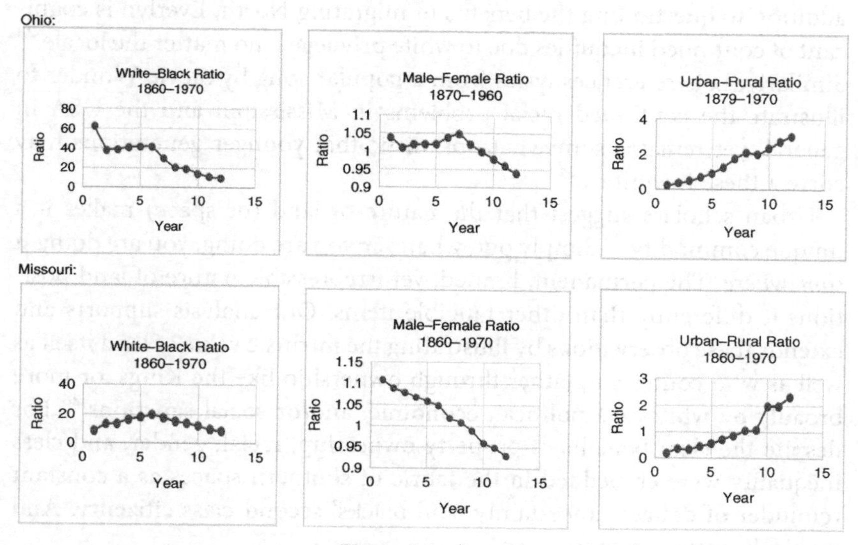

Figure 2.3: (cont.)

to go through the back door . . . and I remember the dogs. But now, you can go eat anywhere. Then you had to get your food and eat on the outside . . . [but] I love Mississippi, you can't take Mississippi out of me. I love the country . . . I don't like the things going on up North . . . They done integrated now, but we still got some prejudiced people down here in Mississippi. They [whites] want everything . . . some of them you can mix with and some of them you can't.

Everlyn provides a two-fold description of segregation that includes negative experiences in predominately white spaces and positive experiences in rural black spaces. The daughter of Irma's brother, Willard Leon, provides a similar account:

I know it was prejudiced. That was a long time ago. Blacks weren't allowed in the white folk's cafés. You were just separated . . . schools . . . they finally integrated, but whites just created private schools. So it's still segregated, but it's better than it was . . . they say, "hard time Mississippi," and it still is. I don't never think there will be equal rights anyway. But there are more problems with older white people. I don't think as many young whites think about that stuff. (Lois, 76 years old)

Segregation in business establishments, police brutality, and daily micro-aggressions enabled whites to exploit blacks economically while disempowering them both economically and non-economically. These negative experiences made black-controlled spaces crucial for group survival. In addition to questioning the benefits of migrating North, Everlyn is cognizant of continued inequities due to white privilege – no matter the locale.[23] Similarly, Lois references lyrics from a popular song by Stevie Wonder to illustrate the continued racial problems in Mississippi and the USA in general, yet remains somewhat optimistic that younger generations may correct these inequities.[24]

Urban scholars suggest that the nature of land (or space) makes it a unique commodity.[25] Simply put, whatever you are doing, you are doing it *somewhere*. The permanent, limited, yet irrepressible nature of land positions it differently than other tangible items. Our analysis supports and extends these observations by illustrating the intrinsic value of land itself as well as who controls it, either through ownership like the Kings, or more broadly by whites via political, economic, and/or social sanctions.[26] For despite the clear benefits of property ownership, racial, gender, and class inequality were embedded in the fabric of southern spaces as a constant reminder of danger, uncertainty, and blacks' second-class citizenry. And no blacks were exempt:

> The first time I really realized racism, my mother worked for this white family. And when I was about six years old, she told me that I couldn't go to work with her anymore. I asked her why and she said because little black boys could not be around little white girls. I didn't understand it at the time, so I felt bad. (James Lewis, 69 years old, steel worker)

James learned at an early age that just as whites used spatial segregation as a means of control, they also endeavored to control bodies. Even young black bodies (and by default, white bodies) were being surveilled and manipulated based on fears and stereotypes that negatively affected both groups, albeit in starkly different ways.[27] Such micro-aggressions could be disaffirming and internally traumatizing for black children; over time, white children were likely to embrace their parents' biases.[28] More specifically, Irma could work as a domestic for this white family, but her small son was considered a threat. This example also illustrates some of the daily tensions she and her family were forced to navigate around whites who likely considered themselves her allies. Irma's daughter-in-law, Deidra, recalls another example of gendered, segregated work roles:

> My mom used to work for this white lady. They would make her call the children, "Ma'am," and "Sir" – to little children as if they were more than my mother was. And that bothered me – why they would call my Mom, a grown woman, by her first name and she had to say, "Miss" and "Mister" to them . . . that was for black women period who worked in white households . . . and my Mom would cook and clean

for them, yet she had to go through the back door. (Deidra, 64 years old, homemaker)

In *Black Feminist Thought*, Collins (1990) assessed the pattern described above: "Making black women work as if they were animals or 'mules of de world' represents one form of objectification. Deference rituals such as calling black domestic workers 'girls' and by their first names enable employees to treat their employees like children, as less capable human beings" (p. 69). Yet Collins also argues that black women engaged in "everyday resistance to this attempted objectification" (p. 70). Applying this premise, like Irma, black women often created affirming self-definitions and subjugated knowledges to push back against such treatment. According to their narratives, King family members concertedly resisted an ecology of segregation in proactive, adaptive, and resilient ways.

Implications of Rural Environmental Injustice

An ecology of segregation helps explain how beliefs and behavior shaped spatial arrangements to reinforce racial oppression. Moreover, these practices and processes extended beyond brick and mortar in the rural South such that even nature was manipulated to support segregation and marginalization. The following comment illustrates how this ecology shaped how spaces were described and controlled over time:

We lived on the country part of town and where we lived there were just black people. And when you were going to the store, then that [land] was white people's . . . their land had been passed down to them by their mothers and fathers . . . It was segregated . . . where they lived, there were no blacks living up there . . . white people owned all the land where blacks lived . . . Mr. Pryor just sold it [some of his land] to My Mul. (Carla, 74 years old)

The segregated infrastructure mirrored the segregated labor market such that whites stymied most blacks from accumulating enough money to purchase land. Even blacks like the Kings who had been able to amass funds were still "relegated" to land that whites did not want. Carla's comment also points to the potent, long-term effects of inheritances, here marked by controlling and "passing down" spaces, as a mechanism by which economic disparity across racial groups is and has been perpetuated.[29] By manipulating "place" and the people who inhabited it, southern whites ensured their control.

Segregation and racism shaped other dimensions of the local terrain and even tainted how flora and fauna might be perceived by southern blacks:

They say, that if you go down that Denville Road, there is a tree that was half alive and half dead . . . they say that's where they use to hang the black people, and they say, that's why the tree was half alive and half dead . . . My Mul and My Daddy never said anything about these rumors, we heard them talking to other grown people. 'Cause back then, grown people did not discuss a lot of stuff with their children . . . [but] Everything is so growed up now. (Carla, 74 years old)

According to Carla in the above memory, land could be the site of violence and unrest in other ways such that trees used as makeshift swings for black children could easily become lynching posts.[30] She also describes her parents' efforts to shield them from the latter activities by encouraging them to engage in the former on the family's homestead. Moreover, studies show that her family's efforts to amass material possessions could make them particularly vulnerable to violence by jealous whites. Bouie (2011) corroborates these dangers, "not only could you be killed for transgressing the nebulous and arbitrary social requirements of Jim Crow, but you could *also* be killed for starting a business, accumulating wealth and otherwise trying to improve your situation" (n.p.). According to Carla, rumors about the "hanging tree" as a palpable place served as reminders to blacks about what could easily become their fate. Yet her closing comment is a reminder of the irrepressible nature of land such that the passage of time has meant that, outside of memories, most tangible signs of the legacy of the Denville Road tree are now overgrown by grass, weeds, and other foliage.

Attempts to completely disempower blacks meant striving to make segregation and other forms of inequality so pervasive that they were considered "natural." However, the following comment about life in and around Gallman, Mississippi, for black youth suggests their cognizance of ill treatment:

I guess I didn't think much about it . . . you could go to the movies, if you went around to the back and you sat in the back. You had to go up these stairs and sit in the back . . . but you still had to pay the same amount as white people to get in . . . I don't remember no restaurants you could go to. They [blacks] had juke joints (dance halls) for people who wanted to dance and drink [but], no sit down and have no church meal . . . The store was in Gallman and you had a post office and the two white churches . . . and that was about it . . . Mr. Roy's store was the gas station . . . You knew there was a difference, because when you went in a store, I'm not talking about Mr. Roy's store, but when you went in a store and got ready to buy something, you definitely weren't going to see no black people on the cash register and the folks didn't want to take your money in their hands. You had to lay your money on the counter and . . . you just knew it was a difference . . . from the way they treated you . . . it was so common and frankly speaking, I don't think blacks wanted to live around them no way because, besides Mr. Roy, they weren't too friendly. (Carla, 74 years old)

As described above, blacks in Gallman created their own protected spaces for spiritual (i.e., churches) and non-spiritual (i.e., juke joints) enjoyment.[31] A few whites attempted to be somewhat less exclusionary, yet they were the numeric minority. Carla also alludes to another subversive dimension of Jim Crowism and segregation that meant blacks were expected to obligingly pay "full-price" for sub-par treatment, goods, and services in white businesses. However, her comment also reminds us that southern blacks enjoyed spaces from which whites were absent. When recalling these types of experiences, it was evident that older members of the King family felt challenged to reconcile their disdain for such whites and the system of segregation with the Christian edict to love unconditionally. Despite direct and indirect sanctions, blacks like the Kings endeavored to amass economic, human, and social capital and create a safe harbor inside the black community.

Yet Martha's reflection below suggests that systemic forces tied to racist values and structures ultimately made southern spaces intolerable for increasing numbers of blacks – even persons like herself whose families were relatively stable economically:

That was the norm back then and you knew what you could do. You knew where you couldn't go. So you just stayed within your boundaries. I can't say how I really felt then. But looking back on it now, I could see the difference in it and knowing what I know now, if I had that same understanding back then, I don't know what I would have done. But one thing I did tell myself when I was growing up down there – they didn't have jobs for blacks other than cleaning and keeping white folk's kids or you could be a nurse's aide. Even if you were a nurse, you could not work in that capacity. You could only be a nurse's aide. So I told myself, if the Lord blessed me to get grown, I wasn't going to stay in Mississippi because I refused to babysit those white folk's kids and clean their houses forever. So I just waited until I got 18 so I could leave to find a better job.

An ecology of segregation meant that "place" was used to control and constrain the life chances and quality of life of southern blacks. As Martha notes, "boundaries" were clear and continually and consistently reinforced.[32] Although farmland was a crucial commodity and helped the Kings experience some semblance of agency and economic stability, it did not fully protect them from the snares of segregation. Nor did it enable even the most industrious blacks to "catch up" with their white middle-class peers economically.[33] Even as landowners, George and Irma had to work multiple jobs to make ends meet as well as to maintain and expand their farm.

Constrained employment for young adults in the King family and other blacks like them meant that sharecropping, becoming domestics or nannies for white families, or working in underpaid service posts were the only positions they could expect. Like Martha, blacks wanted more and they knew that they deserved more. The influence of religious education is also apparent in the above remark. Martha's socialization as a Christian resulted in faith that God's power supersedes that of powerful whites or anyone who would strive to oppress God's people. This stance was common among King family members as they referenced Psalms 110:1 and its counterparts in Luke 20:43 and Acts 2:35 as well as Exodus 3:14 and Matthew 28:18, which confirm the omnipotence, omniscience, and omnipresence of God in the lives of believers.[34] Thus Martha, and family members like her, believed that God would provide them with opportunities to realize their dreams – if they took a chance and migrated.

CONCLUSION: NAVIGATING SEGREGATED RURAL SPACES

An ecology of segregation meant that whites could invade black spaces and if they so chose, black bodies, without permission or repercussions. Comments by King family members illustrate the nature of segregation through the use of terminology such as "up town "or "in town" where whites lived and were dominant versus "out of town" or "away from" town which was largely the domain of blacks. When members of the latter racial group entered the spaces of the former, they did so cautiously, infrequently, and potentially at their own peril. Navigating segregated areas meant knowing that in most instances, "place" referenced physical as well as non-physical barriers established and maintained by values, norms, and discriminatory ideologies. In this way, an *ethos of segregation* created and helped justify and reinforce a palpable chasm referred to here as an ecology of segregation. Some older members of the King family *seemed* to accept these divisions *en face*. However, a deeper analysis of their reflections uncovers tensions, derision, *and* everyday resistance – evident well over half a century later. Additionally, several older family members referenced Ephesians 6:12 that, although blacks navigated segregated places and whites who espoused such beliefs, the actual struggle was ideological in nature – but just as pernicious.[35] They also seemed to be trying to understand and reconcile Christian beliefs in a society that often verbalized similar tenets, but realized them inconsistently.

In addition to describing how place/space was used and periodically shared inter-racially, some family members did not appear to relish spaces shared with whites. Moreover, predominately black places such as churches, homes, and in some instances, juke joints, represented safe spaces where blacks could interact and fellowship outside the purview and control of whites. For these reasons, being able to own one's home represented much more than four walls and a roof; blacks who were forced to live on white-owned property were without a safe space they could call "home." Thus for George and Irma and a cadre of other blacks in Gallman, owning property represented an economic accomplishment, as well as non-economic benefits associated with security, safety, privacy, and temporary *freedom* from subordination.

Certain socio-ecological implications of segregation were subtle. For example, it appears that having the tacit support of influential whites such as Roy meant that Irma, George, and most importantly, their daughters, were excluded from some of the problems with which other blacks contended. This interaction parallels work on weak social ties that suggests "individual economic attainment is related to the extent of one's connections to individuals embedded in positions of power and authority in the marketplace" (Mason 2003: 60). In rural places like Gallman, Mississippi, these weak connections between whites and blacks appeared to be limited, and for this reason, more influential. Although their economic successes were circumscribed by segregation, Irma's domestic ties with an influential white family translated into a certain degree of largess. Roy's support seemed to enable her family to traverse relatively safely in a broader, racially charged social sphere than some other local blacks. It was unclear whether he extended the same *protections* to other blacks. However, it appears that Roy's protection was largely the result of his respect for and interactions with Irma.

Intra-familial decisions also shaped how King family members understood and negotiated segregated spaces. For example, like some parents today, George and Irma attempted to protect their children literally and psychologically. Yet according to our findings, although George and Irma were able to provide some semblance of protection for their family, they could not protect their children from the pervasive ideology that, as Carla described, *things were different*. And this reality meant that seemingly innocuous spaces in nature such as trees or infrequently trod roads could readily become spaces for lynching, rapes, and other physical assaults as a "way to terrorize blacks into acquiescence by brutally killing those who intentionally or accidentally stepped over some invisible and shifting line of permissible behavior" (Ayers 1992: 156–7).

It also appears that Irma's reputation as a stalwart Christian provided her with a certain degree of credibility that, in addition to Roy's support, enabled this family to navigate potentially tumultuous spaces. In this way, certain local whites in Gallman appeared to broadly adhere to Christian tenets that prevented the King family from experiencing some of the more egregious encounters documented in research. However, the statistics included in this chapter confirm socioeconomic disparities that played out daily for black families like the Kings and encouraged many of them to seek opportunities outside the South.

George and Irma purchased their farm in 1944, and gradually begin to improve the house and land. For financial and ecological reasons, it was not until the mid- and late 1950s that they were able to make certain improvements that middle-class whites in Gallman took for granted. Some improvements were slow due to difficulties securing the requisite funds. However, certain constraints were also the result of priorities given to property in and near town owned by whites over property located in the "bottomland" where blacks lived. In this way, ecology was again shaped by segregation and Jim Crowism. According to this family's accounts, although economically able white farmers had running water and electricity in the early 1900s, *structural provisions* were not made for blacks to enjoy these same basic amenities until the mid-1900s.[36]

It is apparent that George and Irma's homestead was a source of family pride as well as an example of what was possible. And because they stayed in Mississippi rather than migrating elsewhere, their homestead represented a socio-ecological haven for other family members and friends. Gallman, Mississippi, was home for Irma. Her childhood wish made so many years before became a reality. Unbeknownst to her, that simple wish placed her family on a path to becoming middle class. Land, a milk cow, family, her church, and friends – all made the choice to remain in Mississippi easy for Irma. By doing do, she continued to represent and provide an anchor for her siblings, their children, and her own children, as they ventured to other places. Through migration spells on the family farm, they always had a safe haven. And annual or bi-annual trips helped maintain kinship ties and enabled them to visit their older sister Irma whose efforts and sacrifices had made many of their accomplishments possible.

3

"Getting to the School on Time": Formal Education and Beyond

Tanya burst into tears as she starred incredulously at her report card – five As and an F! How could she have gotten an F in Advanced English? Her graded papers did not reflect this outcome. And although Mrs. Smith had a reputation for being prejudiced against black students and showing differential treatment toward white students, she had not even hinted that Tanya was not doing well. After several more crying bouts, a petrified Tanya informed her mother. Would Carla believe her accusations of discrimination against this senior white teacher? Carla looked at the grades, listened to her forlorn daughter's explanation, and after a long pause said, "I believe you. Let's talk to Principal Jims." The next day, Carla, Tanya, and the principal meet. In a calm, determined manner Carla presented her daughter's grades and said, "Now I don't have a lot of education, but this seems hard to believe. My daughter has had an A average her entire four years of high school. Now she gets an F – and this grade is not reflected in her graded papers for this class. Something's not right." Carla demanded a formal inquiry that uncovered that Mrs. Smith had manipulated the grades of the top black and white students to favor the latter students. She was severely sanctioned and the students' grades were corrected. Tanya received the A grade she had earned. Disillusioned by her first direct discriminatory experience, Tanya thanked her mother for intervening on her behalf.

Secular education and practical training became central mechanisms by which the Kings acquired the credentials, expertise, and experience required in society to be competitive. This chapter's introductory story reflects their emphasis on education and challenges to that goal. The above subversive scenario was formally attributed to racism by school administrators; it represents the academic environment in which most Kings navigated. It also illustrates the influence of macro-, meso-, and micro-level factors such as school systems, administrators, teachers, and parents, in effecting whether and how students of color are educated, equipped, and empowered in schools. Empirical and experiential information chronicle the influence of secular education and black schools on the King family experiences and outcomes as well as for blacks in general. Particular attention is given to efforts to live out the multi-faceted imperative cultivated by early generations of Kings. Equally important are the threads that stretch across educational domains that shape and reinforce each other, outcomes, and how family members made sense of the varied forms of instruction they experienced. We also examine whether and how the family's ethos withstood migration experiences, specifically, whether and how education influenced their quality of life and aspirations.

EDUCATION AND SEGREGATION: GENERATIONAL TIES THAT BIND

Education in Ecological Context

Formal education and training, coupled with social policies to counter the systemic effects of historic disenfranchisement, have been the primary mechanisms by which blacks have been able to transition into the middle class. Educationally, Mississippi has lagged behind its peers historically; segregation meant blacks were even more disenfranchised without sufficient school resources. Despite the *Brown vs. Board of Education* ruling in 1954, public school segregation continued until the late 1960s. White residents responded by establishing private segregated schools.[1] Whites in Mississippi flaunted their resistance to desegregation, spurred by white supremacy groups such as the White Citizens Council and supported by a cross-section of whites from various class positions, including the Klu Klux Klan.[2] The first school for black youth in Mississippi was not established until 1862. Despite this troubling context, the Kings encouraged family members to earn as much formal education as possible to avoid poverty and improve their lot in life. Reaching these goals would require the combined effects of ecology and ethos in the form of school systems and personal decisions.

Berry and Blassingame (1982) explain the historic challenges blacks faced as they sought an education: "The blacks' passion for education in the 1860s was equaled by the whites' desire to deny and limit the education they received. During the early years of Reconstruction, southern whites burned schools . . . what southern whites feared most was that the education of blacks would destroy white supremacy" (p. 264). According to southern whites, blacks should only be trained to be domestics and farm laborers; higher education was a waste. This mentality resulted in a multi-pronged strategy to squelch educational aspirations among blacks and included dramatically defunding schools; maintaining separate, unequal schools;[3] tactics to force blacks into submission; underpaying black teachers; threatening committed white teachers; limiting the number of black high schools; and, prohibiting blacks from frequenting public libraries.[4] Germaine to the King experience, "in 1940, Mississippi decreed that black and white children would use different textbooks. All books used in black schools had to delete all references to voting, elections, and democracy . . . had it not been for the efforts of blacks and northern philanthropists, there would have been no advance in black education in the 20th century" (Berry and Blassingame 1982: 266).[5]

Tables 3.1 and 3.2 illustrate the educational landscape for blacks and whites during the mid-nineteenth to the late twentieth century. We consider this time frame because it includes seminal social changes in the USA and reflects the period when most of this family's migrating occurred. US school enrollment patterns by race and gender (Table 3.1)

Table 3.1 *School Enrollment Patterns by Race and Gender (1850–1970)*

	Race and Gender (Rates Per 100)			
	Male		Female	
Year	Black	White	Black	White
1970	85.5	89.0	85.2	87.6
1960+	81.7	85.4	81.2	84.2
1950	74.7	79.7	74.9	78.9
1940	67.5	75.9	69.2	75.4
1930*	59.7	71.4	60.8	70.9
1920	52.5	66.6	54.5	65.8
1910	43.1	61.4	46.6	61.3
1900	29.4	53.4	32.8	53.9
1890	31.8	58.5	33.9	57.2
1880	34.1	63.5	33.5	60.5
1870	9.6	56.0	10.0	52.7
1860	1.9	62.0	1.8	57.2
1850	2.0	59.0	1.8	53.3

Key: Taken from US Department of Commerce: Bureau of the Census. 1975. "Historical Statistics of the United States Bicentennial Edition – Colonial Times to 1970" and Snyder, Thomas D (Ed.). 1993. "120 Years of American Education: A Statistical Portrait." National Center for Education Statistics. +Denotes first year figures include Alaska and Hawaii. *Revised to include Mexicans as whites.

Table 3.2 *Median Years of School Completed by Race and Gender (1940–1990) and Percent Illiterate by Race (1870–1979)*

Year	Race and Gender			
Panel A: Median Years of School Completion Rates by Race and Sex (1940–1990)				
	Male		Female	
	Black	White	Black	White
1990	12.4	12.8	12.4	12.7
1985	12.3	12.7	12.3	12.6
1980	12.0	12.6	12.0	12.5
1975	10.7	12.5	11.1	12.3
1970	9.8	12.2	10.3	12.2
1969	9.6	12.2	10.0	12.2
1968	9.2	12.1	9.7	12.1
1967	8.9	12.1	9.8	12.1
1966	8.8	12.0	9.6	12.1
1964	8.7	11.9	9.1	12.0
1962	8.3	11.6	8.9	12.0
1960	7.9	10.6	8.5	11.0
1947*	6.6	9.0	7.2	9.7
1940	5.4	8.4	6.1	8.7
Panel B: Percent Illiterate in the Population, by Race +***(1870 to 1979)***				
	White	Black		
1979	0.4	1.6		
1969**	0.7	3.6		
1959	1.6	7.5		
1952	1.8	10.2		
1947	1.8	11.0		
1940	2.0	11.5		
1930	3.0	16.4		
1920	4.0	23.0		
1910	5.0	30.5		
1900	6.2	44.5		
1890	7.7	56.8		
1880	9.4	70.0		
1870	11.5	79.9		

Key: Taken from U.S. Department of Commerce: Bureau of the Census. 1975. "Historical Statistics of the United States Bicentennial Edition – Colonial Times to 1970" and Snyder, Thomas D (Ed.). 1993. "120 Years of American Education: A Statistical Portrait." National Center for Education Statistics. *Excludes populations for whom school year was not reported. **Denotes first year figures include Alaska and Hawaii. +From 1870 to 1940, data reflect population 10 years old and older; thereafter, they reflect 14 years old and older.

show dramatically different enrollment trends post-emancipation. From 1850 to 1870, whites enrolled at rates between five to twenty-five times those of blacks, regardless of gender. Gradual increases were evident for blacks over time, but by the early 1900s, they still lagged about twenty percentage points behind whites. However, changes included increased enrollment by blacks and relatively slower enrollment rates by whites such that, by the 1930s, blacks lagged about ten percentage points behind whites and the distance was cut in half by 1950 (i.e., 74.7 to 79.7 in 1950). Migration, continued efforts by blacks to earn an education, and policy changes such as *Brown vs. Board of Education* in 1954 ushered in substantial changes in school enrollment. By 1970, blacks only slightly lagged behind whites on this measure. Yet, as King family members described, school segregation and differential educational opportunities continued to occur for blacks. For example, seventy-five-year-old Carla vividly and succinctly recalls her reality in Gallman, Mississippi: "when we were going to elementary school – you were in a two room shack."

Table 3.2 provides school completion and literacy rates. Panel A considers median years of school completion by race and gender. Blacks had fewer years completed in 1940 (i.e., 5.4 for black males, 6.1 for black females, 8.4 for white males, and 8.7 for white females). However, trends show gradual increases over time for all four groups. Black males and females had similar median years completed as did white males and females. Yet median years were slightly higher for females overall. Median years increase noticeably for blacks and whites from 1940 to 1964 and continued consistently for blacks until, in 1990, they were almost comparable to whites (12.4–12.8). Thus by the early 1900s, black enrollment and completion rates were relatively similar.

As presented in Panel B, in 1870, the illiteracy gap between whites and blacks was initially wide – by a factor of seven (i.e., 11.5 for whites and 79.9 for blacks). Blacks experienced a gradual increase in literacy (or decrease in illiteracy) over time. Germane to this study, Janice, George, and Irma's ability to read in the 1900s to 1930s contrasts with relative high illiteracy rates among blacks during that period. However, by 1969, a decline in illiteracy for blacks by a factor of twenty-two is evident. By 1979, differences still existed by a factor of four (0.4 for whites and 1.6 for blacks). These results suggest that since the mid-nineteenth century, blacks enrolled in and completed school at impressive rates, but continued to lag behind whites. The continued upward trends in educational indices for blacks illustrate the positive effects of black community efforts and, to a lesser degree, social policies.[6] Yet scholars suggest that the combination of affirmative action and government anti-discrimination policies as well as the increased education of blacks resulted in reduced racial discrimination and inequality in the labor market between the mid- and late twentieth century.[7] However, these indices do not capture the quality of education across racial groups. This broader educational context informs King family experiences about pursuing education for upward mobility.

Pursuing Education in the Jim Crow South

Whether they lived inside or outside the South, the vast majority of King family members attended predominately black schools.[8] Several common themes emerged from their narratives. The first was the tendency to acknowledge the importance of education followed by personal and/or family outcomes of this ethos. A second more aspirational theme was the deep desire to honor their foreparents by exceeding their educational attainments. Segregation undermined the *material resources* of black students, resulting in sub-par facilities and inadequate funding as compared to their white peers.[9] And correlates are clear between poor public education and long-term wage disparities for blacks. White (2016) suggests, "differences in skills accounted for the most significant portion of the wage disparities in the 1940s. But the root of that skill gap was still racial . . . the persistent inequality of educational opportunities . . . single-handedly cut earnings of black Southern workers by as much as 50 percent" (n.p.). But what were the educational opportunities and constraints in Gallman, Mississippi? What values and efforts were apparent among the Kings?[10] Siddle Walker (1996a) conveys the important, yet incomplete, studies on this subject:

> Historical recollections that recall descriptions of differences in facilities and resources of white and black schools without also providing descriptions of the black schools' and communities' dogged determination to educate African American children have failed to tell the complete story of segregated schools. (P. 5)

Like this author, family narratives help recapture aspects of this lost history. This section takes an emic perspective by assessing King family experiences in segregated educational spaces in the South. And the economic implications of segregation, tangible outcomes, and rationale for these differences were evident:

> In North Carolina in 1945–46 the value of school property per student enrolled was $217 for white pupils and $70 for black pupils . . . these discrepancies were evident

throughout the south . . . this blatant lack of equality in school facilities and resources was, of course, a reflection of the unequal treatment of blacks in all aspects of American life . . . Money for buildings and other things, the priority was white people. Black was secondary . . . it was discrimination. (Siddle Walker 1996a: 2, 25)

Billingsley (1992) provides a counter-narrative about proactive efforts by blacks who understood the relationship between education and upward mobility:

Among all the sources of survival, achievement, and viability of African-American families, education has played a preeminent role . . . Education is the traditional opportunity through which black families find their places in life. And having found it, they replicate their experience again and again through their children. (Pp. 172, 174)

For Billingsley and other scholars who study the black experience, education was a vehicle by which blacks could better navigate a racist system and escape poverty. Moreover, black families like the Kings realized that high school completion, followed by college, could better position them and later generations to live more economically stable lives. However, poverty undermined educational aspirations in Gallman such that most black youth from sharecropping families were forced to tend white-owned crops rather than attend school. For example, 75-year-old Janice provides a direct correlation between property ownership and education in their farming community. She also attributes the ability of her and her siblings to earn a high school diploma, the local educational standard at the time, to the family's farm: "We owned our land and by doing so it allowed us to remain in school. Those families who didn't own their land had to leave school and pick cotton."

Yet the advantage of attending school did not alter limited resources. So make-shift schools in black churches, followed later by Historically Black Colleges and Universities (HBCUs), and colleges and seminaries sponsored by Black denominations sprang up around the country, especially in the South. Most had a religious impetus. Lincoln and Mamiya (1990) describe the post-slavery response to education:

After emancipation, the newly freed people of all ages swamped the schools . . . Sunday schools were often the first places where Black people made contact with the educational process, first hearing, then memorizing, and finally learning to read Bible stories . . . All these Black schools stressed the importance of religion and moral education for the uplift of the race as it was obvious to all parties that socialization was closely tied to the educational process. The molding of young minds in the crucible of education would become determinative of the future options and economic opportunities for African Americans. (Pp. 251–2)

Racism and fear of economic competition from blacks prompted many whites to bar blacks from basic education.[11] Post-Civil War federal and state legislation did not stop these exclusionary practices.[12] Hallinan (2001) notes how Black churches responded: "black communities tried to provide at least some access to education through the creation of Sabbath schools, night schools, and informal learning centers" (p. 51).

Intra-Racial and Intra-Familial Constraints to Advanced Education

Property that enabled them to escape sharecropping and facilitated their children's completion of high school were considered two important facets of middle class for southern farmers. Yet certain systemic and personal decisions inhibited their education-oriented ethos. Recollections by George and Irma's three daughters describe how their advanced education was thwarted by logistics and gender expectations:

> We graduated from 12th grade. All four of us. And the girls wanted to go on to college, but there wasn't a college around where we lived, nothing but Utica Junior College. I went there and took up cosmetology . . . But My Daddy wouldn't let us go off to live in the dorm . . . We didn't have transportation. We could have gone to Jackson State, Alcorn College . . . there were a few colleges around, but you would have had to either stay on campus or drive back and forth. And none of us knew how to drive. (Carla, 74 years old)

Janice and Carla attended Utica Junior College, a historically black community college instituted in 1903 to provide vocational training for blacks, and majored in cosmetology.[13] The approximate nineteen miles from Utica, Mississippi, to Gallman, Mississippi, or twenty-five minute drive meant Irma could transport them to school during the two year periods.[14] However, their aspirations to pursue further education were stymied by distance to other HBCUs; predominately white schools nearby would not grant them entrance. Unable to drive themselves and without their father's support, the 30 miles (33 minutes) to Jackson State in Jackson, Mississippi, and over 60 miles (one hour) to Alcorn State in Lorman, Mississippi, seemed like a lifetime away. [15] Although Carla could not explain her father's reservations, her older sister remembers a reason:

> I wanted to go to college, but we didn't have transportation . . . but at Piney Wood, Mississippi, I had registered over there because I could work my way

through college. I always wanted to be a nurse, but my daddy said I was too young to live on campus. I guess he felt like I wasn't mature enough ... so I couldn't go. (Martha, 77 years old)

Martha finished high school early at the age of 16 years old. George thought her too young to make the 39-mile commute (about 47 minutes) to the HBCU. Martha's aspirations to become a nurse were leveled. The family's hierarchy meant they dare not defy George's decisions. His daughters' could not reconcile his adamancy that they complete high school and community college with his lack of support for additional education. Carla's next remark suggests that the local black middle class was further stratified based on network size, income levels, and more liberal gender beliefs:

You had the Mills family. Now all those children went to Jackson State. But then the older children finished Jackson State and then they stayed there [in Jackson, Mississippi] and so as the others came along, they could stay with their relatives, go to school, and then come home on the weekend. But we didn't have nobody in Jackson that we could've stayed with ... One of My Mul's sisters did stay in Jackson, but My Daddy definitely wouldn't have let us stay with her.

The ability of George to summarily determine his children's educational outcomes reflected the intra-racial gender constraints at the time that marginalized many black women.[16] George's concerns about what could befall his daughters may have over-shadowed his otherwise supportive stance on education. It is likely that George's decision to prevent his daughters from going away to college was informed by his fear of what might happen to them outside his immediate protection and purview. Although he embraced traditional gender roles, his stance may have been informed by concerns that black females could be unapologetically used by white males – and some black males.[17] Regardless of his rationale, George's daughters did not experience the economic benefits of advanced degrees. For example, Darity and Myers (1998) show that:

In 1970, the average white family income for a head with only a high-school degree was $31,238 ... by 1991, the average white family income for a head with only a high-school degree was $30,252 ... in 1970, the average black family income for a head with only a high school degree was $22,876 ... by 1991, the average black family income for a head with only a high-school degree fell to $20,297 ... the black-white family income ratio among high school graduates aged 41 and above was 0.757 in 1970 ... quadrupling of the gap between the racial income inequality among young, educated families and older, educated families ... came about as a result of the worsening position of black families with young, uneducated heads. (Pp. 64, 67)

Similarly, Cotton (1993) finds an inverse relationship between educational attainment and black and white unemployment rates. As adults, George and Irma's children were likely to experience some of the negative effects of these education-based income inequities. According to Carla, George would not entrust them to certain family members who had migrated to Jackson, Mississippi. Furthermore, Carla's comparison of her college experience with peers from another local middle-class black family further nuances the definition, attitudes, and actions of black middle-class residents in Gallman, where, in this instance, the social ties of extended family, weekly migration spells, and supportive paternal views, made the difference between two or four years of college.

Carla and Janice graduated from Utica Junior College. Their formal education in Mississippi ended there. Each sister pursued additional training post-migration. The following comments describe their educational and career trajectories:

> Some of the better jobs I've had were due to education. When I worked for [name] Bank, I had gone to school for data processing, so I got a better job. Before then, I was just a matron at the school, passing out lunches. With a little education, you got a better job and with that same education, I got a job at [company's name] as a computer programmer. I ended up being the manager of the data processing, keypunching area . . . And you needed education at [social services agency from which she retired] because you had to pass meds [distribute medication to clients] and some of these meds were psychotropic drugs. So you had to know what you were doing. (Carla, 74 years old)

Seventy-seven year old Martha details her experiences:

> Once I left Mississippi, when I got to Chicago, I went to a school to take up computer programming and then I went to [technical community college] in Gary for many years. But I was just trying to take classes so I could get training to get a quick job . . . And then with having kids in between, and trying to go to school, it wasn't easy. But I was determined to do something different. I never gave up on going to school, but I didn't get a degree either.

Martha pursued technical training intermittently as she attempted to balance life as a homemaker. Similarly, Carla pursued computer programming and eventually earned an Associate's degree. For Janice, advanced education consisted of a period in nursing school, followed by a manufacturing career that provided substantially higher wages. After a short college stint and period in the military, James Lewis had a career in the steel mill. As illustrated by scholars such as Mincer and Polachek (1974), differential incomes and work lives for these siblings were greatly influenced by childbirth experiences and traditional gender roles that resulted in periods of part-time employment.[18] Yet for Carla and Janice, financial reasons made these intermittent absences from the labor market limited. Each of Irma's children took a different educational path, but ended up living in close proximity to each other in Indiana. Janice's family would be considered solidly upper middle class; the other three families, lower middle class based on their greater relative number of children. And each attempted to instill in their children the importance of formal education or training beyond what their parents had obtained.

AN EDUCATIONAL IMPERATIVE FOR SUBSEQUENT GENERATIONS

It is important to consider whether subsequent generations were similarly affected by the King family's educational ethos. According to their grand- and great-grandchildren, formal and religious education as well as close familial networks continue to be paramount. They also describe some of the factors, both systemic and personal, that affected their decisions and successes, countered leveled aspirations, and promoted educational outcomes for blacks in general today. A careful review of scholarship suggests that despite dramatic economic inequalities, the expressive and instrumental efforts of black parents, teachers, principals, and the black community helped counter what black schools lacked economically.[19] Such schools were considered extensions of the black community and black family. This same sentiment was evident in the King family:

> Education was very important. I used to always say to my daughters, "It's not a matter of *whether* you're going to college, it's *where* [emphasis is hers] you're going to college." (Carla, 75 years old)

> College was the only priority in my family. We were told to learn all we could because "they can take whatever you have, except what's in your head." I'll never forget that because life for me has shown this to be true . . . I always knew I would go to college and actually looked forward to it as my parents were pretty strict . . . Life took twists and turns, but I always made my way back to college and have completed three degrees and a national license. (Lorna, 56-year-old CPA)

These two inter-generational comments inculcate a common theme that parents wanted their children to surpass them educationally and strove to embed this value in them. The ethos itself is not necessarily unique; many black families espouse this goal.[20] Yet for the Kings, this objective seemed to be ideologically driven by faith in and expectations of God more than by the family's immediate socioeconomic or ecological reality.[21]

In another example, according to Barbara, a 44-year-old Program Director, both her parents encouraged advanced education: "education was important because my mother had some college, but she didn't have a degree and I know my father had not completed elementary school, so it was important for them that their children receive a good education." A childhood of sharecropping meant Barbara's father could not read or write; her mother has earned an Associate's degree. Yet following her parents' urging, Barbara earned a Bachelor's degree in Sociology with a minor in Mathematics and a Master's degree in Human Services Administration. Similarly, 49-year-old Manuel recalls a family mantra: "Education has been stressed for as long as I can remember. I got so tired of my mother saying, 'learn, boy, learn,' that I often tease her still today about it. I would hear the stories about how she wished she could have gone to college." Raising five children prevented his mother from completing a computer programming degree. Based on her continued urging, Manuel completed a Bachelor's degree in Law & Society from a top Research I university. He also recalls the effect on his father:

> One of my fondest memories of my dad was when I overheard him telling one of our neighbors that I was going to [college name]. While he never told me how proud he was of me attending [college name], overhearing his proud voice was very heartwarming. I tell everyone about [college name], but up until now no one knew that it was more so because I knew that I made my father proud.

Fifty-three-year-old administrator Tanya, who has earned several graduate degrees, was taught that religious and secular education are intertwined: "Scripture says that people perish for lack of knowledge. At home, in church, and at school, I was always reminded that education would be the way to achieve my goals in life." Paraphrasing Hosea 4:6, Tanya's comment reminds her of the multi-faceted benefits of formal education, temporal and spiritual. King family members often took unexpected paths to the middle class; others were temporarily detoured by imprudent decisions made in youth.[22] And this socioeconomic goal escaped some Kings. Most attribute their accomplishments to God and family support. Yet they contend that it would be incorrect to assume that their educational trajectories were not without challenges.

Countering Leveled-Aspirations

Despite an education-oriented family ethos and linked-fate mentality among the Kings, overcoming various educational challenges and disabilities required concerted interventions by parents, teachers, and community members. They describe individual and systemic impediments to formal education, how they overcome them, and lessons learned. For example, fifty-three-year-old executive director Donna, recalls a problem that had the potential to level her personal and familial aspirations:

I stuttered as a child. I had so much to say, but couldn't get it out. I had speech therapy with Special Ed therapists, had my tongue clipped, etc., but to no avail. I hated it when people would tell me to slow down or look at me as if I was retarded. When I got older, I began to research what other people did to stop. I tried it all . . . When I had my oldest daughter, I prayed and asked God to take it away. I needed to be able to answer the phone, call for help if needed, yell "stop" if she ran into the street. It was a matter of life and death for her and . . . It went away . . . I'm a new creation! Education is very important in our family. It took me ten years to get my Bachelor's Degree in Accounting and I know that I made my mom and dad proud when I finally graduated. I obtained my CPA certification as well. I'm proud of the fact that my sisters and brothers all have college degrees.

The above comment summarizes traumatic childhood experiences that included unsuccessful educational and medical interventions. Donna's concern for her daughter provided intense motivation; yet she credits God with healing her as promised in 2 Corinthians 5:17.[23] The remark also reflects a practical understanding of a passage that guarantees spiritual transformation. Donna's conversion was also physical. Below Constance navigated circumstances that could have derailed her ambitions:

In my early years, I believe I may have had some learning disabilities, as I was held back one grade and the reason for this was never discussed. As I got older, I figured out my own learning style and this has helped me to excel. I've never received special education services, but in hindsight, obviously, there were some concerns that didn't get addressed . . . [E]ducation was and still is a priority in my family. My parents even put money aside for their grandchildren to have the opportunity to attend college. (Constance, 47 year old social worker)

It appears that even Constance's upper-middle-class family could not prevent her learning disability from going unattended. Collins (2009) notes that students of color often "fall through the cracks" of public school systems not designed for their success.[24] Although efforts were made to address Donna's speech impediment, Constance adapted without such interventions. Family members also acknowledge specific black teachers and predominately black schools that were part of a broader supportive *village* that stood in the gap between resource limitations in segregated schools and black student success:[25]

I was dyslexic as a child and would cry and cry because I would write some of my alphabet backward. I later learned that my brain was trying to process information too quickly, but at the time, I was scared and ashamed. My mother worked with me every day. The teachers said I was really smart – I just had to learn to take my time. At every school I attended, black teachers pushed me to excel – white teachers, not so much. But black teachers encouraged me when I needed it and reprimanded me too . . . Once I was placed in advanced classes, I hit my stride . . . My mother and the older women at church helped me in some ways too. School and church were key learning environments for me. (Tanya, 53 years old, school administrator)

Unlike Constance, Tanya suggests that the combined efforts of her mother and a cadre of black teachers addressed her learning disability; church members also provided on-going, long-term support. The above Kings suggest that a combination of formal and religious education helped create a practical, emotional, and psychological safety net. Without such interventions, unchecked challenges of minority youth with special needs can derail their educational outcomes even more quickly than for minority youth in general.[26]

The following experience of one of Irma and George's great-grandchildren, Denny, supports Siddle Walker's (1996b) understanding of an exemplary school system and strides to meet the specific educational needs of black boys. Its details are important as an exemplar on how macro-, meso-, and micro-level dynamics can be effectively harnessed to positively affect educational trajectories:[27]

Throughout elementary, middle and high school Denny experienced some learning and speech challenges and needed extra guidance, speech therapy and support. He was always a smart kid, but needed help to understand and build on his learning style and improve his speech language skills. Despite the struggle, Denny believes in God and has the utmost confidence in himself. He comes from a spiritual family, has had very positive church experiences and made his own decision to join church and be baptized . . . this foundation continues to help and guide him as he completes his bachelor's degree – he is a junior at [name of college] and when he struggles with school or other things, I guide him right back to his faith in God, himself, and prayer. (Vera, 51 years old, preschool transitional social worker)

Without tenets from the Black Church tradition and dedicated parents, it is unlikely that Denny would have completed high school and be matriculating in college.[28] For Vera, both secular and religious education were needed for his success. However, she is particularly beholden to a cadre of committed black teachers and administrators:

There were several African American teachers throughout his elementary, middle and high school years that provided guidance, concern and "parental support" for

Denny – from his third grade elementary teacher who helped to identify ways to support his learning style and fine motor needs to his middle school male teachers and basketball coaches who were excellent role models and who were always imparting valuable information about the value and necessity of advanced education for minorities.

Following guidelines suggested by education scholars such as Ferguson (2001), Hale (2001), and Mehan, Hertweck, and Miehls (1986), Denny thrived based on institutional processes that addressed diverse learning needs.[29] Vera traces the indelible influence of a group of black teachers and school leaders throughout her son's school experience. Their presence provided instrumental and expressive developmental support needed to overcome his learning challenges while also instilling the requisite confidence and high aspirations such that this additional support did not result in internalized shame or a deficit mentality. However, she posits that one black female teacher was particularly instrumental:

> Denny has shared with me that the most influential person was his high school guidance counselor, mentor, teacher and "school mom" – [teacher's name], at [name of high school]. [Teacher's name] followed Denny's school path. She attended all of his educational meetings, guided his class choices to ensure that he would meet all graduation requirements, recommended various supportive services, made sure Denny stayed focused . . . and kept in regular contact with me about his progress. She delayed starting a new job at a different school several months because she was not going to leave Denny's school until he was completely finished. She was so proud and excited at Denny's graduation! The [name] county school system was the best ever for my son.

Using Billingsley's (1992) understanding of the black community's ecology, Vera attributes her son's academic accomplishments to capable, committed black teachers and the broader supportive school environment in which they are employed. Denny benefited from an inclusive school system designed to meet special needs (macro-level support) as well as resourced schools led by prepared administrators (meso-level support) who were accountable to each other, parents, and students. Qualified, committed teachers and guidance counselors provided continual, culturally sensitive instruction and coaching (micro-level support). Yet his success was fueled by Denny's committed parents. In many ways, Billingsley's race-specific typology responds to and extends Bronfenbrenner's generic developmental model.[30]

This commitment level also parallels, Siddle Walker's (1996b) observation about exemplary black schools: "The teachers consistently use the phrase 'their highest potential' when referring to their expectations for students. The term seems to capture their commitment to push students to perform as well as they were intellectually capable of performing. They believe this type of push – giving other children what you would want for your own – was the bases of good teaching and of a good school program" (p. 158). Vera contrasts this overall positive experience with Denny's earlier challenges in a well-resourced, predominately white school district that did not have the diverse, culturally sensitive personnel to appropriately educate and socialize students of color – nor the interest in doing so.[31] The above account also supports studies about a linked fate ethos in effective black schools:

> An environment that was both relational and institutional . . . interested in the "whole child" . . . [and] of developing "multiple intelligences" . . . teaching could not be reduced to a job or an occupation; it was a mission . . . the teacher was thus passing on cultural capital to students, and doing it in a way that the students understood, because they shared a common cultural heritage . . . the schools' major features existed within the framework of a cultural community . . . [that] common religious beliefs, common ways of acceptable communicative patterns, and common beliefs about the appropriate relationship of young people to adults. (Siddle Walker 1986a: 204, 206, 213)

Many of the Kings describe being instructed by individuals with a calling or ecclesiastical vocation to provide academic as well as practical knowledge and expose them to the multi-consciousness needed to successfully navigate a global society.[32] Whether one considers school in a two-room shack in Mississippi or in urban areas, the familial push for higher education, combined with support from parents, teachers, and church members, helped combat leveled aspirations and launch many family members toward the middle class.[33]

Combating Racial Inequality to Earn an Education: Generational Experiences

Quite possibly the most egregious implication of race, class, gender, and *age* inequality for blacks has been the tendency for black youth to struggle for proverbial educational *scraps*. This negative pattern exists in urban and rural spaces.[34] In *Another Kind of Public Education: Race, Schools, the Media and Democratic Possibilities* (2009), Patricia Hill Collins responds compellingly to this chronic dilemma: "because of its history, race has been tightly bundled with the social issues of education and equity in the U.S. context . . . [U.S.] enemies, however, do not include the historically imagined enemy of brown and black youth, more often depicted as

America's problem than its promise" (pp. ix, x). As suggested by Collins and other educators, combating educational inequality will mean directly and proactively addressing politics and pedagogy as well as including adults and youth, rather than simply traditional leaders, in the arduous, but potentially transformative process. In the interim, black youth who make it are applauded for their grit; their less successful peers are usually blamed for their shortcomings.

Successive generations of the King family experienced unfavorable outcomes when marginalization attenuates educational outcomes.[35] Such situations took varied forms, but became part of the social capital that informs their lives:

> I didn't have much exposure to whites and racism as a kid, but I saw that even in the black culture people treated you differently based on your color. The lighter kids received more attention than the darker ones and the ones with "good hair" received more attention than the ones with nappy hair. (Donna, 54 years old)

Her 53-year-old cousin, Tanya, describes specific experiences:

> I have experienced racism from white people since I was in high school. I have also been made fun of because of my dark skin by family members and peers. Even one of our great aunts [Irma's sister] was "color struck." She liked her light skinned nieces better . . . and she was dark skinned! My mother helped when she told me that our ancestors from Africa were dark skinned like me. I learned more about this in college. So that "house slave" and "field slave" mentality still affects some blacks today – even family members. It reminds me of something I heard once in a sermon, "all skin folk ain't kinfolk, and all kinfolk ain't kind." Big Mama and my mother are dark skinned – they made me proud . . . and strong.

Donna and Tanya's remarks are a reminder of the very real intra-racial partiality that still exists in the black community based on phenotypical features such as skin color and hair texture as well as special accommodations often given lighter-skinned blacks. Tanya also details both intra- and inter-racial discrimination in predominately white graduate school and workplaces. Both women suggest that such inequality is often ignored at the peril of both recipients of such privilege and blacks penalized because they lack these ascriptive traits. For Tanya, a seemingly clichéd sermonic reference helped her understand some of these racial complexities. Moreover, she was personally empowered by Irma and Carla as role models as well as by black college professors who taught about the physical, social, emotional, and psychological effects of slavery on many blacks.

Although intra-racial discrimination was mentioned, educational impediments were more often associated with racism and *inter-racial* problems. Barbara recalls negative inter-racial experiences in college:

> When I started my undergraduate degree, I wanted to major in sociology and there was a black admissions officer. I tested really high in math, so he encouraged me to pursue a degree in math because I would get a scholarship. There were only three black math majors on campus . . . and just going through that program was traumatic. I was only one class away from having a math degree. The higher up you got, there were no blacks in your class. You would see the professors reaching out to the white kids, but not to you. It just wasn't the best situation.

The absence of both a critical mass of culturally sensitive white professors, minority professors, and students of color meant black students like Barbara do not have the requisite support system to find success in Science, Technology, Engineering and Mathematics (STEM) fields.[36] So despite being heavily recruited in mathematics, she could not be retained. Unlike Tanya in the introductory story, Barbara did not receive the intervention needed to address the differential treatment she experienced. Moreover, both women were astounded at the unethical lengths some whites would go to maintain their privilege. Barbara has become disenchanted by the returns on her educational investment:

> My education has not taken me where I expected because initially the position that I have required the education level that I have. First they bumped it [requirements for the position] down to a bachelor's degree and this year, they changed that. As long as you don't have to [manage] a group home, all you need is a high school diploma. I manage group homes, so you have to have a degree – that's a state mandate – but you only have to have a bachelor's degree. So based on my chosen field and the state where I live, having a Master's degree is no longer beneficial . . . it is slightly disheartening because part of your bargaining tool is gone. So as far as wages, when you could say, "well, I have this much education *and* this much experience." Now it's cheaper for them to get someone right out of college because they aren't expecting as much [salary] as I am.

Barbara contends that her Bachelor's and Master's degrees did little to guarantee gainful employment.[37] Moreover, her experience suggests that the anti-discriminatory policy changes that resulted in a decrease in labor market discrimination for better educated blacks in the mid- to late 1900s do not necessary account for individual discriminatory experiences; may not be applicable across all labor market sectors; can be undermined by state-specific power; and may have been short-lived.[38] Although she tries to be optimistic, years of being over-worked and under-paid now mean that she calls into question the American Dream that her parents suggested formal education would secure.

In this instance, the implications of labor market demands that do not place the same premium on "soft" as "hard" skilled degrees mean that when Barbara was *forced out* of a mathematics program into the social sciences, her employment prospects and subsequent wages changed drastically.[39] Jacobs (2014) confirms the monetary differences:

> The Department of Education (DOE) report looked at four years of data on college graduates and found that STEM majors – science, technology, engineering, and math – on average earn $65,000, while non-STEM majors earned about $15,500 less. STEM majors were also more likely to be employed and hold only one full-time job, rather than a part-time job or multiple jobs . . . Overall, the information clearly pointed to the advantages of studying a STEM field in terms of employment and salary. (n.p.)

Barbara's concerns are also supported by economists: "African Americans receive a 15 percent wage penalty and nearly half of the raw wage differential is caused by racial discrimination" (Mason 2003: 52). Another relative describes an employment trial:

> I know I am blessed. I have a great career, great salary, and great benefits. But sometimes it really makes me mad to known that I had to have four degrees to get this job and my white colleagues only have two. I know there are people who would love to be in my shoes, but it still makes me angry. And everybody knows I'm a Christian at work. Some whites thought they could just say anything to me and I was going to turn the other cheek and forgive and forget. I gave them a few chances, but then I let them know – being a Christian doesn't mean being a doormat . . . Sometimes I get angry with black churches today. So many horrible things are happening in the black community and they are too silent. It's good to pray, but we should pray – and then act! (Tanya, 53 years old, administrator)

Tanya's advanced degrees have resulted in gainful employment. Yet she was disillusioned to learn that her white peers were not similarly qualified; her concerns were compounded by unprofessional treatment. Thus her advanced education and supposed protections via affirmative action and anti-discriminatory policies have not made her exempt from unequal treatment.[40] Tanya also alludes to both Matthew 18:21–2 and Matthew 5:39 to suggest that her colleagues assumed she would accept negative treatment because she was a Christian.[41] She believes that black Christians, in general, and black churches, have become complacent for not responding collectively to the increased discrimination, racism, and violence against blacks that is taking place in the USA. Studies show that blacks are expected to acquiesce to discriminatory treatment by whites and that the latter group is often dismayed when the former does not comply.[42] Tanya's experience also parallels Cotton's (1993) finding, "in 1980 . . . over one-third of employed black males and over one-fourth of employed black females were overeducated for the jobs they were doing. By comparison less than one-fourth of white males and one-fifth of white females were similarly overqualified" (pp. 193–4). For both Barbara and Tanya, formal education, family support, and an education-oriented ethos were not enough to shield them from discrimination or attenuated educational and employment trajectories. And Barbara wants her children to exceed her educational level, but would rather they avoid "soft skilled" occupations. In contrast, Tanya's position has enabled her to navigate racism despite minimal job support.

Studies seem to have reached an impasse in explaining why most blacks continue to lag behind their white peers in terms of labor market and employment reports, irrespective of educational level.[43] Several of their employment experiences suggest that they were still being sorted into lower-paying occupations. Research points to culprits such as: underfunding; well-meaning, over-worked teachers; outdated curricula; tracking; strained student-teacher relationships; culturally unprepared, non-blacks teachers; and, in some cases, unmotivated students.[44] National studies show the continued value of education among black students and students who beat the odds despite disadvantage.[45] Yet some blacks realize the weak correlation between education and gainful employment in a society fraught with labor market inequities:

> We were all told "go to college, get a degree, and get a good paying job and life will be well." But in this day and age, you can have a degree and still have to work two or three jobs to make it because, in the workplace, they want one person to do the job for two people and to pay you the wages of half a person . . . and if you've gone into a soft field like social services or education, you are underpaid . . . and you have so many people competing for the same jobs and that's why employers are able to underpay and overwork people because they know you can be readily replaced. (Barbara, 44 years old)

Barbara's experiences question human capital explanations for disparate employment outcomes among blacks; she points to a myriad of discriminatory episodes to support claims of differential treatment based on race.[46] Despite her own academic challenges, Barbara recognizes the intrinsic benefits of higher education and family role models:

> I was really proud when my mother got her degree. That was something she wanted to do, just to achieve it because she wanted to . . . she has an Associate's degree in

Computer Science … I'm proud of our family. The way we support each other. The way we band together in a crisis and for the most part, we are responsible citizens and I credit that to where we came from. We are all hard workers and most of us are college educated, even outside of our immediate family … someone before us had to instill that in the generation of our parents for it to have affected us this far.

The following comment by 74-year-old Carla reflects the King's educational ethos:

I wanted all of my children to have a college education – not so they could think they were better than anybody else, but just in case you married somebody and it didn't work out, you didn't have to end up in poverty. You could still take care of yourself and your family, if you had one. Now that was the value I placed on education.

According to the above narratives, formal education was encouraged to improve the likelihood of upward mobility, enhance family outcomes, and prepare one for unexpected challenges such as divorce and other hardships. Like religious education, secular education was deemed the requisite social capital to accomplish these goals. Billingsley (1992) suggests that, because of formal education, blacks were able to transition from working class to middle class in one generation. Whether through college training, on-the-job training, informal skills learned in the military, or wisdom from the college of hard knocks, this pattern is apparent for many of the Kings. Family and generational resources and values were central to this process and parallel Siddle Walker's (1996a) observation: "black parents are shown to be victims of an oppressive system, but are also depicted as agitators to the system, people search for ways to achieve better educational opportunities for their children" (p. 9). It is important to continually delve deeper into the ecology of predominantly black learning spaces and the families and communities that surround them – beyond economic measures. Simply put, substandard resources do not necessarily equate to substandard care and support for black students. However, an absence of systemic and institutional resources and support severely constrain learning opportunities.[47]

Contemporary Educational Attainment and Race and Ethnicity: The Current Black Experience

Scholarship, historical accounts and family narratives illustrate the importance of culturally sensitive, proactive systems, processes, groups, and resources for educational and subsequent employment success. Yet Mason (2003) supports economists who conclude that, "racial inequality from 100 years ago effects individual labor market outcomes today" (p. 66). White (2016) recognizes a key source of these inequities: "school segregation is a major cause of labor inequality in the U.S. – whether it's intentional or not" (n.p.). What educational patterns are evident today?

Table 3.3 presents educational attainment based on race, gender, and age in 2017. Several patterns are apparent. Among completed college degrees, the most common level of educational attainment in 2017 was the Bachelor's degree. In general, across the four racial/ethnic groups, *non-Asians* are more likely to have only earned high school diplomas (ranging from 30.5 percent for white females to 47.4 percent for Hispanic males) or attended college without graduating. Relatively higher percentages of Asians and whites, both males and females, respectively, have earned Bachelor's degrees as compared to their black and Hispanic counterparts. A similar pattern emerges for Master's degrees. However, gender differences exist; relatively higher percentages of Asian males have advanced degrees. Although fewer than 3.0 percent of individuals overall have earned doctoral degrees in 2017, more Asians (5.9 and 3.3 percent) and white (2.2 percent) men have done so. Intra-racial trends are also evident. For example, black males are more likely

Table 3.3 *Educational Attainment by Race, Gender, and Age in 2017*

Educational Level (%)	Race							
	Black		White		Hispanic		Asian	
	M	F	M	F	M	F	M	F
High School	42.1	35.3*	34.1	30.5*	47.4	39.9*	20.9	22.5*
Some College	25.1	26.2	20.5	20.8	24.0	25.4	17.4	14.6*
Associate: Occ.	4.0	4.8	4.7	4.9	3.9	4.9	1.9	2.4
Associate: Academic	5.9	7.1*	5.5	7.4*	5.3	6.9	3.9	5.2*
Bachelors	15.7	16.4	22.6	23.4	14.2	16.4*	29.3	33.7*
Masters	5.6	8.2*	8.5	10.2*	3.9	5.1*	18.5	16.4*
Professional	0.5	0.9	1.9	1.2*	0.6	0.5	2.1	2.0
Doctoral	1.1	1.1	2.2	1.4*	0.8	0.9	5.9	3.3*

Key: Figures taken from 2017 U.S. Census Bureau at https://www.census.gov/ and Snyder, Thomas D (Ed.). 1993. "120 Years of American Education: A Statistical Portrait." National Center for Education Statistics. Table excludes individuals with "no" education and less than a high school diploma or its equivalence. Figures reflect the noninstitutionalized population. M=Male. F=Female. Occ = Occupational. *T-test results that denote significant gender difference for that specific racial category at $p<.05$.

to have high school diplomas, while relatively higher percentages of black females have earned associate academic, Master's, and professional degrees. Yet similar representation is evident for the other educational levels for blacks. Gender differences are also apparent for whites and Hispanics, particularly for "some college" and advanced degrees. Given the direct correlation between advanced education and economic mobility and wealth, Asians, followed by whites, continue to be better positioned for the latter positive outcome.[48] Blacks continue to lag behind on these educational indices.

Findings in Table 3.3 are important because studies show the direct correlation between earnings and educational attainment. In 2015, the median earnings of persons 25–34 years old with a Bachelor's degree ($50,000) were 64 percent higher than their peers who only completed high school ($30,500). Moreover, the median earnings of young adult high school completers were 22 percent higher than their peers who did not complete high school ($25,000). The median income of persons with a Master's degree or higher was $60,000, about 20 percent higher than returns for a Bachelor's degree alone. Thus the lag in educational attainment for blacks has economic implications that influence quality of life and life chances.

The push toward higher education initiated by Irma and George over seven decades earlier was realized and advanced in subsequent generations and meant that all their children completed high school – an accomplishment uncommon among blacks in Gallman, Mississippi, at the time. Their seventeen grandchildren have the following educational outcomes: two earned high school diplomas only (11.8 percent); four had "some college" experience (23.5 percent); one associate degree (5.9 percent); two Bachelor's degrees (11.8 percent); seven Master's degrees (41.2 percent); and one PhD degree (5.9 percent). Occupations are equally diverse and include business administrators, several CPAs, several truck drivers, a college professor, multiple social workers, homemakers, an executive director of a teen mentoring program, a financial analyst, a program director for persons with developmental disabilities, multiple business owners, three ordained ministers, and a carpenter. Like their predecessors, most would be considered middle class, several are solidly upper class. And as illustrated in research by Pattillo-McCoy (1999), some offspring have not experienced the middle-class lifestyles their parents' cultivated.

Like the pattern for blacks in Table 3.3, a greater relative percentage of female grandchildren have earned advanced degrees as compared to males. Unlike results in the table, the King's educational attainment is skewed toward advanced degrees. Some of their employment outcomes support dimensions of labor market segmentation theory where males in this family found themselves in gendered occupations (for example, truck drivers and steel mill workers) that were further bifurcated based on race. Yet, several younger males have forged lucrative careers as entrepreneurs. "Pink collar" sector jobs are common among King women; a substantial number have experienced success as administrators, college educators, and entrepreneurs in predominately male job arenas.[49] In addition to discrimination embedded in labor markets, several other systemic and individual factors have resulted in skewed returns to education for later generations of Kings, including low wage returns for advanced degrees in "soft skilled" fields; unequal access to on-the-job training; and intermittent periods of unemployment for King females to rear children. Darity and Myers (1998) suggest:

> In 1970 the family incomes of blacks with older, better-educated heads was *two and a half* the total of black families with younger uneducated heads. By 1991 it was more than *seven times* [emphasis is theirs] that of black families with younger, uneducated heads . . . the relative difference . . . went from 250 per cent in 1970 to 720 in 1991. The relative difference rose among whites as well, but only from 200 per cent in 1970 to 390 per cent in 1991. (p. 68)

Despite these differential educational returns, their academic accomplishments reflect the King's ethos and honor the sacrifices of their ancestors in Mississippi.

Education in Real Time: Mt. Horeb and Parish High Schools

Research consistently illustrates that blacks value education and understand its role in fostering stability and self-efficacy.[50] Moreover, the school of hard knocks often meant that some form of credentialism via formal education, a trade, informal skills, military stints, or on-the-job training was central to becoming competitive in the labor market. One's livelihood and that of one's family depended upon it. For Irma and George's children, such training took place in segregated southern schools. Carla notes, "Our elementary school was near our church. It went from 1st grade through 8th grade. That was Mt. Horeb School, where the cemetery is now." The embeddedness of religion and secular education in Gallman is apparent in the above quote, as the first school Irma and George's children attended, Mt. Horeb School, was located across from their church. Martha details teachers and process as well as its practical and academic instruction:

The school was across the street from Clear Creek Church No. 1. It was two rooms and went up to the 8th grade. I started when I was five. My first teacher was named Ms. Louis Smith. She took me in like her own child. She was really good. I have a little aluminum roasting pan that she gave me as a gift that I still have. Ms. McClure was another teacher. She was an older lady because she had taught My Daddy [George] when he was in school. Then there was Ms. Webb and Ms. Ketchum and Ms. Vinnie May. Both Ms. Ketchum and Ms. May were kind of young. They were not all there at the same time. We didn't have a principle, just a head teacher. (Martha, 77 years old)

Martha astutely recalls the litany of teachers during her nine years at Mt. Horeb who each had an indelible influence on her. Although the school superintendent was a white male, teachers were black. In several instances, the King children were taught by teachers who had also taught their parents. This inter-generational instruction illustrated the commitment teachers possessed. Although the building was small, she describes its efficient use:

There were four different classes in one room. They divided it up so everybody would get some instruction during the course of the day. Back in the day, we started with inspection to make sure your hair and nails and clothes were clean. Then we said the Pledge of Allegiance, sang a song, and then we had prayer and then class started. School was from about 7:00 am to about 3:00 pm.

Paralleling Siddle Walker's (1996a) research, black children were taught common academic topics such as arithmetic, reading, writing, science, and spelling as well as physical hygiene, civil religion, and religious rituals. The one school meant all black students were taught together, regardless of their class position. James Lewis' next memory alludes to socioeconomic differences:

One year, when I was about 4 or 5 years old, we had to exchange Christmas gifts at school. The boy who pulled my name gave me some apples and oranges, but My Daddy bought the boy whose name I pulled one of those guns and the dart board that you shoot things on and it sticks. I looked at my gift and the one I was supposed to give away – and kept the gift My Daddy bought. But my parents found out about it and made me give it back. (James Lewis, 69 years old, steel worker)

In addition to mentioning his favorite teachers such as Ms. McClure and Ms. Rose Young, James's comical thoughts above about a gift exchange illustrate the differing economic abilities among his peers. Although he was unware of the class distinctions at the time, the types of gifts provided alluded to different families' economic abilities – and his parents' efforts to ensure that he didn't make such distinctions.

The King children were cognizant of the differences segregation made in their learning experiences and contended with sub-par schools led by committed black administrators and teachers. Janice suggests: "We had separate schools. So whites had their schools and the blacks had theirs. We walked to our elementary school. It was two rooms with a wood heater on each side." Martha also acknowledges the differential funding: "We had the hand-me-down books from the white schools all through elementary and most of high school. In high school sometimes we did have some new books." And whites went to great lengths to maintain segregation and Jim Crow:

In '53, they were going to build the whites a new school and they said there were going to give the black kids that building. It was up there in Gallman, in the little community where whites lived . . . before the end of '53, somebody went up there and burned that building down to the ground so the blacks wouldn't be up in their little area – basically in their back yard because that's where the school was. I graduated from high school in '57 – I started high school in '53 . . . when I started high school, they bused us to Crystal Springs because we couldn't go to school with the white kids. (Martha, 77 years old)

The above incidents occurred several months before *Brown vs. Board of Education of Topeka* and illustrate the stark school environment that black children in Gallman faced. Sub-par conditions during elementary and middle school, and continued efforts by vigilante whites to regulate where black students learned in high school, were reminders of who ultimately controlled southern spaces – either by design or force.[51] Yet mechanisms used to control land (and blacks) often emanated differently. Poor and working-class southern whites tended to use violence, coercion, and intimidation (example, burning down the vacant white high school for fear of its use by blacks). In contrast, more economically powerful whites tended to use wealth, political influence, other legal forms of power, or complacency to accomplish the same ends.[52] In this way, Gallman residents were stratified by common factors such as race, class, gender, and age as well as, among whites, by less emphasized factors such as reliance on overt or covert mechanisms to maintain power.

Funds were located to build a high school for black students. Yet it paled in comparison to the new high school for their white peers:

Then when you left there, you went from the 9th to the 12th grade at the high school, Parish High. I was in the first class to graduate from the *new one*. They built us a new school. Now they did have a school in Crystal Springs called Horse Claw Negro High School. It had the word "Negro" in it. But now my school didn't have the word "Negro" in the name, although they had a high school for the white kids and a high school for us . . . The black and white students never went to school together. Matter of fact, our bus passed by the white school to get to the black school. (Carla, 74 years old)

Her sister provides additional details about the differences segregation made:

When you left Mt. Horeb, you went to Crystal Negro High School. It was called that to make the distinction from the white high school Crystal Springs High School. It was a long building like a barn with a long hallway and all classes were on different sides of the hallway. The white school was a big building like schools should be. It had a cafeteria and gym and everything. We didn't have those things, but we had an auditorium and some basketball goals in the back. Our principle was named G.R. Brown. We took the bus to get to high school. (Martha, 77 years old)

Black students welcomed the new learning space, sports teams that fostered interaction with other regional black schools, and advanced classes such as physics and history. Martha notes that the efficient use of human resources meant teachers often held multiple positions: "In high school we had English from Mrs. Brown. She was the librarian too and the principle, Mr. Howard, was the science teacher. Mr. Lilly taught math and physics . . . We also had a music and health education teacher, but I can't remember his name." Sports opportunities were particularly important for several men in this family. James Lewis comments: "I attended Parish High School. I played baseball and football. By the time I got to high school, all my sisters were grown and working. They would send me school clothes and gifts. And I didn't have to work in the field like they did. This was because I was the youngest." Irma's nephew, Fred, segued athletics into two college degrees and a career as an educator:

I graduated from Parish High school in 1969 and went to [name of an HBCU] on a football scholarship . . . Education allowed me to make a decent living with a decent job. It brought me up from the living conditions from my parents and made it better for me and my children. It inspired me to help other children do the same thing. I have taught and coached many of them . . . some went to college and several went pro. Others are lawyers, doctors, some went into education, highway patrol, game warden . . . I loved teaching . . . I loved my students and I coached football too and middle school basketball. The coaching experience was good . . . Sometimes I see my students and they just say "thank you." That makes you feel good. (Fred, 69 years old)

In addition to securing positions that he loved and at which he excelled, Fred's academic success enabled him to experience upward mobility and economic stability. And, like his teachers, he established a strong reputation as a role model and mentor.

Despite its benefits, religious education could be used in controlling ways. Below Martha describes its influence in several settings:

> Back in the day, if a girl got pregnant, she couldn't finish high school unless she strapped herself down. It was considered a disgrace. Parents felt that she might talk other kids into having sex. But they didn't put the boys out . . . that was discrimination, but that's what they did. This was just like in church. If a girl got pregnant, she had to come and apologize to the church, but they didn't ask about the daddy or have him apologize. One of my classmates was pregnant when we graduated, but she was a heavy set girl so the teachers acted like they didn't know.

Martha describes how Christianity and traditional gender expectations influenced religious *and* academic spaces. Adults followed biblical edicts such as John 8 to surveil the sexual activity of youth and squelch inappropriate sexual behavior. Young, pregnant, and unmarried females were sanctioned both at church and in school. Yet their male counterparts were exempt from this same scrutiny and indictment. This tendency troubled Martha as an example of intra-racial discrimination she now condemns. However, overall, she has fond memories of her early academic years:

> I think I got a good education at Mr. Horeb and in high school. I got good grades in all my classes and it was a must that we did our homework when we got home. And we had homework all the time. And you could always ask the teacher for help . . . I hate the fact that I didn't get a college degree. I was a career student.

Despite a strong academic beginning, maternal responsibilities, the death of a spouse, and cancer prevented Martha from completing a college degree. For these reasons, she encouraged her children to complete college and is proud that they all did. Support from teachers and their parents meant Irma and George's children were able to navigate school segregation and have generally favorable experiences that undergirded their parents' push to earn high school diplomas.

Two segregated schools in Gallman were pivotal spaces for local black youth. What they lacked economically, they made up for in committed teachers and administrators. Although not as well appointed as the institutions for white students, King family members suggest that Mt. Horeb School and Parish High focused on their holistic education and socialization and helped realize the King's educational ethos.

CONCLUSION: MULTI-CONSCIOUSNESS AND BEYOND

This chapter documents educational advances among blacks, in general, and the Kings, in particular. Census reports and scholarship show the hard-won academic accomplishments led by teachers in the many Mt. Horeb's of the South and later, in largely segregated schools in the North. And studies illustrate the continued importance of education among black youth.[53] Yet educational strides among blacks do not overshadow the chronic ways in which black youth continue to lag behind most of their non-black peers. And school segregation today is comparable to rates in the 1960s.[54] According to the Urban Institute, "in every state but New Mexico and Hawaii, the average white student attends a school that is majority white" (Jordan 2014: n.p.). Scholars argue for more concerted involvement from the black community to hold political and governmental leaders and groups accountable.[55] Our findings provide a strong argument that improving the economic outcomes for blacks (and thus their ability to effectively compete in the labor market) means acknowledging and confronting the covert, non-traditional forms of racism, sexism, and classism in the USA.[56]

Just as formal education was paramount, it appears that informal instruction and wisdom to navigate inequality were just as crucial for the Kings. Narratives in this chapter describe life lessons and non-traditional forms of social capital that were sobering, disillusioning, and often angering, but that family members now use to navigate society and socialize their children in multi-conscious ways. Such a thought process requires thoughtful instructions around the many facets of being of American descent and African descent, and global citizens. We posit that the difficulty for many family members involves how to prepare themselves and school on their children to navigate spaces that are often disparate in their treatment of blacks, while simultaneously teaching them Christian tenets to be loving, inclusive, optimistic, and forgiving – as modeled by Jesus Christ. The tension between Christ-like ideals and societal reality appears to be the most difficult conundrum more recent generations of the King family face in order to be thoughtfully inclusive in an often exclusive society.

Just as the military provided opportunities beyond what the Kings who enlisted experienced in the rural South, formal education provided opportunities for a broad array of experiences and occupations. So despite the animus some family members suffered in predominately white academic spaces, they consider their educational outcomes generally beneficial. Yet the intrinsic benefits often outweighed the extrinsic ones as well-educated Kings, particularly females, still found themselves hitting glass ceilings and settling for horizontal rather than vertical movement in various employment sectors. In such instances, they contend that gender, race, class, and their intersection continue to eclipse education and expertise for most blacks. Furthermore, because they contend that their educational achievements have not taken them as far as whites and/or white males, they are concerned that their children will experience a similar fate – membership in the *lower* middle class while confronting the daily micro-aggressions required to maintain that precarious class position.

4

"Jesus and the Juke Joint": Blurred and Bordered Boundaries and Boundary Crossing

The small shack was invisible from the main road. Tucked deep in the woods behind a wall of oak trees, it was intentionally hidden from those people who did not know its purpose. Unimpressive from the exterior, it represented a safe space for blacks in need of escape from the severity of segregation. This juke joint (as it would be called today), was run by Howard and Janice. Customers crowded around the makeshift bar and assorted tables, laughed, talked, listened to blues and jazz, and drank homemade moonshine. Because they were usually preoccupied managing the joint, their teen-aged daughter, Irma, manned the bar. Her parents were not yet active believers, but Irma had become a Christian as a child. She felt uncomfortable bartending, but did not dare disobey her parents. Unbeknownst to them, as Irma served drinks to patrons, she also evangelized to them – and watered down the moonshine! Irma believed that she was helping customers avoid drunkenness and the dangers that usually accompanied it. Although they loved her, Janice and Howard were furious upon learning about Irma's proselytizing. They whipped her. But Irma continued to water down the drinks and share the bible.

Irma's Christian convictions were codified early in life – even at the risk of being reprimanded and physically punished by her parents whose decisions were driven by more utilitarian concerns. The above scenario informs readers about another pivotal economic decision made by the Kings. The juke joint provided Janice and Howard with crucial income to supplement their sparse earnings as sharecroppers. Secondary income in the *informal economy* is often associated with life in post-industrial urban cites.[1] Yet a similar dynamic – for similar reasons – was evident among the Kings in early twentieth-century rural Mississippi.

As well as the above introductory example, Chapter 4 relies on several important incidents, some that contradict Christian tenets, to illustrate adaptive practices used by family members to combat challenges. We suggest that the concepts *blurred* and *bordered boundaries* help understand and explain the impetus behind decisions to use non-traditional religious, educational, and practical approaches to address and/or prevent problems. Yet family narratives show that some remedies were still circumscribed by boundaries tied to biblical and personal dictates. These two contrasting concepts provide insight into the complex interplay between spirituality and segregation that often resulted in tensions as well as incongruent beliefs and behavior. We also document the role of the local church, Clear Creek Missionary Baptist Church No. 1, in this family's religious socialization process. Religious education has been presented as the linchpin in the King family's trajectory into the middle class. Moreover, the organized Black Church shaped their lives by providing structured teaching and learning experiences as well as continual spiritual, emotional, and psychological support. Some Kings emphasized religious education more consistently than others. For some, recommitment would come later in life; others did not waver from this ethos. Yet the intrinsic and extrinsic benefits of such education were consistent.

This chapter deviates from other chapters in its emphasis on the *ideological motivation* behind many King family outcomes. Several overarching scriptural references seemed to inform their ethos: "Those who till their land will have plenty of food, but those who follow worthless pursuits have no sense" (Proverbs 12:11 NRSV); "And whoever does not provide for relatives, and especially for family members, has denied the faith and is worse than an unbeliever" (1 Timothy 5:8 NRSV); and "For even when we were with you, we gave you this command: Anyone unwilling to work should not eat" (2 Thessalonians 3:10 NRSV). Individuals candidly describe how such passages undergirded their daily lives to remain steadfast during difficult periods. These same verses also provided the requisite stigma to motivate less conscientious family members whose lackadaisical behavior was believed to undermine their lives, family stability, and the family name. Their religious ethos was linked to people and places, bible verses were directional for adherents, and the socialization process was formalized in the local Baptist church.[2]

These findings illustrate the strategic application of religious education to encourage family members to make prudent decisions such as to pursue formal education; unwavering belief in the Protestant Ethic; a linked fate mentality; and interplay and tensions between beliefs, behavior, and economic outcomes. Content analysis suggests that, in general, family members embraced the secular anecdote, "God helps those who help themselves," as well as Proverbs 10:4 that "A slack hand causes poverty, but the hand of the diligent makes rich" to survive and, in some instances, thrive during chronically harsh conditions.[3] Irma's teenage memory of her parents' clandestine juke joint that was frequented on Friday and Saturday night by many of the same blacks who attended church on Sunday morning provides the context to better understand how her religious beliefs and behavior would eventually become central to the King family's wellbeing.[4] This and other experiences show that some family members synergized Christian and secular beliefs and practices, especially for financial and practical reasons. Such practices were often correlated to family loyalty, self- and group-efficacy and interests, and the imperative to stave off poverty. This chapter is also important because it identifies some of the driving forces – beliefs, values, and a religious institution – that shaped the roles, rituals, and responsibilities from which this family's middle-class outcomes emerged.

BLURRING SPIRITUAL AND SECULAR ROLES, RITUALS, AND RESPONSIBILITIES

The following sections consider some of the effects and implications of socio-ecological structures and the associated ideologies and practices of the Kings and their offspring. We posit that blurred and bordered beliefs and practices are not *enface* bad or good, but rather part and parcel of strategic decisions to accomplish personal and collective goals. Those family members most comfortable engaging in such practices seemed to more effectively navigate many of the macro-, meso-, and micro-level pitfalls that blacks confronted in the segregated South. In addition to examining what family members *did*, we are interested in what they *learned* from these same experiences and shared with each other and subsequent generations. What certain individuals chose to do, avoid, or endure typically informed the subsequent strategies they provided to others. In this way, their own experiences could become formal and informal social capital or educational tools to help other family members improve both their life chances and quality of life. Blurred and bordered socio-ecological boundaries manifested in terms of rituals, roles, and responsibilities associated with family decisions. We present these three thematic foci and representative narratives.

Blurred and Bounded Roles

Blurred boundaries between Christianity and secularism were evident for Janice King, the family's matriarch. As noted earlier, she and her first husband, Howard, managed a juke joint out of necessity. Even after she became committed to Christ, Howard's death meant that she continued to sell moonshine informally to feed her children. Thus the exacting nature of poverty collided with the lofty goals of religiosity. Although likely not described as such then, 74-year-old Carla shows that Janice King responded *entrepreneurially*:

> On My Mul's side of the family, I think it was about twelve of them . . . there were probably about seven girls . . . she [Irma] was the oldest girl, but not the oldest child . . . They were Christians, but Momma Janice [Irma's mother, Janice King] made whiskey. But you have to realize that people, especially black people, had to do whatever they could do in order to survive. Because remember, she lived on the white man's place and they say that Momma Janice would be out there plowing that mule just like a man. But see if you couldn't do the work, then you were going to have to move. And after her husband died . . . where were you going with a lot of kids? So I know she made whiskey because she [Janice King] told me that – made whiskey and sold it. (Carla, 74 years old)

Janice's androgynous personality appeared to initially be born out of necessity, but evolved as she did. Her eventual role as a single-parent mother meant taking on traditionally masculine (such as plowing) and feminine roles to sustain her family. It also meant relying on her oldest daughter, Irma, to take on domestic responsibilities. According to her grandchildren, at this point Janice had also become a staunch Christian, yet, she also relied on economic practices that would likely be considered antithetical to certain scriptural dictates, particularly as interpreted by more conservative Christians. In this instance, ecological conditions resulting from poverty meant Janice's role as primary provider took precedence over her role as a believer. Additionally, Janice's continued practice of selling moonshine, even after the juke joint closed, parallels some of the decisions made in post-industrial urban centers as poor and near-poor persons periodically participate in the informal economy to make ends meet.[5]

Similarly, despite passages such as I Corinthians 6:19–20 that remind believers that their bodies are God's temples and not their own to defile, Janice enjoyed "dipping snuff" or Garrett chewing tobacco.[6] As well as avoiding alcohol, staunch black Christians at the time often understood this passage to warn believers against smoking or using tobacco of any kind. Yet Janice did not appear to believe that enjoying the "sweet-tasting" tobacco undermined her relationship with God. In addition to attending church, Janice loved baseball and also had a large Motorola hand-cranked record player on which she listened to her favorite blues *and* Christian music. According to family members, Janice's identity included gender-blurred domestic and non-domestic roles. One of her more traditional religious rituals involved preparing large Sunday dinners for family and friends during which the following table grace was always recited, *Lord bless us and bump us, keep all of our far away relatives from us. Because when we have something good to eat, they come and eat it up from us.* Janice's favorite grace combines religious, comedic, and practical dimensions as yet another example of her comfort-level blurring spiritual and secular aspects of life and crossing conventional boundaries that influenced many of her children, especially Irma.

As illustrated in earlier chapters, in many ways, George and Irma both embraced traditional gender roles. Yet their roles intersected as a result of the segregation and racism that constrained employment for most black males to low-wage positions and relegated most black females to positions as domestics and nannies. For example, typical annual wages for black women and men in the South during the 1940s ranged from about $300–$600, making it difficult to sustain a family with children. George was considered the family's protector and provider. Yet Irma's income as a domestic was crucial to their economic stability. Moreover, her children consistently suggest that she was largely responsible for ensuring much of the daily maintenance of the family farm. Thus the couple espoused traditional gender roles, however, the reality of everyday life in the rural South meant that their roles were inextricably intertwined:

> She [Irma] did the best she could. He had his own lifestyle and way of doing things and stayed gone a lot. Yes, she did what she was supposed to do to maintain the house and take care of us kids . . . My Daddy did give My Mul money to buy shoes and to buy us clothes . . . [and] we never went hungry and even though our house wasn't in the best of shape, we always had somewhere to stay and My Daddy always had a car or truck or something. (Martha, 77 years old)

When queried about how their mother was able to maintain such a strenuous set of responsibilities inside and outside the home, explanations were inevitably linked to her strong Christian convictions. Carla was decided in her portrayal of Irma:

> The Virtuous Woman . . . because all of us have faults, but as far as My Mul was concerned, she was, in my eyesight, a perfect example of a Christian woman and she tried to help everybody . . . we all have done something, but I didn't know one bad thing about her . . . when I think about it, we had a part in it, kind of making My Mul's life a little miserable . . . because we would say we wanted to go to games and things and My Daddy would say, "you can't go with the boys or by yourself" – although there were three of us [three sisters]. My Mul would ride along and just sit out there in the car while the games and things were going on. And I think about

that now and I get a little sad . . . but she wanted us to be like other kids to be able to go sometimes . . . My Mul was the anchor of the family.

The Virtuous Woman presented in Proverbs 31:10–31 was the most common biblical example associated with Irma by family members – regardless of the generation. Like the woman in the narrative, Irma was consistently described as long-suffering, gentle, hard-working, conscientious, entrepreneurial, and savvy, as well as a source of pride and financial resources for her husband and family. Above, Carla reflects upon her mother's sacrificial nature on behalf of her children as well as the gender constraints many women upheld to help maintain their families. According to Carla, although Irma deferred to George, she chose to chaperone her daughters rather than allow them to miss important childhood experiences. In addition to the Virtuous Woman motif, the secular term *family anchor* was also attributed to Irma by multiple family members and friends. And Carla contends that her immediate family is still experiencing the benefits (particularly her children's academic accomplishments) of Irma's sacrifices so many decades ago.

Reference to the Virtuous Woman passage also alludes to the reality that, like the husband and wife in the narrative, the Kings did not share equal responsibilities for maintaining their household. Readers should note that this exegesis does not diminish George's importance in or contributions to the family, but rather acknowledges the differential roles and responsibilities with which many black women like Irma had to contend at the time.[7] His children did not appear to believe that they were dishonoring George by acknowledging the singular role played by their mother. And common bible interpretations of this type often reinforced traditional gender roles for both husbands and wives about appropriate attitudes and actions at home and in society at large.

Role blurring also occurred from a developmental perspective for George.[8] The comment below by his middle daughter, Janice, summarizes his early years as husband and father: "Poppa was a rolling stone [said jokingly]. My Daddy always made sure we had a way to go to church, but he really didn't start going to church until basically after we left home." She candidly acknowledges the reality that George was not always the Christian, deacon, and staunch church supporter that he became later in life. For him, the effects of religious education and Irma's influence were delayed. His daughter's joking reference to the song by The Temptations also suggests that George was not consistent in his commitment to family life as a young father and husband.[9] However, over time, he appears to have become, like his wife, a model that his children, grandchildren, and many community members emulated:

> He would go to church every now and then, for revival or something like that. But in terms of being a regular church goer, he didn't do that . . . One Sunday at the beginning of the year, I don't know what year it was, but it was the first Sunday and Janice [their middle daughter] came in and asked, "Daddy, you going to church today?" And he said, "I don't know, I might go." Then she said, "It's the first Sunday of the year and I think everybody ought to go to church." And so we went on to church. And a few minutes later, here comes my Daddy. So I say that was an eye opener as far as him becoming faithful in church because he would then go to church and then he started going to Sunday school. He was the Superintendent of the Sunday school and a deacon. And he was faithful after that . . . and then he was an inspiration to the younger kids that were coming along. He would try to encourage them and help them. (Martha, 77 years old)

The above remark is important in its reference to the role of both the local black church, Clear Creek No. 1, and the first Sunday in the Black Church tradition. As a meso-level organization, the local church represented the institutionalized symbol of religious education for most blacks in Gallman. In addition, for many rural black churches, first Sunday was particularly important because it meant they would be graced by a sermon from a circuit pastor, rather than just Sunday school facilitated by deacons; they would often also participate in Communion. The first Sunday *of the year* was imbued with additional significance in its representation of possibility and promise. Thus George's daughter's urging and multiple "firsts" seemed to make it difficult for him to ignore. Equally important, it initiated his religious development during which he took on an increasingly more influential role as a religious leader at church and in his own family.

Just as salient, the multiple, varied roles this family patriarch and matriarch held on their homestead and in the community were reminders of a linked fate and the biblical promises found in passages such as Psalms 6:10, 22:5, 25.3, and 31:1.[10] As described by their granddaughter, Donna, "I loved that my Big Papa and Big Mama were important members of their church. It was important to them that we did things that did not bring shame on the family." A practical expectation of religious living implied that blacks were promised that God would vindicate them against enemies as well as protect them from dishonor and disgrace. Such a promise would be particularly important for believers with a history of marginalization such as southern blacks. Yet adherents were also charged to reciprocate by avoiding behavior that would bring shame and stigma on the community, church, and their family.

Blurring Granddaughters' Gender Expectations

Although a staunch Christian by all accounts, Irma was also considered prudent about the realities of the world, which meant realizing and understanding the humanness in herself, family members, and other people, in general. The following accounts by her granddaughters illustrate how she helped family members overcome personal challenges:

> I had my first child while I was a senior in high school. I remember Big Mama telling me it wasn't all bad and to not let having a baby stop me from making something out of my life. Both my parents were disappointed with me and I felt so embarrassed and lost because I really didn't know what I had embarked upon. I knew early on that for me to grow, I needed to leave [city's name]. There was always a drive within me to want more and I knew I was destined for more. I wanted more for my child. So, one day, when my son was three and I was done with the dead-end jobs, I packed my bags, and moved to Minneapolis and enrolled in college – and the rest is history. (Constance, 47 years old, social worker)

Despite her solidly upper-middle-class parents' concerns and a limited network, with her grandmother's advice and motivation, Constance embarked upon a period of self-discovery during a late twentieth-century migration:

> The move wasn't supported by my parents. I've made some mistakes, took a few detours, but never looked back or regretted my decision to move. That once three-year-old is now married with two children and works for [large automobile company]. I realize now that it has been 25 years since I made that change and looking back, I can say it was only God's grace that protected me because moving here was my first time away from home . . . but the Holy Spirit took care of me – even when I wasn't aware.

According to Constance, her grandmother encouraged her to ignore Christian and societal stigma associated with single parenthood as well as her parents' disappointment to pursue her life dreams by relocating. Moreover, this instance of blurring traditional religious education (specifically verses such as Genesis 2:24, Exodus 22:16, Hebrews 13:4, Matthew 15:19, Mark 7:21, and 1 Corinthians 6:18 that encourage sex after marriage and discourage fornication) from Irma, provided Constance with the requisite motivation. This advice was particularly noteworthy because it deviated from what is usually considered clear biblical standards around appropriate sexual behavior and stigma for sexually active, unmarried *women* (refer to John 4:1–42 about the Samaritan woman).

We posit that Irma's advice did not suggest that she rejected these biblical edicts that encourage childbirth post-marriage, but rather that common sense dictated that such ideological sanctions or shaming would not benefit her granddaughter pragmatically or long-term. Logical advice was needed and crossing this biblical boundary was warranted. And although it took longer than she anticipated, Constance would go on to earn a Master's degree in Health and Human Service Administration and a graduate certificate in marriage/family therapy. Equally important, Constance posits that Irma's unconditional support was evidence of the agape love referenced in passages such as Deuteronomy 6:5, Mark 12: 28–34, Matthew 22:34–46, and Luke 10:25–28. Constance's references to God's unmerited favor, the Holy Spirit as a comforter, and her own faith illustrate the ability of religious education to influence long-term beliefs and outcomes.

Narratives here show that seemingly bounded roles such as wife, domestic, or provider did not prevent some family members from both thinking and making decisions outside the proverbial box. Although religious in their impetus, and family focused, such decisions had a decidedly practical dimension based on the realities of black life. Irma seemed particularly comfortable providing marital advice to her offspring. According to the following account by another granddaughter, her love for George did not mean Irma believed the institution should be embraced by all women:

> I remember some advice Big Mama gave me when I was about 25 years old. I had gone down to visit and we were sitting on her front porch. It was one of the few times I was able to talk to her alone. We were discussing relationships and she spoke fondly about her two marriages. But then she said, "If you're gonna get married, make sure you marry someone who wants as much out of life as you do. If not, he will drag you down. *And you won't have nothing*" [emphasis was hers]. She was not trying to discourage me from getting married, but rather remind me to be smart if I choose a husband. And I took that message to heart and think about it all the time. Although I have been asked to get married, Big Mama's standard has prevented me from making several huge mistakes that would have undermined my life today. They were really nice men, but not right for me. It was hard to say no at the time, but I know I made the right decisions . . . Big Mama saved my life. (Tanya, 53 years old, school administrator)

Irma cautioned Tanya against following societal dictates that often tacitly encourage women to wed, particularly black women – or feel unfulfilled and embarrassed if they don't. The candid conversation about the benefits and possible problems associated with marriage included a reference to 2

Corinthians 6:14 about being incompatible in marriage (i.e., being unequally yoked). Although the bible verse is often referenced when discussing unions between Christians and non-Christians, Irma gave practical advice to only consider partners whose work ethic, ambitions, life goals, temperament *as well as* Christian lifestyles were complementary to Tanya's. In doing so, Tanya suggests that her grandmother was encouraging her offspring, especially females, to strive to be both economically self-sufficient and pursue lofty personal and professional goals. Moreover, Irma's advice held a broader veiled warning that, despite one's accomplishments and financial security, one "poor" decision, such as marrying unsuitably, could result in one falling down into the ranks of the poor.

Certain roles meant being circumspect based on existing expectations around race, gender, class, and place. Blacks in the South who failed to understand this reality could pay with their lives. For the Kings, roles in the family and local black church included certain expectations, but had intrinsic and extrinsic benefits. Moreover, certain roles, especially in predominately black spaces, could afford some latitude and be nuanced to meet individual and group needs. Yet family members, like Irma, had cultivated a reputation for quietly subverting certain traditional roles and expectations.

Blurred and Bounded Rituals

Daily activities of the Kings on their farm, as well as interactions with whites, were informed by spiritual and secular rituals that shaped expectations and behavior. These habits and practices were informed by the King family ethos about what was appropriate across varied ecologies. And over time, such events became tradition.

Rituals and Relationships

Whether intra- or extra-familial in nature, family-inspired rituals fortified individuals in the King family to better navigate societal traditions and norms linked to racism and segregation. The following scenarios illustrate some of the approaches used to negotiate physical and non-physical terrain. Many of these instances focus on George, the family patriarch, given his more direct influence on key family rituals. Applying a Weberian concept to the King family setting, they seemed to routinize the charisma from both their local church and family into adaptive rituals and norms.[11] When considering intra-familial rituals, the memory below evokes customs associated with patriarchy and everyday internal processes:

Each evening when George (my grandfather) came home from work, the ritual was Irma (my grandmother) prepared his meal and had everything in order in the front room where the TV was located. I helped her get things together. The other children had to go outside (unless it was raining and then we had to be quiet – no talking). George would arrive and speak to us. He seemed very happy to see us. He would wash up for dinner and ate while watching the evening news in silence. After dinner, he would call us in around his chair and quiz us with current news or riddles. He seemed very pleased if we got the answer and even more so that we kept trying. He then went outside with the cattle or whatever.

According to the above comment by Lorna, their 56-year-old eldest granddaughter and a CPA, certain rituals structured the King family's evening activities. Seemingly innocuous practices such as when and how George's meals were prepared, the expected ambiance, as well as how other family members, especially children, responded to and in his presence, illustrated a hierarchical, traditional family structure. Although the memory was a positive one for his granddaughter, it also suggested an expectation that part of George's role as protector and provider meant completing certain rituals inside the home before taking care of responsibilities outside on the farm. Moreover, the legacy of patriarchy in this family meant socializing children about their responsibilities when helping Irma prepare for George's arrival, including the post-dinner tradition focused on his grandchildren that contrasted with George's otherwise stoic demeanor. During these latter instances, he appeared comfortable blurring generational boundaries by ignoring the anecdote that children should be seen and not heard.

Scholars who studied black families, such as E. Franklin Frazier and Andrew Billingsley, associated normative lifestyles and economic stability with both the King family structure and middle-class outcomes. According to Darity and Myers (1994): "Frazier linked their comparative wealth to their capacity to form and maintain a patriarchy. The development of family life on an institutional bases was closely tied up with the accumulation of property in these families" (p. 141). As suggested by the above scholars, property and a form of patriarchy undergirded the middle-class status of the Kings and possibly other black families like them. Younger family members who frequented their homestead corroborated certain family rituals suggestive of patriarchy that were common at the time and that created and reinforced specific relationships, interactions, and expectations based on gender and age. Yet Frazier's assessment must be nuanced if applied to this family's decision-making because the combination of military funds and Irma's vision and initiative resulted in the land purchase that would move the Kings toward the middle class. Moreover, their children agree that *Irma's* continued sweat equity largely resulted in gradual improvements to the homestead.

After becoming a committed Christian, new rituals associated with this lifestyle meant that religious and personal hygiene went hand-in-hand for George:

> George was the Chairman of the Deacon Board, so preparation for church had a ritual. We had to memorize a bible verse based on the alphabet letter for the week. Each Saturday was spent finding and learning a verse to recite during Sunday school which we had to present to him that evening ... George also prepped for Sunday service ... he's the only man I knew to dye and hot comb his hair straight each Sunday morning as we listened to the gospel radio station and dressed for church.

In her quotes, Lorna describes six rituals for which George became well known. The first ritual reflects traditional household gender roles that meant dinner was always ready upon his arrival to be eaten in silence as he watched the news. Second, banter around current events, riddles, and learning scripture had practical and spiritual import in terms of religious socialization, honing reading skills, and improved knowledge about society. The third practice involved tending to the livestock and any other farm-related duties that Irma had not or could not perform. Later, bible study and preparation for church became weekend rituals because it was important to prepare one's self spiritually (i.e., fourth, learning bible verses and fifth, listening to a gospel radio station) and physically (i.e., lastly, conking one's hair) for worship service. Their traditional gender roles remained generally bounded throughout Irma and George's relationship. Yet as George aged, personal and religious roles that were initially bounded began to blur significantly, as the latter roles began to shape the former. The influence of Christianity also became more evident in his daily life in terms of both time spent at home and at church.

Irma and George placed high value on how one achieved goals. Church-based customs also meant emphasizing hard work, personal character, and delayed gratification. Ill-gotten gains of any sort were unacceptable:

> I can remember this man who lived next to us who planted corn, Mr. Bunk, and we were down there pulling the little ears of corn that weren't quite ready. We were making dolls out of that silk. He saw us so he came and told My Daddy. And he was just going to stay right there at the house until My Daddy whipped us ... I remember that because My Daddy never did whip us and that was the only time he ever whipped us ... My Daddy told us about telling a story [i.e., lying]. My mother didn't want him to whip my middle sister because she said, "Yes, we

took the silk." But me and my older sister said we didn't do it. And so my mother thought that my father should just whip me and my older sister. But he whipped all three of us . . . My Bunk did that because he was a mean old man . . . Mr. Bunk was black . . . He had land too. He probably had about ten acres. He didn't plant as much stuff as we did . . . we were probably considered a little bit more well off than Mr. Bunk because he didn't have transportation.

Informed by the Ten Commandments and Leviticus 19: 11 ("you shall not steal . . . you shall not lie to one another"), Carla describes a parenting memory that emphasized honesty, the Protestant Ethic, and intra-racial class position.[12] It was also the only time George blurred household gender roles by spanking his children; yet he did so at Irma's insistence. Carla's remark also alludes to how blacks were further stratified in Gallman, as she describes another black landowner. She suggests a hierarchy, even among black landowners, based on acreage and material possessions such as automobiles. The latter types of accouterments further delineated socio-economic status. Carla also alludes to possible intra-racial enmity that may have existed for some blacks as they witnessed their black peers accumulate possessions they did not have. In rural communities, it is not surprising that families that owned land tended to be relatively better off than their peers who did not. Moreover, other factors that delineated class position would also be expected to be influential in how individuals viewed themselves and each other.

Rituals, Religion, and Racism

Christianity was also part of the social fabric of white society in Gallman, Mississippi. However, the racial hierarchy and the beliefs that perpetuated it tended to overshadow religious dictates. The tendency for *whiteness* to trump religiosity for many whites has been documented. Per Emerson (2008), this pattern is indicative of the larger societal stance and even exists in congregational settings: "whites of [church name], like other whites, were firmly planted in a racialized world in which whites expected to be in control and saw themselves as the standard by which others must measure themselves . . . much was needed to overcome the dominant white perspective" (pp. 33, 153). This same author provides a broader explanation: "when people of different cultures come together, the consequence must always be some conflict, for it is more than people coming into contact. It really means that truths come into contact and that means conflict" (Emerson 2008: 147). For the Kings, segregation and Jim Crowism emboldened local whites and made black rituals all the more necessary.

Yet property ownership for some blacks was a precursor to certain intangible assets. It appears that successfully accomplishing such a task (i.e., planning, having the requisite funds, making the land purchase, and sustaining the property) garnered a certain amount of respect from fellow landowners and renters alike. The following comment further explains the King's emphasis on godly living despite systemic reasons to do otherwise – and possible benefits:

I don't know why segregation existed, but in one way, we were blessed because we were on our own land and the other folks [local whites] really didn't bother us, but I saw how other blacks were treated and [relative to them] we were treated very well . . . Land made a difference . . . I also think God blessed our family because our mother was really a praying woman. She brought us up to trust in the Lord and to always tell the truth. She always said, "Tell the truth" and then we would not get a whipping . . . She used to say all the time, "If you lie, you will steal." (Janice, 75 years old)

Janice refers to verses Psalm 41:2, Psalm 59:1, Psalm 64:1, and Job 1:10 to explain God's favor in her family's ability to purchase land as well as the associated protection from the more drastic effects of segregation that land afforded. Such perceived *respectability* among blacks has also been associated with ascriptive traits such as skin color, achievement, and, in this case, property ownership. Additionally, she posits that religious education taught by Irma warned against lying and stealing and encouraged a consistent prayer life as recommended in 1 Samuel 12:23 and 1 Thessalonians 5:17. Thus just as literature shows the importance of prayer in the Black Church, Irma had modeled this ritual for her children.[13] In addition to enabling Irma to communicate with God, they posit that prayer provided validation that enabled her to withstand the difficulties associated with segregation, gender constraints, and the myriad responsibilities as the family anchor.

Janice also provides a practical example of blurred boundaries. Irma's belief that a liar could easily become a thief suggests a possible gateway between seemingly less innocuous, unchecked faults and more serious practices that could follow. The subtlety by which she understood certain scripture and attempted to socialize her children accordingly showed a quiet determinism to forge a different life for them than she had as a child. Beyond Christian dictates, such sanctions had practical import because blacks caught stealing from whites (or even believed to be doing so) faced certain punishment, possibly death.[14] Outside their homes, blacks could not escape racist rituals:

We just knew that there was a difference. At the stores, the black people had to go to the back window to order your food. You couldn't go inside to sit down and eat it. We just knew that that's how it was . . . segregation . . . I never went to school with white people until I got to college. We could go in and cook and clean the white people's houses, but you had to go through the back door. You couldn't sit and eat with them. They ate in the dining room, but you could sit in the kitchen and eat your food. (Martha, 77 year old)

The irony of segregation meant that, despite their dependence on blacks in a myriad of ways, whites manipulated physical location (i.e., opposites such as "front versus back" and "inside versus outside") to codify race-based social location. Seventy-four-year-old Carla, continues this sobering description:

In terms of poor whites interacting with blacks . . . you didn't really interact. You didn't have a reason to interact. You kind of stayed where you were and they stayed where they were. Now if you went in the store and they were in there, you knew that you were going to let Mr. Roy [Williams] wait on them first and then you get waited on. But then you go back to the section of town where you lived and they go back to where they lived. And whites with money treated poor whites bad – just like they treated blacks . . . I believe they were ashamed of them.

Carla describes both racial segregation and intra-racial segregation among whites based on class where ne'er-do-well whites were considered embarrassments by their wealthier peers. Historical accounts also illustrate that poor whites represented a more immediate threat to blacks due to their volatility in response to economic troubles, hatred and jealousy of blacks, perceived white entitlement, and social estrangement from wealthier whites.[15] Other supportive scholarly sources exist:

These big dudes of the white race, they've never showed no care and respect for the poor white man. ... they all white, why don't they hang together in every respect? ... Color don't boot with the big white cats; they only lookin for money ... the poor white man and the poor black man is sittin in the same saddle ... the controlling power, is in the hands of the rich man. That class is standin together and the poor white man is out there on the colored list ... but I don't know how to take poor white people . . . they seemed to have always thought they was a class above the Negro ... There's been many a white man as well as Negro that's been undertrodden. (Rosengarten, 1974: 489)

Like this depiction, the Kings were aware that, despite the economically and socially precarious statuses of poor and working-class whites, they often relied on white privilege to marginalize blacks. Yet, despite class divisions, blacks realized that whites would generally unite against blacks. Rituals in Irma's daily life were affected as well:

She [Irma] let me go to work with her to help clean for Mrs. Dina so she could finish sooner . . . She would wash dishes and put them away, change the bed linen, wash and put them away, dust the living room furniture and floor, clean the bathroom, and mop the kitchen floor, and have me sweep the sidewalks outside. I would tell her, "Big Mama, there's nothing to sweep or dust," but she would say, "Baby, do it anyway and get in the corners." Often, whenever I swept the sidewalk, individuals [white men], most in their cars, would yell words towards me. I felt hurt and stared at them, but I thought, "One day they'll know that's not nice. Maybe they don't go to church!" (Lorna, 56 years old)

In the above comment, Lorna details her grandmother's daily domestic rituals as well as ritualistic racist and sexist behavior. Even as a pre-teen, she was exposed to segregated work and the capricious behavior of certain white males in predominately white spaces. However, following her family's ethos, Lorna attributed these mean-spirited interactions with a lack of religious education. According to this same granddaughter, even *well-heeled* whites expected certain conventions by blacks:

Going to work with Irma allowed me to see racism at an early age. For example, all of the whites called her "Irma" (even the children), but she addressed them all with a title (Miss), including the girls. This burned my heart to no end! How dare you whites disrespect my Grandma! I didn't say anything to Big Mama, but I did think about the situation often. I came to see that they all loved her and she loved them, but that greeting was the "way of the day." Mrs. Dina was good to Big Mama and often gave her nice items as gifts. They were nice to me too, but I noticed after we spoke, they always had somewhere to go with the girls.

As described in Collins' *Black Feminist Thought* (1990), even older blacks were expected to contend with micro-aggressions such as being referred to using diminutive terms (for example, by adding an "a" to the end of one's first name) and expected to welcome hand-me-downs from whites. As Lorna noticed, although friendly, their children could not be her playmates. It is unclear whether Lorna was considered a "baby maid" (Collins 1990: 71).[16] It is also unclear whether genuine relationships were possible across segregated chasms shaped by paternalism, stereotypes, and entitlement. Recollections of these inter-racial ties were tinged with recognition that, no matter how benevolent they were to Irma, the two families could never really be considered *friends* given the racial power dynamics that existed at the time.[17] However, rituals are only sacrosanct when individuals continue to adhere to them. And things were beginning to change societally and developmentally for King family offspring:[18]

When I was older, I told myself one particular morning that I wouldn't say, "Ha do, Ms. Dina?" as Irma told me to say each and every day. When it was time to speak, I said, "How are you, Mrs. Dina?" not knowing what would happen. But I didn't care. Mrs. Dina smiled and said, "Lorna, you are growing up – a smart young lady."

I thought Irma would "get me" when we got in the car to leave, but she never said a word . . . ever, and I continued to greet Mrs. Dina with that phrase.

Increasing numbers of southern blacks were pushing back against poor treatment by whites. For Lorna, life in the 1960s "up North," differing generational expectations, and growing civil rights activism nationwide meant challenging the status quo – both for this white employer and black employee.[19] And in doing so Lorna was agentic during a seemingly innocuous, yet personally empowering conversational ritual.[20] Although a teenager, it was unlikely that Lorna was a threat to Dina given that segregation still prevailed. Yet the change in the nature and structure of their short introduction (including use of correct English rather than the expected vernacular) suggests a conversation between *equals* and likely reflected a milestone for Irma as she witnessed an example of the changing socio-political terrain. Moreover, the exchange illustrated some of the ways ideological boundaries were being challenged by a generation of blacks that would inevitably dismantle Jim Crow.

Family members also posit that rituals around Black Church life helped counter some of the negative effects of southern living then and today as some of the negative impact of racist rituals could be supplanted by positive religious ones. For example, 75-year-old Janice describes important programs: "Sunday school, BTU [Bible Training Union], and choir rehearsal. Religion is very important in my life because I know you can't make it without the Lord." These were multi-purposed places to train black children biblically and socially as suggested in Proverbs 22:6, as well as to teach them strategies around double-consciousness as suggested by DuBois.[21] Martha, concurs:

There was racism everywhere . . . that's just the way it was. I never felt less than they were . . . When I got grown, I thought it was really stupid that they would let us come in and cook for them when we could have done anything to their food, but they wouldn't let us eat with them . . . Religion was very important because My Mul made sure that we went to Sunday school and church . . . we would go back in the afternoon for our girls' group called Red Circle. We had revivals when I was growing up and that's when we would get a chance to see your friends during the week because everybody didn't have cars . . . Religion is still very important in my life today. I still go to church on Sundays. I go to Bible Study and if they started having Sunday school again, I would go.

As suggested in scholarship, rather than diminish their identities, southern blacks like Martha often reflected on the irony of segregation and rejected its potential negative effects on their self-identities.[22] After discussing problems blacks faced outside their community, she details Black Church-related benefits in which she is still involved.

A common theme here is the importance of religious rituals associated with the local black church, Clear Creek Church No. 1. Religious life helped the Kings better navigate the senseless world in which they lived as well as provide intra-racial communal and social benefits. Rituals are important to reinforce beliefs and behavior considered central to maintaining structure and expectations. Just as Jim Crowism was instituted to maintain the status quo and its related benefits for whites, blacks like the King family proactively developed their own rituals, events, programs, and counter-narratives. As suggested by Collins (1990), the self-definitions and subjugated knowledges developed by blacks reflected wisdom about themselves and whites crucial to resistance. Moreover, spaces like black churches, black schools, and for the Kings, their farm, provided shelter, safety, security, pride, and well-being for themselves and their families.

Blurred and Bounded Responsibilities

George and Irma's shared tasks as family providers often manifested as blurred, complementary responsibilities. Their children suggest that they had an understanding of their respective tasks inside and outside the homestead. Minimally, these shared expectations were also functional to simplify their lives. Yet the reality of harsh living in the rural South required a certain amount of relationship fluidity. Despite gender roles that bifurcated certain behavior, according to her children, Irma's role as the family anchor meant additional domestic responsibilities. And her commitment to Christ meant additional duties in the community. A consistent theme has been George and Irma's attempts to encourage formal and religious education for their children. Yet their approach to accomplishing this goal varied:

> Both my parents could read. Education was important to them. My Daddy didn't do too much talking about it. He left everything like that up to my mother. And she wanted her children to be better off than she was. And so she just made sure that you went to school every day and she didn't keep you out of school for nothing. (Martha, 77 years old)

However, prioritizing education in this way was not the case for all family members. According to in-law Deidra, when asked whether education was a family priority, "not really, because during sharecropping time, my daddy would take us out of school to work in the field, so I guess making a living was his priority . . . survival for the family." Although Deidra's father valued education, the exacting nature of farm life meant making difficult decisions to maximize time and human resources on his farm. Their disparate priorities and complex decisions of these two families reflect diversity among black farm owners in Gallman linked to labor and capital.

At the time, completing high school was considered an important milestone for potential economic security in Gallman. Although both parents encouraged this goal, Irma was responsible for ensuring its success. For George, women were responsible for caregiving functions. When they disagreed about their respective roles, particularly if he neglected certain farm duties, problems were addressed non-confrontationally:

> I never heard her and My Daddy – they were married so many years – fuss or argue – although My Daddy wasn't going to argue with her. He would just walk out of the house. She definitely wasn't a violent person. And she didn't drink. She didn't smoke. She said she saw enough of that when she was growing up. I guess the people who would come around to buy whiskey were kind of wild and rowdy. So she said she didn't want to be like that either. (Carla, 74 years old)

Irma's upbringing in an extremely poor sharecropping family had an indelible effect on the type of home she endeavored to create for her own family. This lifestyle meant honoring her husband as the head of household; ensuring, to the best of her ability, that her children were fed, safe, and not exposed to what she considered unsavory aspects of secular life; ensuring that her children complete high school; and maintaining a house without conflict. According to their children and other family members, George understood and respected these rules and endeavored to support them.

Moreover, Irma's disdain for violence resulted in a reputation her eldest daughter, Martha, associates with Proverbs 22:1, "A good name is to be chosen rather than great riches, and favor is better than silver or gold." Irma's calm demeanor and affable spirit resulted in a reputation as a peacekeeper. According to Martha, her mother's personality also harkened to passages such as Romans 12:18 and 2 Corinthians 13:11 that charges believers to live peaceably with each other and to establish and maintain peace. This description of Irma is consistent across the various comments provided by family members and friends. From a religious perspective, it appears that this demeanor enabled Irma to live out Christian tenets that were personally valuable. From a more practical perspective, it enabled her to navigate predominately white spaces without appearing uppity or threatening.

Distinct gender roles and responsibilities were also common in the historic Black Baptist Church. Pastors, ministers, and deacons were male; women held supportive roles as the wives of these leaders or as deaconesses and church mothers. Female evangelists were generally welcome if they did not overstep their boundaries by asserting themselves as ordained ministers or pastors.[23] As a member of the Mother's Board, Irma was considered a church mainstay; parishioners would seek her out for counsel. For example, her 56-year-old granddaughter, Lorna, recalls Irma's community involvement: "I got to go with her to do missionary visits on behalf of the Clear Creek No. 1 Church."

Family narratives detail Irma's other church-related interests as a deaconess, church mother of Clear Creek No. 1, and staunch member of a local black women's lodge, the Heroine of Jericho Court #442 J.C. Gilliam. Collins (1990), Dodson and Gilkes (1987), and Gilkes (2001) posit that such community work, particularly organizations such as lodges, is actually an extension of the black women's activist tradition. Carla, describes additional religious responsibilities for Irma as an "Other Mother":

> She was truly a Christian woman and anybody that she could help, like some of the women, they didn't have cars . . . and she would take them to the store to get food, flour, and meat that you couldn't raise. She was just an all-around wonderful woman. She was a mother of the church and helped with Communion.

Per Carla, the above responsibilities (i.e., serving Communion and helping less fortunate neighbors) reflect a servanthood model inculcated by Jesus Christ throughout the four Gospels; required of diligent believers; and that promises benefits such as steadfast love, compassion, comfort, friendship and vindication.[24] Such godly support would be particularly salient for historically oppressed people. More importantly, and in dramatic contrast to images and practices in white-dominated society, servanthood meant eschewing worldly power and control for acts that empower the disempowered.[25]

The Kings also recognized George's position in the family and community. Particularly after becoming a faithful participate at Clear Creek Church No. 1 and as a mason, George became a church pillar – known as much for his Sunday school involvement as for being a church musician and quartet singer:[26]

> Sometimes, he would come home late and play his guitar on the porch . . . very loudly! We knew that we could count on him being home Sunday morning getting

ready to go to church or later that afternoon going to see a singing group at a visiting church.

Black Christians had a profound effect on Donna: "I respect the saints of old for many reasons. Many of them were uneducated, but served God from the depths of their heart and knew the Word of God." For her, a lack of extensive formal education did not mean the members of Clear Creek No. 1 were not religiously educated and biblically well-versed. Moreover, the communal nature of its membership blurred the roles and responsibilities between biological and fictive kin.[27] Peers also provide insight into the lifestyle and influence of the Kings:

> I been knowing them a long time. I remember them when I was a little girl. We were raised up together. We went to school together . . . I knew Momma Janice [Janice King], Aunt Gina Mae, Fred [Gina's son] . . . I would go over there a lot and we [Irma and her] would talk late at night. And she was comfortable with me and I was comfortable with her. And she be done cooked up some turkey necks and we would eat them. I remember George, and if I had stuff for him to do [repairs], I would always call him about it . . . . And if I go to church and I would sit in the back, she [Irma] would always say, "Come on up to the front." She wouldn't allow me to sit too far in the back. She was sitting on the Mother's Board. I always respected her. (Everlyn Mae, 81 years old)

As their contemporary, Everlyn recalls Irma's cooking skills and George's ability to repair equipment as well as Irma as a role model in the local Baptist church. In addition to making her feel comfortable at church, Irma helped Everlyn's family financially.

Irma also had caretaking responsibilities for a white family – which meant she and her children could be "hired out" to that extended family:

> She was always doing domestic work for this one family, but then if his brother needed somebody, then she would help him out, and then if his mother needed something done, she would help her out . . . basically for Mr. Roy, and anybody in his family that needed help . . . they paid her . . . she cooked for them. She cleaned and they had two little girls and so she would baby sit because she was up there all day . . . wash and iron . . . his brother had about five kids . . . She might not work for the brother every day, 'cause like most of the time if they [Roy's brother] needed someone to babysit, she would send one of us . . . to babysit. They would pay us about two dollars for about four to five hours. (Carla, 74 years old)

Irma's employment boundaries were blurred considerably such that employment for one white family resulted in additional low-paid work for their family members.[28] Although they were paid ($2.00 to babysit in 1950 equates to about $20.99 in 2017), white privilege meant Irma and her children's personal time could be usurped:

The family she used to work [Mr. Roy] for wanted her to shell pecans and it was a bushel of them. And I was like "Big Mama that is not right." They just dropped them off and wanted her to crack every pecan versus just buying some already cracked. And she said that she had been working for them a long time and that she's home and not doing anything and so she could just sit on the porch and do it . . . She said that sometimes it's better to keep the peace than trying to be right. (Dora, 54 years old)

According to her granddaughter, Irma may have felt put upon by the above types of inconveniences, but prudence dictated she choose her battles carefully. Rather than fear of whites, Dora believes that her grandmother ignored her entreaty to refuse such work because Irma was trying to live out the Fruit of the Spirit as described in Galatians 5: 22–23.[29] Yet Roy's relatively positive interactions with local blacks as compared to many of his white peers may have made Irma feel beholden to meet such entitled requests:

Mr. Roy, from what I remember, treated black folks well. He had a store and if you didn't have the money, you could go out there and get stuff that you needed, kind of like a charge card, but you didn't have a charge card. But he kept a list of how much you owed him . . . like your flour, cornmeal, and baking powder, 'cause you didn't raise that on the farm . . . Whites didn't seem to mind the way he treated blacks . . . Mr. Roy was just well liked and he was like a big man in the community. I guess because he had something [wealth]. So nobody ever said anything [about his treatment of blacks] that I knew of. And when he passed, his wife opened up the church so we could go to the funeral, but you had to sit in the back . . . but she let you go to the funeral, which was unheard of back then.

Carla's comment above provides a reminder that their farm enabled them to minimize store purchases, debt, or the need for credit which, following Proverbs 22:7, George strove to avoid.[30] Segregation and Jim Crowism also meant that whites in Gallman such as Roy could *decide* whether and how they interacted with blacks. His upper-class status enabled him to establish inter-racial ties with certain blacks such as Irma and George. Yet by the next account, a racial hierarchy was even maintained in death:

If a nigger walked into a white church, he'd just be driven out, if they didn't kill him. But if a Negro was a servant . . . if they was maids for the white people . . . then they'd . . . accept him to come in and take a seat on the back seat. (Rosengarten 1974: 298)

Although Irma's relationship with Roy and his family blurred certain racial boundaries, it appears that she felt both genuine care, and a certain degree of ambivalence, for them. Moreover, although somewhat altruistic in nature, the storeowner's own interactions with Irma and her family did not absolve him or his wife from systemic inequities, segregation, and racism with which they and other whites were complicit and benefited.

Family members suggest that Irma's enmeshed responsibilities as domestic and *friend* of the Williams's resulted in certain protections:

> I think that it [lack of physical abuse and overt racism] was largely because of Mr. Roy, because you did have some rumors [about lynchings, rapes, and interracial sex], but nobody ever bothered us or said anything to us. Now they say his brother was a drinker, but he never said nothing to us and we were up there babysitting his kids and then ... he would be the one to bring us home. But he never stopped the vehicle to do nothing to us ... you would hear about rumors about white men going with black women. Some of them weren't rumors because I remember this one lady [name] who had about three *white* kids [i.e., mixed-raced]. She wasn't married, but I know she had those white kids. I don't know how she got money. (Carla, 74 years old)

According to Carla, prevailing social norms were used to justify varied forms of abuse of blacks, including one of the more common ways in which inter-racial boundaries were blurred historically – access to black female bodies. The common belief among their children is that Irma's employer's patronage as well as her and George's reputations as upstanding "good Christian blacks" wedded to a Protestant Ethic, family, church, and community, created a unique "place" for them as middle-class blacks in Gallman, Mississippi. Moreover, the above narratives are attempts by family members to explain and, in some instances, reconcile a period in US history for which most explanations do no justice. Despite this conundrum, reliance on God, family ties, self-help, and strategic decisions, seemed to facilitate their ability to navigate southern society.

THE BLACK CHURCH AND RELIGIOUS EDUCATION

Christianity, as a belief system, and the Black Church, and its physical counterpart in the black community, girded the lives and decisions of the earliest generations of the Kings and continue to play enduring roles for most family members. Religious education, as defined in the Introduction, was grounded in the bible. Equally important, the local church, as both an organization and an organism, provided sanctuary for socialization, training, and initiatives needed to educate, equip, and empower blacks.[31] It also afforded a safeguard against the ravages of racism and segregation. Yet scholarship shows that, despite efforts by groups such as Quakers and primitive Methodists, as well as short-lived support among Protestant groups during the Great Awakenings, historically, Christianity has been appropriated by some whites to justify oppression of blacks and to placate them by emphasizing other-worldliness, docility, and paternalism.[32] A similar unfavorable observation of this faith tradition was noted in *All God's Dangers* (Rosengarten 1974), particularly in the segregated South: "the nigger wasn't allowed to have nothing but church services and, O, they like to see you goin' to church" (p. 298). But how did Christianity influence the Kings and their peers? Rather than duplicate the large literature on the influence of the Black Church, we focus on its specific features and effects germane to the Kings.[33]

A strong scholarly argument has been made that, for many black adherents, Christianity could provide social, cultural, economic, and political beneficence beyond its religious impetus.[34] This faith tradition has been linked in research to godly validation and uplift, positive racial identity, socio-political action to combat inequality in its varied forms, resilience, self-help, community service, and skill development. And for the Kings, black churches helped pave the way for the requisite religious and secular education, behavioral tenets, and familial reputation that would lead to middle-class living. Its educational foci has also been documented.[35] Billingsley (1992) provides a strong summary about the holistic nature of religious education: "the Black Church . . . encouraged education, business development, and democratic fellowship beyond its members . . . it represented freedom, independence . . . as well as the opportunity for self-esteem, self-development, leadership, and relaxation" (p. 354).[36] Thus whether one considers DuBois' (1903[2003]) touchstone analysis;[37] Lincoln and Mamiya's (1990) dialectical polarities of *resistance versus accommodation* and conceptualization of the Black sacred cosmos;[38] empirical studies on Black Church cultural tools;[39] youth focused initiatives;[40]community usage of Black Church symbolism;[41] or the intrinsic role of religion in everyday life,[42] these congregations both routinized and activated the lives of many blacks in the North and South. Participant observations and interviews are used here to assess the multi-faceted influence of one church, Clear Creek No. 1 Missionary Baptist Church, on the beliefs and behavior of the King family.

Ecology in Real Time: Religious Life at Clear Creek No. 1 Missionary Baptist Church

The presence of the tiny white structure atop the forked corner grounds the neighborhood.[43] The cemetery where church and community members have been buried for well over a century is visible across the semi-paved road. The same simple signage has been posted there just as long: Clear Creek No. 1 M.B. Church (also refer to Figures 4.1–4.4). Only the three weekly religious services and the pastor's name are listed on the modest frame. Some cars are parked in a row along the small parking area; others sit along the nearby gravel road. Entering the church, one is reminded of the importance of necessity and simplicity. The sparsely decorated sanctuary is immaculate. A deep crimson is prominent and colors the oak-framed pews, carpet, pulpit seats, and draperies that flow along the rear wall. Its dual symbolism for both the sins of humanity and the atoning, redemptive blood of Jesus Christ makes this shade common in many black churches. The sturdy, efficient furniture has stood the test of time. Careful attention has been taken to this place of worship, praise, and education.

Figure 4.1: Clear Creek Missionary Baptist Church No. 1 (Exterior)

Figure 4.2: Clear Creek Missionary Baptist Church No. 1 (Exterior)

Figure 4.3: Clear Creek Missionary Baptist Church No. 1 (Interior)

Figure 4.4: Clear Creek Missionary Baptist Church No. 1 (Interior)

Sunday school is followed by 11 o'clock worship service. It is apparent that individuals who attend the former event will also be present during the latter. Although attendance is sparse, they represent a dedicated cadre of members who are fiercely loyal to Clear Creek No. 1's religious and economic ministries. On this particular Sunday, a total of thirteen persons attend Sunday school. They are mostly older adults, including the pastor, the Sunday school superintendent, several deacons, two women, and a few youth. The Sunday school lesson is based on the theme, "You Can Make It." Although Sunday school has historically been a time for teaching youth, this demographic is noticeably absent. Only two preteen girls and two middle school boys are present. This light attendance parallels research on the *falling away* of increasing numbers of blacks from mainline denominations and the increased competition churches like Clear Creek No. 1 now face from the seven other local black churches and possibly white congregations in Gallman, Hazlehurst, and Copiah county in general.[44] Additional congregants trickle in for worship service. Attendance includes a five-person choir, three deacons, two mothers clad in all white and wearing large white hats, two ushers, and seventeen additional parishioners. Given the ritualistic space, each group sits in a designated area. The pianist and drummer, as well as an elderly song leader, ensure lively music. Gospel songs, spirituals, and hymns are rendered from the American Baptist Hymnal located in the pew stands. Use of this common hymnal guarantees that everyone present can participate in each aspect of the worship service, including singing and reciting scripture collectively.

Other historic vestiges of the Black Baptist Church are evident. For example, although the elderly women on the Mothers' Board provide important guidance and support, only male deacons and ministers are formal church leaders. Several prayers are made followed by a communal scripture reading #581. The sermon, based on Jeremiah 18: 1–4, is entitled, "One Who Made a Difference." The pastor preaches using the traditional charismatic style for which many black Baptist preachers are known. Based on the congregation's responses, he does not disappoint. The combination of straightforward exegesis, practical examples and strategies, and lively call-and-response fosters consistent participation. Other than several out-of-town visitors, only church members are present. Yet another proselytizing ritual, opening the Doors of the Church, occurs to welcome individuals who may want to accept Jesus Christ by letter, Christian experience, or as a candidate for baptism.

Several administrative reports are read by the church secretary. This reading appears to dampen the otherwise upbeat worship; children become antsy and the celebratory atmosphere ebbs. However, she announces the necessity of this interruption because the reports had not been presented during the prior business meeting. Despite the small size of the church and membership, financial donations are impressive. A total of almost $5,000 had been raised during the months of May and June. The secretary proudly reports the church's substantial bank account [amount omitted here]; it too reflects the historic pattern of black philanthropy to churches as well as bible edicts in Malachi 3:8–10, Nehemiah 10:38, and Leviticus 27:30 that encourage tithing and financially supporting God's house.[45]

Her financial update is greeted with loud applause and several "Amens" undoubtedly by members who realize and acknowledge the sacrifices made by this largely working and lower -middle-class congregation to ensure the viability and continued existence of this mainstay religious organization. Like so many other historic black churches, the structure, pomp, and circumstance, and above noted rituals have been part of the Clear Creek No. 1 history since it was chartered. The current leaders are actually the children, grandchildren, and close relatives of past leaders. The primary difference today is the presence of a full-time, rather than itinerate, preacher.[46] A long-line of Kings attended this church. Seventy-four-year-old Carla recalls certain rituals:

When we were coming up, you didn't have all these things in church like they got today. You went to Sunday school and then once a month, you had preaching and then they had BTU (Baptist Training Union) for the children ... Sunday school was your main educational part. Preachers only came once a month. A deacon was appointed over the Sunday school.

In addition to describing common events, her comment points to the religious educational focus on black youth evident in research on the Black Church.[47]

Although several larger congregations with multiple worship services and a larger array of activities dot the growing town, why has this particular congregation withstood the test of time? What is its place in the ecological fabric of this community and for its congregants? And how, as a community educator, did it influence the Kings and their offspring? The above observational summary provides a description of this congregation. Church leaders and members, including King descendants, share the church's prescriptive influence. Church deacon since 1984, Norman Hoskins attempts to summarize Clear Creek No. 1's broad influence:

The church is the bedrock of the community ... foundation of the community. We are a God-fearing community. The church is everything. Other things may come and go, but the church is the bedrock. There are [now] quite a few black churches in the community ... we are one of quite a few.

Per the above leader, the Black Church shapes the religious fabric of the local black community. Repeated words such as *bedrock* and *foundation*, *everything*, and *God-fearing* suggest its enduring influence – regardless of ecological changes. His assessment also substantiates the collective training that occurs at Clear Creek No. 1. This same parishioner specifically addresses his unwavering support:

I've been a member since the early 80s. I'm from [city's name], Mississippi, originally. I was attending school at [name of public college in Mississippi], when I met my wife, who was a member here. I started associating with Clear Creek in about 1980–81. I was married to [wife's name] at this church in 1984. So this church is now my life. I had the pleasure of serving under [prior pastor's name] for a long time. And now I'm serving under [current pastor's name]. So Clear Creek is everything to me. I still reside in [city's name], Mississippi. I usually travel this way twice weekly. Tuesday night for bible class and every

Sunday for Sunday school and worship service. I grew up in a rather modest congregation in my home town. I attended a much larger church in [city's name], but I find that since I've been here at Clear Creek, I enjoy the small church atmosphere.

Norman has been a member of Clear Creek No. 1 for almost forty years after returning post-college. His comment illustrates the two-fold commitment to both the institution and the pastor called to lead it. Just as the church is a community bedrock, Clear Creek No. 1 is Norman's bedrock. He lives in a larger nearby city for employment and amenities, but he and his family make the 45-minute drive twice weekly to church. Lastly, the importance of formal religious education via this church is also apparent, as he references participation in mainstay training events in the Black Church tradition – mid-week bible study and Sunday school.[48]

Beyond its role as a community educator, members like Kevin L. Lloyd, baptized at Clear Creek No. 1 and a deacon there since 1967, recognize its historical ties:

My granddaddy, grandmamma all came through this church. What does it mean to me? It means salvation to me . . . I've been here long enough to know who the Lord is and what He is in my life. Why would I move somewhere else? Everywhere you go, it's Satan everywhere you go, so why would I go anywhere else? [I'd rather] Stay here at my membership church where all my people came through . . . aunts, uncles, all came through this church . . . Like you all, they move away, but the Lord blessed me to be here . . . The church has been here centuries . . . a long time . . . It's the oldest black church in the Gallman community.

Just as Irma was described as a family anchor, Clear Creek No. 1 is considered a community anchor. King family members and other local residents describe the church similarly. Kevin comes from a long line of church members and leaders; he was first introduced to God and Christianity at Clear Creek No. 1. In addition to its spiritual benefits, he suggests that the church's historical place in his family's life is equally important and precipitates his continued participation. Also germane here, he recognizes the black migration that has taken place from Gallman, Mississippi, and, like Irma, his decision to remain.

Several King family progeny detail this church's influence on their family in ways that parallel existing literature on the Black Church:

It was our home church. And you were going to go to church every Sunday. Back then, not only was the church a place to go to be religiously taught, but that's also where you saw everybody. That's when you got caught up on everything. That's where we saw our little friends, 'cause otherwise, you couldn't go over to nobody's house because you lived too far apart. So you just met up at the church because everybody was going to be there and everybody was going to have their kids there ... Sometimes church was an all-day event, especially certain times of the year. (Carla, 74 years old)

Clear Creek No. 1 has and continues to play a multi-faceted role locally.[49] As the primary place for youth to interact, it was the site of another important event for Martha: "I met him [her late husband, Donald Sr.] down South. I never went to school with him, but we were all at the same church. His people were members of Spring Ridge and we were members of Clear Creek No. 1. But we would go to Spring Ridge and they would come to Clear Creek No. 1. So that's how I met him." Thus this weekly religious meeting site could influence dating and mating outcomes. And for other blacks, churches like Clear Creek No. 1 provided safe spaces against various negative influences:

Back in them days, it [the black church] was real religious. But now, society has changed a lot now. But back in them days everything was nice ... Going to church was nice ... I liked those old timey songs and old timey preaching. Now it's that fast stuff and it's a real difference. (Everlyn Mae, 81 years old)

Everlyn critiques what she believes is a focus on cross-over music and questionable theology in contemporary black churches.[50] Similar to Everlyn's comment, Carla points to its sacredness and secularity as a communal meeting place:

Although they were on My Daddy's side of the family, you would see Everlyn, Aunt Denie, and Boy Frank at church . . . and if Uncle Dick came, he probably wouldn't come inside the church. I don't remember him being inside the church. I guess he was a Christian, but you had a lot of men that would come to church, but, say, Uncle Daniel, he drank a lot ... those people never tried to mix in with what I called the "so-called Christians." But if you were having singing and it was sounding real good, they would stick their heads in the window ... but they weren't coming in ... And if they came to the attractive meeting [revival], they would stay on the outside until it was time to eat . . . they respected the church too much.

Most of the persons included in this section were introduced to Clear Creek No. 1 as children. Some left Gallman, but returned after college. Others stayed in the small town and remain faithful to its oldest black church. To some readers, their experiences may seem romanticized. Yet the most recent Pew Research Report (2013) shows that blacks continue to remain faithful to black churches, despite decreased church involvement by whites. Based on the above sentiments, Clear Creek No. 1 serves as a broad-based community educator of both religious and practical knowledge. Individuals here describe spiritual and secular benefits associated with religious hygiene; use of the bible to hone reading skills; teaching tithing and financial management; socializing children to respect adults, each other, and themselves; teaching about self-love, racial uplift, and positive racial identity; instruction around values, morals, and church dedication; and cultivating platonic and romantic ties. Its multi-pronged effects also help explain this congregation's inspirational influence in the lives of King family members and other local blacks.

Broad Benefits of Black Church Life and Clear Creek No. 1

No matter one's class position, involvement in the Black Church meant interacting and worshipping several times weekly with blacks from varied socioeconomic backgrounds. Temporarily blurring statuses, classes, spiritual gifts, and talents helped remind congregants of their linked fate with other blacks.[51] In Gallman, this meant that "Sister Smith," whose weekly occupation was as a domestic for a white household, could hold a singular position in church based on her mezzo-soprano voice or "Brother Jones," an employee at the local chicken plant, could serve as the Chairman of the Deacon Board. And one's reputation as a faithful Christian could translate into important social capital among both blacks and whites.

Similarly, the trials and tribulations due to segregation, discrimination, and other forms of inequality could temporarily take a backseat to the preaching, singing, praying, and other forms of worship - and squelch some of the effects of negative experiences from the prior week. For these reasons, just as the King's homestead was a familial asset, Clear Creek No. 1 represented an institutional *community asset* of which members were extremely proud and beholden. We contend that, the religious ethos and experiences found in such black congregations were (and continue to be) as important and palpable in their effects as more tangible capital such as property and networks. More than coping mechanisms and examples of semi-involuntariness, individuals here *chose to participate* in church. Varied examples of agency are also apparent as spouses and young adults exit church and often return later in life. Specifically, it is suggested that blacks in Gallman attend Clear Creek No. 1 because they want to; there are intrinsic values associated with doing so that have tended to escape the scholarship of non-black researchers.[52] In addition to altruism and providing one's reasonable service as Christians, bible verses such as Luke 6:38 described some of God's economic promises.[53]

Comments made here, informed by participant observations, parallel literature on the continued salience of black congregations and institutionalized religious life for blacks.[54] Moreover, these sentiments support the continued involvement in the Black Church despite other quasi-religious and secular options and call into question the semi-involuntary thesis.[55] The organizational and organismic features of Clear Creek No. 1 were also apparent. As an organization, like other black churches, it provides structure, systemic guidance, processes, and programs that help organize the lives of believers. Yet organizations that survive often become routinized in ways that can undermine their charismatic nature.[56] However, as an organism, Clear Creek No. 1 Baptist Church appears to continue to survive *and* thrive in the life of its members today and in the memories of many of the Kings.

CONCLUSION: SOME IMPLICATIONS OF BLURRING AND BOUNDEDNESS

The concepts *blurred* and *bordered ecological boundaries* provide insight about the complex interplay between context-specific dynamics found in black schools and black churches to potentially combat familial and systemic economic challenges. We contend that these two concepts describe practices used by King family members in two ways – one that is largely ideological and another that is largely structural in nature. First, family members intentionally created and applied a belief system that blurred the role of traditional formal education and Christianity synergistically to foster family economic stability and combat inequality, broadly defined. Second, they used specific strategies to negotiate potentially volatile structural boundaries tied to racism and classism in Mississippi. Narratives illustrate how maneuvers across boundaries, intangible and tangible, were necessary for family safety and beneficence.[57] Moreover, commentary provides a strong argument that the ability for certain family members to experience their personal goals was a result of *boundary crossing* of seemingly entrenched community chasms and race-based self-interests to meet collective and personal objectives.

Scripture provided spiritual power and promise to the Kings who embraced Christianity. These same family members also often expressed a certain comfortability applying passages in practical ways, especially when attempting to navigate segregation, inequality, or personal challenges. Readers are cautioned against using a contemporary lens when assessing decisions made by blacks during that historical period. For Irma and George and other blacks like them, circumspect, cautious living could be the difference between life and death. It should also be noted that boundaries they faced could be physical, practical, or virtual. For example, the family homestead provided a physical boundary between themselves and white encroachment. Similarly, the bible and God's promises provided an emotional and psychological sanctuary against the onslaughts of white supremacy and entitlement, as passages reminded them of God's favor afforded them and other marginalized believers.

In addition, traditional gender roles were embraced ideologically, but often blurred in the face of practicality. This means that "protector/provider" roles became intertwined for most couples in the King family. Traditional gender roles between husbands and wives were subscribed ideologically *en face*, minimally, to keep the peace. However, the harsh realities of black life as well as internal familial needs meant the more adaptive families moved beyond stark traditional expectations and relied on the skills, expertise, experiences, and know-how of family members, regardless of gender. And it appears that families too wedded to traditionalism for its own sake did not fare as well, minimally because they failed to tap into the capacities of female family members or recognize the constraints many male family members were experiencing that prevented them from being protectors and providers in the strictest sense. Despite tensions and uncertainties that can result in the absence of clearly defined roles and expectations, androgyny appears to have served certain King women well and parallels studies about the benefits of more gender fluidity in black relationships based on needs, context, and objectives.[58]

An emergent pattern here suggests that being middle class is as much about one's attitude as their assets. This means that intangible dynamics are just as important as tangible ones. This chapter considered intangible dynamics, more specifically, spiritual and secular beliefs, that influenced King family practices and traditions. We were also interested in the influential effects of Christian ideology from the Baptist Black Church tradition. Moreover, how were beliefs and behavior blurred and/or bordered for a family that espouses Christianity? These concepts extend several important features of socio-ecological theory to explain King family practices. Moreover, they provide a way of understanding how and why certain family members strategically and purposefully synergized aspects of Christian and secular tenets and practices to achieve and maintain middle-class status. Narratives suggest that individuals who were able to maintain their family's adherence to religious education that emphasized consistent church involvement, formal education, prudent decisions, and delayed gratification, were best able to weather the economic challenges in urban spaces, employment discrimination, and other inequities. Still other stories describe accounts of how other family members "found their own way," as they were influenced by the new, unfettered environment in cities and endeavored to shape those spaces to meet their specific needs.

Blurred boundaries reflect the tendency to practice certain *dimensions* of one's belief system to meet specific needs. Like religious education, it tends to be inherently practical and logical. Unlike religious education, designed to engrain a certain ethos, blurring tends to nuance one's perspective based on a specific context, problem, or self-reflection. Blurring boundaries is, by definition, multi-conscious in nature and, for the Kings, was driven by efforts to maintain their family inside and outside Mississippi as well as navigate oppressive structural conditions. In certain instances, it means choosing to delay certain Christian dictates to meet needs – such as one mother's decision to postpone paying her church tithes to buy food for her children and then "double up" on her tithes the following month. She did not consider this decision a betrayal of biblical dictates, but made a logical decision as a caregiver.

In other instances blurred boundaries and boundary crossing require discernment about local economic conditions and making difficult, but prudent, decisions accordingly. For example, although an introvert who appeared to largely adhere to the prevailing gender roles of the time, Irma's decision to purchase the family homestead was unexpected because it took place outside George's purview and challenged prevailing gender norms and bible dictates about submissiveness as described in 1 Peter 3:1. Yet she did so because it would secure her family's socioeconomic stability. A similar rationale explained her later decision to focus on more long-term benefits rather than child-centered luxuries by purchasing electricity rather than bicycles. However, bordered boundaries meant Irma only went so far. The desire to please God and George, and more importantly, take care of her children, were the primary factors that shaped her behavior.

It is also evident that many examples of blurring and crossing boundaries were made by women in the King family. Part of this is due to the reality of controlling images and societal constraints linked to racism, sexism, and classism.[59] This does not mean that men were not actively involved, but rather that certain rigid societal norms and values were being challenged by women, as they made certain context-specific decisions. Several examples are in order. Janice and Howard's decision to operate a juke joint and her decision to continue to sell moonshine after he died; Constance's decision to leave an upper-middle-class family support system and relocate with her son; Katherine Mae's decision to pursue a nursing career despite her initial limited education; Margie Lou's venture with her husband, Amos, to open their own grocery/liquor store; Hana's entrepreneurial role as a rental property owner; and, Irma's decision to disappoint her children and choose lights over bikes – all evidence unexpected choices that outsiders might question – but that helped usher in middle-class lifestyles. The Kings contend that decisions were not happenstance nor were the resulting outcomes serendipitous, but rather reflected specific strategic beliefs, values, and choices on the part of pivotal family members who were models for their progeny. Intestinal fortitude was needed. It appears that these instances reflect being savvy, understanding social, religious, and cultural boundaries, yet, as needed, intentionally and purposefully bending, twisting, and sometimes breaking these same dictates to accomplish familial and personal objectives.

According to national reports, religion, particularly the Black Church, continues to have an indelible place in the lives of many African Americans (Barnes 2004, 2012; Nelson 2005; Pew 2009, 2010). Furthermore, when inter-racial comparisons are made, 2007 results show that, "the [Black] church still serves the primary social, political, and psychic needs of African Americans while white evangelicals have turned to other organizations to respond to these needs" (*Church Leader Gazette* 2013: n.p.). Specifically, per a 2009 Pew Research study, about 56 percent of American adults consider religion to be very important in their lives, about 80 percent of blacks acknowledge this belief, and this level of importance is apparent across the majority of black denominations. About 39 percent of Americans attend religious services at least once weekly; the majority of blacks (53 percent) report doing so. And about 58 percent of Americans pray at least once a day; a significantly higher percentage of blacks (76 percent) do so.[60] Lincoln and Mamiya (1990) refer to blacks as *superchurched* because of their church commitment. The above figures parallel earlier research on black religiosity and reflect behavior for most of the Kings.[61]

According to the narratives and empirical results in this chapter, religious and secular education are intermixed in their influence on the attitudes and actions of the Kings. And a local black church, Clear Creek No. 1, was the original site of religious socialization for this family. Just as Irma and George strove to educate their offspring religiously in the home, Black Church cultural tools such as scripture, prayer, songs, and sermons further reinforced their efforts. As will be evident in the next chapter, members of subsequent generations have continued this tradition, but in nuanced ways.[62]

5

"Keeping God's Favor": Contemporary Black Families and Systemic Change

Dora had reached an impasse in her life. Raised to value marriage and family, she had given it her all, but could no longer endure her husband's abuse. She and her six-year-old son, Kevin, deserved better. Dora wanted to escape, but where? After discussing her dilemma with her mother, she learned that her grandmother Irma was more than willing to provide a safe haven. So Dora packed up her son and all of her possessions in a small U-Haul pulled by her Chevy Monza and headed away from Ft. Worth, Texas, toward Gallman, Mississippi, – but her car stalled an hour outside the city. Irma tapped into her local network and got several close friends to travel to Texas, pull the old car ten hours to a repair shop in Hazlehurst, Mississippi, and transport Dora and Kevin to Irma's home. George had been dead several months, so Irma enjoyed her granddaughter's companionship – and even tolerated her great-grandson's boundless energy. She was also pleased to be able to help Dora get back on her feet. The tiny, temporary family lived together on the farm seven years during which Irma thrived on the camaraderie, and Dora was able to earn both Bachelor's and Master's degrees. Moreover, Kevin blossomed in school. They both also became active members of Clear Creek No. 1 Church and the community. Dora was grateful because Irma's largess enabled her to rebuild her and her son's lives.

Studies such as Wilkerson's (2010) *The Warmth of Other Suns*, Moody's (1968) *Coming of Age in Mississippi* and, *All God's Dangers* by Rosengarten (1974), chronicle the migratory patterns and experiences of blacks who left the South in search of opportunities and to escape Jim Crowism. We have illustrated in prior chapters that this rationale and motivation was no different for the Kings who left Gallman, Mississippi, for places such as St. Louis, Missouri; Jackson, Mississippi; Detroit, Michigan; Toledo, Ohio; Chicago, Illinois; and Denver, Colorado. However, what appears to be different for this family were the effects of who and what they *left behind*. But what were some of the migration implications for later generations, particularly for George and Irma's grandchildren and great-grandchildren?

This chapter continues to detail the migration experiences on which King family members embarked. Although some of their experiences were introduced in earlier chapters, linkages are made here between relocations that occurred in the mid- and late 1900s, with special emphasis on life in urban cities where the vast majority of the Kings eventually settled. Some family members remained in the North; others participated in "migration spells" that enabled them to establish strong footholds in new spaces, respond to varied set-backs, and/or financially assist each other. And still others eventually returned permanently to Mississippi and the familiarity and safety of home. Their varied experiences also illustrate how inequality and segregation continued to follow blacks in new spaces. During many of these transitions, individuals relied on the King family homestead in Gallman, Mississippi, as a temporary domicile.

This chapter also considers the inestimable, non-economic value of *place* beyond its economic import. According to the introductory narrative, Irma's farm (now owned free and clear) and finances (and help from Dora's mother) enabled her to support her granddaughter and great-grandson during an extremely trying time. This meant that Dora, as a single parent, was able to avoid the poverty that often accompanies divorce and debt.[1] Moreover, a supportive kin network helped Dora spiritually, educationally, and financially to eventually transition into the middle class herself.[2] Furthermore, Dora's presence provided a temporary respite for family members who were concerned about an aging Irma living alone after George's death. In this case, a linked-fate mentality informed strategic decisions made across generations of King family members. The King homestead was the destination for many migration spells. So when family members left Mississippi, they always knew they had a place to which they could return because "Irma's house was their house." Thus the family farm represented an emotional and economic safety net. The following sections help explain decisions to leave and/or remain in Mississippi and precipitating factors; implications about in- and out-migration choices; and the influence of industrialization on employment opportunities and constraints in urban locales.

COMING HOME . . . FOR A LITTLE WHILE: MIGRATION SPELLS, THEIR EFFECTS, AND IMPLICATIONS

The initial scenario in this chapter provides an example of a cross-generational migration spell during which one of Irma's grandchildren and great-grandchildren lived with her for almost a decade with mutually beneficial results. Equally important, Irma became a role model for Dora and Kevin; re-socializing them both in religious and practical ways. This section explores some of the implications of migration spells and factors that informed such decisions.

Table 5.1 provides comparative trends of median income by race and gender; unemployment rates; and median income by gender for farmers from 1950–1970, during which many of the Kings made major migratory decisions. According to these figures, median incomes for black males generally exceed that of white and black females, but incomes for these three groups lag considerably behind white males. Yet these incomes also exceed those of farmers and domestics substantially, the two positions George and Irma held. However, products generated from the family farm augmented their formal wages. It is also important to gauge specific periods. For example, in 1959, the median income for white males exceeds that of the five other groups by at least a factor of 1.25. For black females and domestics, incomes differ by almost a factor of five. Similar patterns are apparent in 1950, but black male and white female median incomes are more similar. By 1960, farmers and domestics earn similar amounts ($500 and $473, respectively). But by 1970, the median income for farmers is double that of domestics. In addition, black and white females have similar incomes ($3,285 and $3,870, respectively). Yet the median income for white males ($8,254) still exceeds that of the group closest to them, black males ($5,485), by almost 50 percent. As might be expected, unemployment rates follow a similar pattern, as rates for black males and females consistently exceed those of their peers, in some years, by a factor of two. These types of economic dynamics help explain some of the systemic influences that informed King family migratory decisions and outcomes. Family narratives detail what these factors meant on a daily basis.

Coming Full Circle: An Inter-Generational Migration Experience Across Time

Dora was able to navigate a plethora of economic and non-economic challenges with the assistance of her immediate family, in general, and Irma, in particular. Now a 54-year-old financial analyst, she summarizes her migration spell in Gallman, its outcomes, and her thoughts about Irma:

> I was able to go back to school and finish my undergraduate as well as my graduate degree . . . I earned a B.A. degree with a concentration in Finance and an M.B.A. She [Irma] was there to tend to Kevin when he got home from school, so she helped

Table 5.1 *Economic Statistics by Race and Farm and Domestic Statuses (1950–1970)*

Year	Median Income				Unemployment Rates				Median Income	
	White		Black		White		Black		Male	Female
	M	F	M	F	M	F	M	F	Farmers	Domestics
1950	2,982	1,698	1,828	626	4.7	5.3	9.4	8.4	711	448
1951	3,345	1,855	2,060	781	2.6	4.2	4.9	6.1	482	447
1952	3,507	1,976	2,038	814	2.5	3.3	5.2	5.7	479	433
1953	3,760	2,049	2,233	944	2.5	3.1	4.8	4.1	493	554
1954	3,754	2,046	2,131	914	4.8	5.6	10.3	9.3	597	495
1955	3,986	2,065	2,342	894	3.7	4.3	8.8	8.4	461	502
1956	4,260	2,179	2,396	970	3.4	4.2	7.9	8.9	455	486
1957	4,396	2.240	2,435	1,019	3.6	4.3	8.3	7.3	469	469
1958	4,569	2,364	2,652	1,055	6.1	6.2	13.8	10.8	498	467
1959	4,902	2,422	3,844	1,289	4.6	5.3	11.5	9.4	645	502
1960	5,137	2,537	3,075	1,276	4.8	5.3	10.7	9.4	500	473
1961	5,287	2,588	3,015	1,302	5.7	6.5	12.8	11.8	321	458
1962	5,462	2,630	3,023	1,396	4.6	5.5	10.9	11.0	486	476
1963	5,663	2,723	3,217	1,448	4.7	5.8	10.5	11.2	703	477
1964	5,385	2,841	3,426	1,652	4.1	55	8.9	10.6	710	518
1965	6,188	2,994	3,563	1,722	3.6	5.0	7.4	9.2	696	555
1966	6,510	3,079	3,864	1,981	2.8	4.3	6.3	8.6	1,179	526
1967	6,833	3,254	4,369	2,288	2.7	4.6	6.0	9.1	968	512
1968	7,291	3,465	4,839	2,497	2.6	4.3	5.6	8.3	1,215	546
1969	7,859	3,640	5,237	2,884	2.5	4.2	5.3	7.8	1,151	513
1970	8,254	3,870	5,485	3,285	4.0	5.4	7.3	9.3	1,105	527

Figures from Current Population Reports: Family and Individual Income. Series G 189-204. Median Money Income of Families and Unrelated Individuals in Current and Constant (1967) Dollars, by Race and Head: 1947 to 1970; Series d 87-101. Unemployment Rates for Selected Groups in the Labor Force: 1947 to 1970; Series G 372-415 Median Money Wage or Salary Income of All Workers with Wage or Salary Income, and of Year Round Full-Time Workers, by Sex, Race, and Major Occupation Group: 1939 to 1970.

out in terms of babysitting and I learned . . . domestic things . . . I learned how to cook. I was active in the church . . . During the week, I would go to school and she would wash and cook. Then on the weekend, I'd do those things. So I never told her "no" for anything because if a button fell off my shirt, she was sewing it, if there was a hole in my sock, she was sewing it . . . I remember going to visit her friends and visiting the sick, going to Bumpers (a local hamburger place) . . . She was the nicest person I have ever met.

A formal education dramatically altered Dora's socioeconomic trajectory and resulted in her current middle-class status. For example, according to reports by the Economic Policy Institute, "college graduates, on average, earned 56% more than high school graduates in 2015, . . . up from 51% in 1999 and is the largest such gap in EPI's figures dating to 1973" (Rugaber 2017: n.p.). Ma, Pender, and Williams (2016) found that in 2015, median earnings of bachelor's degree recipients with no advanced degree who work full time were $24,600 or 67 percent higher than earnings of high school graduates. Moreover, individuals with bachelor's degrees took home $17,700 (61 percent) more in after-tax income than high school graduates. Masters degrees tend to garner relatively more economic gains. In addition to formal social capital, Dora learned important domestic skills, became intricately involved in the local community, and received free childcare reflective of inter-generational caregiving well documented in black families.[3]

Kevin, now 37 years old, recalls the benefits of living with and learning from his great-grandmother, Irma:

> I remember being outside with her and trying to get calves back in the fence . . . I remember being on the land. I'd be out there shooing the cows . . . She showed me how to milk cows and pick pecans . . . I know that we came from strong, independent black people, regardless of their education. I know that during those times, you may not have been able to get a lot of education because you had to work . . . I would go to school, come home, watch Thunder Cats, and do my homework. And I would ace everything. It was a good time in my life. I always had the freedom and time to play with my friends . . . I always had toys. I didn't have much, but I was happy.

Irma also modeled biblical tenets associated with peacekeeping and patience (as described in Chapter 4). However, in some instances, her Christian lifestyle was difficult for Dora to comprehend: "How she interacted with people. She didn't say, 'think about other people to keep the peace,' she showed me . . . after a while, now I can understand it better." Dora complained about the seemingly docile way in which her grandmother seemed to respond to requests, particularly from her employer. Yet Irma's decision-making suggested a "big picture" understanding that initially evaded the perception of the younger woman. Almost three decades later, Dora better comprehends her grandmother's wisdom and sacrifices. White–black employer–employee interactions were complex; however, most family members believed that a reciprocally caring relationship did exist between Irma and her employer's family:

> She [Irma] said that they were all one big family. Those people really did love Big Mama. They treated her good. They took money out for her Social Security, just like it was a regular job – because it was a regular job. When I was born, she told

them that she couldn't work for them anymore because she had to keep her grandbaby, and they bought her a baby bed so that I could stay in the baby bed while she worked. And I remember Big Mama telling me that those young girls taught me how to walk.

The above quote is important here because Dora's migration spell that introduced this chapter was not the first time she lived on the King family homestead. As a baby, Irma kept this same granddaughter while her mother, Carla, got acclimated up North. Thus Dora's migration experiences had come full circle.

By all family accounts, the Williams family seemed to care for Irma and her family. Yet their altruism did not erase the inherent paternalism and unequal power dynamics embedded in such interactions, nor the following realities: Irma's salary was central to maintaining the family farm; the Williams's decision to purchase a baby bed was driven by their desire to retain Irma as their domestic; and, regardless of their positivity, genuine relationships can only be forged between people who value each other as equals. Several younger Kings posit that no matter how "good" some southern whites treated certain blacks, until whites dismantled segregation and the culture that fostered it, they were complicit and culpable. Collins (1990) pointedly explains these inter-racial dynamics and the unconventional strength of such black women:

No matter how loved they were by their white "families," black women domestic workers remained poor because they were economically exploited . . . for reasons of economic survival, African American women may play the mammy role in paid work settings. But within African American communities these same women often teach their own children something quite different. (Pp. 72–3)

Yet Irma and George inspired family members:

Because I knew the history, I saw the importance of having your own and not being dependent on living on someone else's land and working on their land and making them money, and you still don't have anything. And then they're giving you almost nothing in pay and watered down milk . . . so I saw the benefits that Big Mama and Big Papa had in having their own land and they could do some farming and taking care of cattle and have chickens. They could make their own money and live off their land . . . I'm really thankful for all the decisions that Big Mama 'nem ["and them" or her ancestors] made back then because it's affecting us now. Purchasing the land, the decision to marry Big Papa, they were good family successes that are continuing today.

In addition to low wages, Dora describes how the Williams's would give Irma hand-me-down clothes, watered down milk, and other items they no longer wanted – rather than simply providing a gainful wage. Now as an adult, Dora has more clarity about a litany of decisions made by her grandparents and other southerners as well as *the big picture* that shaped her ancestors' sacrifices to secure a place in the middle class. Yet she remains certain that her grandmother made dignified concessions to persons who were often undeserving of her largess and kindness.

Irma's House Is Our House: Other Migration Spells

A constant theme among family narratives is the stability represented by the King family homestead. Migration spells provided one of the strongest examples of its beneficence. For example, long before Dora and Kevin's stint, older family members recall what was possibly the first migration spell on the King homestead:

> Ike was George's sister. Today you would say she was autistic. She couldn't talk, but she could motion and make sounds so you could understand. They had a school for special needs kids in Jackson. She was doing well up there. They were teaching her sign language there. She lived there. But then there was a man there, who wanted to marry her, and Big Poppa [George's father] went and got her. And she went down from there. (Martha, 77 years old)

According to Martha and Carla, Ike then came to live with them. Her unpredictable behavior meant she spent time living on their farm, at her father's house – and sometimes in the woods on the homestead. Carla describes the turning point in Ike's migration spell on the King farm:

> We would play with Ike and My Mul would watch her. We were little children . . . we didn't really know anything was wrong with her, except that she couldn't talk. Then one day, Ike got mad at Sul [nickname for Martha] and bit her. My Mul got scared. Ike started getting mad with us a lot, but we were just being children. (Carla, 74 years old)

Ike began to yell, hit, and bit several of Irma's children; her moods became erratic. Irma didn't want Ike to leave, but she was afraid for her to stay. Other local family members were afraid to take Ike into their homes. George's father made the difficult decision to send Ike to a mental institution in Jackson where she lived until her death. Elderly King family members consider Ike's story a troubling reminder of the limits of their family's homestead, Irma's caregiving, and the inadequate knowledge about mental illness in rural black communities at the time.[4]

The daughter of one of Irma's sisters, Robbie, also lived on the King farm long-term. The scenario below summarizes this earlier migration spell:

> When Aunt Robbie went to Jackson, she left her daughter Cora, who was the same age that I was. She left her with My Mul. That girl stayed down there with us – I don't know how long . . . She might not have gotten Cora until she went to Ohio. (Carla, 74 years old)

According to her first cousin, Cora lived with them most of their teenage years as her mother attempted to establish herself in two separate northern cities. The middle class escaped Robbie. She trusted that the King homestead would provide a more stable family life for Cora than she could. Moreover, Carla describes the family network and major migration patterns:

> Momma Janice's [Janice King] children . . . she had two sons who went to Denver, Colorado. Then Aunt Robbie, she went to Jackson, Mississippi, and then Aunt Gert went to Ohio and then Aunt Robbie followed her to Ohio. Aunt Katherine ended up in Chicago . . . Now she was a nurse. She left Mississippi without even a 12th grade education, and ended up working in the operating room. She went to school in Chicago. She went back to Mississippi to visit, but never to live again. That was a success story. Momma Janice had two different husbands, but all of her kids saw themselves as brothers and sisters – no steps. And I don't care [Irma's siblings] when they came back to Mississippi, they were always going to come to see My Mul.

Like Carla's summary, Figure 2.2 (presented in Chapter 2) summarizes some of the outcomes of the family's early migrators. The thumbnail descriptions and examples above illustrate episodic stays in Gallman; intra-family migration spells as Irma's sisters moved to locales where family members had existing networks; and, instances when family members left the South and only returned annually to visit Irma. Katherine and Margie vowed never to return to Mississippi once they had escaped the clutches of Jim Crow; they only returned to visit their older sister. And they knew she would be there:

> My Mul never wanted to leave Mississippi because that's where all of her friends and family were. She didn't even want to leave when we went down there, but she didn't have a choice because everybody had their jobs and homes and families. And while there was more than one of us, it wasn't but one of her. So it was easier to move her up here . . . she never once complained, but I know she didn't want to come up here. (Carla, 74 years old)

With Dora's eventual relocation to begin her career in another state and Irma's failing health, the family matriarch could no longer live alone on the farm. With a hint of sadness in her voice, Carla describes another migration spell – the difficult family decision to relocate their mother to Gary, Indiana, where all of her children now lived. Thus Irma was the final member of the Kings to leave the family homestead. Although the house was sold and relocated to another local family member, the forty acres remain in the family. And Irma's great-grandson, Kevin, understands its importance and promise: "Her [Irma] legacy was the land that they had . . . No one else will get this land. We will keep passing it down through the family . . . and make sure each generation contributes to the next phase . . . and I know that we will be making them [past generations] proud. We are a strong people . . . We should never sell that land."

TIME TO GET OUT! FACTORS THAT PRECIPITATED AND HELPED MAINTAIN FAMILY MIGRATION

Although Irma had no desire to leave Gallman, Mississippi, she did not discourage family members from doing so. It is important to consider instances when migration spells became long-term stays outside of Mississippi. In some cases, Irma emboldened family members to migrate to other areas akin to biblical stories of exodus and empowerment. Both long-term and migration spells resulted – in the following instance, a cross-spatial episode. George's family network in Detroit, Michigan, enabled his eldest daughter to pursue non-domestic employment:

> I graduated in '57 and worked one year in Crystal Spring at the white high school in the cafeteria. We would prepare the food, serve the food to the kids, and then clean up the kitchen. First I went to Toledo and stayed with Aunt Robbie [Irma's younger sister] for a couple of months and then I went to Detroit and stayed with Aunt Hana [George's younger sister] for a few months . . . Aunt Hana got me a job babysitting this Mexican lady's twins and another little child . . . I was okay with that job. The people were nice. I probably would've stayed longer if the lady would not have started adding more stuff. (Martha Ella, 77 years old)

Martha quickly realized both her limited employment prospects in Mississippi as well as her disdain for working as a food server and custodian for white students. Ironically, despite being from a black middle-class family, Martha's job in food services was merely a variation of domestic work. Private sector work was also common for post-migration black women as they competed with a large pool of candidates for low-wage posts.[5] After a two-phased migration process, George's sister offered her a place to live and secured her a job as a nanny:

> At first, all I had to do was take care of these kids and put a load of clothes in the washer when they took their nap and feed them and clothe them . . . after a while, the lady started adding things, "Now Martha, while the kids are down for their nap, could you fold this load of clothes or could you put the dishes in the dish washer and load the dishwasher?" Adding extra stuff, but never said nothing about no

extra money. I told Aunt Hana, "Now I might *look like a fool*, but a fool I'm not. This lady hired me to take care of those kids. She didn't hire me to be no maid."

Martha initially enjoyed her post, but over time, she began to be overworked and under-paid. Collins (1990) challenges this type of exploitation: "as members of a subordinate group, black women cannot afford to be fools of any type, for our objectification as the Other denies us the protections that white skin, maleness, and wealth confer" (p. 208). As this same scholar concludes, Martha quickly made the distinction between knowledge and wisdom and used the latter in decision-making:

> I told Aunt Hana I wouldn't be there too much longer . . . And when I got tired, that one Friday, I left and I told Aunt Hana that I wasn't going back. I didn't even tell the lady I wasn't coming back. I just didn't go back. Aunt Hana tried to talk me into going back and I said, "No, because she knew I was doing extra work when I was there, and she wasn't willing to pay me." And I said, "if I go back, ain't no telling what they might do to me." I decided to go to Chicago . . . I stayed with Donald's brother and his wife [her future in-laws] . . . I was able to get a job.

In response to wage exploitation based on her race, gender, and *age*, Martha's decision meant relocating from Ohio to Chicago where other family networks existed. Beyond the practical benefits of exiting an exploitive workplace, like Collins (1990), we posit that Martha's decision was indicative of the long tradition where, "black women's work remains a fundamental location where the dialectical relationship of oppression and activism occurs" (p. 46). Long-standing ties in Gallman meant Martha was sanctioned to temporarily live with her fiancé's family – but only because they were a nuclear family and could be chaperones. Fueled by heightened expectations, Martha's activist spirit sustained her though various problems and eventually resulted in fair, gainful employment and marriage.

Irma's niece and Gina Mae's oldest child, Ella, initially relocated from Mississippi to Ohio, but a migration spell meant Florida became home. However, she faced challenges as the only female plant supervisor:

> I came to Florida in 1973 because I had a sister down here. I got a job at an orange factory and worked there for 16 years. Whites . . . we got along fine. Only the men, both black and white, didn't want to work for a woman. Until my boss [a white man] told them that I was the boss on the floor and that if they wanted to keep their job, they had to work for me . . . I supervised for 15 years . . . I was the only black woman supervisor in the factory. (Ella, 77 years old)

Ella initially experienced sexism in a predominately male manufacturing arena. However, the hierarchical nature of the plant meant the problem was systematically addressed. Moreover, strong ties meant she had family support.

Although blacks like Martha and Ella wished to get out of Mississippi, overt and covert forms of unrest typically met them. Writers such as Drake and Cayton (1945) as well as Spear (1967) argue that recruiting campaigns to lure blacks from the South to the North existed, but were exaggerated. Yet the draw was infectious:

> The great mass of caste-bound Negroes in the South stirred ... they came in droves – 50,000 of them between 1910 and 1920. And as each wave arrived, the migrants wrote the folks back home about the wonderful North. A flood of relatives and friends followed in their wake. (Drake and Cayton 1945: 58)

In *Black Chicago*, Spear's (1967) explains the route between the King family locale and key final destinations: "as the terminus of the Illinois Central Railroad, Chicago was the most accessible northern city for Negroes in Mississippi" (p. 129). Reception was generally unfavorable. The following quote references the train stop the Kings used and the northern locale where many would make their home:

> Up from Mississippi – where agents were arrested, trains stopped, ticket agents intimidated. And at Brookhaven, a chartered car carrying fifty men and women was deliberately sidetracked for three days . . . newspaper headlines . . . commented in a none too friendly vein: "Half a Million Darkies from Dixie Swarm to the North to Better Themselves" . . . the southern caste system was in the process of profound modification. (Drake and Cayton 1945: 59–61)

Disparaging stereotypes of "big, ignorant, vicious Negroes" (Spear 1967: 37) meant that regardless of their social capital, initiative, and excitement, blacks were considered competition and/or threats on multiple fronts. Ironically, whites were often most concerned about "Negro attempts to secure adequate housing [that] was not restricted to the poor; even the affluent were blocked in their quest for a decent place to live" (Spear 1967: 20, 23) in ways that resulted in congested, predominately black areas that contrasted significantly from the safe haven of the King's homestead.

Chronic economic and ideological constraints meant blacks continued to leave Mississippi.[6] However, post-migration, they often found themselves living in over-crowded, segregated areas in the North; excluded from gainful employment; and, overall, dealing with similar conditions they thought they had escaped in the South. Each of Irma and George's children had periods during which they initially lived in unfavorable conditions, including a housing project, a basement apartment, and, as 53-year-old Donna summarily recalls:

We grew up in the hood . . . We had roaches! I was glad when we moved to [name of city] to be closer to my cousins, but we moved next to a train track. But we had a house, our own house, and that had to stand for something.

Donna's parents' new house next to noisy train tracks was still a dramatic improvement over their original apartment. Carla married and moved her three children out of similarly infested housing. For these families, the ability to purchase houses was part of their trek into the middle class.[7] Yet these initial humble abodes reflected what was available to most urban blacks. Although some scholars have questioned depictions of black kinship networks, it appears that these four siblings avoided poverty by moving in close proximity in the Midwest and supporting each other economically and in-kind.[8]

Reluctant Migration and Non-Traditional Relocation Relations

The pursuit of job opportunities outside the South was not always the motivation for migration for the Kings. Nor were the outcomes always favorable. Other factors precipitated such spells. The following narratives offer instances when a linked-fate mentality undermined agency and positive outcomes. In the first scenario, Irma asked one daughter to temporarily assist her older sisters – with life-changing consequences:

After Utica there wasn't any place to work, no shops. And that's when my sister, Martha was going to have her second child, and My Mul must have gotten sick and couldn't go to Chicago, so she sent me. So that's how I ended up leaving Mississippi . . . [had she stayed in Mississippi] I guess I would've eventually learned how to drive and got me a job. Almost all of the black folks worked at the chicken plant and that's probably what I would've done . . . It was just a job and almost everyone that applied got hired there . . . I never did go back to Mississippi [to live]. I wanted to go back home, 'cause I never wanted to leave home. Janice had left Mississippi and gone to St. Louis to help Aunt Lou [Margie Lou]. They had a store. Janice had gotten another job at a hospital, but Aunt Lou wouldn't let her come through the store – scared somebody was going to steal something. So she had her go through the back door, coming around in the dark to catch the bus to go to work at the hospital. So My Mul told me to go to St. Louis so I could watch Janice as she went out the store to get on the bus. So that's how I ended up in St. Louis. (Carla, 74 year old)

Like her parents, Carla had no desire to leave Mississippi. At twenty-two, she had completed community college with a cosmetology degree. Her initial plans included learning to drive and joining the labor force at Sanders Poultry, the local chicken plant. Yet Carla had not located employment. At her mother's urging and spurred by a linked-fate ethos, she began a two-phased migration spell to assist her sisters in Chicago, Illinois, and then in St. Louis, Missouri. Her experience reflects a more sobering instance of a largely unwanted migration spell. Carla continues:

> Seems like I never wanted to leave Mississippi. I guess I never wanted to leave being around my momma and daddy . . . I never wanted to leave home. And I guess if I had found a job, I wouldn't have left home. But you weren't going to find a job because you didn't know how to drive . . . And I don't know why My Daddy didn't teach us how to drive or why he didn't teach us how to play the piano. He could play the piano and the guitar. And didn't try to teach nobody how to play either one.

Carla did not welcome life away from the family farm. However, limited labor market options after graduation meant she was unemployed and thus *available* for migration spells to help other family members. Several other observations are noteworthy. Traditional gender roles appear to have prevented George from extending a common "rite of passage" to his daughters of learning to drive that he later provided to most of his male grandchildren. Moreover, Carla's belief that playing a musical instrument represented social capital that could also have been parlayed into income informed her later decisions that her own children learn to play instruments, including piano.[9] The intersection of macro-level (i.e., job constraints) and micro-level (i.e., acquiescence to Irma's requests) factors precipitated Carla's northern migration.

Not all King family migration spells had positive outcomes. Although a Christian, Margie Lou (a middle-class store owner in the North) appears to have embraced a different ethos than her older sister Irma (a middle-class farmer in the South), which resulted in less economic and emotional support for Irma's two daughters left in her charge. Additionally, the youngest sister in an aged-based hierarchy meant Carla was expected to assist her sisters. And for Irma, extending such a courtesy was part of their family's value system that Carla respected and obeyed. This same ethos meant Irma eventually intervened on Carla's behalf:

> I often think about Abraham, when the Lord told him to get his family out and to leave in the book of Genesis. That's what My Mul told me about St. Louis. She said, well, I wasn't prospering in St. Louis and so she's the one who told me to leave. And that's how we ended up in [city's name]. So sometimes I think about what would have happened if I had not left, but I kind of think about how Abraham must have felt when the Lord told him to get out.

Based on her daughter's sorrow, Irma encouraged her to relocate near her older sisters. Although Carla waxes nostalgic about possible alternate life outcomes, she takes solace in parallels between her life and that of an Old Testament biblical stalwart who moved into uncharted territory and was spiritually and economically blessed by God. Carla was proud of her unwavering support of her mother's requests. And Irma's directive to migrate from St. Louis parallels Collins' (1990) views, "African American mothers try to protect their daughters from the dangers that lie ahead by offering them a sense of their own unique self-worth" (p. 127).

Although we focus on migration spells in Mississippi, other family members made strategic post-migration decisions that ushered in the middle class. George's sister, Hana, migrated to Detroit in the mid-1900s. A combination of *common* employment for black women and more *uncommon* income-generating efforts for black women were parlayed into economic stability:

> Aunt Hana had a job as a private domestic, but she had also bought a house with three floors. She lived on the first floor and rented out the second and third floors. So she had more income. Her first husband died . . . I remember her last husband that she was renting to . . . he was homosexual. But he had gone to the doctor and got some bad news. So he told Aunt Hana that since she had been so nice to him, why don't they get married and she could get his money when he died. So that's what they did . . . they had just been friends. They [Big Pappa and Aunt Hana] had good business skills. Aunt Hana [she didn't have any children] had already come to Mississippi and took her brother's two children by a lady he wasn't married to. (Carla, 74 years old)

Hana and Irma's close relationship meant Carla had lived "a spell" with her during a medical emergency. There are several parallels between the two women. Like Irma, based on a linked-fate mentality, Hana raised the children of a poor family member. She also spent part of her adult work life as a domestic. And like her sister-in-law, Hana saved funds to purchase property that provided security, safety, and economic stability. However, Hana was much like her brother George in terms of business acumen. Thus Hana created fictive kin in her apartment building and became an "Other Mother" that blurred boundaries relative to gender, class, sexual identity, and mothering.[10]

Yet post-migration, some Kings experienced challenges as single parents. For example, 77-year-old Ella recalls making sacrifices:

> I was always working and trying to take care of my kids because they didn't have nobody but me . . . I am okay with how things turned out . . . except education. I was not able to get an education because I had to take care of my kids . . . When I got married I had just finished 8th grade. And when my husband left, I had to take care of my eight kids.

Based on her childhood as a sharecropper, early marriage, and subsequent divorce, Irma's niece, Lois, found herself in a similar situation:

> My daddy moved us around a lot. He was a sharecropper. He had a piece of land he would farm for the white man. He would keep us out of school to chop, pick, and harvest the cotton. We only got a chance to go to school when it rained or after Christmas. Then I got behind, so when I got big, I dropped out. I got to about sixth or seventh grade. But my brother Charlie didn't get that far because he had to plow. It was daddy's fault, but I forgive him . . . I used to be a little bit mad, but I'm not. He did the best he knew. His parents were slaves . . . I tried to go to night school, but I had all those children and a special needs child too. Some nights I'd be so tired, so I gave that up . . . but I pushed my kids the way my daddy *did not push us* [emphasis is hers] . . . all six graduated from high school . . . It was hard, but I didn't let that bother me . . . God helped me through it. (Lois, 76 years old)

Lois' childhood is a reminder of the cycle of exploitation many southern blacks experienced and the associated deficits for black children, particularly in terms of educational opportunities. Even after emancipation, the damaging effects of slavery still influenced the thoughts and decisions of Lois' parents, particularly her father. In reflection, she has reconciled the less favorable aspects of her upbringing, including her divorce, and acknowledges God's help during the process. As adults, family support helped fortify both Ella and Lois during difficult periods. Both are pleased that their children were still able to complete high school. However, both women are also cognizant of the challenges they could not navigate in the absence of safety nets, such as a homestead or education, common among other King family members. Overall, these diverse migration spells, the relationships and reasons that fostered them, and their diverse outcomes, illustrate how the King family ethos, coupled with systemic effects, altered life chances and socioeconomic trajectories.

NORTHERN STEEL MILLS AND POST-MIGRATION EXPERIENCES

Steel mill positions did for certain King family members what military stints accomplished for others. By the late 1960s, most of Irma and George's children had migrated to the Midwest. This meant steel mills such as United States (U.S.) Steel: South Works in Chicago, Illinois, U.S. Steel: Gary Works, Bethlehem Steel, American Bridge, and Inland Steel mills, all in Gary, Indiana,[11] provided gainful employment, particularly for male family members. But what was the ecological context of these mills and their impact on the socioeconomic outcomes of the Kings?

History and Ecology of a Steel Town

Although they were generally unwelcomed by white workers, blacks were involved during the early US industrialization period. Despite discrimination, blacks and Mexicans began working in steel mills in large numbers in the 1910s and 1920s. Steel work was largely segregated by race/ethnicity and craft. New Deal legislation in the 1930s ushered in the United Steelworkers of America (USWA) in 1942 and attracted minority workers in search of fair treatment and wages. Yet even from 1890 to 1920, blacks held a variety of positions in manufacturing-related industries across the country. For example, in 1890, there were approximately 19,359 blacks in these fields, 24,991 in 1900, and 47,691 by 1910. By 1920, 150,669 blacks worked in manufacturing industries; this represented an impressive increase of 215.7 percent overall. When iron and steel workers are specifically considered, their representation over this five-decade period was as follows: 7,357 in 1890; 13,293 in 1900; and, 33,101 in 1910. By 1920, 129,257 blacks worked in steel and iron industries (Green and Woodson 1930). These same scholars note that, in 1920, there were about 23,616 iron and steel industry semi-skilled black laborers. However, regardless of the city or organization, in the early 1900s, blacks were:

> Virtually excluded from Northern industry both by the hostility of trade unions and white workers ... In a large Western factory, employers and employees actually entered into an agreement in 1892 "not to hire any Negroes or foreign men for twenty years" ... another manufacturer of metal goods in the same state also said, "We have no Negroes in our employ but do not object to them. If we should refuse to employ them it would be solely to avoid the risk of friction between them and white workmen" ... occasionally white employers would hire a Negro for the most dangerous and disagreeable tasks, which white men had refused to perform [and] it was said that these Negroes received smaller pay than the white workers ... the foreman told me that we could never get colored men to grind, because they were afraid of the wheel; we discovered it was not so at all; one of the best grinders we now have is a colored man. (Green and Woodson 1930: 138–40, 142, 329)

Despite discrimination, skilled blacks dispelled stereotypes about their so-called sub-par abilities.[12] Yet both native and immigrant white workers "learned racism ... 'race' has been deeply constitutive of North American economic, cultural, and political life for 300 years" (Williams 1993: 228). Cotton (1993) came to a similar conclusion:

> Capitalist labor markets are characterized for the most part by collusive arrangements between cabals of white workers and white capitalist employers. These

arrangements yield both psychological and material benefits to both parties. White workers have effective control of the best jobs and occupational slots in the economy, and white employers are assured of a relatively docile, system-defending workforce. (P. 203)

Thus impressive work performance by blacks was not likely to dispel the embeddedness of racism and discrimination or their intrinsic and economic value to both white employers and workers.

Competence, strike breaking, and sheer determination meant that by 1910, blacks held skilled posts as die grinders, engineers, plumbers, rammers, shippers, core makers, molders, and in the crucible melting area. As noted by sociologists such as W.E.B. DuBois (1899[1996]) in *The Philadelphia Negro*, many blacks started as unskilled workers; others brought skills with them from manufacturing jobs in the South. Yet, regardless of their acumen, blacks were often demoted from skilled to unskilled posts or even terminated for the sake of white workers. Rather than the result of racial discrimination, these tendencies were often attributed to labor market dynamics or "general consequences of the postbellum deflation from which both white and black suffered" (Green and Woodson 1930: 336).

When referencing apprenticeship opportunities and germane to this book, blacks in various manufacturing positions, "received considerably more wages a month than the average farm hand" (Green and Woodson 1930: 143). Occupational reports confirm:

Earnings of plant workers in iron and steel establishments are among the highest in manufacturing. In April 1959, earnings of production workers in these plants averaged $127 a week or $3.10 an hour. This compares with the $89.87 weekly or $2.23 hourly earnings for production workers in all manufacturing. (*Occupational Outlook Handbook* 1959: 608)

Moreover, in 1958, hourly wage ranges for common steel mill positions were considerably higher than for other posts, such as coke plants ($2.30–$3.03); blast furnaces ($2.23–$2.74); steelmaking ($2.23–$3.57); rolling and finishing mills ($2.16–$3.57); and laborers and janitors ($1.96–$2.03) (*Occupational Outlook Handbook* 1959).[13] Most of these rates should be considered lower bounds because workers were often paid based on how much they produced (i.e., on an incentive basis). Blacks were disproportionately represented in the lower-paying posts, but still noticeably present among skilled artisans. A broad comparison of the above wages to average earnings in general in 1957 ($2.27 per hour for males and $1.45 per hour for females) illustrate the relative economic benefits of steel mill posts.[14]

We focus on steel mill opportunities in the Midwest were many Kings migrated.

When Ecology Meets Entrepreneurship

Most historians agree that the advent of the steel industry in the Chicagoland area was the result of synergy between ecology and entrepreneurship. The Illinois and Indiana regions were replete with high quality natural minerals such as iron ore and coal. Multiple iron ranges were discovered in the mid- to late 1800s that fueled this new industry. Moreover, cheap labor was abundant.[15] In 1900, steel, in particular, was in demand. Several business leaders took advantage of this growth. U.S. Steel was created in 1901 by J.P. Morgan, Elbert Gary, Andrew Carnegie, and Charles M. Schwab. Its South Works location had 11,000 employees in 1910. In 1920, its Gary Works location was the largest steel mill in the USA with 16,000 employees. U.S. Steel's employee base peaked at 340,000 and its Gary location peaked at roughly 30,000 employees. However, the latter location has lost over 80 percent of its employees due to downsizing. Reports show that, in the early 1900s, U.S. Steel was responsible for approximately 66 percent of America's steel production and about 30 percent of international output.

U.S. Steel was not the only employer in this lucrative terrain. Led by Charles M. Schwab and Eugene Grace, by the 1920s, Bethlehem Steel became the second largest steel organization. Its focus was on government contracts for large buildings and bridges. At its height, Bethlehem employed 300,000 workers. The U.S. Steel and Bethlehem locations made Illinois and Indiana important hubs in the country's steel industry in the 1960s as well as locations for severe economic malaise after deindustrialization. Leadership's focus on short-term profits rather than proactively responding to technological innovation; high overhead, particularly employee salaries and benefits; and, increased oversees competition, resulted in bankruptcy in 2001.

Inland Steel was an independent mill in East Chicago, Indiana, designed to reduce iron ore to steel. Headquartered in Chicago and founded in 1893 when Felix Block purchased a failing local mill, Inland had about 2,600 employees by 1910. Steel needs during World War II meant that, by the 1940s, Inland had 14,000 workers and was producing roughly 3.4 million tons of steel annually. Unlike its competitors, enhanced profits in the late 1990s resulted in Inland being acquired by ArcelorMittal. Another local player, American Bridge, was founded in 1900 by J.P. Morgan. However, it became a subsidiary of U.S. Steel in 1902. By 1938, American Bridge had furnished and erected bridges in 69 foreign countries and overseas US territories. However, American Bridge was sold to Continental Engineering Corporation in 1988. The location in Gary, Indiana, was one of the eight original sites and the last to close. Each of these five steel mills summarized above represented job opportunities for blacks like the Kings.

Research suggests that the USA produced between 380,000 and 60 million tons of steel between 1875 and 1920. Yet dramatic reductions in demand, international competition, lax leadership, failure to embrace new technology, changes in demand for imports, mounting overhead, union discord, and worker strikes meant that five decades later, most of these industrial behemoths were either bankrupt, consolidated with other steel corporations, or substantially downsized.[16] Table 5.2 summarizes demographic and economic indicators for the steel mills in which King family members worked.

From the Steel Mills to the Middle Class

Manufacturing jobs represented a path to the middle class for many persons without college educations.[17] A steady increase in black presence in these jobs in places that would eventually be termed *rust-belt cities* mirrored the migration patterns of blacks in the 1920s through the 1960s. Hurley notes (1995):

> Between 1920 and 1930, more than 15,000 migrants, most of them from Mississippi, Alabama, Tennessee, Arkansas, and Georgia arrived in Gary to work

Table 5.2 *Thumbnail Descriptions of Chicagoland Steel Mills*

Name	Founded–Closed	Peak # of Employees	Current Status
American Bridge	1900–1978	About 2,500	Consolidated with U.S. Steel in 1902
Bethlehem	1903–2001	300,000	Bankrupt
Inland: Indiana Harbor	1893–1998	25,000	Acquired by Ispat International
U.S. Steel: Gary Works	1906*–	30,000	5,000 employees (2015)
U.S. Steel: South Works	1901–	340,000	52,500 employees (2000) $2.65 billion earnings (2016)

Key: *Some records show a founding year of 1908.

in the mammoth lakefront factories . . . the following decade, another 20,000 African Americans came to the Steel City to fill industrial positions created by the wartime boom. (Pp. 112–13)

Similar migration patterns were evident in other northern cities. The Kings already had family and fictive kin in Ohio, Michigan, and, later, Illinois that facilitated subsequent relocations once family members became aware of available manufacturing jobs.

Steel mills provided gainful careers for four men from one generation of the King family, George and Irma's son, James Lewis, and his three brothers-in-law. Each of them received on-the-job training; several worked in harsh conditions that included episodic manual labor. As chronicled by Barnes (2005), Caitlin (1993), and Hurley (1995), racial tensions, discrimination, and micro-aggressions by white male workers were cyclic, particularly when blacks were considered competition for *their good jobs*. More broadly, blacks were often barred from competing with whites for higher paying positions and prevented from active union involvement that could help secure their jobs during economic downturns.[18] There were benefits and drawbacks to work in this sector. Pay was steady, benefits were good, and a conscientious black person could learn on the job and earn a solid living. Such was the case for the Kings. James Lewis, now 69 years old, is the surviving family member from this craft and held a position uncommon for blacks:

I worked in the chemical department. I enjoyed my job. You would take one chemical and treat it and get three or four others. We extracted gases and other by-products. I learned on the job . . . just the idea that you could change different chemicals – it was amazing. On the negative side, we learned that some of the by-products had benzene in them. We found out later that it was a cancer-causing agent.

Three of these men would later develop cancer; two would succumb to lung cancer. Carla argues, "I believe that's what killed Horris [her first husband] . . . working all those years in that Coke plant . . . and Janice still gets a small monthly check from U.S. Steel for Calvin . . . for mesothelioma." Her remark parallels existing studies that one of the long-term implications of careers in manufacturing arenas was chronic exposure to sweltering heat, smoke, and carcinogens that may have precipitated lung cancer and death for several Kings.[19] Specific to Carla's husband's experiences, the *Occupational Outlook Handbook* (1959) acknowledges, "much of the work in cokemaking still is physically taxing and requires exposure to heat and dirt" (p. 609).

However, to these same individuals, the benefits of steel mill jobs seemed to outweigh the drawbacks, as they strove to establish their families economically outside of the South. Their steel mill jobs resulted in home purchases, multiple automobiles, vacations, and other accouterments associated with the middle class. Over time, Irma's children transitioned into the middle class; one was solidly upper-middle class. A daughter from the latter family describes her upbringing:

> My brothers and I had an abundance of material possessions. I was involved in the arts, dance, music, and theater. Both my parents had good jobs. We didn't qualify for any services or support meant for families with limited incomes. (Constance, 47 years old, social worker)

Yet according to her 52-year-old brother, Calvin Jr., both parents had to work together to balance the scheduling demands of their jobs:

> One thing that really stood out about my father was that he worked around the clock [swing shifts]. I was always active in school, but he was always there. He always knew and showed up at my games . . . it was special to look up in the stands or at the baseball field and see him there. But that let me know that my mother must have told him. He would not have known otherwise.

Their father, Calvin Sr., had a career at U.S. Steel in Gary, Indiana; their mother, Janice, held an even higher-paying industrial position at Elgin, Joliet and Eastern Railway (referred to simply by family members as EJ&E).[20] One of few women on staff, Janice's career differed dramatically from those of most of her female peers (and sisters) who worked in the "pink collar" sector. She experienced severe overt and covert discrimination based on race and gender as summarized by Collins (1990): "with Northern migration, some black women entered factory employment . . . regardless of their location, African American women faced discrimination" (p. 58). The intersection of several marginalized statuses made Janice particularly vulnerable:

> While working at the railroad [EJ&E], when computers first came out, the white folk didn't want to teach me how to use them. They were only teaching other white folk. I had to teach myself and figure things out. My faith got me through those tough times. (Janice, 75 years old)

Such a coveted position at a predominately white, male-dominated company in a steeply competitive labor market meant Janice was believed to be "taking" a job from a deserving white male. Increased automation followed by significant downsizing only exacerbated tensions. Although Darity and Myers (1994) were considering the plight of young, black males, parallels between their following assessment and Janice's experience in a male workspace are clear: "the real basis for discrimination in the labor market steams from the efforts of groups already in the jobs to guarantee their position. Someone must be part of the surplus since there must be a surplus" (p. 53). And Janice contends that most white male co-workers made concerted efforts to push her out into the surplus labor population. And differential accumulation of *formal* on-the-job training made her more susceptible to unemployment as compared to white males.[21]

Similarly, a steel mill position enabled Martha, Donald Sr., and their five children, to live middle-class lives. Donald Sr. had a reputation for taking advantage of overtime:

> One year when we had a very bad blizzard, my father, who worked in Chicago, went to work anyway. He told us how he didn't have to work, but got paid because even the people who lived in Chicago didn't make it in to work. That created a lifelong work ethic that I still have today. (Manuel, 49 years old)

His father's example as a 30-year veteran of the steel mill has fueled both Manuel's work and entrepreneurial ethics as a businessperson.[22] Yet based on their large family, Martha's domestic skills were also needed to make ends meet. He laughingly describes his mother's role in stretching family resources across five children: "I will never forget how my mother was so frugal that she would cut an apple into five pieces and give us 1/8 of an apple . . . after you took the core out, there was only about 5/8 left."

In addition to non-economic benefits, according to his daughter, Donna, Donald's position as a crane operator provided practical economic perks:

> My dad worked at the steel mill and made decent money, we went on vacation every year, always had food, lights, and gas. I think there were just varying degrees of cheapness and different levels of style . . . Some of our relatives did whatever they wanted and never had money for the things that they needed and others had money, but weren't willing to give it for the things that I thought were important. I loved that my [name of aunt and uncle] always had nice cars and fancy clothes, but they both worked and didn't have as many kids to support as my mom and dad did. I remember that my grandmother's house [Irma] initially had an outhouse and no running water, while my dad's mother had indoor plumbing and a marble coffee table. I just figured that Big Mama wasn't that type of person and wouldn't have spent money on fancy stuff. (Donna, 53 years old)

The above comment describes the following: class variations across King families whose spouses all worked at steel mills; factors such as number of children, family budget, and spending choices that shaped economic outcomes; and common items associated with the middle class.[23] Equally important is the comparative lens between Irma's farm life and those of her in-laws, both middle class, that suggests different priorities around materialism.

However, by the early 1970s, deindustrialization meant the out-migration of the majority of manufacturing positions in rust-belt cities. Most steel mill workers were ill-prepared for these drastic employment downturns:

> Many industrial workers did not predict the dramatic economic changes that were about to occur in the late 1900s ... without foresight, younger men opted for steel mill positions rather than college, and many heads of household were not prepared for the "rainy days" that was postindustrialization ... by the late 1900s, most persons previously gainfully employed in northern industrial cities found themselves either unemployed, underemployed, or working in service occupations for substantially lower pay and reduced benefits ... cities like Chicago, New York, Detroit, and Philadelphia lost over half of their manufacturing jobs during the 2-year period following 1967. (Barnes 2005: 13–14)

Several King family members employed by the bigger steel mills were able to retain their jobs until retirement; jobs in unsafe parts of the mill, combined with seniority, provided a certain degree of security. Flanagan (1999) offers specifics about these employment changes: "From 1963 to 1977 the total number of manufacturing jobs in the central cities of the 25 largest metropolitan areas dropped by 700,000 ... from 1958 to 1972, the more established industrial cities in the North lost between 14 to 18 percent of the manufacturing jobs" (p. 211). In northwest Indiana, American Bridge steel mill experienced sharp downturns which meant Carla's husband, Horris, was let go. His daughter, Vera, describes their family challenges:

> I did not think we were poor ... I was ignorant to the fact. I was always clothed, fed well and cared for ... when our power was out, I thought it was because of the weather, not because we couldn't afford to pay ... I looked forward to a sleepover at my favorite aunt's house because our lights were out. As a teenager, my eyes saw reality in terms of one car for our family, being on the side of the road in that car on occasions because gas was put in it sparingly and the gas gauge was in disrepair ... getting food from government programs ... momma making the hard decision to ask my father who'd never contributed anything to buy my school clothes during my junior year in high school ... Me trying to work part-time at Dairy Queen my senior year for uniforms I needed and trips that I wanted to attend for drill team [a student program] ... but my mom always found a way to get any items I needed – even though it was a struggle. (Vera, 51 years old)

Despite being frugal, five children in this working-class family meant the absence of a financial safety net. They experienced substantial economic hardship for about five years until Horris was able to arrange retirement by combining his career work time at two local steel mills. These financial constraints meant accepting subsidies from a government food program, holding Vera's biological father accountable for child support, part-time jobs for Vera and several of her siblings, and industriousness and sacrifices by their mother. In general, post-industrial dilemmas for the Kings varied based on factors such as at which steel mill one worked, seniority, existing training, and down-sizing decisions outside of their control.

In contrast, other families had a different experience: "we grew up sheltered so I wasn't aware of the violence and poverty around me until I was an adult and had moved out of state. I had a fabulous childhood" (Constance, 47 years old, social worker). For Constance, a combination of her father's seniority at one mill and her mother's position that was only tangentially affected by deindustrialization, meant she and her siblings were oblivious to the dire conditions some of her cousins were facing. Tanya's comment below provides an important retrospective on her parents' ability to navigate these conditions and the linked-fate outcomes they ultimately forged:

> I remember seeing my father come in almost black from soot and smoke. But he continued to work. After he got laid off, I could tell that it really hurt his pride not to be able to take care of us . . . But he collected bottles and cans and stood in line to get government cheese, milk, and rice. My mother also went without things like new clothes to make sure we had pretty normal lives growing up . . . As a teenager, I began to realize that we were struggling and I really began to respect their decisions and see their sacrifices. Even now, I am not wasteful because I remember what my parents went through. My daddy died before we could really do nice things for him, but my mother does not want for anything now . . . I watched them go without for us. They were such godly people. I hope that I am becoming like them.

According to Tanya's comment, even steel mill positions were unstable, no matter one's experience and tenure. Blacks like Horris dealt with being "last hired and first fired" such that historic racial discrimination, coupled with deindustrialization, exacerbated their financial travail.[24] Her family's economic challenges and subsequent choices parallel broader systemic change noted by Mason (2003):

> Tight labor markets and government anti-discrimination and affirmative action policies, along with rapid improvement in the quality and quantity of African American education relative to white education, brought about a reduction of

> inter-racial inequality and racial discrimination in the labor market between 1945 and 1973 . . . during the 1970s and 1980s when inter-racial wage inequality begin to expand, macroeconomic and public policy changes have contributed to an environment than encourages racial discrimination in the labor market. (Pp. 64–5)

Although substantially fewer steel mill posts are available today as compared to during its industrial heyday, their employees still earn relatively high incomes. For example, according to the *Occupational Outlook Handbook* and based on Bureau of Labor Statistics 2017 figures, the average steel mill worker earns a mean annual wage of $49,050. Other common positions and mean annual salaries include: managers ($113,770); building and grounds maintenance workers ($31,640); installation and maintenance jobs ($48,260); production ($43,120); and rolling machine operators ($42,750). Unskilled laborers are the lowest-paying posts such as production helpers ($31,450) and laborers and movers ($35,050).[25] Yet these changes mean, for people like the Kings who had steel mill careers and who were part of the middle class, gone are:

> The $20 hourly wage, introduced on a huge scale in the middle of the last century, [that] allowed masses of Americans with no more than a high school education to rise to the middle class . . . Hourly workers had come a long way from the days when employers and unions negotiated a way for them to earn the prizes of the middle class – houses, cars, college education for their children, comfortable retirements . . . wages are falling below the $20-an-hour threshold – $41,600 annually – that many experts consider the *minimum income* [emphasis is ours] necessary to put a family of four into the middle class. (Uchitelle 2008: 1–2)

Occupations in northern steel mills created the “middle class for blue-collar workers” and provided a bridge by which some King family members could walk a post-migration path into the middle class.[26] For others, it meant traversing a tightrope across financial insecurities based on the realities of deindustrialization that loomed large. Economically, these jobs represented opportunities for many individuals with limited formal education to parlay technical skills and hard work into socioeconomic security. Most started in entry-level positions. Over time, on-the-job training, and in some instances, willingness to perform dangerous jobs, meant relative job security and incomes that often exceeded their peers, both black and white, in other occupations. Yet decades later, health challenges and a line of widows call into question whether, given the limited alternatives, steel mill positions represented viable employment options for blacks or just another example of post-migration disenfranchisement.

POST-MIGRATION EFFECTS ON RELIGIOUS EDUCATION: COMING BACK TO GOD

Migration resulted in employment for most members of the King family and certain difficulties as a result of these same opportunities. Blacks generally contended with worsening employment barriers in unwelcoming northern spaces. Green and Woodson (1899) summarize these grim conditions:

> The lines along most of the avenues of wage earning are more rigidly drawn in the North than in the South. There seems to be an apparent effort throughout the North, especially in the cities, to debar the colored worker from all avenues of highly remunerative labor which makes it more difficult to improve his economic condition even than in the South. (P. 82)

The end results were conditions in many northern areas that paralleled those in the South. How did these new economics problems affect Christian dictates referred to here as religious education? How did the Kings make sense of these situations? Did they continue to espouse the ethos lived by Irma and George or did the unfettered ecology away from Mississippi precipitate new ways of thinking about religiosity and lifestyles?

Blacks in the South who espoused Christianity were often skeptical of whites who embraced this same faith tradition, but directly or indirectly perpetuated segregation and racism. Attitudes and actions of such whites contradicted the Great Commandment in Matthew 22: 34–40 to treat others as you want to be treated, Luke 6:31, to do unto others as you would have them do unto you, and the plethora of other bible verses that encourage unconditional love, altruism, community-building, and selflessness.[27] The primary character in *All God's Dangers* (1974) describes this conundrum: "God's got some good people here [in the South] and He's got some here ain't fittin to go to hell . . . and you don't know none of em's ways unless you watch 'em. The Bible tells you to watch as well as pray" (p. 489). Disillusioned by continued oppression by southern whites, this caustic indictment suggests both a weak correlation between attitudes and behavior as well as the importance of prayer and discernment described in passages such as Matthew 26:41, Mark 13:33, Mark 14:38, Ephesians 6:18, and Colossians 4:2.[28]

Despite the dramatically different ways that many whites and blacks in the historic South understood and appropriated scripture, the Kings were staunch Christians. Narratives presented in earlier chapters illustrate their attempts to socialize their children, and by proxy, other progeny, to embrace this same faith tradition. Applying developmental phases from ecological theory enable us to consider whether and how the Kings changed over time in terms of their religious ethos. Some family members describe a consistent Christian lifestyle. However, it is equally common for others to describe periods during which they "got off track" and were less committed to Christianity and the Black Church, yet, like the Prodigal son in Luke 15:11–32, returned and recommitted. This latter tendency was typically linked to age and migration experience.

For example, James Lewis, a 69-year-old steel worker, remembers George and Irma's training: "Religion was very important growing up. We went to church on Sunday and respected our elders ... it is still the same. Matter of fact, I spend more time in church now that I used to 'cause I understand it better." James acknowledges his family's emphasis on church attendance and respecting adults. He also alludes to his personal Christian development that included inactivity as a young man followed by hyperactivity as an older one. His wife, Deidra, describes a period when her husband had distanced himself from his parents' religious ethos:

> It [Christianity] helped me through a lot of hard times because when I first got married, my husband wasn't saved and he was in the street all night long and I had to keep praying and asking the Lord to see him safely home. So the more I prayed, the closer I got to the Lord. And I was praying for deliverance for him. And God answered my prayer.

Like his father, George, James was not an active Christian during his early adult years. Post-migration life in the North provided varied diversions. And like his father, James re-committed himself later in life. Yet, as promised in Proverbs 22:6, this *home training* stayed with him.[29] Moreover, Deidra describes her own commitment to God, family, a local church, and the Black Church prayer tradition – and eventually witnessing her husband's spiritual transformation. James Lewis is now a deacon, usher, and Sunday school superintendent. Deidra also describes the source of her long-suffering faith, and dedication that suggests the importance of Christianity for blacks in Gallman:

> Religion was my father's top priority. He would make sure we sang songs, read scripture and prayed every night ... The thing I'm most proud of happened between me and my father. After I had grown up and came back [to Mississippi] and I went to Sunday school with him one Sunday and they asked me to review the lesson [summarize it to the congregation]. And the way I reviewed the lesson, I could just see the pride on his face. That was a lasting memory. And when I got home, he asked me how did I learn all about that and I told him how he had taught me about the bible. It made me proud because it made him happy ... It is still top

priority. I live based on the same values my father passed on to me. (Deidra, 64 years old homemaker)

For Deidra, religious education has been invaluable and enabled her to weather personal and familial challenges. This training came full circle during one of her visits to Gallman. Her ability to exegete scripture is a direct result of religious education as a child that she and her spouse now provide to their family. Fred, Gina Mae's son, specifically references Matthew 7:7 and Luke 11:9 when describing a godly intervention:[30]

People say "God is good," but they don't really know unless they ask God to do something and God gives it to him. When I was injured I started drinking. I told my daddy, "I'm gonna stop drinking." He said, "You've been saying that for years" . . . He [God] said, "If I make one step, he'll make two." A couple of weeks later, I realized I didn't want anything and I haven't drank since December 1982. So I just asked. I have a close bond there [with God] because of the things He has done for me. I tell anybody, "If you ask, He will do the same for you." If you ask and really want it. You just have to be sincere about it. (Fred, 69 years old)

After self-medicating for several years after a chronic back injury, Fred believes that God miraculously cured him from drinking. Sober for 34 years and because God is not a respecter of persons, Fred is confident that other people's needs can be similarly met. Martha also correlates religious and secular education and family support: "I'm proud of the fact that we all try to stay connected and close to one another and we stress education. We encourage our family members who are not saved to get saved and to accept the Lord as their personal Savior" (Martha, 77 years old).

Although described in diverse ways, the importance and nature of religious education provided during childhood emerged as a common theme; subsequent generations of Kings recall benefits and, in some instances, challenges associated with Christianity for them and blacks today. In support of the introductory scenario to this chapter, for 51-year-old Vera, home life included seemingly constant exposure to Christianity that she now imparts to her 22-year-old son:

Church was the foundation of everything! We practically lived in church – preaching, praying, testimonies and spiritual songs until I left for college. My faith and direction were definitely realized and reinforced from church. And I found that I have instilled these same things in Denny.

She provides a more detailed, somewhat comical childhood account:

The family ritual before we ate was long. First we all had to go around the table and recite a bible verse. I had learned to say, "Jesus wept," because it was quick and I was trying to speed things along so we could eat . . . Let's get to the food that mama cooked! Often fried chicken, greens, macaroni and cheese, cornbread! ... My daddy always prayed before every family meal – and his prayers *were so long* [emphasis is hers]. I mean he would pray for at least twenty minutes. And while he was praying, we were supposed to have our eyes closed and head bowed. But about fifteen minutes in, I would always open one eye and look around the table – and my sisters would be doing the same thing [laughs]. I would slowly sneak some food from my plate while he was praying ... Sometimes he prayed so long that the food would be cold by the time he finished. My Daddy was known for his long prayers [laughs again].

The above comment reflects two facets of Vera's family's religious education – the importance of scriptural knowledge and prayer. Her reference to Jesus' response to the death of Lazarus in the book of John 11 reflects a *mini-bible lesson* her father routinely provided prior to meals – as well as her attempts to expedite such experiences. As a deacon, Vera's father also had a reputation for lengthy prayers during worship services which he continued at home. Vera's childhood response differs from her current understanding of prayer, but this story is told often during large family gatherings.

Family members make the distinction between a personal relationship with God that reflects abiding belief in biblical tenets and consistent participation in church. Many of George and Irma's grandchildren admitted to being inactive in church as young adults, yet never wavering in their faith, and eventually re-committing themselves to Christ and a local congregation. Below Deidra describes a challenge and her faith that, through prayer, her youngest son will also follow this same path:

My personal challenge was my third son, Conrad. For some reason he was adventurous and he would get into a lot of trouble, so I had to stay prayed up. He is one of the reasons that I have a personal relationship with the Lord now because he kept me on my knees praying for the situations he got into ... My Mom taught me cleanliness. My Dad taught me about the Lord, and my children taught me about love.

Deidra attributes both her spiritual and personal hygiene to her parents. Just as they trained her, she hopes that her children will also rely on the godly model she and her husband provide. Moreover, she attributes her maternal role as the source of her own development about unconditional love.

Intra-Generational Doubts and Destinies

It was more common for members of later generations to question the rigidity of religious education. Other family members did not stray from

Christianity, but now rely on a more critical exegetical lens. For example, Donna is critical of the *silk stocking*, upper-class black church she was required to attend as a child, and expresses appreciation for the small black church that she would eventually join as an adult:

We went to a large church with a pipe organ and lots of snooty people. I'm sure they believed in God, but they were very mean and hateful. There was a stark difference between that church and the members of [church name]. That's where my Aunt Dee [Carla] and Uncle Horris went to church. It was a small church with no parking and probably only twenty seats including the choir stand. But, they sang with their whole hearts, prayed, and received God's favor. I needed that level of dedication in my life at that time. Sitting under their ministry, I believed that there was nothing that God couldn't do. I loved that they were willing to put their faith into action ... They nourished my soul. Things weren't the same when the older saints passed on, but the legacy they left was absolutely amazing! I praise God for Uncle Horris and Aunt Dee, Sister Parkins, Brother and Sister Raberry, Pastor and Sister Jimson, Sis McMurry and Elaina on the piano. (Donna, 53 years old)

Another family member describes a similar experience at a class-conscious black church:

For about two years, I was a member of a good-sized Baptist church in [city's name]. I joined because of their commitment to the community. They provided low-income housing, a weekly feeding program, and other social justice programs. But after getting involved, I became disillusioned by the politics and negativity. Yes, they did things in the community, but looked down on the people they were helping – just a lot of middle- and upper-class blacks who did not get along and who put on a lot of airs. There was a lot of infighting and very little love in that church. It was very stressful. I eventually left. I don't ever want to attend a church like that again. (Tanya, 53 years old)

Like their predecessors, the vast majority of Kings attend Baptist churches. This tradition is known for its intra-denominational diversity. Based on their experiences, both Donna and Tanya rejected the pomp and circumstance of churches frequented by doctors, lawyers, teachers, and other influential blacks. The former experience occurred in the Chicagoland area, the latter in a southern city, yet both parallel scholarship about black middle-class proclivities common in the Black Belt.

As noted in Drake and Cayton's (1945) *Black Metropolis*, education, occupation, family habits and structure, recreational activities, and *religious behavior* delineate socioeconomic class position:

The middle class is marked off from the lower class by a pattern of behavior expressed in stable family and associational relationships, in great concern with "front" and "respectability" . . . the middle class is subdivided by effective income and (to a lesser extent) by occupation and education. (Pp. 661–2)

These same scholars posit that respectable activities often took place in church. Donna contrasts her experience with the more folksy worship and close-knit fellowship she experienced after joining the Baptist church that her Aunt Carla attended. She describes specific spiritual benefits and a litany of *saints* who helped transform her life.[31] Although Donna avoids questioning another person's salvation, she is suspicious of Christians whose behavior does not reflect God's central command in I John 4:8 to love one another.[32] Similarly, Tanya became both disenchanted and traumatized by her time in an upper-class black church that appeared more focused on paternalism and materialism than agape love and community empowerment.

The thoughts above illustrate increased circumspection and self-reflection around Christianity as well as religious accountability and expectations. Tanya acknowledges her commitment to Christianity *and* respect for other faith traditions:

I'm glad our family continues to be Christian. My Christianity is unwavering, yet I still respect and appreciate other religious beliefs, especially African religions, that actually pre-date Christianity. I don't think that honoring those practices contradicts what it means to be a Christian. As I've gotten older, I am much more bothered by people who use Christianity to shame and hurt. Jesus was all about love. So I don't support interpreting the bible in ways that hurt or devalue gays and lesbians, women who have children out of wedlock, poor people, or anyone. A lot of people may think my version of Christianity is too liberal . . . God bless them too. (Tanya, 53 years old)

Tanya wishes more Christians were inclusive of diversity and focused on love. Moreover, she has great concerns about traditional biblical interpretations that have historically been used to vilify certain groups. These reflections among later generations illustrate another way in which migration, exposure to multicultural spaces, and secular education have influenced how Christianity is understood and appropriated by certain Kings. Clear Creek No. 1, by all accounts, is a small country church. Some Kings went on to attend larger, *silk-stocking* black churches and had varied experiences – positive and negative. Their experiences are reminders of the diversity inherent in black churches based on factors such as class make-up, size, theology, demographics, and leadership type and style.[33] In addition to following their family tradition, younger Kings appear to be drawn to the independence of the Baptist tradition.

Yet not all family members have continued to embrace this tradition. Several other generational differences are apparent. Gina Mae's daughter was raised Baptist, but is no longer associated with this tradition:

I'm a Jehovah's Witness . . . for about 11 years. I got baptized in 2002 . . . First I was a Baptist. I didn't care too much for the Baptists. The preachers and deacons were going with women and married people in the church . . . they weren't right and they didn't think nothing of it . . . I met black and white people who were Jehovah's Witness and I like their lifestyles . . . Blacks have so many little churches on every corner instead of being together. (Ella, 77 years old, retired manager)

For Ella, a litany of incongruences caused her exodus from her family's faith tradition including infidelity by church leaders and lack of convictions for such sins. In addition, she suggests that the feature for which Baptist churches are heralded – congregational independence – is problematic for collective mobilization. Lastly, Ella was drawn to the inter-racial environment among Jehovah's Witnesses.

Although reared in a black Baptist church, 37-year-old Kevin now embraces spirituality rather than Christianity. He notes how Christianity has been used to marginalize blacks and justify social injustice:

I was raised in church, it was a big part of my life. And I still love God, but as I began to learn about my culture and my people, I started to wonder about some of the things black people believe about the bible. I learned that a lot of things in the bible have been found in African religions long before the bible was even written! We come from a rich culture and our people don't realize that – we don't realize that Christianity does not represent us. As a proud black man and the things I learned about Africa really connected with me . . . taking care of my family, being in nature . . . we [black people] need to know these things.

Kevin contends that blacks should question the appropriateness of Christianity as a model for their race based on its ethos that encourages docility, its historic application that oppressed minorities, and its inconsistent stance on diversity. He questions organized religions and Christianity's connections to colonialism:

We went to church every Sunday. I believe in universal laws. I don't feel it is right or just to take from, harm, or hinder any living being . . . I try not to hang around people who try to be so focused on being a Muslim or Christian. I pray to the God of our ancestors. I am the same person, I just don't consider myself a Christian. I know the bible . . . Spirituality is very, very important to me. If your spirit is negative, I can't vibe with you . . . I don't judge them, but I choose not to be around them closely . . . religion has no part in my life, but growing spiritually and being morally and spiritually sound is part of my daily life.

As he got older, Kevin had more questions than answers:

I understand the history of slavery, and even when I understood part of slavery, I accepted Christianity. But then when I learned on a different level, I checked my sources and history. The more I found out about slavery, I saw that we were spiritual before that. Then I learned about Constantine. And how people who did not accept Christianity were killed ... Who did we pray to before [Christianity]? We were good before then and before that. ... religion was given to us and it wasn't what we initially had ... I pray to my Creator. Christianity has been used in opposite ways from African religions. Everything is male dominated in our society and women are degraded, in African religions, women were powerful.

Kevin points to African faith traditions that pre-date Christianity as more appropriate exemplars for Africans in the Diaspora.[34] He still loves God, but encourages blacks to consider spiritual paradigms that empower them, in general, and black women, in particular. Like Kevin, Nate, the protagonist in *All God's Dan*gers (1974), predicts increased self-reflection and self-efficacy among blacks:

[T]he bottom rail will come to the top someday ... the poor generation on earth will banish away their toils and snares. But won't nobody do it for em but themselves ... I'm a colored man, I've come in the knowledge of what it feels like to move out of this back yonder 'ism'; and I'm confident all of my race will someday move out from under earthly bondage. (P. 7)

Similarly, 53-year-old Tanya questions strict adherence to a biblical model of the nuclear family given challenges among blacks to realize this standard:

Christian women are told to be submissive to their husbands as family leaders. Even some really educated black men still want to follow this tradition. Yet they forget the rest of the passage about husbands being willing to sacrifice themselves for their wives. We have so many men who are attracted to the independent, educated, and successful women in our family, but they become intimidated and threatened by the very things that attracted them. Even when women in our family try to be submissive, it's still not enough. That conflict is still there and sometimes their men treat them badly ... They say blacks raise our daughters and love our sons. I think this happens a lot in our families. We try to protect our boys, but we sometimes enable them. We are trying to live based on bible verses that don't reflect our families and the black experience.

Tanya was taught bible versions such as Genesis 2:24, Matthew 19:5, Mark 10:7, Ephesians 5:31, and Ephesians 5:23 that encourage Christian women to desire marriage as central to their identities and lives.[35] To her, black churches and black families often send a message that marriage is the ideal, despite the reality that the biblical model, Jesus Christ, was single. Her concern is also informed by labor market inequities, differential rearing and educational outcomes based on gender, and other systemic problems that mean many black men are unable to enact a protector/provider role tied to this biblical edict.[36] Per Tanya, the result has been a mismatch between formally educated, gainfully employed black females and black males who are Christians, but not marriageable. Both Kevin and Tanya's views suggest that certain Christian dictates conflict with secular, systemic, and historic realities for blacks. According to Tanya, these tensions have resulted in as many divorces as marriages. Divorced couples among the Kings were not exempt from childcare challenges, budgetary constraints, and other logistics associated with raising children on one income rather than two. However, the advanced education and employment of most of the mothers, financial and social support from kin, and second marriages appear to have pre-empted the following common outcomes suggested by Darity and Myers (1994):

> The black middle class displayed a pattern of increasingly solid and stable family life until the late 1950s and early 1960s. But thereafter the pattern of marital dissolution and instability that has characterized the black underclass also penetrated the black middle class. (P. 221)

Despite concerted interventions by family networks that have largely staved off downward mobility predicted by scholars such as Berry and Blassingame (1982) when nuclear families dissolve, the influence of the King's ethos means that marriage remains the preferred arrangement – even among divorced family members.

Lastly, several family members candidly describe more extreme deviations from the King family ethos and life-changing experiences akin to the Prodigal Son in Luke 15:11–32. Although 56-year-old Donald Jr. is now a pastor and administrator in the Salvation Army, he describes a path away from his middle-class, Christian upbringing:

> I went to college in Industrial Education . . . got a job, then I went into the Marine Corp. I got caught up and went to jail for possession to sell in 1985 . . . Then I landed in prison 1987–97 – going in and out of prison. In 1997, I went to a sober living program in San Diego. I got a job and then my life changed . . . God took me through what He took me through to place me where I am today . . . now I can help others. I've been homeless, doing drugs . . . not knowing what to do . . . God brought me out of that and placed me where I need to be to help those who need help.

An administrative oversight meant Donald Jr. left college after his freshman year. Drug-related decisions resulted in a military discharge and years of incarceration. Yet he attributes his current successes, including marriage and ministry, with a holistic understanding of God and Jesus Christ as well as God's purpose for his life to help re-acclimate incarcerated men:

I thank God for pulling me up out of the muck and mire . . . He placed me on the right road when I surrendered my life to Him . . . God spoke to me when I was about 11 or 12 . . . He told me he wanted me to preach His word. But I ran into the world . . . then 40 years later, God spoke to me again . . . I truly know God today. I have seen Him orchestrate and work in my life firsthand . . . When my mother was sending me to church, knowing Jesus is one thing, but having a relationship and understanding Him is a totally different thing. When I was a kid, I just heard about Jesus . . . now I have an understanding of Him . . . I have ninety-two men in our program . . . I oversee the men's residence, donations, the warehouse . . . the entire operation. We try to help them get back on their feet and get back into society.

Donald paraphrases Psalms 69:1–3 to contrast his past problems with his present transformation. Calvin Jr. also offers a "before and after" narrative of worldly living during his youth that deviated dramatically from his upper-middle-class upbringing:

I knew about God as a teenager, but I ran in the opposite direction . . . I found myself running with the wrong crowd. I never was a drug dealer, but I hung around people in that world. You can get addicted to that lifestyle. Most of the time when I went out, I had a gun or a knife . . . but when that excitement was fading, I found myself asking, "Calvin, why are you here? You know God has a purpose for you." Then I asked God to save me . . . the last time I asked, He said, "I already have, but you keep going back to it" [his past life] . . . I've been a minister 22 years . . . That's why I ran at first . . . but when I made the commitment, I was serious about it. By growing up in church, I knew it wasn't nothing to play with, but I knew God had a great purpose for me. (Calvin Jr., 52 years old)

Although Calvin Jr. was able to avoid criminal activity, he was drawn to the entitled lives of such persons – until he realized he had strayed from his family's religious ethos:

God put people in my life who mentored me. They told me to be true to who you are and who God called you to be . . . my previous pastor once said, "every good idea is not a *God idea*" [emphasis is his] . . . be true to what God gave you . . . I was fortunate that I had people in my life early in my ministry that helped me so I didn't have to stumble and make a lot of mistakes. Let God open doors . . . I've been a truck driver about six years . . . The money is great, but I know that God has more for me . . . ministry-related things, but we don't know his timing.

Now a minister, Calvin's current trucking position represents stable employment he believes is a precursor to a return to the pastorate. Both Donald Jr. and Calvin Jr. describe "running away" from their early upbringing, but now having a deeper appreciation for religious education and its importance in their lives. Their developmental processes include several common factors: an existential question about their purpose in life; parallels to an Old Testament story in Jonah chapter 1; disillusionment with secular trappings; recollection of their childhood religious training; and a return to Christianity.

For some family members, negative church experiences meant leaving Christianity for short periods or re-locating to smaller black churches more akin to their spiritual and non-spiritual needs. Other Kings strayed away from their early religious upbringing, but are grateful for the opportunity to return later in life. Several consider themselves more spiritual than religious. And others continue to espouse Christian dictates, but have become disenchanted by a routinized Black Church that they believe is failing to proactively respond to contemporary needs and concerns in the black community. Their views and practices are nuanced based on factors such as age, formal education, parenthood, personal experiences, and discriminatory encounters.[37] Yet the vast majority of Kings suggest that adhering to the religious education espoused by their foreparents has enabled them to more adeptly and positively navigate society.

CONCLUSION: MIGRATING OUT OF THE SOUTH AND INTO THE MIDDLE CLASS

A group of King family members left Gallman, Mississippi, in search of opportunities and to escape the yolk of segregation. Just as the South represented a litany of limitations, losses, and lies about the American Dream, the North symbolized a place of promise and prosperity expected by God's chosen. Many of Irma's siblings migrated in the mid- and late 1940s; her children would follow suit in the late 1950s and early 1960s. Over time, most would establish themselves as members of the middle class in the places where they settled; some struggled as working-class families, but were able to position their children firmly among the middle and upper middle class. Yet several were unable to pass on their middle-class sensibilities and status to their offspring.

Given the centrality of property in establishing the Kings as middle class in this analysis, all of George and Irma's four children purchased homes that are now paid for; 50 percent of their grandchildren are also homeowners.[38] And the King homestead in Gallman, Mississippi, still remains in the family. Migration spells on the family farm, as an important mechanism by which other generations of King's often experienced the middle class, helped some family members stay on track; it also helped others get back on track. These transition periods are in the spirit of scholarship by Oliver and Shapiro (1997) on how many white families often help launch their children. And unlike many whites, a self-help tradition has been associated with the black community, regardless of class position.[39] As was the case in Hazlehurst and Gallman, urban blacks were broadly stratified based on race, class, and gender. As in the South, property influenced class position. Unlike the South, the Kings suggest that, despite racial and economic challenges, job opportunities were more likely and chronic overt oppression less frequent. However, not every well-off family member welcomed migration spells or availed relatives of their networks and weak social ties. And not all migration spells were favorable. Yet family members who embraced the King family ethos and related religious practices were more apt to participate in these linked-fate initiatives.

Conclusion

"What Would Big Mama Do?" Activation and Routinization of a Black Family's Ethos

WHAT OF THE TWENTY-FIRST-CENTURY BLACK MIDDLE-CLASS FAMILY?

The objective of this historical analysis was to study varied ways of conceptualizing and understanding the black middle class. One southern black family's story was examined across time. A socio-ecological lens focused on systemic, group, and individual dynamics that enabled the King family to both become middle class in Gallman, Mississippi, as well as support relatives and friends in need. Early generations of Kings were slaves, then sharecroppers; land acquisition and farming positioned their children for the middle class. But in many ways, the Kings led unconventional middle-class lives. Their experiences illustrated the multi-faceted nature of the black middle class that includes material possessions, but equally important, certain values, aspirations, and expectations informed by an agrarian context. The King's story also differs from common portraits of the "middle class" - particularly for their earlier generations - based on their emphasis on religious education and largely disdain for conspicuous consumption. George and Irma were Mississippi farmers. They only wanted an agrarian lifestyle - but an economically stable one for themselves and their children. Family narratives suggest that Irma activated most of the efforts that resulted in their socioeconomic successes. We contend that their experiences help expand scholarship on the black middle class.

Like the Kings, studies show that it continues to take more time, effort, as well as familial and community resources for middle-class blacks to maintain their economic foothold. Just as George and Irma walked an economic tightrope most of their lives, many black middle-class families continue to do so today.[1] The goal here was to inform and continue the conversation around a seminal question – what of the twenty-first-century black middle-class family?

In addition to the import of religious and practical education, the King family's communal ethos included family connectedness, a linked-fate mentality, delayed gratification, and community involvement. Yet their lives illustrated that grit and a religious orientation alone are insufficient when compared to the wash of systemic forces that are capitalistic and often culturally pejorative in nature. Just as Mason (2003) suggests that economically, blacks are still dealing with the negative repercussions of racial inequality one hundred years later, we argue that black families today are still attempting to navigate many of the negative effects of historic inequities, post-industrialization, the advent of belief in reverse racism, a white racial frame – and more recent upsurges in old fashioned racism. Cotton (1993) provides a more sobering perspective:

> Racial employment and wage differences have persisted over time because labor market discrimination has persisted over time. And labor market discrimination has persisted because racism is still an American social affliction, the true "tangle of pathology." (P. 202)

A combination of these socioeconomic and ideological challenges mean that many black families – even middle-class black families – continue to expend an inordinate amount of time and resources to combat dynamics that are often ignored or downplayed in the broader society. Economic indices illustrate these challenges.

According to census data, the 2016 median household incomes for Asians, non-Hispanic whites, Hispanics, and blacks (provided from greatest to least) were $81,431; $65,041; $47,675; and $39,490. Average household income for all races was $59,039. Non-Hispanic whites (2.0 percent), Hispanics (4.3 percent), and blacks (5.7 percent) have experienced increases in household income between 2015 and 2016, yet the latter two groups continue to lag considerably behind whites and Asians.[2] From 1967 to 2016, continued disproportionately higher real earnings are apparent for Asians and whites as well as the tendency for blacks to consistently have the lowest median household income. These differences are driven by factors such as lower relative earnings and single-parent households among blacks. White (2016) further illustrates continued inter-racial challenges: "median wages for black male workers during the fourth quarter of 2015 were only 72.4 percent of those of their white counterparts. And unemployment among black workers is around 8.8 percent, while for whites, it's closer to 4 percent. And for workers lower down on the totem pole of skills, the gaps are even more troubling" (n.p.).

Wage differentials are directly correlated with poverty. The overall official poverty rate in 2016 was 12.7 percent or 40.6 million people.[3] Rates based on race during that same year were: 22.0 percent or 9.2 million blacks; 19.4 percent or 11.1 million Hispanics; 10.1 percent or 1.9 million Asians; and, 8.8 percent or 17.3 million non-Hispanic whites.[4] The poverty rates have slightly declined for blacks (from 24.1 percent) and Hispanics (from 21.4 percent) in 2015, yet are still considerably higher than those of their peers. Figure C.1 graphically illustrates black–white household income ratios from 1990 to 2016. Ratios range from a low of 0.56 in 1992 to a high of 0.65 in 2000. Despite fluctuations across time, ratios gradually increased between 1999 and 2008, but have declined since then. Overall, median black household incomes have been *substantially below those of whites each year* under consideration. So although blacks have experienced increases in real household income and a slight decline in poverty, they continue to lag behind non-blacks for common economic indices – suggesting on-going systemic problems.[5] Austin (2016) provides the following contrast about black economic outcomes:

> The earnings gap between African American men and white men is the same now as it was 60 years ago for the median worker ... [yet] the wage gap [is] closing swiftly at the top of the earnings scale, suggesting that targeted, race-specific policies have been effective in expanding access to top-tier educational

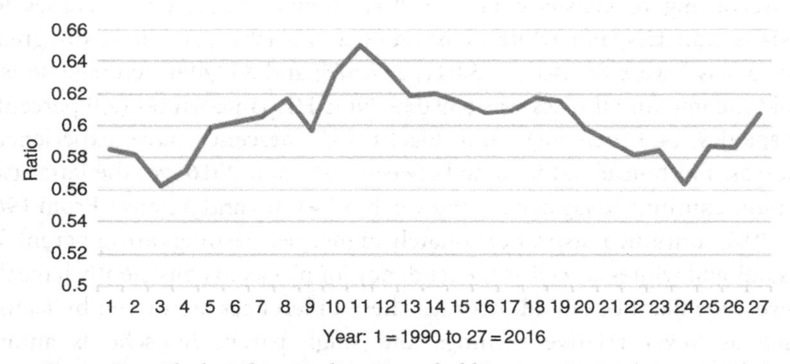

Figure C.1: Black–White Median Household Income Ratios (1990–2016)
Key: 1 = 1990 and 27 = 2016. Each data point represents a one year increment. Data taken from the US census.

opportunities and high-wage professions ... it's astounding that, in terms of economic rank, a black man in the middle of his economic distribution is no closer to his white counterpart in terms of earnings than was his grandfather ... the labor market for low-skilled workers has basically collapsed. (n.p.)

This earning bifurcation provides a glimmer of optimism for blacks at the top of the socioeconomic ladder and less hope for those below. As this study has illustrated, the economic conditions many of their grandchildren and great-grandchildren face today do not differ dramatically from those that Irma and George encountered.

Our results suggest that the Kings, especially their matriarch and patriarch, believed that they were emotionally, psychologically, and literally sustained by their religious beliefs and protections via God's promises.[6] Land and livestock were tangible evidence of God's favor. Findings here suggest multiple benefits of property ownership for the Kings. According to their recollections, the family homestead was the primary mechanism by which they were able to establish themselves as a middle-class family. Equally important, landownership enabled them to actively participate in what we referred to here as "migration spells" where family members could periodically seek refuge on their farm for economic and non-economic reasons.

In addition to spiritual, emotional, and psychological support, food stuffs and other resources generated on the family farm helped enhance the quality of life of other local blacks. The influence of property beyond its common association with wealth, but rather as a community asset, was apparent and reflects an understudied line of inquiry. The appropriation of a *village* form of family support could not have been possible without the family homestead. Thus these findings are also important, as they illustrate how property and homes can serve as meso-level resources. Although the actual farm no longer exists, the land remains in the family as an example of sacrifices and as a symbol of what the Kings were able to achieve.[7] Just as scholarship emphasizes the importance of weak social ties, we consider the King family homestead a *familial social tie* with bonding and bridging effects.

Bonds reflected the family's ability to remain connected across generations and locations. Bonding also enabled certain individuals to build upon the economic stability of Irma, George, and other middle-class family members. Similarly, bonding enabled certain economically challenged family members to avoid the long-term consequences literature suggests occur when one experiences poverty. For some family members, the homestead was the difference between bouncing back rather than being set back by problems.[8] When bridging is concerned, the family ethos informed by religious education was an intangible tool to guide beliefs and behavior. Religious education also encouraged family members to pursue formal education – which also facilitated the transition to the middle class for many of them.[9] And just as religious education provided practical motivations, family members believed that it provided them with crucial guidance, encouragement, biblical success stories, and, sometimes, social sanctions.

The family ethos was directly influenced by a key institution and individual, initially, Clear Creek Baptist #1 Church and the family matriarch, Irma. This church and several others, including New Sardis Missionary Baptist, Miller First Baptist, Bethlehem Baptist, Christ Baptist, and Trinity Missionary Baptist represent meso-level community educators for this family and its offspring. This particular finding is important because it directly correlates both meso-level organizations and an intangible cultural ethos linked to Christianity to the subsequent socioeconomic well-being of this black family. Studies about the importance of the Black Church for black families are not new. We have referenced many of them in this book. However, we show the singular thread that a nuanced understanding of religious education can play in the lives of blacks. However, Christianity as a sole motivator was insufficient; family accomplishments required a strategic combination of religious, secular, and practical education and decisions as modeled by seminal family members. And in doing so, seemingly traditional rituals, roles, and responsibilities were blurred, especially by female family members, to secure certain socioeconomic outcomes. Furthermore, blurring boundaries enabled individuals to navigate macro-level impediments through the strategic use of meso- and micro-level group and individual capital.

Later generations of Kings have had considerable educational accomplishments, especially females. Most are middle class as defined in this study. However, their education did not necessarily translate into high incomes based on their tendency to earn advanced degrees in "soft sciences." Additionally, divorced mothers among the Kings are not necessarily benefiting from consistent child support or the financial stability associated with dual income households.[10] So although these caregivers have been able to avoid the feminization and juvenilization of poverty,[11] their families continue to have economic challenges. Yet the King's ethos values and encourages marriage and nuclear families. Black families like these must grapple with the implications of diverse family forms, especially marriage, and its expected economic and emotional support in light of the following: the likelihood of meeting marriageable males, black male incarceration rates, inter-racial marriage rates between black men and non-black women, and the elusiveness of this institution for better educated black females.

Extending Socio-Ecological Theory in Black Middle-Class Spaces

This study has also endeavored to illumine the broader ecological landscape in which the King family navigated. Informed by ecological work by Bronfenbrenner (1977, 1979, 1986) and Billingsley (1992), this study illustrates that middle-class black families like the Kings were a product of micro-, meso-, and macro-level structural influences. Micro-level dynamics were driven by a familial ethos that encouraged education in multiple forms and other energizing values. Meso-level influences were connected to the Black Church and local under-serviced black schools. Moreover, the King family homestead also represents a meso-level asset in its use as a multi-dimensional, intra-familial and inter-generational haven.

Middle-class aspirations alone did not result in the outcomes documented in this book. Another important finding here reflects the implications of racialized employment patterns. King family members who became middle class did so as a result, in part, of employment opportunities linked to the military, manufacturing jobs, and labor-market-driven dynamics due to international military conflicts that opened up employment for blacks. The combination of these individual and systemic effects are important, as we consider the black middle class moving forward. It was common for husbands to be employed via military stints or in steel mills. Although gainful employment, these were often positions that put their lives in danger or would later result in health problems. Yet these wages alone could not sustain their families. Spouses, usually wives, held crucial, often lower-paying, positions that augmented their partner's incomes – while also juggling domestic responsibilities. Most wives had some formal education; most husbands did not, but had acquired valuable on-the-job training.

This economic formula seemed to work for many of the Kings, but presents a contemporary conundrum. What are the equivalent manufacturing and/or government-related occupations or arenas were conscientious blacks, particularly men, who lack formal education can find gainful employment? What are the present day systemic alternatives or substitutes? Can blacks expect this *dual-family employment formula* to be replicated? Can the current global economy support such needs? In addition, social policies that forced integration in the military, and later, housing and educational arenas, helped cultivate a fertile context for this family's similar successes. It remains to be seen whether future policies and practices that result in structural change will advance or turn back similar possibilities for blacks today. Similarly, military stints may not be as attractive for black millennials who have come to expect more or embrace more entitled attitudes. Or will belief in a post-racial society mean young blacks are emotionally and psychologically ill-prepared to navigate covert forms of racism, classism, and discrimination that exist.[12] Also specific to King family strategies, it will be important to consider property ownership by blacks where contemporary versions of "migration spells" could occur. For this family, property ownership and the person of Irma were critical anchors and stabilizing forces. Do similar individuals and organizations exist today? These and other questions are imperative as we contemplate the future of the black family in general; unchecked challenges will greatly influence the continued presence of the black middle class.

More than Grit

Continued references to Irma as the "family anchor" are indicative of her singular role. However, this analysis illustrates a myriad of ways that the King's history is more than a tale of grit and intestinal fortitude. Readers should be cautious against attributing their "triumphs over trials" to exceptionalism or grit. Their candid, often solemn comments, as well as the often tumultuous socio-ecological context in which they lived, belies a one-dimensional understanding of their experiences as they navigated systemic forces. A grit-based exemplar is reductionist because it does not respond to the following types of questions. Why did the Kings have to work *so hard* to accomplish their dreams? Why were Irma and George's dual jobs necessary?

Moreover, most individuals followed the family ethos and worked tirelessly, and yet some of them did not escape poverty. And paralleling findings in Pattillo-McCoy's (1999) *Black Picket Fences*, some middle-class family members were unable to pass their class position on to their children. These constraints reflect another effect of systemic inequities that grit cannot remedy. Yet grit tacitly positions individuals as herculean in their energies, abilities, and wherewithal – no matter the context. This was not the case for Irma or other family members. Exceptionalism does not do justice to the physicality and energies needed to fight for middle-class entrance in a society fraught with dynamics designed to prevent this goal linked to racism, segregation, Jim Crowism, industrialization, post-industrialization, and contemporary forms of racism and systemic inequities that exist today.

However, an emphasis on grit has been historically used to divert attention from the structural forces acting upon individuals who are encouraged and expected to exhibit grit. This tendency is the most problematic outcome of the use of simplistic, often-ethnocentric approaches to examine the experiences of historically oppressed peoples. We have intentionally attempted to avoid this paradigmatic and methodological trap. Moreover, grit exonerates the multitude of people and groups in power who have historically oppressed other people, as well as supposed allies of minority groups who have directly or indirectly been complicit and benefitted from the same ends. In addition to navigating macro-level dynamics, our results show that the King's outcomes were the result of a plethora of intersecting beliefs, behavior, mistakes, strategic decisions, and sometimes serendipity, that cannot be adequately captured in scholarship.

IF IT WEREN'T FOR THE WOMEN: THE ROLE OF IRMA AND OTHER MOTHERS

Upon embarking on this study, we had no idea that one individual would emerge as the linchpin in this family's story. Yet Irma's name was continually mentioned in narratives and decision-making across generations and among local blacks in Gallman, Mississippi. Without discounting the influence and input of a plethora of other family members, individuals thought it not robbery to acknowledge Irma's hand in what ultimately became a black family's middle-class ethos and lifestyle. Most respondents describe Irma as the King family matriarch. They refer to her respectfully and affectionately as "Big Mama." Irma's ideology about religion, education, family, and hard work undergirded the family's positive outcomes. Moreover, our findings suggest that her perspective – that combined religious and practical teachings – was a common, albeit varied, thread through many family narratives. Stories describe a woman whose quiet demeanor hid the savviness she possessed to navigate precarious, potentially dangerous, predominately southern white spaces.

Because Irma was a "country girl," it is unclear whether she was necessarily interested in becoming "middle class" per se, but rather had her own organic understanding about what constituted stability for herself and, more importantly, for her children. Her beliefs and behavior around what is typically considered "middle class" replaced conspicuous consumption with Christian dictates. Her appropriation of Christianity may have seemed quite traditional at first glance. However, a deeper examination of family narratives suggests that she appropriated key biblical tenets to reflect the contextual realities of black life, particularly for women and children. For Irma, the exegetical process was informed by ecology. In this way, her worldview was akin to those of black women documented in research by scholars such as Patricia Hill Collins, Cheryl Townsend Gilkes, Stacey Floyd-Thomas, and Emily Townes. However, we posit that Irma's story differs based on her spiritual and practical discernment of the southern economic, social, and political terrain that enabled "her own house and a milk cow" to translate over time into a southern middle-class reality.

It is equally important to consider how this process that led to upward mobility unfolded. Without much fanfare, Irma's basic beliefs about God, family, and community were tacitly embedded into this family's attitudes and actions – largely based on her actions. Family members suggest that she was extremely transparent in her views, desires, and interests – even regarding traditional gender roles. Whites endeavored to politically, economically, and socially disempower blacks in Gallman, Mississippi. Yet it was reputational and relational power rather than positional, coercive, or institutional power that made Irma, and later George, most influential. Additionally, this matriarch appeared to have embraced personal power based on her relationship with God and the knowledge and wisdom she believed it afforded. Her lived experiences seemed to evoke a certain amount of trust and care among her family and peers, including local whites in Gallman.[13]

In addition to being a mother to her children and subsequent generations, Irma's role as an "Other Mother" in the community meant that blacks who were suffering even more than she could receive various forms of support. The mothers and other mothers depicted here are described by Collins (1990): "black women's vulnerability to assaults in the workplace, on the street, and at home has stimulated black women's independence and self-reliance. In spite of differences created by historical era, age, social class, sexual orientation, or ethnicity, the legacy of struggle against racism and sexism is a common thread binding African American women" (p. 22). We suggest that King women such as Janice, Irma, and their female offspring engaged in nontraditional forms of activism that, "reflected black women's refusal to relinquish control over their self-definitions. While they pretend to be mules and mammies and thus appear to conform to institutional rules, they resist by creating their own self-definitions and self-evaluations in the safe spaces they create among one another" (Collins 1990: 142). The King homestead was considered such a safe space.

Yet property ownership alone could not foster upward mobility without the corresponding linked-fate motivations that inspired migration spells. Not every family member was open to hosting migrations spells. This is another way in which Irma represented an Other Mother. She could not turn anyone away. An etic perspective might frame her as docile and over-accommodating, particularly relative to gender roles. However, an emic frame suggests a discerning, shrewd black women driven by a daily desire to live out biblical tenets, particularly as a model for her family and the community. Over time, such actions became routinized into the family's ethos. And in addition to the intrinsic benefits of servanthood, God's promises in verses such as Luke 6:38 ensured that her work would not be in vain.[14] More practically, these efforts also meant that each of her children achieved the educational milestone that symbolized socioeconomic stability at the time – a high school diploma. This educational standard may seem minimal today, but was uncommon in segregated agrarian communities.

Although revered and esteemed, women in the King family agree that mothering and other mothering is exhausting. Family members are clear – Irma bore the brunt of the work to rear her younger siblings, and later, to establish and maintain their homestead. Must contemporary black women replicate Irma's behavior and sacrifices? Given contemporary gender dynamics, expectations, and needs, is such behavior likely? Do black women want to embrace these gender roles? Should they? And how should today's black men and gendered power dynamics at various levels figure into black female caregiving experiences and expectations? Scholarship on the black experience, gender dynamics, and socioeconomic outcomes are challenged to thoughtfully and holistically consider the implications of black women as "anchors" in the black community and society overall as well as the effects of carrying such *weight*.

Irma's role as the family anchor and community servant was readily acknowledged and appreciated – but at what price? Did Irma and other black women like her have other aspirations, desires, wishes, and concerns beyond family and farm that were unstated or unaddressed? Did Irma have other unmet dreams that were tabled for altruistic efforts to help others and due to deeply ingrained gender expectations? What are the spiritual, economic, emotional, and psychological ramifications for black women who live similarly – for them, families, communities, labor markets, and society in general? Although the economic and non-economic costs seem incalculable, they demand consideration. In the spirit of the *Me Too* and *Black Lives Matter* movements, and simply because of the need to do the right thing, it is imperative to consider contemporary, new, personal and professional horizons black women wish to pursue as well as how black men, the black community, and majority members, broadly defined, must be involved in these transformative efforts.

FORWARDING A NEW MILLENNIUM DUBOISIAN MODE OF INQUIRY

W.E.B. DuBois performed the first comprehensive sociological study in the USA and it focused on the black experience. This project reflects the spirit of his work and endeavors to extend the processes he employed. *A new millennium DuBoisian mode of inquiry* represents a broad-based research paradigm and process to more thoughtfully and holistically examine social phenomena. Introduced by the first author in an earlier work, it reflects a multifaceted, multidisciplinary, and methodologically mixed model relevant to both the black experience and other historically disenfranchised groups.[15] As defined below:

> This *new millennium DuBoisian inquiry* will reflect both the spirit and rigor of his original efforts – applied to a contemporary global context . . . [it] moves the best of past academic practice into uncharted territory. It requires us to broaden our queries to consider international correlates, unexpected implications . . . as well as subjects and sites that have been heretofore rarely investigated . . . [researchers must engage in] revitalizing and re-appropriating historical Black inquiry . . . broad-based, traditional research should continue in order to compare and contrast quality of life indices of majority and minority groups; only by continuing to do so will scholars keep the larger society abreast of changes in the life chances of the populace, particularly for groups disproportionately affected by social problems such as poverty, classism, sexism, and health inequities. And just as large-scale studies can inform us about lived experiences, ethnographic work, theoretical projects, and other qualitative endeavors provide the singular ability to give voice to those who are often voiceless. (Barnes et al 2014: 190–1)

Black family studies that re-appropriate DuBois' scholarly lens in a multicultural, global context are challenged to consider: internationally-informed definitions of the black family; differing middle-class configurations; multi-consciousness, especially in terms of educational arenas; meaning making from religious and spiritual perspectives; comparative experiences beyond blacks and whites; and, intrinsic and extrinsic dynamics.[16] Additional intra-family analyses focused on religious and non-religious factors will also help develop more holistic research on the black experience.

This expanded paradigm also recommends the use of intersectionality, culturally sensitive lenses, counter-narratives as initial points of departure, and use of emic and etic perspectives. Practical and policy-focused research is equally important, particularly given the continued socio-political challenges black families face. Thus the scholarship we propose would reflect the spirit of DuBois' research to continue to: assess social problems that disproportionately impact black and brown people, the poor, and other marginalized groups; examine systemic to micro-level dynamics; illumine everyday forms of resistance and new forms of subjugated knowledges; and be widely disseminated in academic outlets as well as contemporary avenues such as blogs and newsprint. Specifically, for studies on the black middle class, this means considering contemporary views and appropriation of religiosity, spirituality, and evangelicalism by black religious consumers; multi-racial families and relationships; and global diasporic implications of class distinctions.[17] The overall objective is to perform studies that are "on the whole enough reliable to matter to some as the scientific basis of further study, and of practical reform" (DuBois 1899: 4). Trans-disciplinary research would result.

Finally, earning an education was the primary way later generations of Kings experienced upward mobility. Innovative studies on educational outcomes for black youth are needed to foster upward economic mobility based on their existing class positions. Because the King's cultural tool kit included religious, secular, and practical education, research would benefit from meso- and macro-level studies focused on multicultural educational programs that originate in religious institutions, culturally-sensitive instruction to cultivate diverse capacities across socioeconomic levels, and school metrics and black student successes across time and class.[18]

Members of the King family were able to survive and thrive as a result of attitudes and behavior informed by religious and secular education, as they availed themselves of systemic changes that temporarily kept some of the more deleterious effects of inequality at bay. Family narratives across time suggest that they believed Romans 8:31 applied to them: "What shall we then say to these things? If God be for us, who can be against us?" (KJV 2017). Such confidences buoyed their beliefs and efforts. But these same narratives described a plethora of challenges, many of which still exist for blacks today. Thus contemporary black families, in general, and black middle-class families, in particular, are challenged to proactively identify and respond to similar societal challenges now confronting them as they, like the Kings, strive for security, safety, and economic stability.

Appendix

A Mixed-Methodological Approach: Capturing King Family Voices and Experiences

PROJECT OVERVIEW

This project was an historical analysis of one particular rural southern black family (the Kings), and their relatives who migrated elsewhere, based on quantitative and qualitative data sources and spanning approximately from 1900 to 2016. Seven generations of Kings are referenced in the study. Profiles and individual perceptions were collected about some of the mechanisms individuals used to combat various forms of inequality to become middle class. Moreover, experiences across various generations were linked to earlier decisions by the family matriarch and patriarch *as well as* the realities of black life in rural and urban spaces. Interview results helped uncover mediating effects of dynamics associated with, among other factors, military involvement, family and peer support systems, churches, and local schools.[1] Although this study does not do justice to the complexities apparent in the black experience across time, in general, and for the Kings and their offspring, in particular, the analysis helped glean valuable insight into some of the primary structural, social, familial, and individual mechanisms and decisions salient to this family's trek into the middle class.[2]

Qualitative Analyses

Narratives were based on in-depth interviews with a sample of 28 King family members (n = 23) and peers (n = 5).[3] Interviews were provided across four generations of family members; their comments and family documents also enabled information to be gleaned about three earlier generations. Interviews lasted 30 to 120 minutes and were audiotaped and transcribed by the authors. Multiple interviews were required to further probe comments by four of the older family members and three of their children. Eighteen females and ten males were included in the sample. After the initial round of seventeen interviews, purposive sampling was used to ensure that at least one child from each of Irma and George's four living children was included. Interviews with two peers who were George and Irma's contemporaries, as well as three members of the local church they attended (Clear Creek No. 1), provided additional sample diversity and insight. A total of 20 questions and probes were asked about life in the South, migration experiences, religious education, secular education and training, labor market experiences, connections and thoughts about the King family, and any other information individuals wished to share about their lives. They also answered questions about their individual and family demographics, views, hardships, and victories (interview questions provided upon request).

In addition to descriptions of societal changes over time, content analysis was used to identify meanings, common emergent themes, and patterns (Denzin and Lincoln 2005; Krippendorf 1980). During this qualitative process, data were categorized to illumine and identify cogent trends. Respondents' views were systematically examined using two primary processes: open-coding, in which broad concepts from verbiage were labeled and categorized, and axial coding, in which connections between these concepts and themes were analyzed (Strauss and Corbin 1990). An additional level of open coding took place using NVivio 11 to identify response patterns that might have been overlooked. Patterns that resulted from content analysis and NVivio coding were compared and contrasted to confirm the reliability of the themes identified from narratives. We then included emergent themes, corresponding quantitative analyses, and representative quotes in the book. Information from family obituaries, local newspapers, and courthouse data was also included to further augment these data. In several instances, the authors were required to identify and confirm the biblical "address" of certain passages older family members recited or paraphrased because their ages meant that these details learned in youth now escaped them.

Quantitative Analyses

During the quantitative analyses, bivariate tables, percentages, and/or averages of national, local, and individual-level data were calculated to compare and contrast certain economic indices over time to consider some of the broad effects of macro-economic and public policy changes on King family economics. These statistics informed discussions around topics such as: labor market differences by race; race-based economic disparities across time; and employment experiences and problems among family members.[4] Demographics associated with farming by race were used to discuss employment and wage differentials due to racial discrimination.[5] Moreover, family status as a key determinant of economic status was compared to decisions and changes in the King family and resulting class effects. In addition, demographics in cities where key family members migrated were used to discuss migration options and opportunities. We also considered the implications of social changes linked to military involvement and black educational attainment that affected labor markets from the mid-1900s moving forward.[6] The analysis also assessed some of the implications of systemic changes such as government anti-discrimination and affirmative action policies as well as greater educational attainment among blacks that reduced racial inequality and racial discrimination in the labor market between the mid- and late 1900s.[7]

Data Sources

The book relied on the following sources; census data, courthouse data, family records, deeds, newsprint, direct observation, and extensive field notes collected over a four-year period. The ethnographic portion of the study (i.e., mini-histories) was built on in-depth interviews with family members and colleagues across four generations based on factors such as age, gender, socioeconomic status, and education level. In this way, thick descriptions and narratives, augmented by inter- and intra-group statistical data, informed our multi-disciplinary lens. Moreover, the book moved beyond emphasis on family members to include the experiences of fictive kin, clergy, and other peers linked to the Kings as middle-class Mississippians and their offspring who migrated.

Census archival and current data provided national, state, and local demographic information across time for variables associated with class such as race, employment rates, labor force participation, poverty rates, and educational attainment, when available, for whites and blacks. When available and appropriate, such statistics were broadly compared with figures for King family members across generations, particularly the matriarch's siblings (who were central to the narrative), as well as for subsequent offspring to assess indicators typically associated with the middle class.[8]

In addition to national statistics, the study focused on the following cities and states, where many family members migrated circa 1940 and after: Gallman, Mississippi; Jackson, Mississippi; Chicago, Illinois; St. Louis, Missouri; Detroit, Michigan; Toledo, Ohio; and Denver, Colorado. These comparisons enabled us to assess inter- and intra-group differences for indicators commonly associated with class. Additionally, we collected available courthouse data from Copiah County where the earliest generations of family members lived. These data enabled us to gauge dynamics such as land and vehicle ownership in the county and in Gallman, Mississippi. Unfortunately, beyond interview results, data do not exist to allow us to empirically examine inter- and intra-racial religious stratification in the community. A family tree and migration mapping summarized family outcomes. Equally important was the way persons understood their class status and how they acquired possessions such as farm land and livestock.

Additional Data Collection

Land role data, purchase receipts, and tax documents for purchases such as automobiles were collected July 12–17, 2017, at the Chancellor Clerk Office, 122 South Lowe St., Hazlehurst, MS 39083.

Notes

Introduction: A Black Family from Mississippi as a Socio-Ecological Phenomenon

1. We retain use of this family's actual last name (i.e., King). All other names have been changed to provide confidentiality to participants.
2. During Reconstruction, the federal government attempted to enforce civil rights post-slavery via the 13th, 14th, and 15th Amendments to ensure that blacks had freedom, the ability to vote, and equal protection under the law. However, these assurances were short-lived as southern whites used political, economic, and physical means to re-establish power (Rosengarten 1974; Sterling 1994; Thurman 1965; Van Wormer et al 2012; Wilkerson 2010). Also refer to Chafe et al 2014; DuBois 1903[2003]; Ginzburg 1988; Moody 1992. In particular, *On Lynchings* (2014) is a compilation of Wells-Barnett's three pamphlets on lynchings in the United States. (*Southern Horrors, A Red Record*, and *Mob Rule in New Orleans).*
3. Ibid; Darity, Hamilton, and Stewart (2015).
4. Billingsley (1992); Lincoln and Mamiya (1990); Van Wormer et al (2012); Wilkerson (2010).
5. Wilkerson (2010: 10).
6. Darity et al (2015).
7. Blau and Duncan (1978); Darity and Myers (1994, 1998); Darity, Hamilton, and Stewart (2015); Duncan (1992); Frazier (1937); Newman (1989); Oliver and Shapiro (1997); Pattillo-McCoy (1999); Shapiro (2004).
8. Billingsley (1992); Blumer (1958); Collins (2009); Frazier (1964); Lincoln and Mamiya (1990); Mays and Nicholson (1933).
9. Barnes (2002, 2010); Collins (1990); Rosengarten (1974).
10. Barnes (2002, 2010); Billingsley (1992); Blumer (1958); Collins (2009); Frazier (1964); Lincoln and Mamiya (1990).
11. The complete verse in 2 Thessalonians 3:10 in two common translations is: "For even when we were with you, this we commanded you, that if any would not work, neither should he eat" (King James Version or KJV 2017) and "For even when we were with you, we gave you this command: Anyone unwilling to work should not eat" (New Revised Standard Version 1989).

12. Chafe et al (2014); Moody (1992); Rosengarten (1974); Wilkerson (2010).
13. Van Wormer, Jackson, and Sudduth (2012).
14. Children whose families were renting from whites were required to farm the land throughout the year. Because the King family owned their own land by 1944, their children attended school throughout the entire academic year. This gave them an educational advantage as compared to their black and white peers whose families were renters.
15. Other studies such as Frazier's (1937) *Black Bourgeoisie.*
16. Rosengarten (1974); Wilkerson (2010).
17. Billingsley (1992); Lincoln and Mamiya (1990); Mays and Nicholson (1933).
18. Billingsley (1992); Drake and Cayton (1945); Spear (1967).
19. Ecological theory suggests that individual and collective histories, social institutions, societal values, norms, systemic forces, and social networks shape individual behavior (Bronfenbrenner 1977, 1979, 1986; Darity et al 2015). Although the model includes five dimensions (microsystem, mesosystem, ecosystem, macrosystem, and chronosystem), our analysis focuses on three of them.
20. Migration reflected the exodus of an estimated six million black Southerners to the North and Midwest (Billingsley 1992; Chafe et al 2014; Lincoln and Mamiya 1990; Mays and Nicholson 1933; Moody 1992; Rosengarten 1974; Van Wormer and Jackson 2012; Wilkerson 2010). Speaking to racial tensions, see Blumer (1958) and Darity et al (2015).
21. Irma did not participate in the Great Migrations. However, as her health deteriorated, she reluctantly moved to Indiana to be cared for by her children and died there soon after.
22. Copiah County was first formed in 1823 from Choctaw land relinquished from the Choctaw Indians to the United States through the Treaty of Doak's Stand (National Parks Service and Michael Fazio 1985).

1 "My Own Land and a Milk Cow": Race, Space, Class, and Gender as Embedded Elements of a Black Southern Terrain

1. Newman (1989).
2. Cotton (1993) provides a thoughtful critique of economists' and theorists' attempts to explain continued racial discrimination in a supposedly unbiased marketplace. The author questions explanations linked to employer "tastes", human capital differences, cultural factors, employers' risk-aversive behavior, or "race" as a cost-free signal to screen out blacks. He illustrates that profit-seeking white employers have logical reasons for rejecting racial discriminatory hiring practices. Cotton challenges his peers to acknowledge the prevalence of racism and appropriately include discrimination as an "endogenous as well as exogenous labor market variable" in their research (p. 202).
3. The original Native American presence in Mississippi consisted of tribes such as the Natchez, Choctaw, and Chickasaw. They experienced relocation west of the Mississippi River by the Indian Removal Act of 1830 and overall displacement. Spanish and French explorers arrived in the mid-sixteenth and late seventeenth centuries, respectively.

4. From 1817 to 1860, Mississippi was the largest cotton-producing state in the USA. The highest acreage recorded was in 1930 (4.2 million acres) and the highest production year was 1937 (2.7 million bales produced over 3.4 million acres) (Bruchey 1967).
5. U.S. Department of Commerce: Bureau of the Census (1975).
6. This conservative religious emphasis appears to continue today. For example, according to one study, since 2011, Mississippi has been ranked as the most religious state in the country (Newport 2014).
7. Mississippi led the South in politically disfranchising blacks constitutionally since 1890. In 1890, Mississippi also enacted a poll tax to disenfranchise blacks who had committed petty crimes and required that voters be able to read and understand a section of the Constitution. Between 1890 and 1910, most southern states disenfranchised blacks in this way. Voter registration barriers, other discriminatory practices, and violence meant blacks (and some poor and working-class whites) were excluded from the political process until after the passage of the Voting Rights Act of 1965. Grassroots activism secured registration and fostered voting among blacks. However, a one-party system dominated by white Democrats was supplanted in the mid-1900s by Republican domination (Berry and Blassingame 1982; Lincoln and Mamiya 1990; Quarles 1987).
8. Solomon (1999). For example, before the Civil War, Mississippi was the fifth-wealthiest state in the USA, largely due to slave labor on cotton plantations. By 1860, slaves comprised 55 percent of the state's population. During this same year, property in Mississippi was valued at over $500 million, of which about $218 million (or 43 percent) was the value of slaves. However, by 1870, the state's total assets had decreased to about $177 million (DuBois 1935).
9. According to the 1933 U.S. Bureau of Census, "In 1930 of the more than 2.3 million black families in the United States, 541,000 were headed by females. This was nearly 24 percent of black families with female heads in contrast with the mere 12 percent recorded for white families" (Darity and Myers 1994: 119). These same scholars provide an equally troubling correlate for black males; "In 1950 the mortality rate for young black males 15 to 25 years of age was 289.7 per 100,000 … The 1950s generation of young black males were barred legally from full and active participation in virtually every aspect of social and economic life in America: they were constrained to a limited set of job opportunities; they were limited in how far they could progress in the nation's most prestigious schools and academies; they were restricted in where they could live" (Darity and Myers 1994: 170). A similar statement could be made about their female counterparts.
10. According to 2010 census figures, the majority of residents in Hazlehurst were African American (2,997 or 74.8 percent), followed by Whites (717 or 17.9 percent), and Hispanics (276 or 6.9 percent) (Population Demographics for Hazlehurst, Mississippi in 2016 and 2017). The 4,009 people include 1,594 households and 1,121 families. Over one-third (34.3 percent) of households include children under the age of 18 years old, 36 percent of households are married couples, and 28 percent are female head of household (DADS 2010). The median age of residents is 35 years. Median household income is $25,008; median family income is $26,081. And per capita income is $11,839. Lastly, about 24.0 percent of families and 26.3 percent of the population live below the poverty line, including

35.8 percent of residents under age 18 years old and 26.7 percent of those age 65 years old or over (DADS 2010).

11. Foster (n.d.); Great American Stations (n.d.). From the early to mid-twentieth century, "non-white" in the census included blacks (referred to as "Negros"), Indians, Chinese, Japanese, and all of the non-white races. So when considering the black experience, such figures reflect an upper bound. However, the majority of members of this category were black. The median age in Hazlehurst was 29.9 years, 8.8 percent of the population was 65 years old or older, and 49.8 percent of the population was non-white. When education is considered, for persons 25 years old and older, the median number of school years completed was 9.5 years. For persons 14 years old and older, 75.6 percent of males were in the labor force, unlike 31.7 percent of their female peers. The median family income was $1,366 and 67.8 percent of people in the city had incomes less than $2,000 a year.
12. United States Geological Survey (n.d.).
13. This historic district includes a ten-building section of Gallman built from the late 1800s to the early 1900s (National Parks Service and Michael Fazio 1985).
14. "Hazlehurst, MS Census Data" (n.d.); Geonames (n.d.); National Parks Service and Michael Fazio (1985). Also, in 1858, Gallman became a stagecoach stop between Hazlehurst and Crystal Springs, Mississippi, on the New Orleans, Jackson and Great Northern Railroad.
15. Ibid.
16. During the same three years, tomato shipments from Crystal Springs were 1,553 (cars), 790, and 1,047. Amounts in Hazlehurst were 1,399, 448, and 649 (Crook 1996).
17. The town of Wesson was named after Union Colonel James Madison Wesson and incorporated March 31, 1864. A sawmill, cotton mills, and other manufacturing enterprises fueled its early growth. Its mills, including the Mississippi Mills, reportedly employed about 1,200 in 1887 and fostered other industries. However, mill closings in the mid-1900s resulted in substantial economic decline and out-migration.
18. In 1900, about 8,833,994 blacks lived in the USA. And about 89.7 percent of blacks living in the USA lived in the South; more specifically, about 31.4 percent of them lived in Georgia, Mississippi, or Alabama. More than three fourths of blacks lived in the country rather than in cities as compared to their white peers. Additionally, in 1900, the value of farmland in Mississippi totaled $114,856,660–$152,007,000 (land without buildings). The average size of all farms in the USA that same year was about 146.6 acres. Refer to Willcox (1904).
19. Per this same source, by 1900, that number had increased 28.2 percent to 757,822. The numbers of black farmers in 1910 and 1920 were 879,600 and 927,533, respectively.
20. When we consider the types of crops that were raised by the King family, in that same year, the average value of farms that grew cotton was $1,033, and $3,508 if the crops were vegetable. Moreover, the average value of land alone used to grow cotton was about $653 per farm or $7.82 per acre. Thus this crop had become less lucrative post-slavery (but had improved from its low of $0.08 per pound in 1886). The average value of land for vegetable farms was $2,325 or $35.69 per acre. The average value for

vegetable farms in general was $2,285 per farm or $15.59 per acre – relatively higher than the average value of all farms in the United States. (Crook 1996).

21. Black farmers operated farms that were 20 to 50 acres in size; white farmers tended to operate farms that were 100 to 175 acres. The majority of white farmers had income $100–$1,000; black incomes were about half that. Additionally, "The average area of farms cultivated by white farmers was 160.7 acres; by Negros, 51.2 acres; by Indians, 172.5 acres; by Chinese, 63.8 acres; by Japanese, 37.1 acres; and by Hawaiians, 996.4 acres" (U.S. Census 1975: xcvi). During that period, "of the Negro farmers in the United States, 70.5 percent derived their principal income from cotton; 12.4 percent, from miscellaneous products; 6.9 percent, from hay and grain; 4.1 percent, from livestock, and 2.6 percent, from tobacco" (U.S. Census 1975: cvi).
22. Croppers were at the bottom of the agricultural ladder and many of them were displaced by the mechanization of southern farms and left farming for urban employment, while others became wage hands.
23. The effects of the Great Depression were evident. No more than about 25 percent of whites were croppers, except in the 1930s. Moreover, substantially more whites were tenants and croppers during this same period than in other time frames.
24. Wesson, Mississippi, is about 45 miles south of Jackson, MS, and 145 miles north of New Orleans.
25. Barnes (2004, 2005a, 2014a); Billingsley (1992); DuBois, 1903[2003], 1953[1996]; Frazier (1964); Gilkes (1998, 2001); Lincoln and Mamiya (1990); Mays and Nicholson (1933); McRoberts (2003); Wilmore (1994).
26. Danny (born October 1908), Colvin (born January 1896), Bella (unknown), Mindy (September 1878), and Janice (born September 1890). They also had three half siblings; Marny (born October 1883), Nina (born February 1887), and L. King (born January 1898).
27. *The Holy Bible Containing the Old and New Testaments Translated Out of the Original Tongues and with Former Translations Diligently Compared and Revised, Self-Pronouncing Edition*. 1910. Nashville, TN: The Southwestern Company Publishers and Booksellers (conformable to the edition of 1611, commonly known as the authorized or King James Version). In addition to aids to bible study; the Hebrew calendar; maps of Palestine, Jerusalem and the journeys of Christ; bible stories for children; a self-pronouncing dictionary of proper names and foreign words contained in the bible; black-and-white and color pictures; a concordance; and, an index of leading biblical doctrines, the bible includes marriage certificates as well as lists of names and dates of marriages, births, and deaths.
28. At age 12, Janice joined the Mount Salem Missionary Baptist Church under the pastorate of Rev. Jake Ray. She later joined the Western Star Missionary Baptist Church, and later, St. Paul's Rock Missionary Baptist Church. After moving to St. Louis, MO, in 1968, Janice joined the Jerusalem Missionary Baptist Church.
29. Poppa Sam: Jan. 16, 1892–May 11, 1968.
30. Genesis 1:28 (NRSV 1989).
31. Two siblings, Leddie and Lily, died as children.
32. Farmers would make arrangements for their cows to periodically mate with the bulls of other local farmers such that calves could be born.
33. George was her second husband. Irma had a brief May–December marriage to James Rae Smith. It is said that she divorced him upon learning that he had

somehow tricked her into marriage. Irma admitted that he had been kind, but she divorced him because of their age difference. Although she loved George, Irma seemed to regret her hasty first divorce.

34. Janice Mae was named after her grandmother, Janice King.
35. Refer to scholars such as Cannon (1998); Baker-Fletcher (2006); Douglas (2006); Collins (1990); Floyd-Thomas (2006); Gilkes (2001); Grant (1989); Thomas (2004); and Townes (2006).
36. This report reflects females 14 years old or older; 261 non-white females were "live in" domestics. According to this same report, in 1950, only 1.1 percent of white women in Mississippi (0.9 percent in urban areas, 1.6 percent on rural non-farms, and 0.9 percent on rural farms) held these jobs. However, 35.4 percent of non-white females in Mississippi (52.5 percent, 48.6 percent, and 9.7 percent in urban, rural non-farm, and rural farm areas, respectively) were domestics. Representation for males in this occupation, regardless of race, was well under 1.0 percent. The greatest relative percent of non-white females worked in this arena, followed by farm laborers and foremen (18,656 persons) and non-domestic service workers (11,813 persons) such as cooks and waitresses.
37. This concept was popularized in Hochschild's (1990) *The Second Shift*.
38. Annual incomes taken from U.S. Census Table D. 739–764 "Average Annual Earnings Per Full-Time Employee, by Industry: 1900–1970."
39. Grannoveter (1973, 1993).
40. Scholars such as Newman (1989) suggest that a "middle-class" mindset fosters such decisions. Thus Irma's mindset and beliefs about the importance of land ownership occurred *first* and fostered her subsequent behavior. Other middle-class beliefs, such as concerns about downward mobility, disdain for the less industrious, and support for "middle-class" values were also apparent among family members. Conspicuous consumption, another expected trait, was not apparent in Irma and George's generation, but was evident among some persons in subsequent ones.
41. In 2015, an average acre of land in Mississippi was worth about $5,600. This amount is less than half the value of an average acre across the contiguous USA. In addition, less than 9 percent of land in Mississippi is federally owned, notably less when compared to the nearly 24 percent of the country owned by the US government (Frohlich and Kent 2015).
42. We include titles here, such as "Mr. Roy" and "Ms. Dina" not out of deference, but to reflect the exact language used by respondents.
43. Collins (1990).
44. Ibid.
45. Ibid.
46. Scholarship by Collins (1990), Rollins (1985), and Gilkes (2001) provide exemplary depictions of such black women.
47. However, another source suggests that, in 1920, there were black 9,290 mechanics "not otherwise specified" in the United States. (Green and Woodson 1930: 335). Regardless which figure one considers, blacks were under-represented in this skilled trade.
48. Collins (1990); Rosengarten (1974); Wilkerson (2010).
49. According to the ground-breaking study by W.E.B. DuBois, in 1944, almost 4,000 black businesses existed in 12 US cities. He also noted that in 1929, blacks operated about 25,000 stores with sales of $101 million. By 1950, the number of black

businessmen had increased to 42,500. However by 1960 the number of a black businessmen declined to about 32,400 (Berry and Blassingame 1982).

50. According to Tillitt (1944), the average unmarried American private soldier earned a base pay of $50.00 ($30.00 prior to September 1942) per month, or $600.00 before any spending per year. That would translate to a net annual pay of about $3,600.00 "civilian" dollars in 1944. And in 1944, an annual income of $3,600.00 was more than 80 percent of all single workers in the USA earned. Also refer to Dansby and Landis (2001); Elder (1986, 1987); Lutz (2008); Sampson and Laub (1996).
51. Dansby and Landis (2001); Lutz (2008).
52. Per Berry and Blassingame (1982): "WACs made the same charge of discrimination in assignments as those made by blacks and other branches of the Army. Black women were being sent to cooks' and bakers' schools instead of being assigned to higher technical jobs" (p. 325).
53. Salary comparisons between Navy and Mariners show the following in 1943 for Seaman first class versus an Ordinary seaman ($1,888 versus $1,897) and between Petty officer second class versus Able seaman ($2,308 versus $2,132) (American Merchant Marine at War 2003).
54. Biderman and Sharp (1968); Broom and Smith (1963); Dansby and Landis (2001); Sampson and Laub (1996).
55. These same authors calculated income differentials for black, white, and Mexican veterans and non-veterans that support their results that include mean net gains for black and Mexican veterans ($163 and $387, respectively) in general and income benefits for black, white, and Mexican veterans over non-veterans, in general, for occupations such as operatives, service, and laborers (Browning et al 1973). Additionally, income benefits were also apparent for blacks in other occupations such as clerical and craftsmen. Other studies also support the possible economic benefits of the military for disadvantaged individuals [Elder (1986, 1987) and Sampson and Laub (1996)].
56. Earlier studies suggested that military service resulted in *income penalties* for individuals in the form of lower relative pay as compared to civilians because of interruptions in important work-related experiences such as higher education, apprenticeships, and on-the-job training [refer to Browning et al (1973) for a list of these scholars]. Browning and his colleagues found that such penalties did not bear out for most black and Mexican enlisted men, but rather, the exact opposite.
57. Non-rural wages tended to be higher, in general, but were still dependent on the ability of blacks to locate gainful employment.
58. Moskos (1986) contends that military service benefits blacks. Per his somewhat romanticized view, "In recent years several factors have made a spell in the Army increasingly attractive … the availability (since 1982) of GI Bill-style educational benefits, and the generous pay earned by new recruits. A buck private receives a base pay of $7,668 a year, in addition to room and board, medical care, pension, and other benefits. He may receive an enlistment bonus of up to $8,000" (n.p.).
59. Refer to contrasting research that home ownership does not typically translate to wealth for blacks in the same way that it does for whites, largely because black homes are often located in less desirable areas, can be affected by red-lining, and may be in neighborhoods that are not as well maintained (Oliver and Shapiro 1997; Pattillo-McCoy 1998).

60. Berry and Blassingame (1982); Billingsley (1992); Dansby and Landis (2001); Elder (1986, 1987); Lutz (2008); Quarles (1987); Sampson and Laub (1996); Tillitt (1944).
61. Billingsley (1992).
62. Billingsley (1992); Browning et al (1973); Butler (1976); Quarles (1987).
63. Dansby and Landis (2001); Elder (1986, 1987); Feagin (2006, 2010); Feagin and Feagin (1978); Lutz (2008); Sampson and Laub (1996).
64. In 1948, President Harry Truman issued Executive Order 9981 that stated: "It is hereby declared to be the policy of the President that there shall be equality of treatment and opportunity for all persons in the Armed Forces without regard to race, color, religious, or national origin. This policy shall be put into effect as rapidly as possible, having due regard to the time required to effectuate any necessary changes without impairing efficiency or morale" (Lutz 2008: 172). Although this statement was applauded as an example of crucial systemic change, it included caveats that enabled white military leaders to delay proactive implementation such that, despite certain improvements, blacks today still complain about challenges in the military.
65. Butler (1973).
66. Billingsley (1992); Leigh and Berney (1971); Quarles (1987).
67. U.S. Department of Defense (2015).
68. Browning et al (1973); Butler (1976); Dansby and Landis (2001); Elder (1986, 1987); Lutz (2008); Moskos (1986); Moynihan (1965); Quarles (1987); Sampson and Laub (1996). For example, Moskos (1986) suggests that "in 1985, 95.4 percent of black men joining the Army had high school diplomas, in comparison with 87.6 percent of whites" (n.p.). Lutz (2008) provides a similar comment; "research shows that the all-volunteer force continues to see over-re-presentation of the working and middle class, with fewer incentives for upper class participation" (p. 185).
69. In response to continued segregation in the military, a black civilian group formed on October 10, 1947, Committee Against Jim Crow in Military Service and Training. Eight months later, President Truman issued an order calling for both the end of segregation and equality of opportunity for all servicemen and constituted the Committee on Equality of Treatment and Opportunity in the Armed Services to keep him abreast of implementation. In May 1950, the committee recommended that "every vestige of segregation be removed in the Army, Navy, and Air Force" (Quarles 1987: 273). After the Korean conflict, all segregated units had been abolished. Moreover, by 1963, there were about 3,000 black officers in the Army, 2,200 black officers in the Air Force, and 300 black officers in the Navy (Quarles 1987).
70. Under the administration of Franklin D. Roosevelt, black government employees increased from 50,000 to 200,000 between the periods of 1937 to 1946.
71. Margo (1995) also contends that wage compression occurred that resulted in racial wage convergence in the 1940s. This same author suggests that internal migration, occupational shifts, and diminishing racial differences in schools also helped narrow the black–white wage gap between 1940 and 1950. Yet other research, many included in this book, counter this study.
72. Yet blacks were summarily excluded from skilled posts in the defense industry as well as in most automobile plants, save the Ford plant in Detroit, Michigan. Despite riots, union discrimination and other exclusionary tactics by whites, 1940–1944

saw some gains in black skilled posts for males. However, in 1965, only 12 percent of white families lived in poverty, but 43 percent of all black families earned less than $3,000 annually. In other words, blacks were more than three times as likely as whites to live in poverty. After World War II, many blacks lived in a state of constant depression (Berry and Blassingame 1982).

2 "Bikes or Lights": Familial Decisions in the Context of Inequality

1. Barnes (2005a, 2012); Lincoln and Mamiya (1990); Mays and Nicholson (1933); Morris (1984); Pattillo-McCoy (1998).
2. According to Series S 108–119. Growth of Residential Service, and Average Prices for Electric Energy: 1902 to 1970, the following percentage of US residential farms that had electricity and the average monthly price were: 1930 (10.4 percent, $5.00); 1940 (32.6 percent, $4.06); 1950 (77.7 percent, $3.76); and 1956 (95.9 percent, $3.88). Usage figures for farms after 1956 were not available. These figures are over-represented by white residents.
3. Series K 538–549. Rice, Sugarcane, Sugar Beets, and Peanuts–Acreage, Production, and Price: 1895 to 1970-C.
4. Series K 550–563. Hay, Cotton, Cottonseed, Shorn Wool, and Tobacco–Acreage, Production, and Price: 1790 to 1970.
5. Series K 564–582. Livestock–Number, Value Per Head, Production, and Price: 1867 to 1970.
6. Ibid.
7. Series K 609–623. Poultry and Eggs–Number, Production, and Price: 1909 to 1970.
8. Newman (1989).
9. For example, based on an early survey, only three percent of black mothers were recipients of such aid (Bell 2004; also see Roberts 2012).
10. Ibid.
11. Berry and Blassingame (1982); Billingsley (1992); Moody (1992); Quarles (1987); Rosengarten (1974); Wilkerson (2010).
12. Newman (1989).
13. Rosengarten (1974); Sterling (1994); Van Wormer et al (2012); Wilkerson (2010).
14. Barnes (2005); Berry and Blassingame (1982); Golden (1983); Lincoln and Mamiya (1990); Quarles (1987); Rosengarten (1974); Van Wormer et al (2012); Wilkerson (2010).
15. Robbie Lee would later migrate to Toledo, Ohio, to live with her sister Gertha Mae.
16. Refer to Granovetter (1973, 1993) regarding the benefits of weak social ties. See also Wilkerson (2010) for race-specific examples.
17. Refer to Blumer's (1958) insights on how perceptions of group position can influence prejudices.
18. Refer to Benbow (2010) and Rubio (2010) about race and the history of the U.S. Postal Service.
19. Wilson (1996).
20. DuBois (1899[1996]); Van Wormer et al (2012); Wilkerson (2010).
21. Spear is referencing the original comment by Gunnar Myrdal.
22. Using Weber's terminology found in Barnes (2004, 2005b).

23. Feagin (2006, 2010); Gillispie (2003); McIntosh (1989); Messner (2011); Samuels, Ferber, and O'Reilly Herrera (2003).
24. "Living for the City" is a 1973 single written by Stevie Wonder from his *Innervisions* album. Verse 1 begins, "A boy is born in hard time Mississippi."
25. Logan and Molotch (1987); Squires (1994).
26. Using "space" above land, such as building (i.e., skyscrapers) and traveling vertically (i.e., flying), are examples of practical responses to the limited nature of this commodity.
27. Foucault (1970); Foucault and Rabinow (1984).
28. Collins (1990); Feagin (2006, 2010); Feagin and Feagin (1978).
29. Oliver and Shapiro (1997). Shapiro (2004) notes that "family inheritance and continuing racial discrimination in crucial areas like homeownership are reversing gains earned in schools and on jobs and making racial inequality worse" (p. 2).
30. Rosengarten (1974); Sterling (1994); Van Wormer et al (2012); Wells-Barnett (2014); Wilkerson (2010).
31. Cone (1992).
32. Rosengarten (1974); Sterling (1994); Van Wormer et al (2012); Wilkerson (2010).
33. "Among whites, parental status and ethnic group status have a strong direct impact on inter-generational mobility. Children of high-income and high-education parents, neighborhoods, and white ethnic groups have a leg up on other children in attaining high income and an advanced education. However, we are unable to distinguish the manner in which class background matters. Is it because superior class position creates an advantage in skills acquisition or is it because superior social status increases access to persons embedded in positions of power and authority?" (Mason 2003: 66). Scholars, particularly economists, continue to try to untangle this dynamic.
34. All three verses are, "The Lord says to my lord, 'Sit at my right hand until I make your enemies your footstool'" (Acts 2:35, NRSV 1989). The complete verse in Exodus is, "God said to Moses, 'I am who I am.' He said further, 'Thus you shall say to the Israelites'" (Exodus 3:14, NRSV 1989) and describes Moses' burning bush encounter with God. The complete verse in Matthew is, "And Jesus came and spake unto them, saying, All power is given unto me in heaven and in earth." (Matthew 28:18, KJV 2017).
35. This common family theme is based on the following verse, "For our struggle is not against enemies of blood and flesh, but against the rulers, against the authorities, against the cosmic powers of this present darkness, against the spiritual forces of evil in the heavenly places" (NRSV 1989).
36. Electrical wires had to be installed along long gravel roads; trenches had to be built such that water could be piped in. Interested blacks were responsible for making arrangements either through payment or sweat equity as well as secure monthly payments to continue such amenities.

3 "Getting to the School on Time": Formal Education and Beyond

1. Bolton (2005).
2. Quarles (1987).

3. For example, although blacks comprised "31.6 percent of the school age population in the South in 1899, they received only 12.9 percent of public school funds ... in 1930 the South spent an average of $44 for white pupils and only $12 per black pupil annually" (Berry and Blassingame 1982: 265–6).
4. Berry and Blassingame (1982); Collins (2009); Davis (2004); Delpit (2006); Ferguson (2001); Hale (2001); Hale-Benson and Hilliard (1982); Hallinan (2001); Ladson-Billings (2009); Porter (1998); Siddle Walker (1996a); Woodson (2013).
5. Barnes (2010); Billingsley (1992).
6. Although we acknowledge the influence of social policies on educational outcomes for blacks, researchers such as Siddle Walker (1996a) caution against over-crediting the benefits of such initiatives because they are/were only as effective as the individuals and groups required to implement them – usually whites – who have been historically inconsistent in their support. In contrast, she suggests the importance of black efforts toward group empowerment.
7. Darity Jr. and Deshpande (2003); Darity Jr. and Myers (1994, 1998); Mason (2003).
8. Billingsley (1992).
9. Anderson (1988); Ashmore (1954); Bullock (1967); Irvine and Irvine (1983); Jones (1981); Sowell (1976).
10. Post-Reconstruction, the Democrat-run legislature in Mississippi severely reduced the already minimal funding for public schools. However, northern philanthropists such as the Anna T. Jeanes Foundation (via the Negro Rural School Fund, started in 1907) funded basic education for rural southern blacks, including vocational training, and later in the 1940s, traditional academic subjects. In addition, the General Education Board (funded by the Rockefeller Foundation and the Rosenwald Fund) helped construct over 5,000 schools. Northern churches supported denominational colleges (Krause 2003).
11. Women, Jews, and other minorities were also similarly mistreated. Ibid.
12. Franklin et al (1991).
13. In 1982, Utica Junior College merged with Hinds Junior College under federal court order as part of a class action racial discrimination lawsuit.
14. Utica is a town of 3.0 square miles in Hinds County, Mississippi, named in 1837, by its postmaster, Ozias Osborn, who had relocated from Utica, New York. The town was incorporated in 1880. In 1907, its population was about 1,000; in 2016 it had 825 residents. In 2016, the racial makeup was 82.9 percent black, 17.1 percent white, and less than 1.0 percent Hispanic, Asian, and Native American (U.S. Census American Fact Finder 2016).
15. Alcorn State University is an HBCU located in Lorman, Mississippi. It was founded in 1871 during Reconstruction to provide higher education for freed persons. It is the first black land grant institution in the USA. Jackson State University is an HBCU located in Jackson, Mississippi. It was founded during Reconstruction in 1877 in Natchez, Mississippi, as Natchez Seminary by the American Baptist Home Mission Society of New York City. The Society moved the school to the capital, Jackson, in 1882. One of the largest HBCUs in the country, Jackson State became a state-supported public institution in 1940 and is now classified as a research university of almost 10,000 students.
16. Collins (1990).

17. Collins (1990); Rosengarten (1974); Sterling (1994); Van Wormer et al (2012); West (1993); Wilkerson (2010).
18. Other scholars who note differential employment outcomes and job challenges for women, especially black women, based on their childrearing roles include Brown (1997); Hochschild (1990); Hofferth (1984); and Moen (1992). Dual labor market or labor market segmentation theory can also inform the discourse on women's differential job outcomes (Dickens and Lang 1993).
19. Billingsley (1992); Bullock (1967); Irvine and Irvine (1983); Jones (1981); Siddle Walker (1996a); Sowell (1976).
20. Billingsley (1992); Frazier (1964); Lincoln and Mamiya (1990); Mays and Nicholson (1933).
21. Ibid.
22. Barnes (2005).
23. The complete passage is, "So if anyone is in Christ, there is a new creation: everything old has passed away; see, everything has become new" (NRSV 1989).
24. Also refer to Ferguson (2001); Hale (2001).
25. Barnes and Streaty-Wimberly (2016).
26. Collins (2009); Ferguson (2001); Hale (2001).
27. Collins (2009); Kunjufu (1986); Porter (1998); Wimberly (2005); Woodson (2013).
28. Barnes (2002, 2004, 2005a); Billingsley (1999); Collins (2009); Frazier (1964); Lincoln and Mamiya (1990); Mays and Nicholson (1933).
29. Suggestions such as Individualized Educational Plans (IEP) implemented by culturally sensitive, accountable administrators and teachers are the most effective processes to teach students of color (Ferguson 2001; Hale 2001). Moreover, Hale (2001) contends that black children would benefit from educational practices that are innovative and reflective of the "culture of power" evident in the best prepared black schools.
30. Bronfenbrenner (1977, 1979, 1986).
31. Barnes (2008, 2010, 2014b); Collins (2009); Davis (2004); Delpit (2006); Ferguson (2001); Hale (2001); Hale-Benson and Hilliard (1982); Hallinan (2001); Kunjufu (1986); Ladson-Billings (2009); Nwosu and Barnes (2014); Porter (1998); Siddle Walker (1996a); Wimberly (2005); Woodson (2013).
32. Billingsley (1992); Nwosu and Barnes (2014).
33. National Center for Education Statistics (2015).
34. Barnes (2010); Collins (2009); Davis (2004); Delpit (2006); Ferguson (2001); Hale (2001); Hale-Benson and Hilliard (1982); Hallinan (2001); Kunjufu (1986); Ladson-Billings (2009); Porter (1998); Siddle Walker (1996a); Woodson (2013).
35. Her comments parallel assessments in studies by scholars such as Collins (1990) and Rosengarten (1974) about intra-racial socio-emotional and psychological damage that can result from decades of systemic oppression.
36. McGee (2015); McGee and Stoval (2015). Some remedies include increasing the pipeline of students of color who are encouraged and better prepared for STEM fields by improving K–12 science and mathematics education as well as attracting and retaining instructors trained to teach them in these areas.
37. Uchitelle (2008).
38. Darity Jr. and Myers (1998); Deshpande and Darity (2003); Mason (2003).
39. Darity Jr. and Myers (1998); Darity, Jr. et al (2015).

40. Ibid.
41. Matthew 5:39 reads, "But I say unto you, that ye resist not evil: but whosoever shall smite thee on thy right cheek, turn to him the other also" (KJV 2017). Matthew 18:21–22 reads, "Then came Peter to him, and said, Lord, how oft shall my brother sin against me, and I forgive him? Till seven times? Jesus saith unto him, I say not unto thee, until seven times: but, until seventy times seven" (KJV 2017).
42. Billingsley (1992); Lincoln and Mamiya (1990); Quarles (1987); Wilmore (1994).
43. *Journal of Blacks in Higher Education* (2010).
44. Davis (2004); Delpit (2006); Hale (2001); Hale-Benson and Hilliard (1982); Porter (1998).
45. Ainsworth-Darnell and Downey (1998); Hale (2001); Kunjufu (1986); Ladson-Billings (2009); Porter (1998); Siddle Walker (1996a); Woodson (2013).
46. Darity Jr. (1993).
47. Collins (2009); Siddle Walker (1996a).
48. Blau and Duncan (1978); Newman (1989); Oliver and Shapiro (1997); Shapiro (2004).
49. Refer to work by Averitt (1968), Darity (1993), and Heckman and Sadlacek (1985) on labor market segmentation theory. Dickens and Lang (1993) provide a summary of research on this theory.
50. Barnes (2008, 2010, 2014b); Berry and Blassingame (1982); Billingsley (1992); Collins (2009); Davis (2004); Delpit (2006); Ferguson (2001); Frazier (1964); Hale (2001); Hale-Benson and Hilliard (1982); Hallinan (2001); Kunjufu (1986); Ladson-Billings (2009); Lincoln and Mamiya (1990); Mays and Nicholson (1933); Porter (1998); Quarles (1987); Siddle Walker (1996a); Wimberly (2005); Woodson (2013).
51. Anderson (1988); Davis (2004); Irvine and Irvine (1983); Siddle Walker (1996a).
52. Berry and Blassingame (1982); Quarles (1987).
53. Ainsworth-Darnell and Downey (1998); Barnes (2008, 2010); Collins (2009).
54. Collins (2009); Jordan (2014): Nazaryan (2018). Also refer to the Civil Rights Project at https://civilrightsproject.ucla.edu/research/k-12-education/integration-and-diversity/brown-at-60-great-progress-a-long-retreat-and-an-uncertain-future.
55. Barnes (2008, 2010, 2014b); Barnes and Wimberly (2016); Collins (2009); Davis (2004); Delpit (2006); Ferguson (2001); Frazier (1964); Hale (2001); Ladson-Billings (2009); Porter (1998); Siddle Walker (1996a); Wimberly (2005); Wimberly et al (2013); Woodson (2013).
56. Bonilla-Silva (2010); Feagin (2010, 2006); Feagin and Feagin (1978).

4 "Jesus and the Juke Joint": Blurred and Bordered Boundaries and Boundary Crossing

1. Refer to *When Work Disappears* by Wilson (1996) about the benefits of employment in the informal economy for the poor and working class.
2. Barnes (2010).
3. Content analysis was used to identify common biblical and practical themes noted by family members. This approach included a close reading and analysis of narratives to identify meanings; common, emergent themes; and response

frequencies often overlooked by other methods (Denzin and Lincoln 2005; Krippendorf 1980).

4. Cone (1992).
5. Wilson (1996) describes urbanites who may sell marijuana part-time. Other practices such as working in their homes as cosmetologists or providing lawn care services for cash have legal implications if taxes are not paid on these wages. Rather than debate such decisions, it is important to consider structural constraints that result in a lack of gainful employment and foster such practices.
6. Levi Garrett is a brand of chewing tobacco produced by the American Snuff Company.
7. Refer to Collins (1990) and Gilkes (2001) about gender differentials for black women historically. Williams (1993) also notes that in the early 1900s, black women were disproportionately concentrated in agriculture and continually excluded from mill positions and administrative and sales positions.
8. Bronfenbrenner (1977, 1979, 1986).
9. "Papa Was a Rollin' Stone" was written by Norman Whitfield and Barrett Strong. The song was remade by The Temptations in 1973.
10. The exact passages are "All my enemies shall be ashamed and struck with terror; they shall turn back, and in a moment be put to shame" (Psalm 6:10 NRSV 1989); "To you they cried, and were saved; in you they trusted, and were not put to shame" (Psalm 22:5 NRSV 1989); "Do not let those who wait for you be put to shame; let them be ashamed who are wantonly treacherous" (Psalm 25:3 NRSV 1989); and "In you, O Lord, I seek refuge; do not let me ever be put to shame; in your righteousness deliver me" (Psalm 31:1 NRSV 1989).
11. Weber's terminology referenced in Barnes (2004, 2005b).
12. Exodus, chapter 20 or Deuteronomy, chapter 5 (*The Holy Bible: New Revised Standard Version* 1989).
13. Barnes (2002, 2004, 2005a); Billingsley (1992); Frazier (1964); Lincoln and Mamiya (1990); Mays and Nicholson (1933).
14. Rosengarten (1974); Sterling (1994); Van Wormer, Jackson, and Sudduth (2012); Wells-Barnett (2014); Wilkerson (2010).
15. Roediger (1991).
16. Domestic employment by black females was expected. Collins' (1990) notes that a young white girl referred to a black female baby as follows, "Oh look, Mommy, a baby maid" (p. 71).
17. The complicated relationship between the two families did not preclude Dina's children from traveling from Mississippi to Gary, Indiana, to visit Irma in the hospital before she died. Emotionally distraught, they commented that Irma had been a mother to them.
18. Bronfenbrenner (1977, 1979, 1986).
19. Cotton (1993).
20. Morris (1984); Wilmore (1994).
21. Billingsley (1992); DuBois (1899[1996], 1903[2003], 1935, 1953[1996]).
22. Rosengarten (1974).
23. Barnes (2006, 2010); Wilmore (1994).
24. Several specific references to Jesus Christ as a servant are: "Here is my servant, whom I have chosen, my beloved, with whom my soul is well pleased. I will put my

Spirit upon him, and he will proclaim justice to the Gentiles – referring to Jesus" (Matthew 12:18 NRSV 1989), and "For I tell you that Christ has become a servant of the circumcised on behalf of the truth of God in order that he might confirm the promises given to the patriarchs" (Romans 15:8 NRSV 1989). Servanthood is a common biblical charge for adherents. Examples include: "Let your steadfast love become my comfort according to your promise to your servant" (Psalm 119:76 NRSV 1989); "Deal with your servant according to your steadfast love, and teach me your statutes" (Psalm 119:124 NRSV 1989); "For the LORD will vindicate his people, and have compassion on his servants" (Psalm 135:14 NRSV 1989); and "I do not call you servants any longer, because the servant does not know what the master is doing; but I have called you friends, because I have made known to you everything that I have heard from my Father" (John 15:15 NRSV 1989).

25. Several common passages include: "It will not be so among you; but whoever wishes to be great among you must be your servant" (Matthew 20:26 NRSV 1989), and "But it is not so among you; but whoever wishes to become great among you must be your servant" (Mark 10:43 NRSV 1989).
26. Known for their four-part harmony, black quartet singers became most popular in the early to mid-twentieth century and continue to perform today. Typically known for gospel renditions, they are also often adept at spirituals and gospelized hymns. Famous black quartets include the Brooklyn Allstars, the Dixie Hummingbirds, the Five Blind Boys (several groups), the Mighty Clouds of Joy, the Pilgrim Travelers, the Sensational Nightingales, the Soul Stirrers, and the Swan Silvertones. George sang with a local quartet and frequently visited churches on Sunday evenings where such singing took place.
27. Billingsley (1992) details the existence and positive influence of fictive kin for the black family.
28. Collins (1990).
29. The passage is "But the fruit of the Spirit is love, joy, peace, forbearance, kindness, goodness, faithfulness, gentleness and self-control. Against such things there is no law" (KJV 2017).
30. The passage is, "The rich rule over the poor, and the borrower is the slave of the lender."
31. Barnes (2004, 2005b, 2010).
32. Lincoln and Mamiya (1990); Wilmore (1994).
33. See DuBois (1903[2003]) and Lincoln and Mamiya (1990) for historical accounts of the Black Church, as well as the references in Barnes (2004, 2005b) for additional resources on this topic.
34. Lincoln and Mamiya (1990); Wilmore (1994).
35. Barnes (2004, 2005b, 2010); Lincoln and Mamiya (1990).
36. Siddle Walker (1996).
37. In addition to information on Black Church life not previously collected, DuBois' (1903[2003]) *The Negro Church* chronicled denomination-specific schools and efforts to reach the young. Unique even for today, the book includes interviews with 1,339 Black public school children in Atlanta about their religious beliefs, church attendance, and views about the church. Other topics include training for ministers and candid commentary about the character and educational credentials of clergy.

38. In addition to priestly and prophetic dictates, also refer to Lincoln and Mamiya (1990) for details on the remaining five polarities: other-worldly versus this-worldly; universalism versus particularism; communal versus privatistic; charismatic versus bureaucratic; and resistance versus accommodationist. They contend that a priestly function focuses on worship, personal religious living, and other-worldly issues. In contrast, a prophetic function focuses on proactive efforts to transform society politically, economically, and socially often via activism (Lincoln and Mamiya 1990; Morris 1984).
39. Barnes (2004, 2005b, 2010).
40. Barnes and Streaty-Wimberly (2016); Wimberly, Barnes, and Johnson (2013); Wimberly (2005).
41. Pattillo-McCoy (1998).
42. Barnes (2012); Frazier (1964); Lincoln and Mamiya (1990); Marx (1971); Mays and Nicholson (1933); Nelson (2005); Wilmore (1994).
43. Participant observations were performed by the first author.
44. Barnes (2010, 2012, 2014a); Gilkes (1998); Glaude Jr. (2010); Hunt and Hunt (1999); Olson and Perl (2000).
45. Barnes (2010); Burnette (2001).
46. Billingsley (1990); Frazier (1964); Lincoln and Mamiya (1990); Mays and Nicholson (1933); Wilmore (1994, 1995).
47. Ibid. Also note Barnes (2008, 2014b); Barnes and Streaty-Wimberly (2016); Wimberly et al (2013).
48. Barnes (2002, 2005a); Frazier (1964); Lincoln and Mamiya (1990); Mays and Nicholson (1933).
49. Billingsley (1990); Frazier (1937); Lincoln and Mamiya (1990); Mays and Nicholson (1933); Wilmore (1994, 1995).
50. Barnes (2012, 2014a); Glaude Jr. (2010).
51. Although in rural settings, pastors, blacks who could read, less impoverished blacks, or blacks who had curried favor from local whites, often wielded more power than their peers.
52. Ellison and Sherkat (1995).
53. This well-known verse is, "Give, and it will be given to you. A good measure, pressed down, shaken together, running over, will be put into your lap; for the measure you give will be the measure you get back" (KJV 2017).
54. Barnes (2010, 2012, 2014a); Billingsley (1990); DuBois (1953[1996], 1903[2003]); Frazier (1937, 1964); Gilkes (1998); Hunt and Hunt (1999); Mays and Nicholson (1933); Olson and Perl (2000); Pattillo-McCoy (1998); Wilmore (1994, 1995).
55. Ellison and Sherkat (1995).
56. Coined by Max Weber [refer to Henderson and Parsons (1947) and Parsons (1937)].
57. Barnes (2004, 2005, 2005a); Drake and Cayton (1945); Frazier (1939); McRoberts (2003); West (1993).
58. Franklin and Pillow (1982).
59. Collins (1990).
60. Yet religious affiliation and commitment for blacks vary by gender (women more than men) and age (older persons more than younger ones) (Pew 2009).

61. Billingsley (1992); DuBois (1903[2003]); Lincoln and Mamiya (1990); Mays and Nicholson (1933).
62. Barnes (2005, 2006); DuBois (1903[2003]); Lincoln and Mamiya (1990); Pattillo (1998).

5 "Keeping God's Favor": Contemporary Black Families and Systemic Change

1. Barnes (2005); Edin and Lein (1996); Hays (2003); Hogan, Hao, and Parish (1990); Jarrett (1994); Wilson (1987).
2. According to Kurtzleben (2014), "among millennials ages 25 to 32, median annual earnings for full-time working college-degree holders are $17,500 greater than for those with high school diplomas only. … Indeed, median annual earnings for full-time working 25- to 32-year-olds with bachelor's degrees grew by nearly $6,700 to $45,500 from 1965 to 2013" (n.p.). According to the article, "What is the Value of a College Degree?", "In sheer monetary terms, a college graduate with a bachelor's degree is expected to earn twice as much as a college dropout. This translates into about 1.6 million USD more in lifetime earnings than workers with a high school diploma" (n.p.).
3. Billingsley (1992); Staples (1998).
4. Logan, Denby, and Gibson (2007).
5. Collins (1990).
6. Census data from 1960–2010 show population ranges from 2.2 to 3.0 million residents. The state continues to lag behind economically. Since 1970, it has experienced a five-fold increase in unemployment, but high school graduation and home ownership rates have exceeded 60 percent. In 2014–2016, poverty rates in Mississippi as compared to the USA in general are 13.7 percent and 2.08 percent, respectively; median incomes are $41,754 and $57,617, respectively (taken from https://www.statista.com/statistics/205481/poverty-rate-in-mississippi/).
7. Williams (1993) notes housing discrimination that made it difficult for blacks to secure home loans. These practices were legally codified. He suggests, "at the federal level, public expenditures and programs had race-specific impact. The 1949 National Housing Act subsidized the suburbanization of whites, but not blacks" (p. 221). Affordable housing remains out of reach for many US residents. Per the National Low Income Coalition report, "the hourly wage rate needed for a 'modest' two-bedroom rental is more than double the federal minimum wage of $7.25 per hour in all but four states … minimum wage hasn't kept up with inflation: In 1968, the federal minimum wage was equivalent to $10.90 in 2015 dollars – nearly $4 higher than today's actual deferral minimum wage" (n.p.). Hourly wages needed to afford such housing in the eleven states where the Kings either migrated and/or currently live are; CA ($30.92), CO ($21.97), GA ($16.79), IL ($20.87), IN ($15.17), MI ($16.24), MN ($18.60), MO ($15.67), MS ($14.84), OH ($15.00), and TN ($15.34). Mississippi has the lowest needed hourly wage; California has the highest. The report suggests that the USA has the resources to address the affordable housing crisis, but is politically stymied from doing so (Strutner 2017: n.p.).
8. Darity and Myers (1994) note challenges about depictions of kinship ties, "given the economic deprivation of urban black families, the impersonality of the urban environment, the cultural push towards individualism, the preference shown by

urban institutions for the nuclear family structure, and the difficulty of maintaining strong kinship ties in the urban environment, the reciprocal exchange system (of the extended family) is not as effective as Stack suggests ... Stack fails to discuss the strains, setbacks, and contradictions within urban kinship networks" (p. 125).

9. One daughter has been employed as a pianist for multiple church choirs for decades.
10. Billingsley (1992).
11. The first author profiled the steel mill experience in Gary, Indiana, in *The Cost of Being Poor* (2005). Other writers such as Drake and Cayton (1945) and Spear (1967) describe black migration to the Chicago area and work in steel mills.
12. Also refer to Haynes (1921).
13. Per this same source, workers 65 years of age with 30 years of service could receive a minimal pension of $188 monthly. Most workers could also receive supplemental unemployment benefits for up to 52 weeks as well as $90 per month for disability and pension. Additionally, most workers received vacation pay from 1 to 3.5 weeks based on their tenure. Working conditions varied, but in 1957, steel mills had an injury frequency rate (injuries per million hours of work) of 4.0 compared with 11.4 for all manufacturing posts. That same year, about 85 percent of iron and steel plan workers were members of the United Steelworkers of America union.
14. According to the 1978–1979 *Occupational Outlook Handbook*, hourly incomes for these same types of steel mill positions ranged from $6.55 to $8.90 (or $68,120–$74,048 annually).
15. Bensman and Wilson (2005).
16. This summary of the US steel industry was informed by the following studies: American Bridge (2017); Barnes (2005); Bensman and Wilson (2005); Caitlin (1993); D'Costa (1999); Hurley (1995); Industrial Steel Construction (2017); Mohl and Betten (1986); Stoddard (2015); and Warren (2001, 2008).
17. Austin (2016).
18. Ibid; Mohl and Betten (1986).
19. Caitlin (1993); Hurley (1995).
20. Elgin, Joliet, and Eastern Railway was a Class I railroad that opened January 1, 1889. Headquartered in Gary, Indiana, it operated primarily to Waukegan, Illinois, to service the U.S. Steel South Works in Gary. By the 1970s, EJ&E operated almost 170 miles of track and carried millions of tons of revenue. In December 2012, Canadian National completed the merger between EJ&E and Wisconsin Central Ltd. On January 1, 2013, EJ&E ceased to exist, exactly 124 years after it was founded (Caitlin 1993; Hurley 1995; Sanders 2003).
21. Field and Goldsmith (1993).
22. According to Donald's wife, Martha, during that 30-year period, he was only laid off for a short period and also took part in a labor strike for about six months. By the end of his career, a decline in jobs meant that his work hours were sparse. He vowed not to give them any more than 30 years of service; when he reached that milestone, he retired.
23. Blau and Duncan (1978); Newman (1989); Oliver and Shapiro (1997); Pattillo-McCoy (1999); Shapiro (2004).
24. Blau and Kahn (1981) suggest a higher likelihood of layoffs based on race (blacks more than whites) and gender (females more than males), respectively, after

controlling for personal productivity. These differences are, in part, attributed to discrimination.

25. Strauss (2017). Common annual salaries at U.S. Steel in 2018 range from $42,869 to $56,098 for positions ranging from utility person to maintenance electrician (www.glassdoor.com/Salary/United-States-Steel-Salaries-E1251.htm).
26. Writers such as Uchitelle (2008: 3) attribute these problems to decisions such as buy outs; the development of a second-tiered wage scale that compresses earnings; outsourcing; and, employees who are reluctantly accepting wage cuts for fear of losing their jobs.
27. The complete passage in Matthew is: "When the Pharisees heard that he had silenced the Sadducees, they gathered together, and one of them, a lawyer, asked him a question to test him. 'Teacher, which commandment in the law is the greatest?' He said to him, 'You shall love the Lord your God with all your heart, and with all your soul, and with all your mind. This is the greatest and first commandment. And a second is like it: You shall love your neighbor as yourself. On these two commandments hang all the law and the prophets'" (NRSV 1989).
28. The complete passages (KJV 2017) are "Watch and pray, that ye enter not into temptation: the spirit indeed is willing, but the flesh is weak" (Matthew 26:41); "Take ye heed, watch and pray: for ye know not when the time is" (Mark 13:33); "Watch ye and pray, lest ye enter into temptation. The spirit truly is ready, but the flesh is weak" (Mark 14:38); "Praying always with all prayer and supplication in the Spirit, and watching thereunto with all perseverance and supplication for all saints" (Ephesians 6:18); and, "Continue in prayer, and watch in the same with thanksgiving" (Colossians 4:2).
29. The complete verse is, "Train children in the right way, and when old, they will not stray" (NRSV 1989).
30. The two complete passages are: "Ask, and it will be given you; search, and you will find; knock, and the door will be opened for you" (Matthew 7:7 KJV 2017) and "So I say to you, Ask, and it will be given you; search, and you will find; knock, and the door will be opened for you" (Luke 11:9 KJV 2017).
31. Barnes (2012); Frazier (1964); Lincoln and Mamiya (1990); Marx (1971); Mays and Nicholson (1933); Wilmore (1994).
32. The complete passage is, "Whoever does not love does not know God, for God is love" (KJV 2017).
33. Barnes (2002, 2006, 2014a).
34. Collins (2009); Hale-Benson and Hilliard (1982); Kunjufu (1986).
35. The complete passages (KJV 2017) are: "Therefore shall a man leave his father and his mother, and shall cleave unto his wife: and they shall be one flesh" (Genesis 2:24); "And said, for this cause shall a man leave father and mother, and shall cleave to his wife: and they twain shall be one flesh?" (Matthew 19:5); "For this cause shall a man leave his father and mother, and cleave to his wife" (Mark 10:7); "For this cause shall a man leave his father and mother, and shall be joined unto his wife, and they two shall be one flesh" (Ephesians 5:31); and "For the husband is the head of the wife, even as Christ is the head of the church: and he is the savior of the body" (Ephesians 5:23).

36. Franklin and Pillow (1982) suggest that black men and women should develop relationship expectations based on their specific needs and goals rather than societal dictates.
37. Barnes and Wimberly (2016); Wimberly et al (2013).
38. After her husband's death, Carla sold their home and now lives with one of her daughters.
39. Billingsley (1992) suggests that blacks are more likely to provide such support to their family members than are whites.

Conclusion: "What Would Big Mama Do?" Activation and Routinization of a Black Family's Ethos

1. Oliver and Shapiro (1997); Shapiro (2004).
2. 2016 median household incomes by region are: $64,390 (Northeast); $64,275 (West); $58,305 (Midwest); and, $53,861 (South). Although a 3.9 percent increase was evident between 2015 and 2106, the lowest relative incomes are still found in the South.
3. This translates to 2.5 million fewer persons than in 2015 and 6.0 million fewer than in 2014.
4. Non-Hispanic whites represented 61.0 percent of the US population and 42.5 percent of the people in poverty; 2016 poverty rates by region are: 14.1 percent (South); 12.8 percent (West); 11.7 percent (Midwest); and, 10.8 percent (Midwest). The highest poverty rates are in the South.
5. Darity Jr. and Myers (1994, 1998); Mason (2003); Oliver and Shapiro (1997); Shapiro (2004).
6. Mattis (2002).
7. The family now rents the property to local hunters annually during hunting season and sells lumber off the land as well.
8. Barnes (2005) finds that, due to factors such as discrimination and limited social networks, poor and near-poor blacks tend to be less likely to bounce back from economic challenges and thus spend longer periods in poverty after making imprudent decisions. Foster (1993) suggests that disparate ideological views about poverty are part of US culture as well; "in periods where the poor are seen as strapped by constraints, the generosity of poverty programs increases ... in periods where the poor are perceived as deviant (e.g., as preferring to 'cheat' the system rather than work), welfare policy is reshaped and restricted" (p. 271).
9. Billingsley (1992).
10. Oliver and Shapiro (1997); Shapiro (2004).
11. Barnes (2005).
12. Bonilla-Silva (2010); Feagin (2006, 2010); Feagin and Feagin (1978).
13. Bolman and Deal (2008).
14. The complete passage is: "Give, and it shall be given unto you; good measure, pressed down, and shaken together, and running over, shall men give into your bosom. For with the same measure that ye mete withal it shall be measured to you again" (KJV 2017).
15. Barnes et al (2014).
16. Nwosu and Barnes (2014).

17. Barnes (2010); Berger (1963, 1969).
18. Barnes (2010); Collins (2009); Davies (2008); Delpit (2006); Hale (2001); Ladson-Billings (2009); Woodson (2013).

Appendix: A Mixed-Methodological Approach: Capturing King Family Voices and Experiences

1. Alex-Assensoh and Assensoh (2001); Anderson (1997); Barnes (2005); Barnes and Jaret (2003); Drake and Cayton (1945); Massey and Denton (1993); Spear (1967); Wilson (1987, 1996).
2. Blau and Duncan (1978); Frazier (1937); Newman (1989); Oliver and Shapiro (1997); Pattillo-McCoy (1999); Shapiro (2004). Census archival data were also gathered from sources such as "A Half-Century of Learning: Educational Attainment 1940 to 2000 Tables", "The Changing Shape of the Nation's Income Distribution: 1947–1998", "Historical Poverty Tables: People and Families - 1959 to 2015", and "Poverty Status of Families, by Type of Family, Presence of Related Children, Race, and Hispanic Origin: 1959 to 2015".
3. The small sample size and non-random selection process preclude generalizability; yet this is not the project goal. We contend that the study's strength lies in the ability to provide an in-depth examination of some of the contextual and familial dynamics that explain socioeconomic outcomes among family members.
4. Darity Jr. and Deshpande (2003); Darity Jr. and Myers (1994).
5. Darity, Jr. (1993); Darity Jr. and Deshpande (2003); Darity Jr. and Myers (1994, 1998); Darity, Jr., et al (2015).
6. Darity Jr. and Deshpande (2003).
7. Darity Jr. and Deshpande (2003); Darity Jr. and Myers (1998); Mason (2003).
8. Blau and Duncan (1978); Darity Jr. and Myers (1994, 1998); Darity, Jr. Hamilton and Stewart (2015); Duncan (1992); Frazier (1937); Newman (1989); Oliver and Shapiro (1997); Pattillo-McCoy (1999); Shapiro (2004).

Bibliography

Ainsworth-Darnell, James W., and Douglas B. Downey. 1998. "Assessing the Oppositional Culture Explanation for Racial/Ethnic Differences in School Performance." *American Sociological Review* 63(4): 536–53.

Alex-Assensoh, Yvette and A.B. Assensoh. 2001. "Inner-City Contexts, Church Attendance, and African-American Political Participation." *The Journal of Politics* 63(3): 886–901.

American Bridge Co. 2017. "American Bridge History." Retrieved Feb. 25, 2018, from www.americanbridge.net/about/history/#1930s.

American Merchant Marine at War. 2003. "Truth About Salaries, the Draft, Unloading Ships, and Court Martials." *Barron's National Business and Financial Weekly* (April 24, 1944). Retrieved Feb. 2, 2018, from www.usmm.org/salary.html.

Anderson, Elijah. 1997. *Streetwise: Race, Class, and Change in an Urban Community*. Chicago: The University of Chicago Press.

Anderson, James. 1988. *The Education of Blacks in the South, 1860–1935*. Chapel Hill: University of North Carolina Press.

Ashmore, Harry. 1954. *The Negro and the Schools*. Chapel Hill: University of North Carolina Press.

Austin, Elizabeth. 2016. "Black-White Earnings Gap Remains at 1950s Levels for Median Worker." *University of Chicago News* (Dec. 14). Retrieved Feb. 8, 2018, from https://news.uchicago.edu/article/2016/12/14/black-white-earnings-gap-remains-1950s-levels-median-worker.

Averitt, Robert. 1968. *The Dual Economy*. New York: Norton.

Ayers, Edward. 1992. *The Promise of the New South: Life After Reconstruction*. New York: Oxford University Press.

Baker-Fletcher, Karen. 2006. "A Womanist Journey." Pp. 158–75 in *Deeper Shades of Purple: Womanism in Religion and Society*, edited by Stacey Floyd-Thomas. New York: New York University Press.

Barnes, Sandra. 2002. "Then and Now: A Comparative Analysis of the Urban Black Church in America." *Journal of the Interdenominational Theological Center* 29(1): 137–56.

2004. "Priestly and Prophetic Influences on Black Church Social Services." *Social Problems* 51(2): 202–21.

2005a. *The Cost of Being Poor: A Comparative Study of Life in Poor Urban Neighborhoods in Gary, Indiana*. New York: State University Press of New York.
2005b. "Black Church Culture and Community Action." *Social Forces* 84(2): 967–94.
2006. "Whosoever Will Let *Her* Come: Gender Inclusivity in the Black Church." *Journal for the Scientific Study of Religion* 45(3): 371–87.
2008. "'The Least of These': Black Church Children's and Youth Outreach Efforts." *Journal of African American Studies* 12:97–119.
2010. *Black Megachurch Culture: Models for Education and Empowerment*. New York: Peter Lang Press.
2011. "Black Church Sponsorship of Economic Programs: A Test of Survival and Liberation Strategies." *Review of Religious Research* 53(1): 23–40.
2012. *Live Long and Prosper: How Black Megachurches Address HIV/AIDS and Poverty in the Age of Prosperity Theology*. New York: Fordham University Press.
2014a. "The Black Church Revisited: Toward a New Millennium DuBoisian Mode of Inquiry." *Sociology of Religion: A Quarterly Review* (doi:10.1093/socrel/sru056).
2014b. "To Educate, Equip, and Empower: Black Church Sponsorship of Tutoring or Literary Programs." *Review of Religious Research* (doi: 10.1007/s13644-014-0173-2).
Barnes, Sandra and Charles Jaret. 2003. "The 'American Dream' in Poor Urban Neighborhoods: An Analysis of Home Ownership Attitudes and Behavior and Financial Saving Behavior." *Sociological Focus* 36(3): 219–39.
Barnes, Sandra L. and Anne Streaty-Wimberly. 2016. *Empowering Black Youth of Promise: Education and Socialization in the Village-Minded Black Church*. New York: Routledge Press.
Barnes, Sandra, Zandria Robinson, and Earl Wright III (Eds). 2014. *Re-Positioning Race: Prophetic Research in a Post-Racial Obama Age*. New York: State University Press of New York.
Bell, Derrick. 2004. *Race, Racism and American Law*. New York: Aspen Publishers.
Benbow, Linda. 2010. *Sorting Letters, Sorting Lives: Delivering Diversity in the United States Postal Service*. Lanham, MD: Lexington Books.
Bensman, David and Mark R. Wilson. 2005. "Iron and Steel." *The Electronic Encyclopedia of Chicago: Chicago Historical Society*. Retrieved Jan. 15, 2018, from www.encyclopedia.chicagohistory.org/pages/653.html.
Berger, Peter. 1963. A Market Model for the Analysis of Ecumenicity. *Social Research* 30: 77–93.
1969. *The Sacred Canopy*. New York: Anchor Books.
Berry, Mary and John Blassingame. 1982. *Long Memory: The Black Experience in America*. New York: Oxford University Press.
Best College Values. "What Is the Value of a College Degree?" 2015. Retrieved Dec. 15, 2017, from www.bestcollegevalues.org/what-is-the-value-of-a-college-degree/.
Biderman, Albert and Laura Sharp. 1968. "The Convergence of Military and Civilian Occupational Structures." *American Journal of Sociology* 73: 381–99.
Billingsley, Andrew. 1992. *Climbing Jacob's Ladder: The Enduring Legacy of African-American Families*. New York: A Touchstone Book.
Blau, Peter M. and Otis Dudley Duncan. 1978. *The American Occupational Structure*. New York: Free Press.

Blau, Francine and Lawrence Kahn. 1981. "Causes and Consequences of Layoffs." *Economic Inquiry* 19(2): 270–96.

Blumer, Herbert. 1958. "Race Prejudice as a Sense of Group Position." *Pacific Sociological Review* 1(1): 3–7.

Bolman, Lee and Terrence Deal. 2008. *Reframing Organizations: Artistry, Choice and Leadership* (4th Edition). San Francisco: Jossey-Bass.

Bolton, Charles C. 2005. *The Hardest Deal of All: The Battle Over School Integration in Mississippi, 1870–1980*. Jackson: University Press of Mississippi.

Bonilla-Silva, Eduardo. 2010. *Racism Without Racists: Color-Blind Racism and the Persistence of Racial Inequality in the United States*. Lanham, MD: Rowman & Littlefield Publishers.

Bouie, Jamelle. 2011. "Violence and Economic Mobility in the Jim Crow South." *The Nation*. (July 29). Retrieved Dec. 31, 2017, from www.thenation.com/article/violence-and-economic-mobility-jim-crow-south/.

Bronfenbrenner, Urie. 1977. "Toward an Experimental Ecology of Human Development." *American Psychologist* 32: 513–31.

1979. *The Ecology of Human Development: Experiments by Nature and Design*. Cambridge: Harvard University Press.

1986. "Ecology of the Family as a Context for Human Development: Research Perspectives." *Developmental Psychology* 22: 723–42.

Brook, Tom Vanden. 2017. "Black Troops as Much as Twice as Likely to Be Punished by Commanders, Courts." *USA TODAY* (June 7). Retrieved April 8, 2018, from www.usatoday.com/story/news/politics/2017/06/07/black-troops-much-twice-likely-punished-commanders-courts/102555630/.

Broom, Leonard and J. Smith. 1963. "Bridging Occupations." *The British Journal of Sociology* XIV: 321–34.

Brown, Irene. 1997. "Explaining the Black–White Gap in Labor Force Participation among Women Heading Households." *American Sociological Review* 62: 236–252.

Browning, Harley, Sally Lopreato, and Dudley Poston, Jr. 1973. "Income and Veteran Status: Variations Among Mexican Americans, Blacks, and Anglos." *American Sociological Review* 38: 74–84.

Bruchey, Stuart. 1967. *Cotton and the Growth of the American Economy: 1790–1860*. New York: Random House.

Bullock, Henry. 1967. *A History of Negro Education in the South*. Cambridge: Harvard University Press.

Burnette, Alice. 2001. "Giving Strength: Understanding Philanthropy in the Black Community." *Philanthropy Matters* 11(1): 3–5.

Butler, John S. 1976. "Inequality in the Military: An Examination of Promotion Time for Black and White Enlisted Men." *American Sociological Review* 41: 807–18.

Caitlin, Robert. 1993. *Racial Politics and Urban Planning: Gary, Indiana 1980–1989*. Lexington : The University of Kentucky Press.

Cannon, Katie G. 1988. *Black Womanist Ethics*. United Kingdom: Scholar's Press.

Census of Population and Housing: 1950. "Census of Agriculture – Special Reports: Section I Land Quantity and Value-Method of Holding." Retrieved Feb. 23, 2018, from www.census.gov/prod/www/decennial.html.

Chafe, William, Raymond Gavins, and Robert Korstad (Eds.). 2014. *Remembering Jim Crow: African Americans Tell About Life in the Segregated South*. New York: The New Press.

Church Leader Gazette. 2013. "Pew Research Report Says Loyalty to Black Protestant Churches Remains Strong." *Church Leader Gazette* (May 8, 2013). Retrieved May 9, 2013, from http://churchleadergazette.com/clg/2013/05/pew-research-report-says-loyal.html.

Collins, Pat Hill. 1990. *Black Feminist Thought: Knowledge, Consciousness, and the Politics of Empowerment*. New York: Routledge.

2009. *Another Kind of Public Education: Race, Schools, the Media and Democratic Possibilities*. Boston: Beacon Press.

Cone, James H. 1992. *The Spirituals and the Blues*. Maryknoll: Orbis Books.

1997. *God of the Oppressed*. Maryknoll: Orbis Books.

Cotton, Jeremiah. 1993. "Labor Markets and Racial Inequality." Pp. 183–208 in *Labor Economics: Problems in Analyzing Labor Markets*, edited by William Darity, Jr. Boston: Kluwer Academic Publishers.

Crook, Brenda. 1996. "Historic and Architectural Resources of Copiah County." *Mississippi: National Register of Historic Places. United States Department of the Interior National Park Service* (1–75). Retrieved April 23, 2018, from www.apps.mdah.ms.gov/t_nom/Historic%20and%20Architectural%20Resources%20of%20Copiah%20County,%20Mississippi.pdf.

Dansby, Mickey and Dan Landis. 2001. "Intercultural Training in the United States Military." Pp. 9–28 in *Managing Diversity in the Military*, edited by M.R. Dansby, J.B. Stewart, and S.C. Webb. New Brunswick, NJ: Transaction Publishers.

Darity, Jr., William (Ed). 1993. *Labor Economics: Problems in Analyzing Labor Markets*. Boston: Kluwer Academic Publishers.

Darity. Jr., William and Ashwini Deshpande. 2003. *Boundaries of Clan and Color: Transnational Comparisons of Inter-group Disparity*. New York: Routledge.

Darity, Jr., William and Samuel L. Myers. 1994. *The Underclass: Critical Essays on Race and Unwantedness*. New York: Garland Publishing, Inc.

1998. *Persistent Disparity: Race and Economic Inequality in the United States Since 1945*. New York: Garland Publishing, Inc.

Darity, Jr., William, Darrick Hamilton, and James B. Stewart. 2015. "A Tour de Force in Understanding Intergroup Inequality: An Introduction to Stratification Economics." *The Review of Black Political Economy* 42: 1–6.

Data Access and Dissemination Systems (DADS). 2010. "Geographic Identifiers: 2010 Demographic Profile Data (G001): Hazlehurst city, Mississippi." American FactFinder. Retrieved July 8, 2017, from https://factfinder.census.gov/faces/ tableservices/jsf/pages/productview.xhtml?src=bkmk.

Davis, Donna. 2004. "Merry-Go-Round: A Return to Segregation and the Implications for Creating Democratic Schools." *Urban Education* 39(4): 394–407.

D'Costa, Anthony P. 1999. *The Global Restructuring of the Steel Industry: Innovations, Institutions, and Industrial Change* London: Routledge Press.

Delpit, Lisa. 2006. *Other People's Children: Cultural Conflict in the Classroom*. New York: The New Press.

Denzin, Norman and Yvonna Lincoln. 2005. *The Sage Handbook of Qualitative Research*. Thousand Oaks: Sage.

Deshpande, Ashwini and William Darity, Jr. 2003. "Boundaries of Clan and Color: An Introduction." Pp. 1–13 in *Boundaries of Clan and Color: Transnational Comparisons of Inter-group Disparity*, edited by William Darity, Jr. and Ashwini Deshpande. New York: Routledge.

Dickens, Williams and Kevin Lang. 1993. "Labor Market Segmentation Theory: Reconsidering the Evidence." Pp. 141–80 in *Labor Economics: Problems in Analyzing Labor Markets*, edited by William Darity, Jr. Boston: Kluwer Academic Publishers.

Dodson, Jualyne and Cheryl Townsend Gilkes. 1987. "Something Within: Social Change and Collective Endurance in the Sacred World of Black Christian Women." Pp. 80–130 in *Women and Religion in America, Vol. 3: 1900–1968*, edited by Rosemary Reuther and R. Keller. New York: Harper and Row.

Douglas, Kelly Brown. 2006. "Twenty Years a Womanist: An Affirming Challenge." Pp. 145–57 in *Deeper Shades of Purple: Womanism in Religion and Society*, edited by Stacey Floyd-Thomas. New York: New York University Press.

Drake, J.G. St. Clair and Horace R. Cayton. 1945. *Black Metropolis: A Study of Negro Life in a Northern City*. New York: Harcourt, Brace and Company.

DuBois, W.E.B. 1953[1996]. *The Souls of Black Folk*. New York: Random House.

1899[1996]. *The Philadelphia Negro: A Social Study*. Pennsylvania; University of Pennsylvania Press.

1935[1998]. *Black Reconstruction in America, 1860–1880*. New York: Harcourt Brace (Reprint New York: The Free Press, 1998).

1903[2003]. *The Negro Church*. Walnut Creek, CA: Altimira Press.

Duncan, Greg. 1992. *W(h)ither the Middle Class?: A Dynamic View*. Syracuse University: Metropolitan Studies Program, The Maxwell School of Citizenship and Public Affairs.

Edin, Kathryn and Laura Lein. 1996. Work, Welfare, and Single Mothers' Economic Survival Strategies. *American Sociological Review* 61: 253–66.

Elder, Glen. 1986. "Military Times and Turning Points in Men's Lives." *Developmental Psychology* 22: 233–45.

1987. "War Mobilization and the Life Course: A Cohort of World War II Veterans." *Sociological Forum* 2(3): 449–72.

Ellison, Christopher G. and Darren E. Sherkat. 1995. "The 'Semi-Involuntary Institution' Revisited: Regional Variations in Church Participation Among Black Americans." *Social Forces* 73(4): 1415–37.

Emerson, Michael. 2008. *People of the Dream: Multiracial Congregations in the U.S.* Princeton: Princeton University Press.

Farmworker Justice. 2014. Retrieved March 11, 2018, from www.farmworkerjustice.org/sites/default/files/NAWS%20data%20factsht%201-13-15FINAL.pdf.

Feagin, Joe R. 2006. *Systemic Racism: A Theory of Oppression*. New York: Routledge.

2010. *The White Racial Frame: Centuries of Racial Framing and Counter-Framing*. New York: Routledge.

Feagin, Joe R. and Clairece B. Feagin. 1978. *Discrimination American Style: Institutional Racism and Sexism*. Englewood Cliffs: Prentice Hall.

Ferguson, Ann Arnet. 2001. *Bad Boys: Public Schools in the Making of Black Masculinity*. Ann Arbor: University of Michigan Press.

Field, Alfred and Arthur Goldsmith. 1993. "The Impact of Formal on-the-Job Training on Unemployment and the Influence of Gender, Race, and Working Lifecycle Position on Accessibility to the On-the-Job Training." Pp. 77–116 in *Labor Economics: Problems in Analyzing Labor Markets*, edited by William Darity, Jr. Boston: Kluwer Academic Publishers.

Flanagan, William. 1999. *Urban Sociology: Images and Structures* (3rd edition). Boston: Allyn and Bacon.

Floyd-Thomas, Stacey. 2006. "Introduction: Writing for Our Lives: Womanism as an Epistemological Revolution." Pp. 1–14 in *Deeper Shades of Purple: Womanism in Religion and Society*, edited by Stacey Floyd-Thomas. New York: New York University Press.

Foster, Daphine G. (n.d.). "Hazlehust History." Hazlehurst History. Retrieved July 8, 2017, from www.hazlehursthpc.com/history.html.

Foster, E. Michael. 1993. "Labor Economics and Public Policy: Dominance of Constraints or Preferences?" Pp. 269–94 in *Labor Economics: Problems in Analyzing Labor Markets*, edited by William Darity, Jr. Boston: Kluwer Academic Publishers.

Foucault, Michel. 1970. *The Order of Things: An Archaeology of the Human Sciences.* New York: Random House.

Foucault, Michel and Paul Rabinow. 1984. *The Foucault Reader*. New York: Pantheon.

Franklin, Clyde and Walter Pillow. 1982. "The Black Males Acceptance of the Prince Charming Ideal." *Black Caucus* 13: 3–7.

Frazier, E. Franklin. 1937. *Black Bourgeoisie*. New York: Free Press.

1939. *The Negro Family in Chicago*. Chicago: University of Chicago Press.

1964. *The Negro Church in America*. New York: Schocken Books.

Frohlich, Thomas C. and Alexander Kent. 2015. "What US Land Is Really Worth, State by State" (July 3). Retrieved Dec. 15, 2017, from www.msn.com/en-us/money/generalmoney/what-us-land-is-really-worth-state-by-state/ar-BBkTfZS#image=BBkGo1K|5.

Geonames (n.d.). "Gallman." Geographic Names Information System. Retrieved July 8, 2017, from https://geonames.usgs.gov/apex/f?p=gnispq% 3A3%3A%3A%3ANO%3A%3AP3_FID%3A670303.

Gilkes, Cheryl Townsend. 1998. "Plenty Good Room: Adaptation in a Changing Black Church." *Annals of the American Academy of Political and Social Science* 558: 101–21.

2001. *If It Wasn't For the Women*. Maryknoll: Orbis Books.

Gillespie, Diane. 2003. "The Pedagogical Value of Teaching White Privilege Through a Case Study." *Teaching Sociology* 31(4): 469–77.

Ginzburg, Ray. 1988. *One Hundred Years of Lynchings*. Baltimore: Black Classic Press.

Glaude Jr., Eddie. 2010. "The Black Church Is Dead." *Huffington Post* (Feb. 24). www.huffingtonpost.com/eddie-glaude-jr-phd/the-black-church-is-dead_b_473815.html.

Golden, Marita. 1983. *Migrations of the Heart*. New York: Ballantine Press.

Granovetter, Mark. 1973. "The Strength of Weak Ties." *American Journal of Sociology* 78: 1360–80.

1983. "The Strength of Weak Ties: A Network Theory Revisited." *Sociological Theory* 1: 201–223.

Grant, Jacquelyn. 1989. *White Women's Christ and Black Women's Jesus: Feminist Christology and Womanist Response*. Atlanta: Scholars Press.

Great American Stations. (n.d.). "Hazlehurst, MS (HAZ)." *Great American Stations*. Retrieved July 8, 2017, from www.greatamericanstations.com/stations/hazlehurst-ms-haz/.

Green, Lorenzo J. and Carter G. Woodson. 1930. *The Negro Wage Earner*. Washington DC: The Association for the Study of Negro Life and History.

Hale, Janice. 2001. *Learning While Black: Creating Educational Excellence for African American Children*. Baltimore: The Johns Hopkins University Press.

Hale-Benson, Janice and Asa Hilliard. 1982. *Black Children: Their Roots, Culture, and Learning Styles*. Baltimore: The Johns Hopkins University Press.

Hallinan, Maureen. 2001. "Sociological Perspectives on Black-White Inequalities in American Schooling." *Sociology of Education* 74: 50–70.

Haynes, George. 1921. *The Negro at Work During the World War and the Reconstruction*. Washington, DC: U.S. Department of Labor.

Hays, Sharon. 2003. *Flat Broke with Children: Women in the Age of Welfare Reform*. New York: Oxford University Press.

Heckman James J. and Guilherme Sadlacek. 1985. "Heterogeneity, Aggregation and Market Wage Functions: An Empirical Model of Self Selection in the Labor Market." *Journal of Political Economy* 93: 1077–125.

Henderson, A.M. and Talcott Parsons (translators). 1947. *The Theory of Social and Economic Organizations*. New York: The Free Press.

Hochschild, Arlie Russell. 1990. *The Second Shift*. New York: Avon Books.

Hofferth, Sandra. 1984. "Kin Networks, Race, and Family Structure." *Journal of Marriage and the Family*: 791–806.

Hogan, Dennis P., Ling-xin Hao, and William Parish. 1990. "Race, Kin Networks, and Assistance to Mother – Headed Families." *Social Forces* 68(3): 797–812.

The Holy Bible Containing the Old and New Testaments Translated Out of the Original Tongues and with Former Translations Diligently Compared and Revised, Self-Pronouncing Edition. 1910. Nashville: The Southwestern Company Publishers and Booksellers.

The Holy Bible: King James Version. 2017. Nashville: Thomas Nelson.

The Holy Bible: New Revised Standard Version. 1989. Nashville: Thomas Nelson.

Hunt, Larry and Matthew Hunt. 1999. "Regional Patterns in African American Church Attendance: Revisiting the Semi-Involuntary Thesis." *Social Forces* 78 (2): 779–91.

Industrial Steel Construction, Inc. 2017. "Gary, IN, Facility." Retrieved Feb. 25, 2018, from www.iscbridge.com/location-and-history/gary-in-facility.

Irvine, Russell and Jackie Irvine. 1983. "The Impact of the Desegregation Process on the Education of Black Students: Key Variables." *Journal of Negro Education* 52(4): 410–22.

Jacobs, Peter. 2014. "Science and Math Majors Earn the Most Money After Graduation." *Business Insider* (July 9). Retrieved Jan. 5, 2018, from www.businessinsider.com/stem-majors-earn-a-lot-more-money-after-graduation-2014-7.

Jarrett, Robin. 1994. "Living Poor: Family Life among Single Parent, African-American Women." *Social Problems* 41(1): 30–49.

Jones, Faustine. 1981. *A Traditional Model of Educational Excellence: Dunbar High School of Little Rock, Arkansas.* Washington: Howard University Press.

Jordan, Reed. 2014. "America's Public Schools Remain Highly Segregated." *Urban Institute.* Retrieved May 6, 2018, from www.urban.org/urban-wire/americas-public-schools-remain-highly-segregated.

The Journal of Blacks in Higher Education. 2010. "College Graduation Rates: Where Black Students Do the Best and Where They Fare Poorly Compared to Their White Peers." Retrieved March 1, 2013, from www.jbhe.com/features/65_gradrates.html.

Krause, Bonnie J. 2003. "The Jeanes Supervisor: Agent of Change in Mississippi's African American Education." *Journal of Mississippi History* 65(2): 127–45.

Krippendorf, Klaus. 1980. *Content Analysis: An Introduction to Its Methodology.* Beverly Hills: Sage Publications.

Kunjufu, Jawanza. 1986. *Developing Positive Self-Images & Discipline in Black Children.* Chicago: African American Images.

Kurtzleben, Danielle. 2014. "Income Gap Between Young College and High School Grads Widens." U.S. News and World Report (Feb. 11, 2014). Retrieved Dec. 15, 2017, from www.usnews.com/news/articles/2014/02/11/study-income-gap-between-young-college-and-high-school-grads-widens.

Ladson-Billings, Gloria. 2009. *The Dreamkeepers: Successful Teachers of African American Children.* San Francisco: Jossey-Bass.

Leigh, Duane and Robert Berney. 1971. "The Distribution of Hostile Casualties on Draft-Eligible Males with Differing Socioeconomic Characteristics." *Social Science Quarterly* 51: 932–40.

Lincoln, C. Eric and Lawrence H. Mamiya. 1990. *The Black Church in the African-American Experience.* Durham: Duke University Press.

Logan, Sadye, Ramona Denby, and Priscilla Gibson (Eds.). 2007. *Mental Health Care in the African-American Community* (1st Edition). New York: Routledge.

Logon, John R. and Harvey L. Molotch. 1987. *Urban Fortunes: The Political Economy of Place.* Berkeley: University of California Press.

Lutz, Amy. 2008. "Who Joins the Military? A Look at Race, Class, and Immigration Status." *Journal of Political and Military Sociology* 36(2): 167–88.

Ma, Jennifer, Matea Pender, and Meredith Williams. 2016. "Education Pays 2016: The Benefits of Higher Education for Individuals and Society." College Board: Trends in Higher Education Series. Retrieved Dec. 16, 2017, from https://trends.collegeboard.org/sites/default/files/education-pays-2016-full-report.pdf.

Margo, Robert. 1995. "Explaining Black-White Wage Convergence, 1940–1950." *Industrial and Labor Relations Review* 48(3): 470–80.

Marx, Gary. 1971. "Religion: Opiate or Inspiration of Civil Rights Militancy?" Pp. 150–60 in *The Black Church in America*, edited by Hart Nelsen, Raytha Yokley, and Anne Nelsen. New York: Basic Books.

Mason, Patrick. 2003. "Understanding Recent Empirical Evidence on Race and Labor Market Outcomes in the USA." Pp. 52–69 in *Boundaries of Clan and Color: Transnational Comparisons of Inter-group Disparity*, edited by William Darity, Jr. and Ashwini Deshpande. New York: Routledge.

Massey, Douglas S. and Nancy A. Denton. 1993. *American Apartheid: Segregation and the Making of the Underclass.* Massachusetts: Harvard University Press.

Mattis, Jacqueline S. 2002. "The Role of Religion and Spirituality in the Meaning-Making and Coping Experiences of African American Women: A Qualitative Analysis." *Psychology of Women Quarterly* 26: 308–20.

Mays, Benjamin and Joseph Nicholson. 1933. *The Negro's Church*. New York: Institute of Social and Religious Research.

McGee, Ebony. 2015. "Why Black Students Struggle in STEM Subjects: Low Expectations." *New Republic* (May 1, 2015) Retrieved Jan. 5, 2018, from https://newrepublic.com/article/121693/why-black-males-struggle-stem-subjects.

McGee, Ebony and David Stovall. 2015. "Reimagining Critical Race Theory in Education: Mental Health, Healing, and the Pathway to Liberatory Praxis." *Educational Theory* 65(5): 491–511.

McIntosh, Peggy. 1989. "White Privilege: Unpacking the Invisible Knapsack." *National SEED Project on Inclusive Curriculum*. Retrieved Dec. 10, 2017, from https://nationalseedproject.org/white-privilege-unpacking-the-invisible-knapsack.

McRoberts, Omar M. 2003. *Streets of Glory: Church and Community in a Black Urban Neighborhood*. Chicago: The University of Chicago Press.

Mehan, Hugh, Alma Hertweck, and J. Lee Miehls. 1986. *Handicapping the Handicapped: Decision-Making in Students' Educational Careers*. Stanford: Stanford University Press.

Messner, Michael. 2011. "The Privilege of Teaching About Privilege." *Sociological Perspectives* 54(1): 3–14.

Mincer, Jacob and Solomon Polachek. 1974. "Family Investments in Human Capital: Earnings of Women." *Journal of Political Economy* 82(2): 576–88.

Moen, Phyllis. 1992. *Women's Two Roles: A Contemporary Dilemma*. New York: Auburn House.

Mohl, Raymond and Neil Betten. 1986. *Steel City: Urban and Ethnic Patterns in Gary, Indiana, 1906–1950*. New York: Holmes and Meier.

Moody, Anne. 1992. *Coming of Age in Mississippi: The Classic Autobiography of Growing Up Poor and Black in the Rural South*. New York: Dell Books.

Morris, Aldon D. 1984. *The Origins of the Civil Rights Movement: Black Communities Organizing for Change*. New York: The Free Press.

Moskos, Charles. 1986. "Success Story: Blacks in the Army." *Atlantic Monthly*: 64–72.

Moynihan, Daniel. 1965. *The Negro Family: The Case for National Action*. Washington DC: U.S. Government Printing Office.

National Center for Education Statistics. 2015. "Fast Facts: Income of Young Adults." Retrieved Dec. 15, 2017, from https://nces.ed.gov/fastfacts/display.asp?id=77.

National Parks Service and Michael Fazio. 1985. *Gallman Historic District | National Register of Historic Places Inventory-Nomination Form*. Retrieved July 8, 2017, from www.apps.mdah.ms.gov/nom/dist/29.pdf.

Nazaryan, Alexander. 2018. "School Segregation in America Is as Bad Today as It Was in the 1960s." *Newsweek* (March 22). Retrieved May 6, 2018, from www.newsweek.com/2018/03/30/school-segregation-america-today-bad-1960-855256.html.

Nelson, Timothy. 2005. *Every Time I Feel the Spirit: Religious Experience and Ritual in an African American Church*. New York: New York University Press.

Newman, Katherine. 1989. *Falling from Grace: The Experience of Downward Mobility in the American Middle Class*. New York: Vintage.

Newport, Frank. 2014. "Mississippi Is Most Religious U.S. State, Vermont and New Hampshire Are the Least Religious States." Retrieved Oct. 5, 2017, from Gallup .com/poll.

Nwosu, Oluchi and Sandra Barnes. 2014. "'Where Difference Is the Norm': Exploring Refugee Student Ethnic Identity Development, Acculturation, and Agency at Shaw Academy." *Journal of Refugee Studies* (doi: 10.1093/jrs/fet050).

Oliver, Melvin and Thomas Shapiro. 1997. *Black Wealth/White Wealth*. New York: Routledge.

Olson, Daniel and Paul Perl. 2000. "Religious Market Share and Intensity of Church Involvement in Five Denominations." *Journal for the Scientific Study of Religion* 39 (1): 12–31.

Parsons, Talcott. 1937. *The Structure of Social Action*. New York: The Free Press.

Pattillo-McCoy, Mary. 1998. Church Culture as a Strategy of Action in the Black Community. *American Sociological Review* 63(6): 767–84.

1999. *Black Picket Fences: Privilege and Peril Among the Black Middle Class*. Chicago: University of Chicago Press.

Pew Forum on Religion & Public Life. 2013. "U.S. Religious Landscape Survey." Retrieved April 1, 2017, from http://religions.pewforum.org/affiliations.

Pew Research. 2009. "A Religious Portrait of African-Americans" (Jan. 30, 2009). Retrieved from www.pewforum.org/2009/01/30/a-religious-portrait-of-african -americans/.

"Population Demographics for Hazlehurst, Mississippi in 2016 and 2017." 2017. Retrieved Sept. 25, 2017, from https://suburbanstats.org/population/mississippi/ how-many-people-live-in-hazlehurst.

Porter, Michael. 1998. *Kill Them Before They Grow: Misdiagnosis of African American Boys in American Classrooms*. Chicago: African American Images.

Quarles, Benjamin. 1987. *The Negro in the Making of America*. New York: MacMillan.

Roberts, Dorothy E. 2012. "Welfare and the Problem of Black Citizenship," *105 Yale Law Journal* 1563–1602, 168–1572 (April, 1996). Retrieved Dec. 3, 2017, from https://academic.udayton.edu/race/04needs/welfare01b.htm.

Roediger, David. 1991. *The Wages of Whiteness: Race and the Making of the American Working Class*. New York: Verso Books.

Rollins, Judith. 1985. *Between Women: Domestics and Their Employers*. Philadelphia: Temple University Press.

Rosengarten, Theodore. 1974. *All God's Dangers: The Life of Nate Shaw*. Chicago: University of Chicago.

Rubio, Philip. 2010. *There's Always Work at the Post Office: African American Postal Workers and the Fight for Jobs, Justice, and Equality*. Chapel Hill: University of North Carolina Press.

Rugaber, Christopher S. 2017. "Pay Gap Between College Grads and Everyone Else at a Record." *USA Today* (Jan. 12, 2017). Retrieved Dec. 15, 2017, from www .usatoday.com/story/money/2017/01/12/pay-gap-between-college-grads-and -everyone-else-record/96493348/.

Sampson, Robert and John Laub. 1996. "Socioeconomic Achievement in the Life Course of Disadvantaged Men: Military Service as a Turning Point, Circa 1940–1965." *American Sociological Review* 61(3): 347–67.

Samuels, Dena, Abby Ferber, and Andrea O-Reilly Herrera 2003. "Introducing the Concepts of Oppression and Privilege into the Classroom." *Race, Gender, and Class* 10(4): 5–21.

Sanders, Craig. 2003. *Limiteds, Locals, and Expresses in Indiana, 1838–1971*. Bloomington: Indiana University Press.

Shapiro, Thomas. 2004. *The Hidden Cost of Being African American: How Wealth Perpetuates Inequality*. New York: Oxford University Press.

Siddle Walker, Vanessa. 1996a. "Can Institutions Care? Evidence from the Segregated Schooling of African American Children." Pp. 211–26 in *Beyond Desegregation: The Politics of Quality in African American Schooling*, edited by M.J. Shujaa. Thousand Oaks: Corwin Press.

1996b. *Their Highest Potential: An African American School Community in the Segregated South*. Chapel Hill: University of North Carolina Press.

Snyder, Thomas D (Ed.). 1993. "120 Years of American Education: A Statistical Portrait." National Center for Education Statistics. Retrieved March 25, 2018, from https://nces.ed.gov/pubs93/93442.pdf.

Solomon, John. 1999. *The Final Frontiers, 1880–1930: Settling the Southern Bottomlands*. Westport: Greenwood Press.

Sowell, Thomas. 1976. "Patterns of Black Excellence." *Public Interest* 43: 26–58.

Spear, Allan. 1967. *Black Chicago: The Making of a Negro Ghetto 1890–1920*. Chicago: University of Chicago Press.

Squires, Gregory D. 1994. *Capital and Communities in Black and White*. New York: State University of New York Press.

Staples, Robert. 1998. *The Black Family: Essays and Studies* (6th Edition). Independence, KY: Cengage Learning.

Sterling, Dorothy (Ed). 1994. *The Trouble They Seen: The Story of Reconstruction in the Words of African Americans*. New York: Da Capo Press.

Stoddard, Brooke. 2015. *Steel: From Mine to Mill, the Metal that Made America*. Minneapolis: Zenith Press.

Strauss, Anselm L. and Juliet M. Corbin. 1990. *Basics of Qualitative Research: Grounded Theory Procedures and Techniques*. Newbury Park: Sage.

Strauss, Eric. 2017. "Salary of Steel Plant Workers." Retrieved Feb. 25, 2018, from http://work.chron.com/salary-steel-plant-workers-2958.html.

Strutner, Suzy. 2017. "The Hourly Income You Need to Afford Rent Around the U.S." *Huffington Post* (Nov. 19). Retrieved April 8, 2018, from www.huffingtonpost.com/entry/how-much-you-need-for-rent_us_5942cc92e4b0f15cd5b9e2ee.

Thomas, Linda. 2004. *Living Stones in the Household of God: The Legacy and Future of Black Theology*. Minneapolis: Augsburg Fortress Press.

Thurman, Howard. 1965. *The Luminous Darkness*. Richmond: Friends United Press.

Tillitt, Malvern H. 1944. "Army-Navy Pays Tops Most Civilians': Unmarried Private's Income Equivalent to $3,600." *Barron's National Business and Financial Weekly*. Retrieved Feb. 7, 2018, from www.usmm.org/barrons.html.

Townes, Emily. 2006. *Womanist Ethics and the Cultural Production of Evil*. New York: Palgrave MacMillian.

Uchitelle, Louis. 2008. "The Wage That Meant Middle Class." *The New York Times* (April 20): 1–5. Retrieved Feb. 25, 2018, from www.nytimes.com/2008/04/20/weekinreview/20uchitelle.html.

U.S. Bureau of Census. 2016. "Income and Poverty in the United States." Washington, DC: U.S. Government Printing Office. Retrieved April 15, 1917, from www.census.gov.

U.S. Bureau of Labor Statistics. 2011. "Highlights of Women's Earnings in 2010: Report 1031." July: 1–91. Retrieved March 10, 2018, from www.bls.gov/opub/reports/womens-earnings/archive/womensearnings_2010.pdf.

U.S. Department of Commerce: Bureau of the Census. 1975. "Historical Statistics of the United States Bicentennial Edition – Colonial Times to 1970: Series K 538–549. Rice, Sugarcane, Sugar Beets, and Peanuts Acreage, Production, and Price: 1895 to 1970-C." Retrieved Feb. 15, 2018, from www.census.gov/history/pdf/histstats-colonial-1970.pdf.

U.S. Department of Commerce: Bureau of the Census. 1975. "Historical Statistics of the United States Bicentennial Edition – Colonial Times to 1970: Series K 550–563. Hay, Cotton, Cottonseed, Shorn Wool, and Tobacco-Acreage, Production, and Price: 1790 to 1970." Retrieved Feb. 15, 2018, from www.census.gov/history/pdf/histstats-colonial-1970.pdf.

U.S. Department of Commerce: Bureau of the Census. 1975. "Historical Statistics of the United States Bicentennial Edition – Colonial Times to 1970: Series K 564–582. Livestock-Number, Value Per Head, Production, and Price: 1867 to 1970." Retrieved Feb. 15, 2018, from www.census.gov/history/pdf/histstats-colonial-1970.pdf.

U.S. Department of Commerce: Bureau of the Census. 1975. "Historical Statistics of the United States Bicentennial Edition – Colonial Times to 1970: Series K 609–623. Poultry and Eggs-Number, Production, and Price: 1909 to 1970." Retrieved Feb. 15, 2018, from www.census.gov/history/pdf/histstats-colonial-1970.pdf.

U.S. Department of Commerce: Bureau of the Census. 1975. "Historical Statistics of the United States Bicentennial Edition – Colonial Times to 1970: Series S 108–119. Growth of Residential Service, and Average Prices for Electric Energy: 1902 to 1970." Retrieved Feb. 15, 2018, from www.census.gov/history/pdf/histstats-colonial-1970.pdf.

U.S. Department of Commerce Bureau of the Census. 1976. "Table F: Income Characteristics of Farm and Nonfarm Families, by Race: 1974." Current Population Reports. Farm Population Series 27. 47 (Sept.): 1–27.

U.S. Department of Defense. 2015. "2015 Demographics Profile of the Military Community." Retrieved Feb. 25, 2018, from http://download.militaryonesource.mil/12038/MOS/Reports/2015-Demographics-Report.pdf.

United States Department of Labor. 1957. *Occupational Outlook Handbook*. Washington DC: US Government Printing Office.

1959. *Occupational Outlook Handbook*. Washington DC: US Government Printing Office.

1978–1979. *Occupational Outlook Handbook*. Washington DC: US Government Printing Office.

United States Geological Survey. "Hazlehurst, MS Census Data." (n.d.). *U.S. Beacon*. Retrieved July 8, 2017, from www.usbeacon.com/economic/Mississippi/Hazlehurst.html.

Van Wormer, Katherine, David W. Jackson III, and Charletta Sudduth. 2012. *The Maid Narratives: Black Domestics and White Families in the Jim Crow South*. Baton Rouge: Louisiana State Press.

Warren, Kenneth. 2001. *Big Steel: The First Century of the United States Steel Corporation, 1901–2001*. Pittsburgh: University of Pittsburgh Press.

2008. *Bethlehem Steel: Builder and Arsenal of America*. Pittsburgh: University of Pittsburgh Press.

Wells-Barnett, Ida B. 2014. *On Lynchings*. Mineola, NY: Dover Books.

West, Cornel. 1993. *Race Matters*. Boston: Beacon Press.

White, Gillian. 2016. "Searching for the Origins of the Racial Wage Disparity in Jim Crow America." The Atlantic (Feb. 9). Retrieved Feb. 8, 2016, from www.theatlantic.com/business/archive/2016/02/the-origins-of-the-racial-wage-gap/461892/.

Wilkerson, Isabel. 2010. *The Warmth of Other Suns: The Epic Story of the Great Migration*. New York: Random House.

Williams, Rhonda. 1993. "Racial Inequality and Racial Conflict: Recent Developments in Radical Theory." Pp. 209–35 in *Labor Economics: Problems in Analyzing Labor Markets*, edited by William Darity, Jr. Boston: Kluwer Academic Publishers.

Willcox, Oliver. 1904. *The Negro Population*. 1–89. Retrieved Feb. 24, 2018, from www2.census.gov/prod2/decennial/documents/03322287no8ch1.pdf.

Willis, John C. 2000. *Forgotten Time: The Yazoo-Mississippi Delta After the Civil War*. Charlottesville: University of Virginia Press.

Wilmore, Gayraud S. (Ed.). 1994. *Black Religion and Black Radicalism: An Interpretation of the Religious History of Afro-American People*. New York: Orbis Books.

1995. *African-American Religious Studies: An Interdisciplinary Anthology*. Durham: Duke University Press.

Wilson, W.J. 1987. *The Truly Disadvantaged: The Inner City, the Underclass, and Public Policy*. Chicago: University of Chicago Press.

1996. *When Work Disappears: The World of the New Urban Poor*. New York: Alfred A. Knopf.

Wimberly, Anne Streaty. 2005. *Keep It Real: Working with Today's Black Youth*. Nashville: Abingdon.

Wimberly, Anne Streaty, Sandra Barnes, and Karma Johnson. 2013. *Claiming Hope: Youth Ministry in the Black Church*. New York: Judson Press.

Woodson, Carter G. 2013. *The Mis-Education of the Negro* (re-print). New York: Tribeca Books.

Index